Optical Communication Networks

The McGraw-Hill Series on Computer Communications (Selected Titles)

ISBN	AUTHOR	TITLE
0-07-005147-X	Bates	*Voice and Data Communications Handbook*
0-07-005560-2	Black	*TCP/IP and Related Protocols*, 2/e
0-07-005590-4	Black	*Frame Relay Networks: Specifications and Implementation*, 2/e
0-07-011769-1	Charles	*LAN Blueprints: Engineering It Right*
0-07-011486-2	Chiong	*SNA Interconnections: Bridging and Routing SNA in Heirarchical, Peer, and High-Speed Networks*
0-07-016769-9	Dhawan	*Mobile Computing: A Systems Integrator's Handbook*
0-07-018546-8	Dziong	*ATM Network Resource Management*
0-07-020359-8	Feit	*SNMP: A Guide to Network Management*
0-07-021389-5	Feit	*TCP/IP: Architecture, Protocols and Implementation with IPv6 and IP Security*, 2/e
0-07-024563-0	Goralski	*SONET: A Guide to Synchronous Optical Networks*
0-07-024043-4	Goralski	*Introduction to ATM Networking*
0-07-031382-2	Huntington-Lee/Terplan/Gibson	*HP's OpenView: A Manager's Guide*
0-07-034249-0	Kessler	*ISDN: Concepts, Facilities, and Services*, 3/e
0-07-035968-7	Kumar	*Broadband Communications*
0-07-041051-8	Matusow	*SNA, APPN, HPR and TCP/IP Integration*
0-07-060362-6	McDysan/Spohn	*ATM: Theory and Applications*
0-07-044435-8	Mukherjee	*Optical Communication Networks*
0-07-044362-9	Muller	*Network Planning Procurement and Management*
0-07-046380-8	Nemzow	*The Ethernet Management Guide*, 3/e
0-07-051506-9	Ranade/Sackett	*Introduction to SNA Networking*, 2/e
0-07-054991-5	Russell	*Signaling System #7*
0-07-057724-2	Sackett/Metz	*ATM and Multiprotocol Networking*
0-07-057199-6	Saunders	*The McGraw-Hill High Speed LANs Handbook*
0-07-057639-4	Simonds	*Network Security: Data and Voice Communications*
0-07-912640-5	Simoneau	*Hands-On TCP/IP*
0-07-060363-4	Spohn	*Data Network Design*, 2/e
0-07-069416-8	Summers	*ISDN Implementor's Guide*
0-07-063263-4	Taylor	*The McGraw-Hill Internetworking Handbook*
0-07-063301-0	Taylor	*McGraw-Hill Internetworking Command Reference*
0-07-063639-7	Terplan	*Effective Management of Local Area Networks*, 2/e

Optical Communication Networks

Biswanath Mukherjee

Department of Computer Science
University of California, Davis

McGraw-Hill

New York San Francisco Washington, D.C. Auckland Bogotá
Caracas Lisbon London Madrid Mexico City Milan
Montreal New Delhi San Juan Singapore
Sydney Tokyo Toronto

Library of Congress Cataloging-in-Publication Data

Mukherjee, Biswanath.
Optical communication networks / Biswanath Mukherjee.
p. cm.—(McGraw-Hill series on computer communications)
Includes index.
ISBN 0-07-044435-8
1. Optical communications. I. Title. II. Series.
TK5103.59.M84 1997
621.382'7—dc21 97-16145
CIP

McGraw-Hill

A Division of The McGraw-Hill Companies

1 2 3 4 5 6 7 8 9 0 FGR/FGR 9 0 2 1 0 9 8 7

ISBN 0-07-044435-8

The sponsoring editor for this book was Steven M. Elliot, the editing supervisor was David E. Fogarty, and the production supervisor was Tina Cameron.

Printed and bound by Quebecor/Fairfield.

This book was printed on recycled, acid-free paper containing a minimum of 50% recycled, de-inked fiber.

To my family and friends

ABOUT THE AUTHOR

Dr. Biswanath Mukherjee is a Professor of Computer Science at the University of California-Davis and has consulted with many companies on the deployment and installation of optical networks. He received his Ph.D. in Electrical Engineering from the University of Washington, Seattle, in 1987. He currently serves on the editorial boards of IEEE/ACM *Transactions on Networking*, *IEEE Network*, and *The Journal of High-Speed Network*.

Preface

The Topic

The basic premise of our subject, optical communication networks, is that, as more users start to use our data networks, and as their usage patterns evolve to include more bandwidth-intensive networking applications such as data browsing on the world wide web, Java applications, video conferencing, etc., there emerges an acute need for very high-bandwidth transport network facilities, whose capabilities are much beyond those that current high-speed asynchronous transfer mode (ATM) networks can provide. Given that a single-mode fiber's potential bandwidth is nearly 50 terabits per second (Tbps), which is nearly four orders of magnitude higher than electronic data rate, every effort should be made to tap into this huge opto-electronic bandwidth mismatch. Wavelength-division multiplexing (WDM) is an approach that can exploit this bandwidth mismatch by requiring that each end-user's equipment operates only at electronic rate, but multiple WDM channels from different end-users may be multiplexed on the same fiber.

Research and development on optical WDM networks have matured considerably over the past few years, with a number of commercial WDM products already available. A number of different experimental prototypes have been and are being deployed and tested in the U.S., Europe, and Japan, with significant support from their respective government agencies and telecommunication providers. There now exists a large and expanding interest on this topic to better understand the issues and challenges in designing such networks. It is anticipated that the next generation of the Internet will employ WDM-based optical backbones.

Intended Audience

Many electrical engineering, computer engineering, and computer science programs around the world have started to offer a graduate course on this topic. That is, research and development on optical communication networks have matured significantly to the extent that some of these principles are being moved from the research laboratories to the formal (graduate) classroom setting. However, there are no textbooks to fill this emerging void, except for those that deal mainly with the physical-layer issues of optical communications, e.g., [Gree93]. These observations led me to the decision to write a textbook on this topic.

I expect that instructors will find the book useful for teaching a graduate course on this subject. The book has been "class tested" in a one-quarter graduate course, ECS 289I, "Optical Communication Networks," which I taught at my home institution, University of California, Davis, during Winter Quarter 1997. As a result, the book has been "debugged" significantly, including the creation of most of the end-of-chapter exercises.

Since the major developments in optical communication networks have started to capture the imagination of the computing, telecommunications, and opto-electronic industries, I expect that industry professionals will find this book useful as a well-rounded reference. Through my own consulting work, I find that there exists a large group of people who are experts in physical-layer optics, and who wish to learn more about network architectures, protocols, and the corresponding engineering problems in order to design new state-of-the-art optical networking products. This group of people is also who I had in mind while developing this book.

Organization Principles and Unique Features

Writing a book on this topic is not easy since such a book has to cover material that spans several disciplines ranging from physics to electrical engineering to computer science, and also since the field itself is still evolving.

This book is *not* intended to cover in any detail the physical-layer aspects of optical communications; readers should consult an appropriate book, e.g., [Gree93], for such material. We summarize these "enabling technologies" in a single chapter (Chapter 2). Our treatment of the material in this chapter should allow us to uncover the unique strengths and limitations of the appropriate technologies, and then determine how the characteristics of the physical devices may be exploited in pragmatic network architectures, while compensating for the device limitations or "mismatches".

An important organizing principle that I have attempted to keep in mind while developing this book is that research, development, and education on optical communication networks should allow tight coupling between network architectures and device capabilities. To date, research on optical network architectures has taught us that, without a sound knowledge of device capabilities and limitations, one can produce architectures which may be unrealizable; similarly, research on new optical devices, conducted without the concept of a useful system, can lead to sophisticated technology with limited or no usefulness.

The book is organized into four parts. Part I introduces WDM and its enabling technologies. Parts II and III are devoted to WDM local-area and WDM wide-area network architectures, respectively. Part IV discusses a number of problems that address tight coupling between device capabilities/limitations and the corresponding networking challenges; Part IV also includes a chapter that briefly discusses optical time-division multiplexing (TDM) and optical code-division multiplexing (CDM). The appendices on "where to learn more" and a "Glossary of Important Terms" should also be beneficial to many readers. More details on the book's organization can be found in Section 1.7.

The most unique feature of this book is its timeliness to fill a void in an important and emerging networking topic. Other major salient aspects of the book are its breadth of coverage, depth of analytical treatment, clear identification of recent developments and open problems, review of testbeds and implementation efforts, an extensive bibliography (292 references), an extensive number of end-of-chapter exercises (a total of 229), an extensive set of illustrations (207 figures, 26 tables), etc.

A solutions manual is currently being developed. Instructors may obtain a copy from the publisher. Please see additional web-based features below.

Web Enhancements

This book is "web-enhanced," i.e., we have (or plan to have) material such as the following available through the web (see author's web address below):

1. table of contents,
2. list of figures,
3. color versions of some figures – some of the figures were drawn originally in color, and the color versions may be more informative than the black-and-white versions appearing in the book,
4. Chapter 1 in its entirety,
5. introductory material for the other chapters,
6. posting of corrections, revisions, updates, etc.,
7. easy mechanism for readers to provide feedback to the author and the publisher, and
8. if and when possible, additional material as deemed useful, e.g., simulations for some single-hop protocols, multicasting, virtual-topology embedding, etc.; allowing animation and remote operation of simulations via parameter selection by the user using Java.

Acknowledgements

Although my name is the only one to appear on the cover, the combination of efforts from a large number of individuals is required to produce a quality book.

Much of the book's material is based on research that I have conducted over the years with my graduate students and research scientists visiting my laboratory, and I would like to acknowledge them as follows: Dhritiman Banerjee for the wavelength-routing material (Chapters 8-10, 12); Mike Borella for the enabling-technologies review (Chapter 2); Debasish Datta for optical TDM (Chapter 16); Jason Iness for GEMNET, sparse wavelength conversion, optical-cycle elimination, and amplifier placement (Chapters 6, 11, 14, 15); Jason Jue for enabling technologies, RAINBOW analysis, and optical CDM (Chapters 2, 4, 16); Byrav Ramamurthy for wavelength conversion, cycle elimination, and amplifier placement (Chapters 2, 11, 14, 15); S. Ramamurthy for wavelength routing and WDDI (Chapters 8, 12, 13); and Srini Tridandapani for channel-sharing and multicasting (Chapter 7).

A number of additional individuals who I have collaborated with over the years, who have enabled me to better understand the subject matter, and who I would like to acknowledge are the following: Krishna Bala, Subrata Banerjee, Chuck Brackett, Kamal Goel, Jon Heritage, Feiling Jia, and Behzad Moslehi.

I would like to thank the following Winter97 ECS 289I students who "debugged" the book and who provided many of the end-of-chapter problems: Steven Cheung, Bill Coffman, Jason Jue, Arijit Mukherjee, Raja Mukhopadhyay, Vijoy Pandey, Byrav Ramamurthy, S. Ramamurthy, Laxman Sahasrabuddhe, Feng Wang, and Raymond Yip. Murat Azizoglu and Debasish Datta also contributed several problems, while Jason Jue and Laxman Sahasrabuddhe "organized" these exercises; their efforts are highly appreciated.

Special thanks are due to Byrav Ramamurthy for the book's "electronic organization," several figure drawings, and answers to my LaTeX2e questions; and to Ken Victa for developing the graphical user interface (GUI) to visualize and draw the results of our optimization studies.

This book wouldn't have been possible without the support of my research on optical communication networks from several funding agencies as follows: National Science Foundation (NSF) (Grant Nos. NCR-9205755, NCR-9508239, and ECS-9521249); Defense Advanced Research Projects Agency (DARPA) (Contract Nos. DABT63-92-C-0031 and DAAH04-95-1-0487); Air Force Office of Scientific Research (AFOSR) (Grant No. 89-0292); Pacific Telesis; Optivision, Inc.; and University of California MICRO Program.

Quality control for a book can be ensured through independent technical reviews. In this regard, the following reviewers are deeply appreciated for their time, effort, suggestions for improvement, and feedback on the book's manuscript and/or its initial proposal: Murat Azizoglu, Andrea Fumagalli, Anura Jayasumana, Walter Kahn, and Arun Somani.

I wish to acknowledge the people at McGraw-Hill who I interfaced with – Steven Elliot and David Fogarty – for their encouragement and attention to detail during the book's production. The book's camera-ready version was produced by the author using the LaTeX2e document-processing system.

Finally, I wish to thank my family members for their constant encouragement and support: my father Dhirendranath Mukherjee, my mother Gita Mukherjee, my wife Supriya, my daughters Bipasha and Suchitra, and my brother Amarnath. I couldn't have done it without you!

I welcome email from readers who wish to provide any sort of feedback: errors, comments, criticisms, and suggestions for improvements.

Biswanath Mukherjee
`mukherje@cs.ucdavis.edu`
`http://networks.cs.ucdavis.edu/~mukherje`

Davis, California, U.S.A.
June 1997

Contents

Preface vii

List of Figures xxi

List of Tables xxxi

I Introduction 1

1 Optical Networking: Principles and Challenges 3
1.1 Need + Promise = Challenge! 3
1.2 xDM vs. xDMA 5
1.3 Wavelength-Division Multiplexing (WDM) 5
1.3.1 A Sample WDM Networking Problem 7
1.4 WDM Networking Evolution 8
1.4.1 Point-to-Point WDM Systems 8
1.4.2 Wavelength Add/Drop Multiplexer (WADM) 9
1.4.3 Fiber and Wavelength Crossconnects – Passive Star, Passive Router, and Active Switch 10
1.5 WDM Network Constructions 14
1.5.1 Broadcast-and-Select (Local) Optical WDM Network . 14
1.5.2 Wavelength-Routed (Wide-Area) Optical Network . . . 15
1.6 WDM Economics 17
1.7 Road Map – Organization of the Book 18
Exercises 20

2 Enabling Technologies: Building Blocks 23
2.1 Optical Fiber 24
2.1.1 Optical Transmission in Fiber 25
2.1.2 Single-Mode vs. Multimode Fiber 28
2.1.3 Attenuation in Fiber 30
2.1.4 Dispersion in Fiber 31
2.1.5 Nonlinearities in Fiber 32

2.1.6 Couplers 35
2.2 Optical Transmitters 37
2.2.1 How a Laser Works 37
2.2.2 Tunable and Fixed Lasers 40
2.2.3 Optical Modulation 43
2.2.4 Summary 44
2.3 Optical Receivers and Filters 44
2.3.1 Photodetection 45
2.3.2 Tunable Optical Filters 46
2.3.3 Fixed Filters 50
2.4 Optical Amplifiers 52
2.4.1 Optical Amplifier Characteristics 53
2.4.2 Semiconductor Laser Amplifier 54
2.4.3 Doped-Fiber Amplifier 55
2.5 Switching Elements 57
2.5.1 Fiber Crossconnect Elements 58
2.5.2 Nonreconfigurable Wavelength Router 62
2.5.3 Reconfigurable Wavelength-Routing Switch 66
2.5.4 Photonic Packet Switches 66
2.6 Wavelength Conversion 69
2.6.1 Wavelength Conversion Technologies 73
2.6.2 Wavelength Conversion in Switches 78
2.7 Designing WDM Networks: Systems Considerations 79
2.7.1 Channels 82
2.7.2 Power Considerations 83
2.7.3 Additional Considerations 85
2.7.4 Elements of Local-Area WDM Network Design 85
2.7.5 WDM Wide-Area Network Design Issues 89
2.8 Experimental WDM Lightwave Networks 91
2.8.1 Local Area Network Testbeds 91
2.8.2 Wide-Area Network Testbeds 92
Exercises 97

II Broadcast (Local) Optical Networks 107

3 Single-Hop Networks 109
3.1 A Passive-Star-Based Local Lightwave Network 109
3.2 Characteristics of a Single-Hop System 113

3.3 Experimental WDM Systems 115
3.3.1 LAMBDANET 116
3.3.2 Rainbow 116
3.3.3 Fiber-Optic Crossconnect 117
3.3.4 STARNET 117
3.3.5 Other Experimental WDM Systems 118
3.4 Other Non-Pretransmission Coordination Protocols 118
3.4.1 Fixed Assignment 119
3.4.2 Partial Fixed Assignment Protocols 120
3.4.3 Random Access Protocols I 121
3.4.4 Random Access Protocols II 122
3.4.5 The PAC Optical Network 123
3.5 Pretransmission Coordination Protocols 123
3.5.1 Partial Random Access Protocols 123
3.5.2 Improved Random Access Protocols 126
3.5.3 Extended Slotted-ALOHA and Reservation-ALOHA Protocols 129
3.5.4 Receiver Collision Avoidance (RCA) Protocol 132
3.5.5 Reservation Protocols 133
3.5.6 The TDMA/N-Server Protocol 137
3.6 Special Case: Linear Bus with Attempt-and-Defer Nodes . . . 138
3.6.1 AMTRAC 139
3.6.2 Multichannel Probabilistic Scheduling 139
3.7 Summary 139
Exercises 141

4 Single-Hop Case Study: IBM Rainbow Protocol 149
4.1 Introduction 149
4.2 Rainbow Protocol 151
4.3 Model 152
4.4 Analysis 155
4.5 Illustrative Examples 159
4.6 Summary 166
Exercises 167

5 Multihop Networks 171
5.1 Characteristics of a Multihop System 171
5.2 Topological Optimization Studies 174
5.2.1 Flow-Based Optimization 174

5.2.2 Delay-Based Optimization 175
5.3 Regular Structures 176
5.3.1 ShuffleNet 176
5.3.2 de Bruijn Graph 181
5.3.3 Torus (Manhattan Street Network) 183
5.3.4 Hypercube 186
5.3.5 Other Regular Multihop Topologies 188
5.4 Near-Optimal Node Placement on Regular Structures 188
5.4.1 Flow-Based Heuristics 189
5.4.2 Delay-Based Heuristics 192
5.4.3 Dynamic Load Balancing 192
5.5 Shared-Channel Multihop Systems 192
5.5.1 Channel-Sharing in ShuffleNet 193
5.5.2 Channel-Sharing in the Manhattan Street Network . . . 193
5.5.3 Channel-Sharing in the Generalized Hypercube 194
5.5.4 Multihop Systems Based on Subcarrier Multiplexing . . 195
5.6 Summary 195
Exercises 197

6 Multihop Case Study: GEMNET 203
6.1 Introduction 203
6.2 GEMNET Architecture 206
6.2.1 Interconnection Pattern 206
6.2.2 Routing 206
6.3 GEMNET Properties 208
6.3.1 Routing Algorithms for Balancing Link Loading 208
6.3.2 Bounds on the Average Hop Distance 209
6.3.3 Which Configuration of GEMNET Is the Best? 211
6.4 Scalability – Adapting the Size of a GEMNET 213
Exercises 217

7 Channel-Sharing and Multicasting 219
7.1 Introduction 220
7.2 Background 222
7.2.1 Channel-Sharing 222
7.2.2 Multicasting 224
7.3 Shared-Channel Multihop GEMNET 226
7.4 Performance Evaluation 229
7.5 Illustrative Examples: Unicast Traffic 237

7.5.1 Twelve-Node Example: Various Delay Components . . 238
7.5.2 Twelve-Node Example: Delay Behavior with Variation in λ 240
7.5.3 Delay Behavior in Larger Networks 243
7.6 Illustrative Examples: Multicast Traffic 246
7.7 Summary 251
Exercises 253

III Wavelength-Routed (Wide-Area) Optical Networks 257

8 Elements of Virtual Topology Design 259
8.1 Introduction 260
8.2 System Architecture 261
8.2.1 Motivation 261
8.2.2 General Problem Statement 263
8.2.3 An Illustrative Example 265
8.3 Formulation of the Optimization Problem 271
8.3.1 Given: 271
8.3.2 Variables: 272
8.3.3 Constraints: 272
8.3.4 Objective: Optimality Criterion 274
8.3.5 Explanation of Equations 274
8.4 Algorithms 275
8.4.1 Subproblems 275
8.4.2 Previous Work 275
8.4.3 Our Solution Approach 276
8.4.4 Simulated Annealing 277
8.4.5 Flow Deviation Algorithm 278
8.5 Experimental Results 278
8.5.1 Physical Topology as Virtual Topology (No WDM) . . 280
8.5.2 Multiple Point-to-Point Links (No WRS) 281
8.5.3 Arbitrary Virtual Topology (Full WDM) 282
8.5.4 Comparisons 282
8.5.5 Effect of Nodal Degree and Wavelength Requirements . 285
8.6 Summary 287
Exercises 288

9 Virtual Topology: LP, Cost, Reconfiguration **291**
9.1 Introduction 292
9.2 Problem Specification 293
9.2.1 Linear Formulation 295
9.2.2 Simplifying Assumptions 299
9.3 Heuristic Approaches 302
9.4 Network Design: Resource Budgeting and Cost Model 304
9.4.1 Resource Budgeting 304
9.4.2 Network Cost Model 305
9.5 Virtual Topology Reconfiguration 308
9.5.1 Reconfiguration Algorithm 309
9.6 Illustrative Examples 310
9.7 Summary 316
Exercises 317

10 Routing and Wavelength Assignment **321**
10.1 Introduction 321
10.2 Problem Formulation 325
10.2.1 Solution Approach 326
10.2.2 Problem Size Reduction 326
10.2.3 Randomized Rounding 328
10.2.4 Graph Coloring 330
10.3 Illustrative Examples 332
10.3.1 Static Lightpath Establishment (SLE) 332
10.3.2 Dynamic Lightpath Establishment (DLE) 336
10.4 Summary 338
Exercises 339

11 Wavelength Conversion **341**
11.1 Introduction 341
11.2 Basics of Wavelength Conversion 344
11.2.1 Wavelength Converters 344
11.2.2 Switches 345
11.3 Network Design, Control, and Management Issues 347
11.3.1 Network Design 347
11.3.2 Network Control 348
11.3.3 Network Management 349
11.4 Benefit Analysis 350

11.4.1 A Probabilistic Approach to Wavelength-Conversion Benefits Analysis 351
11.4.2 A Review of Benefit-Analysis Studies 352
11.5 Benefits of Sparse Conversion 356
11.5.1 Goals 356
11.5.2 Simulator 357
11.5.3 Single Optical Rings 357
11.5.4 NSFNET 358
11.6 Summary 363
Exercises 364

12 Additional Topics on Wavelength Routing 367
12.1 Introduction 367
12.2 Circuit-Switched Approaches 369
12.2.1 LDC-Based Approach 369
12.2.2 Lightpath-Based Approach 370
12.2.3 Multihop Virtual Circuits 370
12.2.4 Routing of Session Traffic 371
12.2.5 Bounds for the RWA Problem 371
12.2.6 Least-Congested-Path (LCP) Routing 372
12.2.7 Wavelength-Conversion-Based Routing 373
12.2.8 Latin-Router-Based Routing 373
12.2.9 Theoretical Results 374
12.3 Packet-Switched Approaches 375
12.3.1 Logical Topologies for Electronic Packet-Switched Networks 376
12.3.2 Deflection Routing Networks 377
12.3.3 Optical Packet-Switch Design 377
12.4 Reconfiguration in WDM Networks 377
12.4.1 Passive-Star-Based (LAN) Algorithms 378
12.4.2 WAN Algorithms 380
12.5 WDM Network Control and Management 381
12.5.1 State Information 382
12.5.2 Connection Setup 384
12.5.3 Connection Takedown and Update 384
12.5.4 Fault Recovery 384
12.6 Amplification-Related Issues 385
12.7 Systems Design Considerations 386
12.8 Testbed Proposals 387

12.8.1 AT&T/MIT-LL/DEC All-Optical Network (AON) Architecture . . . 387
12.8.2 Bellcore's Optical Network Technology Consortium (ONTC) . . . 389
12.8.3 RACE MWTN Project . . . 391
12.8.4 Multiwavelength Optical Networking (MONET) Project . . . 391
12.9 Summary . . . 393
Exercises . . . 394

IV Potpourri 395

13 Multiwavelength Ring Networks 397
13.1 Introduction . . . 398
13.2 System Architecture and Assumptions (Model) . . . 401
13.3 Illustrative Examples . . . 405
13.4 Optimization Criteria . . . 406
13.4.1 MIN-CROSS . . . 406
13.4.2 MIN-DIFF . . . 407
13.4.3 MIN-DELAY . . . 407
13.5 Flow-Based Algorithms . . . 407
13.5.1 MIN-CROSS Algorithms . . . 408
13.5.2 MIN-DIFF Algorithm . . . 412
13.6 Delay-Based Algorithms . . . 415
13.6.1 Performance Analysis . . . 415
13.6.2 Partitioning Algorithm . . . 417
13.7 Illustrative Examples . . . 418
13.7.1 Network Description . . . 418
13.7.2 Delay vs. N Characteristics (Two Partitions) . . . 419
13.7.3 Two or Greater Partitions . . . 425
13.8 Summary . . . 427
Exercises . . . 429

14 All-Optical Cycle Elimination 431
14.1 Introduction . . . 432
14.1.1 Wavelength Crossconnect Switches . . . 433
14.2 Network Assumptions . . . 436
14.3 Overview of Solution and Algorithms . . . 439

14.4 Details of Algorithms 440
14.4.1 NETWORK GENERATOR (Module 1) 441
14.4.2 SWITCH PORT LABELER (Module 2) 442
14.4.3 Λ CYCLE ELIMINATOR (Module 3) 443
14.4.4 CONNECTION SETUP (Module 4) 444
14.4.5 Λ CYCLE ELIMINATOR (Module 5) 445
14.4.6 (STATIC) NETWORK CONFIGURATION ANALYZER (Module 6) 446
14.5 Illustrative Examples 447
14.5.1 Dynamic Analysis 447
14.5.2 Static Analysis 453
14.6 Summary 458
Exercises 458

15 Optimizing Amplifier Placements in an Optical LAN/MAN 463
15.1 Introduction 464
15.1.1 Network Environment 464
15.1.2 Problem Definition 467
15.1.3 Amplifier Gain Model 469
15.2 Solution Approach 471
15.2.1 Formulation 472
15.2.2 Solver Strategies 477
15.2.3 Amplifier-Placement Module 479
15.3 Illustrative Examples 479
15.4 Open Problems 486
15.4.1 Switched Networks 486
15.4.2 Gain Model 488
15.5 Summary 488
Exercises 488

16 Optical TDM and CDM Networks 493
16.1 Optical TDM Networks 494
16.1.1 Basics of TDM 494
16.1.2 Optical TDM 496
16.1.3 Optical Sources 497
16.1.4 Modulation and Multiplexing 498
16.1.5 Transmission of Ultrafast OTDM Signal Using Soliton 499
16.1.6 Demultiplexing and Clock Recovery 501
16.1.7 Optical Processing 504

16.1.8 Optical TDM Network Architectures and Proposals . . 504
16.2 Optical CDMA Networks 507
16.2.1 Basics of CDMA 507
16.2.2 Spread Spectrum 509
16.2.3 Code Sequences 510
16.2.4 CDMA Example 512
16.2.5 Optical CDMA 514
Exercises 520

Appendix A: Further Reading **523**
A.1 General Resources and Publications 523
A.2 Enabling Technologies 524
A.3 Tutorials/Surveys/Reviews 525

Appendix B: Glossary of Important Terms **527**

Bibliography **533**

Index **565**

List of Figures

1.1 The low-attenuation regions of an optical fiber. 6
1.2 A four-channel point-to-point WDM transmission system with amplifiers. 10
1.3 A Wavelength Add/Drop Multiplexer (WADM). 11
1.4 A 4 × 4 passive star. 12
1.5 A 4 × 4 passive router (four wavelengths). 12
1.6 A 4 × 4 active switch (four wavelengths). 13
1.7 A passive-star-based local optical WDM network. 14
1.8 A wavelength-routed (wide-area) optical WDM network. 16
1.9 A wavelength-routed WDM network. 22

2.1 The low-attenuation regions of an optical fiber. 24
2.2 Single-mode and multimode optical fibers. 26
2.3 Light traveling via total internal reflection within a fiber. 27
2.4 Graded-index fiber. 28
2.5 Numerical aperture of a fiber. 28
2.6 Splitter, combiner, and coupler. 35
2.7 A 16 × 16 passive-star coupler. 36
2.8 The general structure of a laser. 38
2.9 Structure of a semiconductor diode laser. 39
2.10 Free spectral range and finesse of a tunable filter capable of tuning to N different channels. 47
2.11 Cascading filters with different FSRs. 48
2.12 Structure of a Mach-Zehnder interferometer. 49
2.13 A semiconductor optical amplifier. 54
2.14 Erbium-doped fiber amplifier. 55
2.15 The gain spectrum of an erbium-doped fiber amplifier with input power = −40 dBm. 56
2.16 2 × 2 crossconnect elements in the cross state and bar state. . . 58
2.17 Schematic of optical crosspoint elements. 60
2.18 A 2 × 2 amplifier gate switch. 62
2.19 A 4 × 4 nonreconfigurable wavelength router. 63
2.20 The waveguide grating router (WGR). 64

2.21 A P $\times$ P reconfigurable wavelength-routing switch with M wavelengths. . . . 67
2.22 The staggering switch architecture. . . . 68
2.23 The CORD architecture. . . . 69
2.24 The HLAN architecture. . . . 70
2.25 An all-optical wavelength-routed network. . . . 70
2.26 Wavelength-continuity constraint in a wavelength-routed network. 71
2.27 Functionality of a wavelength converter. . . . 72
2.28 An opto-electronic wavelength converter. . . . 73
2.29 A wavelength converter based on nonlinear wave-mixing effects. 74
2.30 A wavelength converter using co-propagation based on XGM in an SOA. . . . 76
2.31 An interferometric wavelength converter based on XPM in SOAs. 76
2.32 Conversion using saturable absorption in a laser. . . . 77
2.33 A switch which has dedicated converters at each output port for each wavelength. . . . 79
2.34 Switches which allow sharing of converters. . . . 80
2.35 The share-with-local wavelength-convertible switch architecture. 81
2.36 Architecture which supports electronic wavelength conversion. . 81
2.37 Broadcast-and-select WDM local optical network with a passive-star coupler network medium. . . . 86
2.38 Lightpath routing in a WDM WAN. . . . 90
2.39 MONET New Jersey Area Network. . . . 93
2.40 ONTC testbed. . . . 95
2.41 The AT&T/MIT-LL/DEC AON testbed architecture. . . . 96
2.42 Critical angle in a step index fiber. . . . 98
2.43 Critical angle in a graded index fiber. . . . 98
2.44 Two architectures for wavelength convertible routers: (a) share-per-node, (b) share-per-link. . . . 103
2.45 T=Transmitter, R=Receiver. All connections begin at transmitters and end at receivers . . . 105

3.1 A broadcast-and-select WDM network. . . . 110
3.2 Alternative physical topologies for a WDM local lightwave network. . . . 111
3.3 Architecture of the PAC optical packet network (the dashed lines are used to detect energy on the various channels from the "main" star). . . . 124
3.4 The ALOHA/ALOHA protocol. . . . 125

3.5 Bimodal throughput characteristics of the slotted-ALOHA/delayed-ALOHA protocol for $L = 10$ slots per data packet and N = number of data channels. 127
3.6 Nonmonotonic delay characteristics of the slotted-ALOHA/delayed-ALOHA protocol for L=10 slots per data packet, N=3 data channels, zero propagation delay, and different values of the backoff parameter K. 128
3.7 The extended slotted-ALOHA protocols: (a) simple case, (b) higher concurrency to reduce channel wastage. Note: λ is the control channel and $\lambda, \lambda, \ldots, \lambda$ are data channels. 130
3.8 The dynamic time-wavelength division multiple access (DT-WDMA) protocol. 135
3.9 The multichannel bus network: (a) network structure with tunable transmitters and fixed receivers, and (b) a cycle in AMTRAC for N=4 and M=4. 140
3.10 Classification of single-hop network architectures. 142

4.1 Passive-star topology for Rainbow. 150
4.2 State diagram for the Rainbow model. 154
4.3 Timing for connection setup. 156
4.4 Throughput vs. arrival rate. Slot = 1 μs, $N = 32$, $R = 50$ μs, τ = 1 ms, $1/\rho = 100$ ms. 160
4.5 Throughput vs. message size. Slot = 1 μs, $N = 32$, $R = 50$ μs, $\tau = 1$ ms, $\sigma = 10 - 4$ msg/slot, $\phi = 10$ ms. 161
4.6 Throughput vs. timeout duration. Slot = 1 μs, $N = 32$, $R = 50$ μs, $\tau = 1$ ms, $1/\rho = 100$ ms. 161
4.7 Timeout probability vs. timeout duration. Slot = 1 μs, $N = 32$, $R = 50$ μs, $\tau = 1$ ms, $1/\rho = 100$ ms. 162
4.8 Throughput vs. timeout duration for different message lengths. Slot = 1 μs, $N = 32$, $R = 50$ μs, $\tau = 1$ ms, $\sigma = 10 - 4$ msg/slot. 163
4.9 Delay vs. throughput with parameter ϕ. Slot = 1 μs, $N = 32$, $R = 50$ μs, $\tau = 1$ ms, $\sigma = 10 - 4$ msg/slot, $1/\rho = 100$ ms. . . . 164
4.10 Delay vs. throughput with parameter σ. Slot = 1 μs, $N = 32$, $R = 50$ μs, $\tau = 1$ ms, $1/\rho = 100$ ms, $\phi = 10$ ms. 165
4.11 Delay vs. throughput for varying message size. Slot = 1 μs, $N = 32$, $R = 50$ μs, $\tau = 1$ ms, $\sigma = 10 - 4$ msg/slot, $\phi = 10$ ms. 165
4.12 Throughput vs. timeout duration for different number of stations. Slot = 1 μs, $R = 50$ μs, $\tau = 1$ ms, $\sigma = 10 - 4$ msg/slot, $1/\rho = 100$ ms. 166

5.1 An example four-node multihop network: (a) physical topology, (b) logical topology. 172
5.2 A $(2,2)$ ShuffleNet. 177
5.3 A $(2,3)$ de Bruijn graph. 182
5.4 A 4×6 Manhattan Street Network (MSN) with unidirectional links. 184
5.5 A $(2,4)$ ShuffleNet. 185
5.6 An eight-node binary hypercube. 187
5.7 A linear dual-bus network. 189
5.8 Classification of multihop network architectures. 196
5.9 The (10-node) Peterson graph. 199
5.10 A bidirectional ring network. 199

6.1 A 10-node (2,5,2) GEMNET. 205
6.2 Bounds and average hop distance for a $P = 2$, 64-node GEMNET with different values of K. 212
6.3 Growing a (1,6,2) GEMNET by one node. 216

7.1 Delay behavior of a shared-channel, WDM, multihop network with changes in load λ and number of available wavelengths w. . . 221
7.2 Logical assignment of wavelengths in an eight-node network arranged as a ShuffleNet with $K = 2$ and $P = 2$: (a) nonshared case where $w = 2N = 16$; (b) shared case where $w = 8 < 2N$. . . 223
7.3 Twelve-node SC_ GEMNETs, along with corresponding timing diagrams, for various values of w: (a) $w = 12$; (b) $w = 6$; (c) $w = 4$; (d) $w = 3$; and (e) $w = 2$. (Note: unless otherwise shown, all links are directed from left to right.) 228
7.4 A (3,2) complete Moore graph. 233
7.5 Effects of sharing on multicasting in a (3,2) complete, Moore graph. The source node is node 1 and the destination nodes are nodes 2, 3, 6, and 7: (a) nonshared case; (b) shared-channel case. 234
7.6 Various delay components vs. number of wavelengths in a 12-node network with $R = 2$ and $\lambda = 0.05$. 239
7.7 Average delay vs. number of wavelengths in a 12-node network with $R = 2$ and various load values. Marked points indicate values of admitting equal sharing of the available bandwidth among the 12 nodes when each node has a single transmitter-receiver pair. 241

7.8 Upper (w_H) and lower (w_L) bounds on the number of channels admitting stability and optimal number of channels (w^*) vs. the load in a 12-node network. Note that the curves for $R = 1$ and $R = 0$ are identical; therefore only one of them is shown. 242

7.9 Average delay, T, vs. number of wavelengths, w, for various values of propagation delay in a 120-node network: (a) $\lambda = 0.001$; (b) $\lambda = 0.005$; (c) $\lambda = 0.01$; (d) $\lambda = 0.05$. 244

7.10 Average delay, T, vs. number of wavelengths, w, for various values of load in a 120-node network: (a) $R = 0$; (b) $R = 1$; (c) $R = 10$; (d) $R = 100$. 245

7.11 Upper (w_H) and lower (w_L) bounds on the number of channels admitting stability and optimal number of channels (w^*) vs. load in a 120-node network. 246

7.12 Delay vs. load for Systems A through F: (a) $m = 1$; (b) $m = 5$; (c) $m = 20$; (d) $m = 36$. 251

7.13 Ratio of maximum throughput at a given m to the maximum throughput at $m = 1$. 252

7.14 A (2,2) complete Moore graph. 255

7.15 A (4,2) incomplete Moore graph. 255

8.1 NSFNET T1 backbone, 1991. (©Merit Network, Inc.) 262

8.2 Modified physical topology. 266

8.3 A 16-node hypercube virtual topology embedded on the NSFNET physical topology. 267

8.4 Details of the Utah (UT) node. 268

8.5 The physical topology with embedded wavelengths corresponding to an optimal solution (more than one transceiver at any node can tune to the same wavelength). 269

8.6 The physical topology with embedded wavelengths corresponding to an optimal solution (all transceiver pairs at any node must be tuned to different wavelengths). 269

8.7 Delay vs. throughput (scaleup) characteristics with no WDM, i.e., physical topology as virtual topology. 283

8.8 Delay vs. throughput (scaleup) characteristics with WDM used on some links, but no WRSs, i.e., multiple point-to-point links are allowed on the physical topology. 284

8.9 Delay vs. throughput (scaleup) characteristics with full WDM on some links and a WRS at each node, i.e., arbitrary virtual topologies are allowed. 284

8.10 Delay vs. throughput (scaleup) characteristics for different virtual topologies. 285
8.11 Distributions of the number of wavelengths used in each of the 21 fiber links of the NSFNET for the virtual topology approach with nodal degree $P = 4$, 5, and 6. 287

9.1 NSFNET T1 backbone. 293
9.2 Optimal solution for a two-wavelength and a five-wavelength network. Each physical link consists of two unidirectional fibers carrying transmissions in opposite directions (hence, each wavelength may appear twice on any link in the diagrams; their signal propagation directions are opposite to each other in such cases). Wavelength 0 is used to embed the physical topology over the virtual topology, so the Wavelength-0 lightpaths are not shown explicitly in these diagrams to preserve clarity. Note: $\circ$ = transmitter; $\bullet$ = receiver. 294
9.3 Transport node in the RACE WDM optical network architecture. 306
9.4 Average packet hop distance for the optimal solution. 311
9.5 Average transceiver utilization for the optimal solution. 312
9.6 Average wavelength utilization for the optimal solution. 312
9.7 Comparison of heuristic algorithms for a four-wavelength network. 314
9.8 Reconfiguration statistics. 316
9.9 Physical network topology. 318
9.10 Physical network topology. 319

10.1 Lightpath routing in an all-optical network. 322
10.2 Effect of nodal degree d (for $K = 2$ alternate paths) on wavelength routing. 335
10.3 Effect of number of connections on link congestion. 338
10.4 Connection requests. 340

11.1 An all-optical wavelength-routed network. 342
11.2 Wavelength-continuity constraint in a wavelength-routed network. 343
11.3 Functionality of a wavelength converter. 344
11.4 A switch which has dedicated converters at each output port for each wavelength. 345

11.5 Switches which allow sharing of converters. 346
11.6 Wavelength conversion for distributed network management. . . 350
11.7 Blocking probabilities for different loads in a 10-node optical ring with sparse nodal conversion. 358
11.8 NSFNET with the number of convertible routes shown. A number on a link indicates how many source-destination paths passed through the previous node and possibly could have been converted. A number next to a node indicates how many source-destination paths pass through the node and can possibly be wavelength converted. 359
11.9 Blocking probabilities in the NSFNET for optimal and heuristic placement of wavelength converters (30 ERLANG load). 360
11.10 Comparison of blocking probabilities in the NSFNET when using full conversion and no conversion in the network with the Best-Fit algorithm. 362
11.11 Percent gain in the NSFNET from using full-conversion at every node as opposed to no conversion in the network. 362
11.12 Distribution of the number of wavelength converters utilized at node 2 in the NSFNET (30 ERLANG load). 363
11.13 Network with uniform loading. 366

12.1 Implementation of the circuit-migration sequence. 381
12.2 Network architecture for distributed control and management. . 382
12.3 The AT&T testbed architecture. 388
12.4 The Bellcore all-optical network architecture. 390
12.5 Transport node in the RACE architecture. 392

13.1 The WDDI ring network. 400
13.2 WDDI node and bridge interfaces. 403
13.3 Delay versus N characteristics for two-server traffic ($W = 2$). . . 420
13.4 Delay versus N characteristics for clustered traffic, $c = 2$, $k = 5$ ($W = 2$). 422
13.5 Delay versus N characteristics for pseudo-random traffic ($W = 2$). 423
13.6 Delay versus throughput characteristics ($W = 2$). 424
13.7 Delay vs. throughput characteristics for multiple partitions for MIN-DIFF-based algorithms. 426
13.8 Delay vs. number of partitions for MIN-DIFF-based algorithms. 426

13.9 Delay vs. throughput characteristics for multiple partitions for MIN-CROSS-based algorithms. 428
13.10 Delay vs. number of partitions for MIN-CROSS-based algorithms. 428

14.1 EDFA gain curve. 433
14.2 Wavelength routing using AOTF. 434
14.3 Wavelength-routed network with Λ cycles. 436
14.4 An example five-station, five-switch subgraph of the NSFNET T3 backbone. This network is used for the example "static" analysis results. 437
14.5 A (random) four-station, eight-switch network generated by Module 1. Note that this network contains Λ cycles (as indicated by dashed and dotted lines) when all switches are in BAR state. . . 438
14.6 Flow chart of modules. 439
14.7 Network after elimination of Λ cycles using Module 3. 448
14.8 Network after establishing two connections – heavy lines – (from station 2 to station 4 and from station 4 to station 3) using Module 4. However, a new connection – dashed heavy line – (from station 3 to station 1) is causing a Λ cycle – dashed light line. . 448
14.9 Network after elimination of Λ cycles using Module 5. 449
14.10 Fraction of unblocked calls vs. M for the 5-station, 17-link network. 452
14.11 Probability of resource blocking vs. M for the 5-station, 17-link network. 452
14.12 Probability of crosstalk blocking vs. M for the 5-station, 17-link network. 453
14.13 Fraction of unblocked calls vs. M for the 14-station, 56-link network. 454
14.14 Probability of resource blocking vs. M for the 14-station, 56-link network. 454
14.15 Probability of crosstalk blocking vs. M for the 14-station, 56-link network. 455
14.16 Network for Problem 14.1. 459
14.17 Network for Problems 14.2 and 14.3. 460
14.18 Network for Problem 14.4. 461

15.1 Example of a passive-star-based optical metropolitan-area network (slightly modified version of the one used in [LTGC94]). . 464

15.2 Example of a nonreflective star. 465
15.3 Two examples of powers on three wavelengths passing through a fiber. 467
15.4 Simple two-star network that needs no amplifiers to operate. . . 468
15.5 Original amplifier gain model approximations used in previous studies [LTGC94]. 470
15.6 More-accurate amplifier gain model used in this study. 471
15.7 Mid-sized tree-based network needing no amplifiers to function. 481
15.8 A possible MAN network. 482
15.9 A scaled-up version of the MAN network in Fig. 15.8. 485
15.10 A scaled-down version of the MAN network in Fig. 15.8. 485
15.11 A sample switched network. 487
15.12 A cascade of amplifiers along a link. 489
15.13 Network for Problems 15.5 and 15.6. 490
15.14 A portion of a network. 491
15.15 A distribution network. 492

16.1 A TDM link and multiplexer. 495
16.2 Generation of the OTDM signal: packet compression. 498
16.3 Evolution of a nonhyperbolic secant pulse in a fiber. 502
16.4 Evolution of a hyperbolic secant pulse in a fiber. 502
16.5 All-optical clock recovery system (BPF = band-pass filter, PC = polarization controller) [BCHK96]. 503
16.6 The HLAN architecture. 505
16.7 The staggering switch architecture. 506
16.8 The CORD architecture. 507
16.9 A pseudo-random sequence generator. 511
16.10 A CDMA receiver. 512
16.11 Original data streams and coded, transmitted streams. 513
16.12 Combined signals on data channel. 514
16.13 Decoded sequence consisting of original signal and pseudo-noise. 515
16.14 Decoded sequence with varying number of overlapping signals. . 516
16.15 Implementation of a CDMA coder and decoder based on optical splitters and combiners. 517
16.16 Optical time-spreading CDMA. 519
16.17 CDMA codes. 521

List of Tables

2.1 Tunable optical transmitters and their associated tuning ranges and times. 45

2.2 Tunable optical filters and their associated tuning ranges and times. 51

2.3 Amplifier characteristics. 57

2.4 Requirements for EDFA applications. 84

6.1 Comparison of GEMNET (GN), ShuffleNet (SN) and de Bruijn (DB) graph. (Link loads are computed under the assumption of one unit of flow between every source-destination pair and different routing schemes. Also in this table, $\overline{h}$ is averaged over all the individual nodes' average hop distance ($\overline{h_i}$).) 214

7.1 Optimal number of wavelengths $w*$ for $N = 120$ and various values of λ and R. 247

7.2 w^*, w_L, w_H, and $w_{\max}$ for $R = 10$ and various values of N and λ. 248

7.3 Delay, T, in slots, for the seven example systems. 249

8.1 Traffic matrix (in bytes per 15-minute interval). 279

8.2 Summary of experimental results. 281

8.3 Virtual topology for nodal degree $P = 4$ and best scaleup (106). 282

8.4 Traffic scaleups for different nodal degrees. 286

9.1 Cost of upgrading the NSFNET using WDM. 308

9.2 Average packet hop distance for different virtual topology establishment algorithms. 315

10.1 Sample numerical results for static lightpath establishment (SLE) on a 100-node random network. 334

10.2 Sample numerical results for dynamic lightpath establishment (DLE) using the LCP routing scheme on the same 100-node random network as in the static case. Same set of lightpaths as in SLE is considered, but lightpaths are made to arrive randomly. Results in this table are averaged over 10 random arrival patterns of lightpaths. 337

12.1 The CST at node A. 383

13.1 Traffic matrix no. 1, showing two communities of interest. . . . 405
13.2 Traffic matrix no. 2, showing balanced traffic flows. 405
13.3 Traffic matrix no. 3, with each nondiagonal entry being an uniformly distributed random number between 0 and 7. 410

14.1 Static analysis via exhaustive search on the network with five access stations, 5 switches, and 22 links. 456
14.2 Static analysis with an expanded/contracted number of switches. 457

15.1 Important parameters and their values used in the amplifier-placement algorithms. 466
15.2 Number of amplifiers needed for the various amplifier-placement schemes. (Note that N = number of stations and M = number of stars for the lower bound computation. A "*" in column 4 indicates that the NLP solver could not perform better than the LP solution, even when it was given multiple feasible starting points, including the solutions found in [LTGC94] and [RaIM96].) 480
15.3 Exact amplifier placements for the network depicted in Fig. 15.8. 483
15.4 Transmitter and receiver powers for the network depicted in Fig. 15.8. 484

Part I

Introduction

Chapter 1

Optical Networking: Principles and Challenges

1.1 Need + Promise = Challenge!

Life in our increasingly information-dependent society requires that we have access to information at our finger tips *when we need it, where we need it, and in whatever format we need it.* The information is provided to us through our global mesh of communication networks, whose current implementations – such as the present-day Internet and asynchronous transfer mode (ATM) networks – unfortunately, don't have the capacity to support the foreseeable bandwidth demands.

Enter fiber optic technology which can be considered our "saviour" for meeting our above need because of its potentially limitless capabilities [Coch95]:

1. huge bandwidth (nearly 50 terabits per second (Tbps),

2. low signal attenuation (as low as 0.2 dB/km),

3. low signal distortion,

4. low power requirement,

5. low material usage,

6. small space requirement, and

7. low cost.

While fiber optics has been a *curiosity technology* for the past few decades, our challenge now is to turn its promise into reality to meet our information networking demands of the next decade (and well into the next century!).

Thus, the basic premise of this book's topic – viz., optical communication networks – is that, as more and more users start to use our data networks, and as their usage patterns evolve to include more and more bandwidth-intensive networking applications such as data browsing on the world wide web (WWW), java applications, video conferencing, etc., there emerges an acute need for very high-bandwidth transport network facilities, whose capabilities are much beyond those that current high-speed (ATM) networks can provide. Another term that we are increasingly hearing today from anyone who uses the Internet is "network lag," or "net lag" for short. That is, the network is taking longer to accomplish a task today, e.g., to access a WWW server and display a picture, relative to how long it took to perform the same task a few days back. There is just not enough bandwidth in our networks today to support the exponential growth in user traffic!

Given that a single-mode fiber's potential bandwidth is nearly 50 Tbps, which is nearly four orders of magnitude higher than electronic data rates of a few gigabits per second (Gbps), every effort should be made to tap into this huge opto-electronic bandwidth mismatch.

Realizing that the maximum rate at which an end-user – which can be a workstation or a gateway that interfaces with lower-speed subnetworks – can access the network is limited by electronic speed (to a few Gbps), the key in designing optical communication networks in order to exploit the fiber's huge bandwidth is to introduce concurrency among multiple user transmissions into the network architectures and protocols. In an optical communication network, this concurrency may be provided according to either wavelength or frequency [wavelength-division multiplexing (WDM)], time slots [time-division multiplexing (TDM)], or wave shape [spread spectrum, code-division multiplexing (CDM)].

Optical TDM and CDM are somewhat futuristic technologies today. Under (optical) TDM, each end-user should be able to synchronize to within one time slot. The optical TDM bit rate is the aggregate rate over all TDM channels in the system, while the optical CDM chip rate may be much higher than each user's data rate. As a result, both the TDM bit rate and the CDM chip rate may be much higher than electronic processing speed, i.e., some part of an end user's network interface must operate at a rate higher than electronic speed. Thus, TDM and CDM are relatively less attractive than WDM, since WDM – unlike TDM or CDM – has no such requirement.

Specifically, WDM is the current favorite multiplexing technology for optical communication networks since all of the end-user equipment needs to operate only at the bit rate of a WDM channel, which can be chosen arbitrarily, e.g., peak electronic processing speed. Hence, most of this book's material (Chapters 1 through 15) will concentrate on WDM networks; but Chapter 16 will be devoted to optical TDM and CDM.

1.2 xDM vs. xDMA

We have introduced the term xDM where x = {W, T, C} for wavelength, time, and code, respectively. Sometimes, any one of these techniques may be employed for multiuser communication in a *multiple access* environment, e.g., for broadcast communication in a local-area network (LAN) (to be examined in Section 1.5.1).[1]

Thus, a *local optical network* that employs wavelength-division multiplexing is referred to as a *wavelength-division multiple access (WDMA) network*; and TDMA and CDMA networks are defined similarly.

In this book, we will rarely use the term xDMA; instead, we will use the term *xDM* to refer to the multiplexing strategy, and employ the term *xDM network* to refer to the corresponding network, either for local-area or for wide-area (switched) applications. (Corresponding example WDM networks will be introduced in Sections 1.5.1 and 1.5.2, respectively.)

1.3 Wavelength-Division Multiplexing (WDM)

Wavelength-division multiplexing (WDM) is an approach that can exploit the huge opto-electronic bandwidth mismatch by requiring that each end-user's equipment operate only at electronic rate, but multiple WDM channels from different end-users may be multiplexed on the same fiber. Under WDM, the optical transmission spectrum (see Fig. 1.1) is carved up into a number of non-overlapping wavelength (or frequency) bands, with each wavelength

[1] For example, Ethernet is a *multiple access* protocol employed on a broadcast bus.

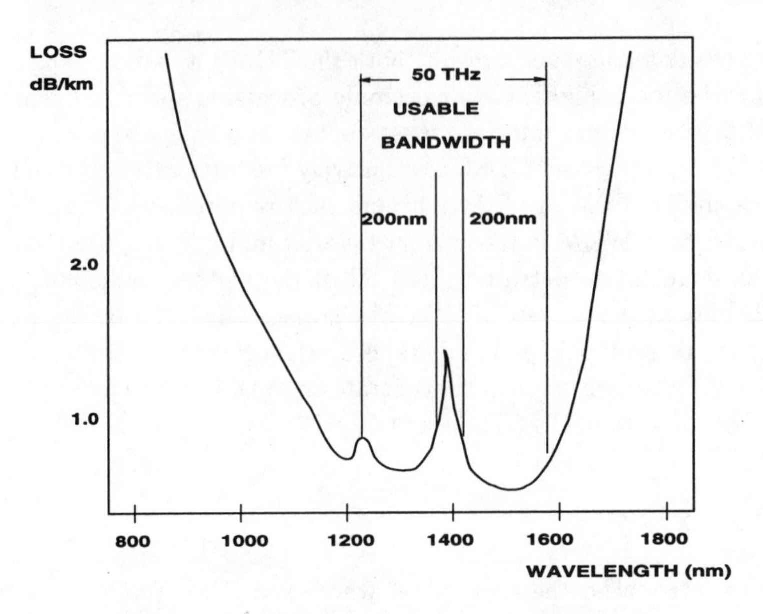

Figure 1.1 The low-attenuation regions of an optical fiber.

supporting a single communication channel operating at whatever rate one desires, e.g., peak electronic speed. Thus, by allowing multiple WDM channels to coexist on a single fiber, one can tap into the huge fiber bandwidth, with the corresponding challenges being the design and development of appropriate network architectures, protocols, and algorithms. Also, WDM devices are easier to implement since, generally, all components in a WDM device need to operate only at electronic speed; as a result, several WDM devices are available in the marketplace today, and more are emerging.

Research and development on optical WDM networks have matured considerably over the past few years, and they seem to have suddenly taken on an explosive form, as evidenced by recent publications [JLT96, JSAC96, JLT93, JSAC90, JHSN95] on this topic as well as overwhelming attendance and enthusiasm at the WDM workshops during recent conferences: Optical Fiber Communications (OFC '97) conference and IEEE International Conference on Communications (ICC '96). A number of experimental prototypes have been and are currently being deployed and tested mainly by telecommunication providers in the U.S., Europe, and Japan. It is anticipated that the next generation of the Internet will employ WDM-based optical backbones.

Current development activities indicate that this sort of WDM network will be deployed mainly as a backbone network for large regions, e.g., for nationwide or global coverage.[2] End-users – to whom the architecture and operation of the backbone will be transparent except for significantly improved response times – will attach to the network through a wavelength-sensitive switching/routing node (details of which will become clearer later in the book). An end-user in this context need not necessarily be a terminal equipment, but the aggregate activity from a collection of terminals – including those that may possibly be feeding in from other regional and/or local subnetworks – so that the end-user's aggregate activity on any of its transmitters is close to the peak electronic transmission rate.

Let us examine below a sample networking problem on such a WDM network.

1.3.1 A Sample WDM Networking Problem

End-users in a fiber-based WDM backbone network may communicate with one another via *all-optical (WDM) channels*, which are referred to as *lightpaths*. A lightpath may span multiple fiber links, e.g., to provide a "circuit-switched" interconnection between two nodes which may have a heavy traffic flow between them and which may be located "far" from each other in the physical fiber network topology. Each intermediate node in the lightpath essentially provides an all-optical bypass facility to support the lightpath.

In an N-node network, if each node is equipped with $N-1$ transceivers [transmitters (lasers) and receivers (filters)] and if there are enough wavelengths on all fiber links, then every node pair could be connected by an all-optical lightpath, and there is no networking problem to solve. However, it should be noted that the network size (N) should be scalable, transceivers are expensive so that each node may be equipped with only a few of them, and technological constraints dictate that the number of WDM channels that can be supported in a fiber be limited to W (whose value is a few tens today, but is expected to improve with time and technological breakthroughs). Thus, only a limited number of lightpaths may be set up on the network.

Under such a network setting, a challenging networking problem is that, given a set of lightpaths that need to be established on the network, and given a constraint on the number of wavelengths, determine the routes over which these lightpaths should be set up and also determine the wavelengths that should be assigned to these lightpaths so that the maximum number

[2] Optical WDM networks for local applications are also being researched and prototyped, e.g., Rainbow, STARNET, etc.; the corresponding network architectures, protocols, algorithms, and challenges are topics of this book's Part II (Chapters 3 through 7).

of lightpaths may be established. While shortest-path routes may be most preferable, note that this choice may have to be sometimes sacrificed, in order to allow more lightpaths to be set up. Thus, one may allow several alternate routes for lightpaths to be established. Lightpaths that cannot be set up due to constraints on routes and wavelengths are said to be blocked, so the corresponding network optimization problem is to minimize this blocking probability.

In this regard, note that, normally, a lightpath operates on the same wavelength across all fiber links that it traverses, in which case the lightpath is said to satisfy the *wavelength-continuity constraint*. Thus, two lightpaths that share a common fiber link should not be assigned the same wavelength. However, if a switching/routing node is also equipped with a *wavelength converter facility*, then the *wavelength-continuity constraints* disappear, and a lightpath may switch between different wavelengths on its route from its origin to its termination.

This particular problem, referred to as the Routing and Wavelength Assignment (RWA) problem, will be examined in detail in Chapter 10, while the general topic of wavelength-routed networks will be studied in Part III (Chapters 8 through 12).

Returning to our sample networking problem, note that designers of next-generation lightwave networks must be aware of the properties and limitations of optical fiber and devices in order for their corresponding protocols and algorithms to take advantage of the full potential of WDM. Often a network designer may approach the WDM architectures and protocols from an overly simplified, ideal, or traditional-networking point of view. Unfortunately, this may lead an individual to make unrealistic assumptions about the properties of fiber and optical components, and hence may result in an unrealizable or impractical design. The goal of this book is to clarify the properties of WDM optical components and present the WDM networking architectures and challenges.

1.4 WDM Networking Evolution

1.4.1 Point-to-Point WDM Systems

WDM technology is being deployed by several telecommunication companies for point-to-point communications. This deployment is being driven by the increasing demands on communication bandwidth. When the demand exceeds the capacity in existing fibers, WDM is turning out to be a more cost-effective alternative compared to laying more fibers. A recent study [MePD95] compared the relative costs of upgrading the transmission capacity of a point-to-point transmission link from OC-48 (2.5 Gbps)[3] to OC-192 (10 Gbps) via three possible solutions:

1. installation/burial of additional fibers and terminating equipment (the "multifiber" solution);

2. a four-channel "WDM solution" (see Fig. 1.2) where a WDM multiplexer (mux) combines four independent data streams, each on a unique wavelength, and sends them on a fiber; and a demultiplexer (demux) at the fiber's receiving end separates out these data streams; and

3. OC-192, a "higher-electronic-speed" solution.

The analysis in [MePD95] shows that, for distances lower than 50 km for the transmission link, the "multi-fiber" solution is the least expensive; but for distances longer than 50 km, the "WDM" solution's cost is the least with the cost of the "higher-electronic-speed" solution not that far behind.

WDM mux/demux in point-to-point links is now available in product form from several vendors such as IBM, Pirelli, and AT&T [Gree96]. Among these products, the maximum number of channels is 20 today, but this number is expected to increase soon.

1.4.2 Wavelength Add/Drop Multiplexer (WADM)

A Wavelength Add/Drop Multiplexer (WADM) is shown in Fig. 1.3. It consists of a demux, followed by a set of 2×2 switches – one switch per wavelength – followed by a mux. The WADM can be essentially "inserted" on a physical fiber link. If all of the 2×2 switches are in the "bar" state, then all of the wavelengths flow through the WADM "undisturbed." However, if one of the 2×2 switches is configured into the "cross" state (as is the case for the λ_i switch in Fig. 1.3) via electronic control (not shown in Fig. 1.3), then the signal on the corresponding wavelength is "dropped" locally, and a new

[3]The terminology OC-n is a widely used telecommunications jargon. "OC" stands for "optical channel" which is unfortunate since it has almost nothing to do with our type of *optics* research and development; it simply specifies electronic data rates. "OC-n" stands for a data rate of $n \times 51.84$ megabits per second (Mbps) approximately; so OC-48 and OC-192 correspond to approximate data rates of 2.5 Gbps and 10 Gbps, respectively. OC-192 is the next milestone in highest achievable electronic communication speed.

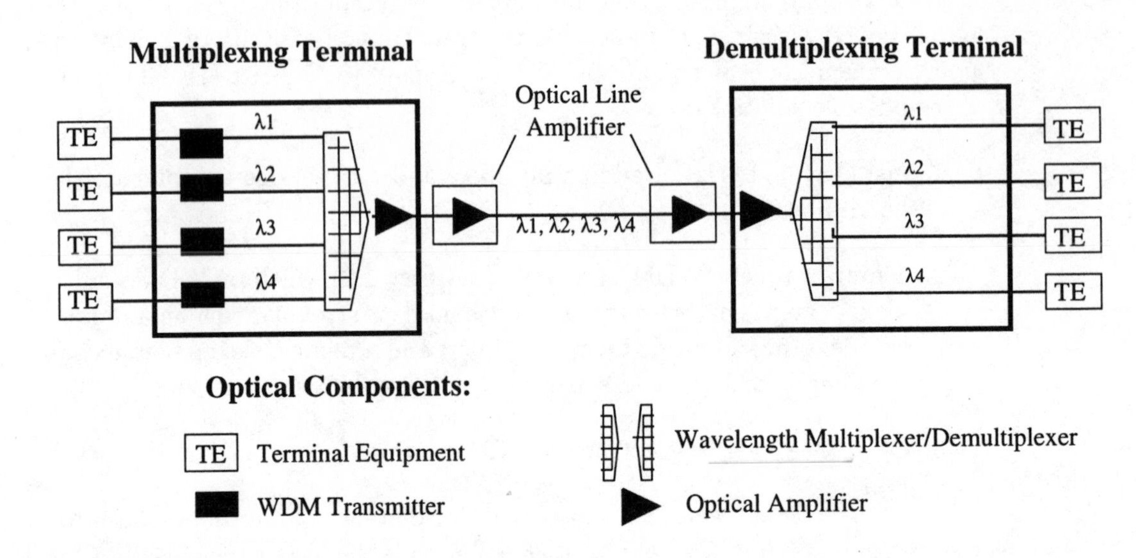

Figure 1.2 A four-channel point-to-point WDM transmission system with amplifiers.

data stream can be "added" on to the same wavelength at this WADM location. More than one wavelength can be "dropped and added" if the WADM interface has the necessary hardware and processing capability.

1.4.3 Fiber and Wavelength Crossconnects – Passive Star, Passive Router, and Active Switch

In order to have a "network" of multiwavelength fiber links, we need appropriate fiber interconnection devices. These devices fall under three broad categories:

- passive star (see Fig. 1.4),
- passive router (see Fig. 1.5), and
- active switch (see Fig. 1.6).

The passive star is a "broadcast" device, so a signal that is inserted on a given wavelength from an input fiber port will have its power equally divided among (and appear on the same wavelength on) all output ports. As an example, in Fig. 1.4, a signal on wavelength λ_1 from Input Fiber 1 and another

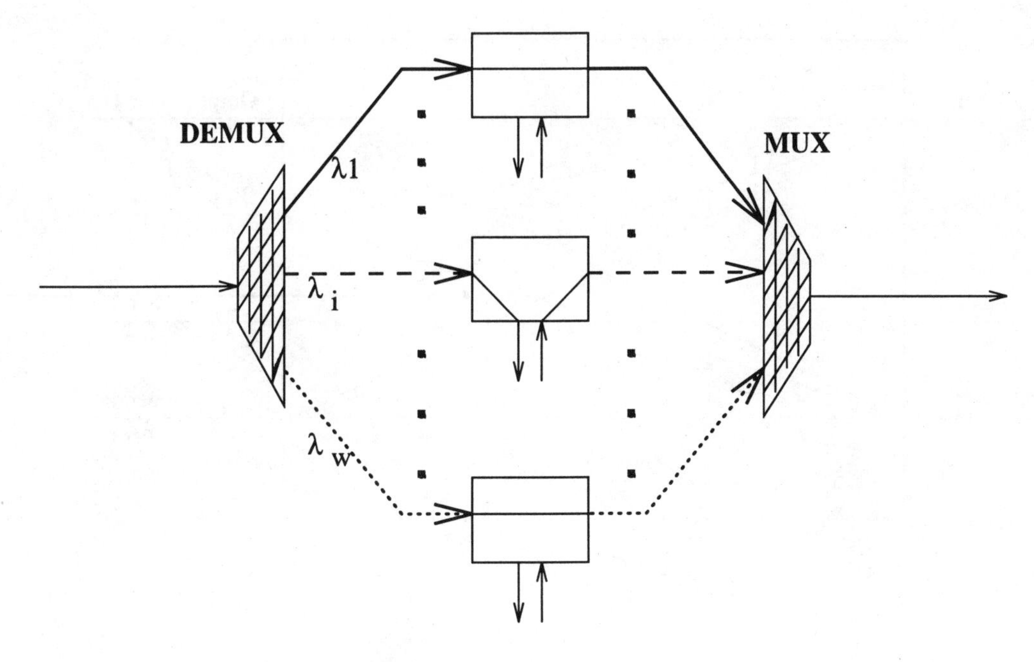

Figure 1.3 A Wavelength Add/Drop Multiplexer (WADM).

on wavelength λ_4 from Input Fiber 4 are broadcast to all output ports. A "collision" will occur when two or more signals from the input fibers are simultaneously launched into the star on the same wavelength. Assuming as many wavelengths as there are fiber ports, an $N \times N$ passive star can route N simultaneous connections through itself.

A passive router can separately route each of several wavelengths incident on an input fiber to the same wavelength on separate output fibers, e.g., wavelengths λ_1, λ_2, λ_3, and λ_4 incident on Input Fiber 1 are routed to the same corresponding wavelengths to Output Fibers 1, 2, 3, and 4, respectively, in Fig. 1.5. Observe that this device allows *wavelength reuse*, i.e., the same wavelength may be spatially reused to carry multiple connections through the router. The wavelength on which an input port gets routed to an output port depends on a "routing matrix" characterizing the router; this matrix is determined by the internal "connections" between the demux and mux stages inside the router (see Fig. 1.5). The routing matrix is "fixed" and cannot be changed. Such routers are commercially available, and are also known as Latin routers, waveguide grating routers (WGRs), wavelength routers (WRs), etc.

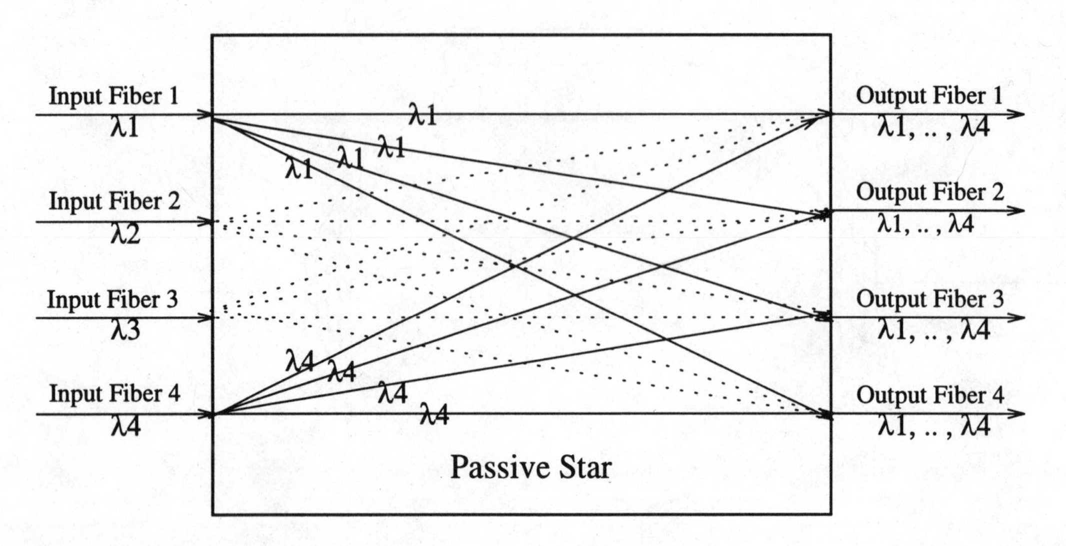

Figure 1.4 A 4 × 4 passive star.

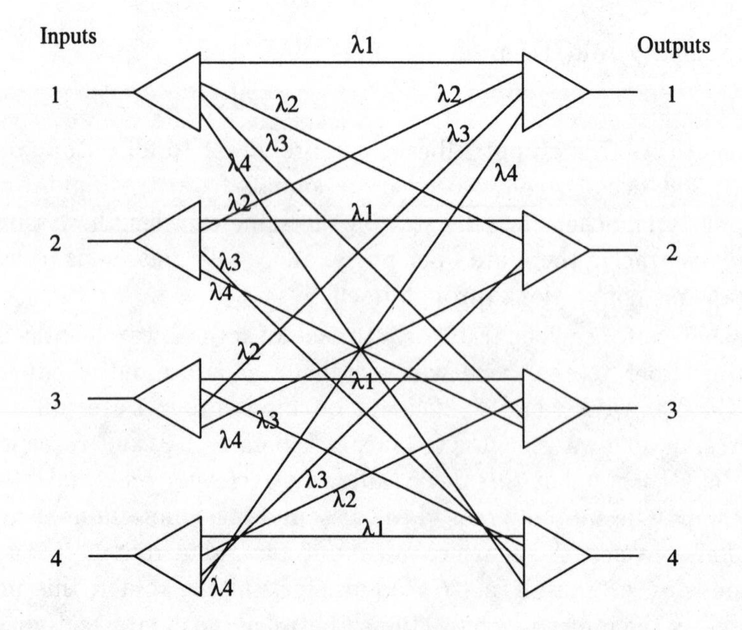

Figure 1.5 A 4 × 4 passive router (four wavelengths).

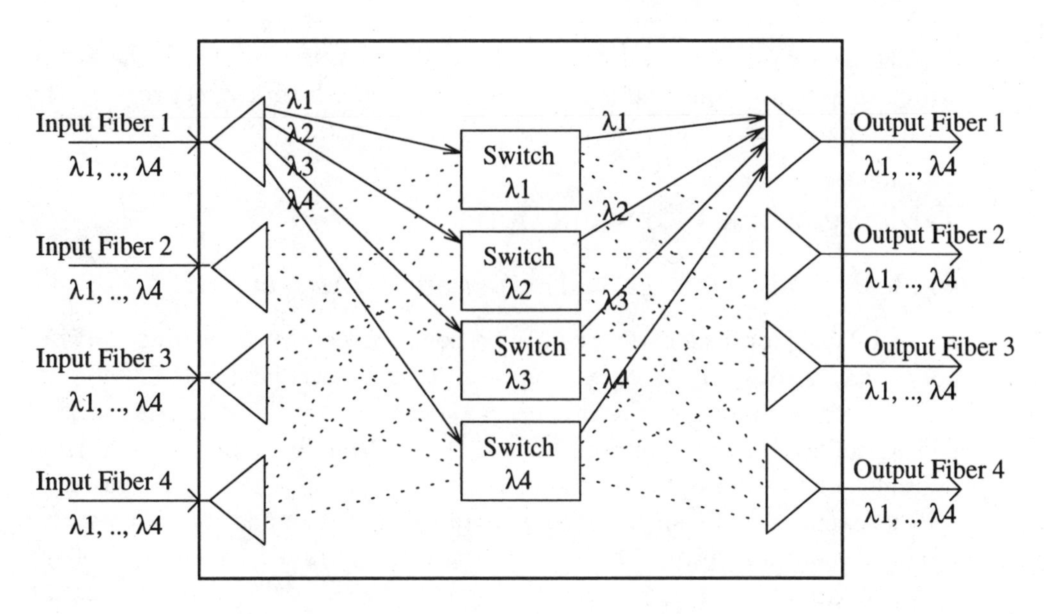

Figure 1.6 A 4 × 4 active switch (four wavelengths).

Again, assuming as many wavelengths as there are fiber ports, a $N \times N$ passive router can route N^2 simultaneous connections through itself (compared to only N for the passive star); however, it lacks the broadcast capability of the star.

The active switch also allows *wavelength reuse*, and it can support N^2 simultaneous connections through itself (like the passive router). But the active star has a further enhancement over the passive router in that its "routing matrix" can be *reconfigured* on demand, under electronic control. However the "active switch" needs to be powered and is not as fault-tolerant as the passive star and the passive router which don't need to be powered. The active switch is also referred to as a wavelength-routing switch (WRS), wavelength selective crossconnect (WSXC), or just crossconnect for short. (We will refer to it as a WRS in this book.)

The active switch can be enhanced with an additional capability, viz., a wavelength may be converted to another wavelength just before it enters the mux stage before the output fiber (see Fig. 1.6). A switch equipped with such a wavelength-conversion facility is more capable than a WRS, and it is referred to as a wavelength-convertible switch, wavelength interchanging crossconnect (WIXC), etc.

The passive star is used to build local WDM networks, while the active switch is used for constructing wide-area wavelength-routed networks. The passive router has mainly found application as a mux/demux device.

1.5 WDM Network Constructions

1.5.1 Broadcast-and-Select (Local) Optical WDM Network

A local WDM optical network may be constructed by connecting network nodes via two-way fibers to a passive star, as shown in Fig. 1.7. A node sends its transmission to the star on one available wavelength, using a laser which produces an optical information stream. The information streams from multiple sources are optically combined by the star and the signal power of each stream is equally split and forwarded to all of the nodes on their receive fibers. A node's receiver, using an optical filter, is tuned to only one of the wavelengths; hence it can receive the information stream. Communication between sources and receivers may follow one of two methods: (1) single-hop, or (2) multihop (which will be studied in Part II of this book). Also, note that, when a source transmits on a particular wavelength λ_1, more than one receiver can be tuned to wavelength λ_1, and all such receivers may pick up the information stream. Thus, the passive-star can support "multicast" services.

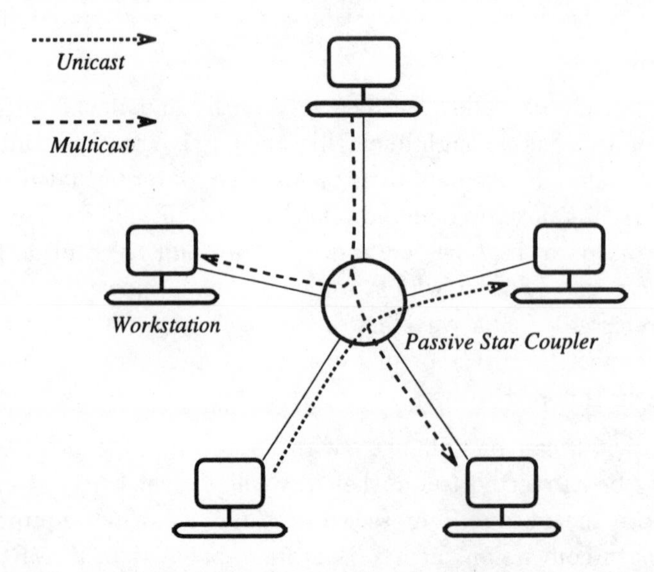

Figure 1.7 A passive-star-based local optical WDM network.

Passive-Star-Based Optical WDM LAN vs. Centralized, Nonblocking-Switch-Based LAN

Consider the passive-star-based optical WDM LAN in Fig. 1.7. If there are N nodes in the system and as many wavelengths as nodes, and also if the bit rate of each WDM channel (and hence of each electronic interface) is B bps, then the aggregate information-carrying capacity of the LAN is upper-bounded by $N \times B$ bps.

Now consider the same network topology as in Fig. 1.7, but the passive star is replaced by a centralized, nonblocking, space-division switch, where the notion of WDM does not exist. Each of the N nodal interfaces still operate at B bps, all on the "same wavelength," so that the aggregate system capacity is still $N \times B$ bps, due to "space-division multiplexing" at the nonblocking switch. So, how does this architecture compare with the passive-star-based WDM LAN solution?

While the passive-star WDM solution cannot boast any capacity enhancement, it nevertheless has the following advantages:

1. In the space-division-switch solution, the "switching intelligence" is centralized. However, the passive star relegates the switching functions to the end nodes (in terms of wavelength tunability at the nodal transmitters and receivers). If a node is down, the rest of the network can still function. Hence, the passive-star solution enjoys the fault-tolerance advantage of any distributed switching solution, relative to the centralized-switch architecture, where the entire network goes down if the switch is down.

2. Another advantage of the WDM passive-star solution is that it allows multicasting "for free." There are some processing requirements with respect to appropriately coordinating the nodal transmitters and receivers. Centralized coordination for supporting multicasting in a switch (also referred to as a "copy" facility) is expected to require more processing.

3. Also, the passive-star solution can be potentially much cheaper since it is purely glass with very little electronics.

1.5.2 Wavelength-Routed (Wide-Area) Optical Network

A wavelength-routed (wide-area) optical WDM network is shown in Fig. 1.8. The network consists of a *photonic switching fabric*, comprising "active switches" connected by fiber links to form an arbitrary physical topology. Each end-user is connected to an active switch via a fiber link. The combination of an end-user and its corresponding switch is referred to as a network node.

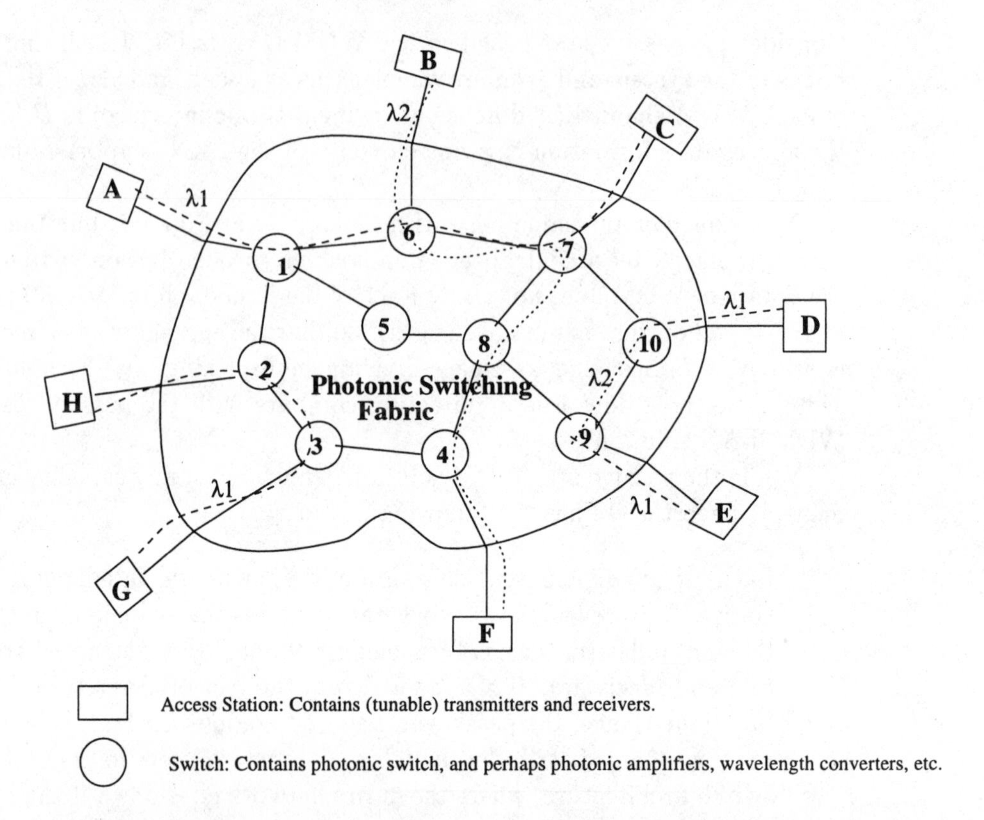

Figure 1.8 A wavelength-routed (wide-area) optical WDM network.

Each node (at its access station) is equipped with a set of transmitters and receivers, both of which may be wavelength tunable. A transmitter at a node sends data into the network and a receiver receives data from the network.

The basic mechanism of communication in a wavelength-routed network is a *lightpath.* A lightpath is an all-optical communication channel between two nodes in the network, and it may span more than one fiber link. The intermediate nodes in the fiber path route the lightpath in the optical domain using their active switches. The end-nodes of the lightpath access the lightpath with transmitters and receivers that are tuned to the wavelength on which the lightpath operates. For example, in Fig. 1.8, lightpaths are established between nodes A and C on wavelength channel λ_1, between B and F on wavelength channel λ_2, and between H and G on wavelength channel λ_1. The lightpath between nodes A and C is routed via active switches 1, 6, and 7. (Note the wavelength reuse for λ_1.)

In the absence of any wavelength conversion device, a lightpath is required to be on the same wavelength channel throughout its path in the network; this requirement is referred to as the *wavelength continuity* property of the lightpath. This requirement may not be necessary if we also have wavelength converters in the network. For example, in Fig. 1.8, the lightpath between nodes D and E traverses the fiber link from node D to switch 10 on wavelength λ_1, gets converted to wavelength λ_2 at switch 10, traverses the fiber link between switch 10 and switch 9 on wavelength λ_2, gets converted back to wavelength λ_1 at switch 9, and traverses the fiber link from switch 9 to node E on wavelength λ_1.

A fundamental requirement in a wavelength-routed optical network is that two or more lightpaths traversing the same fiber link must be on different wavelength channels so that they do not interfere with one another.

1.6 WDM Economics

One study on the economic benefit of WDM in a point-to-point system [MePD95] was highlighted in Section 1.4.1.

Another recent study provides a quantitative analysis of the economic benefits of using WDM in a telephone company's Local Exchange Carrier (LEC) network under traffic demands projected for the year 2000 [Bala96]. An aggressive traffic demand model was used which multiplied the normal traffic projection by 10, assuming that heavy deployment of high-speed networking technology will occur by the year 2000. Synchronous Optical Network (SONET) terminal equipment was assumed to operate between OC-3 rate (155 Mbps) and OC-48 rate (2.5 Gbps).

The cost of introducing WDM in the terminal electronics – i.e., the cost of WDM transmitters and receivers – was assumed to be negligible. Costs for electronic equipment was averaged over several vendor-supplied data. The following costs were modeled: optical amplifier cost = 35% of a OC-48 terminal's cost, WADM cost = 40% of a OC-48 terminal's cost. Only a moderate cost decrease was modeled for WDM devices over the next few years. Two fiber cost models were employed: one considered cost for fiber material and cabling only; the other assumed structure exhaust and computed costs for new conduits and trenching, whenever required.

The above model, further details of which can be found in [Bala96], was applied to three metropolitan-area networks belonging to various telephone companies. Using fiber cost model 1 (no conduit exhaust), *the cost savings of using WDM (vs. no WDM) ranged from 16% to 36%, with the actual "dollar values" of the savings ranging from $86,000,000 to $151,000,000.* Fiber cost model 2 (which models extra cost for conduit exhaust) was applied to only one of the three networks, and it yielded a *cost savings of 33% for using WDM (relative to no WDM), with the actual "dollar values" of the savings equalling $224,000,000.*

It is therefore safe to predict that WDM is here to stay! WDM standardization efforts, e.g., to set up a standard set of wavelengths to facilitate WDM equipment interoperability, are currently in progress under the watch of the International Telecommunications Union (ITU-T Study Group 15/WP4 Q.25) [HaSo96].

1.7 Road Map – Organization of the Book

This book is organized into four parts – Introduction, Broadcast (Local) Optical Networks, Wavelength-Routed (Wide-Area) Optical Networks, and Potpourri – and each part contains between two and five chapters. Each part, as well as each chapter, has been organized as a "stand-alone" entity so that the interested and/or advanced reader can directly go to the part or chapter that is of interest. The contents of each chapter include discussion of theories as well as highlights on prototype implementations/demonstrations, wherever appropriate.

Part I consists of this chapter and the next which deals with enabling technologies, viz., the building blocks for constructing WDM networks. Chapter 2 is essentially a summary of the physical aspects of WDM systems, and is written from the point of view of a computer scientist/engineer. For further details on WDM device technologies, the reader should consult a book that exclusively deals with these topics, e.g., [Gree93]. Readers familiar with device technologies and wishing to study WDM network architectures could skip Chapter 2.

Part II – Broadcast (Local) Optical Networks – contains five chapters. Its first two chapters – Chapters 3 and 4 – examine single-hop network architectures and protocols. While Chapter 3 provides *breadth* of discussion on single-hop networks, a particular single-hop protocol – the IBM Rainbow protocol, which has also been prototyped – is discussed and analyzed in *depth* in Chapter 4. The corresponding *breadth* and *depth* treatment of multihop networks is provided in Chapters 5 and 6, respectively; the multihop network analyzed in depth in Chapter 6 is called *GEMNET*. Chapter 7 examines a novel concept called "channel sharing," and studies how it can be used to support multicasting – the ability to transmit information from a single source node to multiple destination nodes – in a passive-star-based optical network.

Part III – Wavelength-Routed (Wide-Area) Optical Networks – also consists of five chapters. Section 1.3.1 introduced the concept of a *lightpath*. The set of lightpaths in a wavelength-routed network forms a "virtual topology," (1) which may be operated as a "virtual Internet," viz., a *packet-switched electronic overlay* on top of the *WDM optical layer*; (2) which may be optimized based on prevailing traffic conditions; and (3) which may be reconfigured on demand when the pattern of offered traffic changes. Such topics related to optimal virtual topology design are studied in Chapters 8 and 9. Chapter 10 deals with the solution to the Routing and Wavelength Assignment (RWA) problem outlined earlier in this chapter (Section 1.3.1). Chapter 11 examines the issues related to wavelength conversion, reviews several approaches, and reports on a simulation-based study on "sparse wavelength conversion" under which conversion capabilities (which are costly) are sparsely sprinkled through the network. Chapter 12 reviews other work and results on wavelength-routed networks.

The organizing principle behind Part IV – Potpourri – is less strong than those for the other parts of this book, i.e., the contents of Part IV can be considered to be "other important material in WDM networks that did not fit elsewhere." The first chapter (out of four) in Part IV considers a fiber-optic ring network such as a SONET ring or a Fiber Distributed Data Interface (FDDI) network, and examines how to incorporate "growth capability" in such a network by operating it over multiple wavelengths; corresponding optimized partitioning problems to operate the network over multiple subnetworks are defined and analyzed. The next two chapters in Part IV – Chapters 14 and 15 – however, do possess a reasonably cohesive theme, viz., they attempt to address topics at the "device-network interface"; specifically, these chapters *examine the not-so-desirable properties of optics, and try to correct for these "mismatches" using intelligent networking algorithms.* The final chapter – Chapter 16 – deals with optical TDM and CDM networks.

Problem sets are included at the end of each chapter. There are two appendices containing (1) information on further reading, and (2) a glossary of important terms. Finally, all citations are included in a common bibliography at the end of the book.

Exercises

1.1. What are the advantages of fiber optic technology in communication systems?

1.2. In order to take full advantage of the huge bandwidth available on fiber, various multiplexing techniques such as WDM, TDM, and CDM can be used which allow multiple users to share the bandwidth on a single fiber. Compare and contrast these multiplexing techniques. Why is WDM currently the most promising choice for optical communication networks?

1.3. Figure 1.1 shows two regions, 1200-1400 nm and 1450-1650 nm, which are capable of providing up to 50 THz of bandwidth. Calculate the actual bandwidth provided by each region. (Hint: Use the identity $f = v/\lambda$ where $v = 2.0 \times 10^8$ m/s.)

1.4. What is the bandwidth of a 1 nm signal at 1500 nm? At 1350 nm? Give an approximate relation for the bandwidth of a $\Delta\lambda$ nm signal at λ nm.

1.5. Consider the three solutions for upgrading the transmission capacity of a link from OC-48 to OC-192. Suppose the cost of installing additional fiber is $100 per meter, the cost of each transciever is $1000, and the cost of a WDM multiplexer/demultiplexer is $10,000. Determine the maximum length for which you would want to use the multi-fiber solution.

1.6. Give the advantages and disadvantages of the following wavelength crossconnects:
(a) passive star,
(b) passive router, and
(c) active switch.

1.7. Consider the passive star of Fig. 1.4, the passive router of Fig. 1.5, and the active switch of Fig. 1.6. Which of these devices can support the following simultaneous connections? (Assume that TDM is not used, but multicasting is allowed.)

(a) Wavelength λ_1 from input fiber 1 to output fiber 1,
Wavelength λ_1 from input fiber 1 to output fiber 2,
Wavelength λ_2 from input fiber 2 to output fiber 1.

(b) Wavelength λ_2 from input fiber 1 to output fiber 2,
Wavelength λ_2 from input fiber 2 to output fiber 1,
Wavelength λ_3 from input fiber 3 to output fiber 1.

(c) Wavelength λ_1 from input fiber 1 to output fiber 1,
Wavelength λ_2 from input fiber 2 to output fiber 1,
Wavelength λ_3 from input fiber 1 to output fiber 3.

1.8. The routing matrix for an $N \times N$ passive router is called an $N \times N$ Latin Square.
(a) Identify which of the following 2×2 matrices below are Latin Squares.

$$\Lambda_1 = \begin{bmatrix} \lambda_1 & \lambda_2 & \lambda_3 & \lambda_4 \\ \lambda_2 & \lambda_3 & \lambda_4 & \lambda_1 \\ \lambda_4 & \lambda_1 & \lambda_3 & \lambda_2 \\ \lambda_3 & \lambda_4 & \lambda_2 & \lambda_1 \end{bmatrix}$$

$$\Lambda_2 = \begin{bmatrix} \lambda_1 & \lambda_2 & \lambda_3 & \lambda_4 \\ \lambda_2 & \lambda_3 & \lambda_4 & \lambda_1 \\ \lambda_3 & \lambda_4 & \lambda_1 & \lambda_2 \\ \lambda_4 & \lambda_1 & \lambda_2 & \lambda_3 \end{bmatrix}$$

(b) How many distinct 3×3 Latin Squares are there?

1.9. Consider the active switch of Fig. 1.6.
(a) What is the size of each switching element in the center?
(b) Is it possible to construct a 4×4 switch out of 2×2 switches?

1.10. Consider the network of Fig. 1.7. Suppose we replace the passive star by a TDM switch. Compare this new architecture with the previous architecture.

1.11. Consider the simple wavelength-routed optical WDM network shown in Fig. 1.9. Two connections have been established: A-B on wavelength λ_1, and C-B on wavelength λ_2. Establish the connections D-B and C-D while using the minimum number of wavelengths. How would your solution change if wavelength conversion is available at each node?

1.12. Suppose the photonic switching fabric in Fig. 1.8 is replaced by a passive-star coupler. What is the minimum number of wavelengths required to maintain the connections shown in the figure?

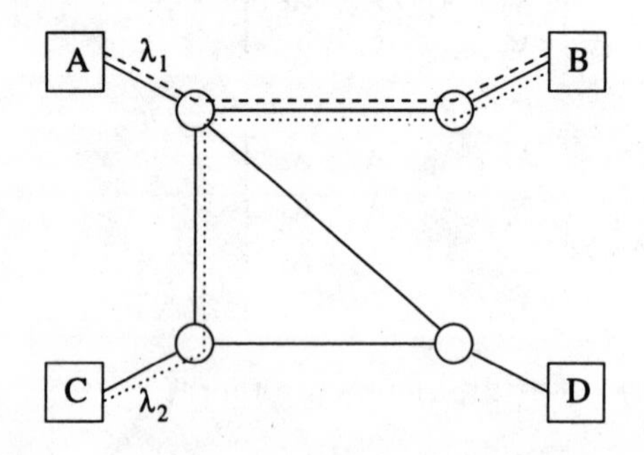

Figure 1.9 A wavelength-routed WDM network.

Chapter 2

Enabling Technologies: Building Blocks

This chapter serves as an introduction to WDM device issues. No background in optics or advanced physics is needed. For a more advanced and/or detailed discussion of WDM devices, we suggest that the reader refer to the references cited in this section. Highly recommended are [Gree93, Henr85, Brac90, Hech92, Hech93].

This chapter presents an overview of optical fiber and devices such as couplers, optical receivers and filters, optical transmitters, optical amplifiers, optical routers, and switches. The chapter attempts to condense the physics behind the principles of optical transmission in fiber in order to provide some background for the nonexpert. WDM network design issues are then discussed in relation to the advantages and limits of optical devices. Finally, we demonstrate how these optical components can be used to create broadcast networks for local networking applications and wavelength-routed networks for wide-area deployment. The chapter concludes with a note on the current status of optical technology and how test networks have used some of the optical devices described in this chapter with a reasonable amount of success.

2.1 Optical Fiber

Fiber possesses many characteristics that make it an excellent physical medium for high-speed networking. Figure 2.1 shows the two low-attenuation regions of optical fiber [Gree93]. Centered at approximately 1300 nm is a

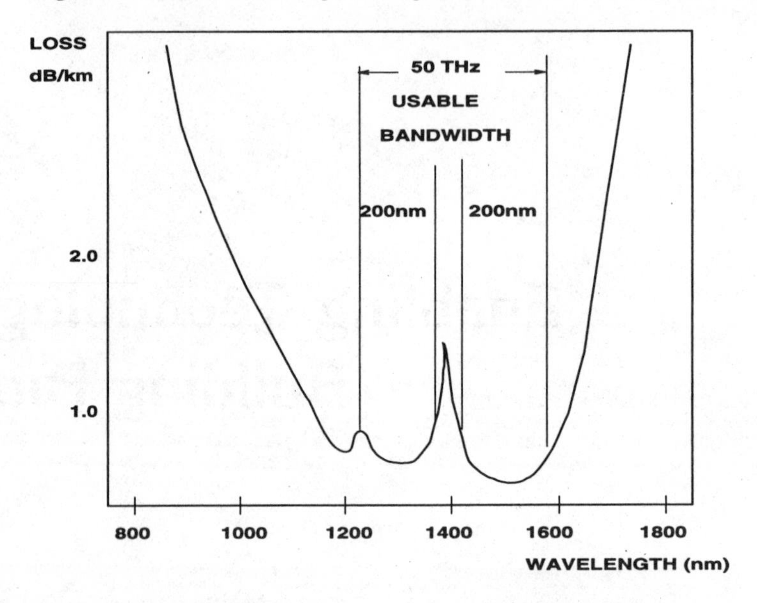

Figure 2.1 The low-attenuation regions of an optical fiber.

range of 200 nm in which attenuation is less than 0.5 dB/km. The total bandwidth in this region is about 25 THz. Centered at 1550 nm is a region of similar size, with attenuation as low as 0.2 dB/km. Combined, these two regions provide a theoretical upper bound of 50 THz of bandwidth.[1] The dominant loss mechanism in good fibers is Rayleigh scattering, while the peak in loss in the 1400 nm region is due to hydroxyl ion (OH^-) impurities in the fiber. Other sources of loss include material absorption and radiative loss.

By using these large low-attenuation areas for data transmission, the signal loss for a set of one or more wavelengths can be made very small, thus reducing the number of amplifiers and repeaters needed. In single-channel long-distance experiments, optical signals have been sent over hundreds of km without amplification. Besides its enormous bandwidth and low attenuation, fiber also offers low error rates. Fiber optic systems typically operate at bit error rates (BERs) of less than 10^{-11}.

[1]However, usable bandwidth is limited by fiber nonlinearities (see Section 2.1.5).

The small size and thickness of fiber allows more fiber to occupy the same physical space as copper, a property which is desirable when installing local networks in buildings. Fiber is flexible, difficult to break, reliable in corrosive environments, and deployable at short notice (which makes it particularly favorable for military communication systems). Also, fiber transmission is immune to electromagnetic interference, and does not cause interference. Finally, fiber is made from one of the cheapest and most readily available substances on earth, viz., sand. This makes fiber environmentally sound, unlike copper.

2.1.1 Optical Transmission in Fiber

Before discussing optical components, it is essential to understand the characteristics of the optical fiber itself. Fiber is essentially a thin filament of glass which acts as a waveguide. A waveguide is a physical medium or a path which allows the propagation of electromagnetic waves, such as light. Due to the physical phenomenon of *total internal reflection*, light can propagate the length of a fiber with little loss. Figure 2.2 shows the cross section of the two types of fiber most commonly used: multimode fiber and single-mode fiber. In order to understand the concept of a mode and to distinguish between these two types of fiber, a diversion into basic optics is needed.

Light travels through vacuum at a speed of $c = 3 \times 10^8$ m/s. Light can also travel through any transparent material, but the speed of light will be slower in the material than in a vacuum. Let c_{mat} be the speed of light for a given material. The ratio of the speed of light in vacuum to that in a material is known as the material's *refractive index* (n), and is given by $n_{\text{mat}} = c/c_{\text{mat}}$.

When light travels from material of a given refractive index to material of a different refractive index (i.e., when refraction occurs), the angle at which the light is transmitted in the second material depends on the refractive indices of the two materials as well as the angle at which light strikes the interface between the two materials. Due to Snell's Law, $n_a sin\theta_a = n_b sin\theta_b$, where n_a and n_b are the refractive indices of the first substance and the second substance, respectively; θ_a is the angle of incidence, or the angle with respect to normal that light hits the surface between the two materials; and θ_b is the angle of light in the second material. However, if $n_a > n_b$ and θ_a is greater than some critical value, the rays are reflected back into substance a from its boundary with substance b.

Looking again at Fig. 2.2, we see that the fiber consists of a core completely surrounded by a cladding (both of which consist of glass of different refractive indices). Let us first consider a *step-index fiber*, in which the change

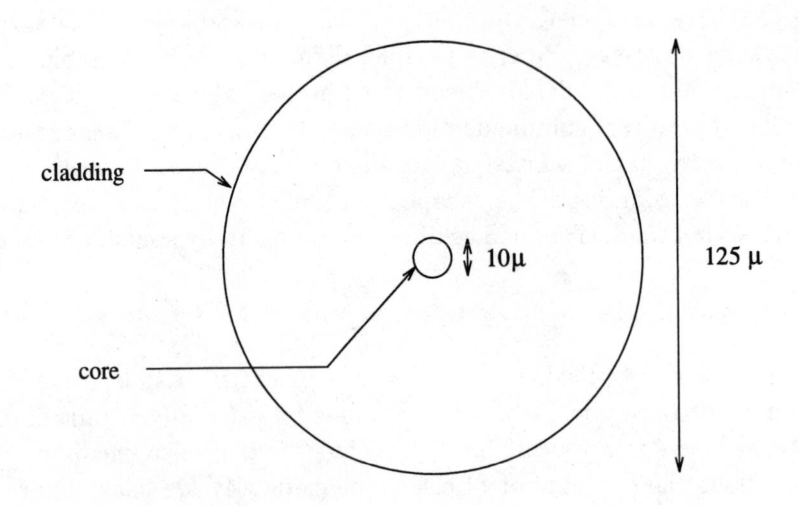

(a) Single-Mode Optical Fiber

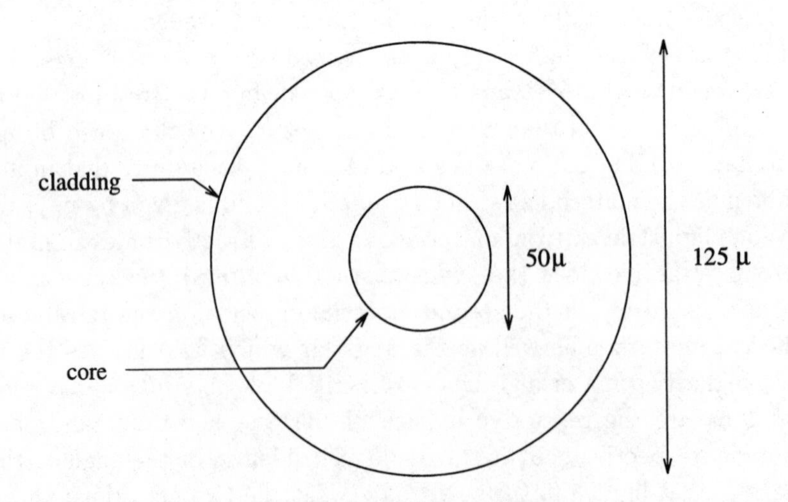

(b) Multimode Optical Fiber

Figure 2.2 Single-mode and multimode optical fibers.

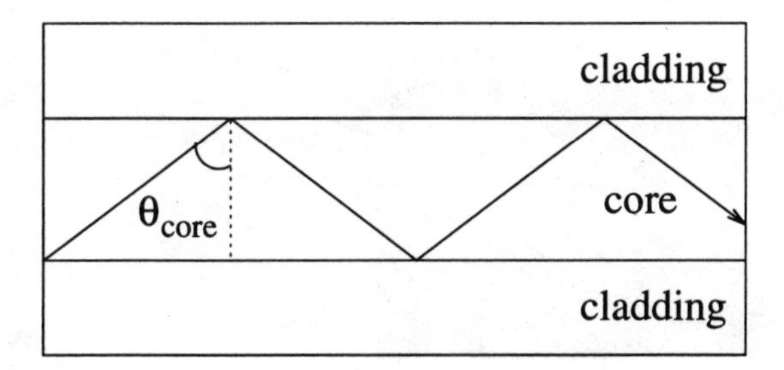

Figure 2.3 Light traveling via total internal reflection within a fiber.

of refractive index at the core-cladding boundary is a step function. If the refractive index of the cladding is less than that of the core, then *total internal reflection* can occur in the core, and light can propagate through the fiber (as shown in Fig. 2.3). The angle above which total internal reflection will take place is known as the *critical angle*, and is given by θ_{core} which corresponds to $\theta_{\text{clad}} = 90°$. From Snell's Law, we have

$$\sin\theta_{\text{clad}} = \frac{n_{\text{core}}}{n_{\text{clad}}}\sin\theta_{\text{core}}$$

The critical angle is then

$$\theta_{\text{crit}} = \sin^{-1}\left(\frac{n_{\text{clad}}}{n_{\text{core}}}\right). \tag{2.1}$$

So, for total internal reflection, we require

$$\theta_{\text{crit}} > \sin^{-1}\left(\frac{n_{\text{clad}}}{n_{\text{core}}}\right)$$

In other words, for light to travel down a fiber, the light must be incident on the core-cladding surface at an angle greater than θ_{crit}.

In some cases, the fiber may have a *graded index* in which the interface between the core and the cladding undergoes a gradual change in refractive index with $n_i > n_{i+1}$ (Fig. 2.4). A *graded-index* fiber reduces the minimum θ_{crit} required for total internal reflection, and also helps to reduce the *intermodal dispersion* in the fiber. Intermodal dispersion will be discussed in the following sections.

In order for light to enter a fiber, the incoming light should be at an angle such that the refraction at the air-core boundary results in the transmitted light being at an angle for which total internal reflection can take place at the

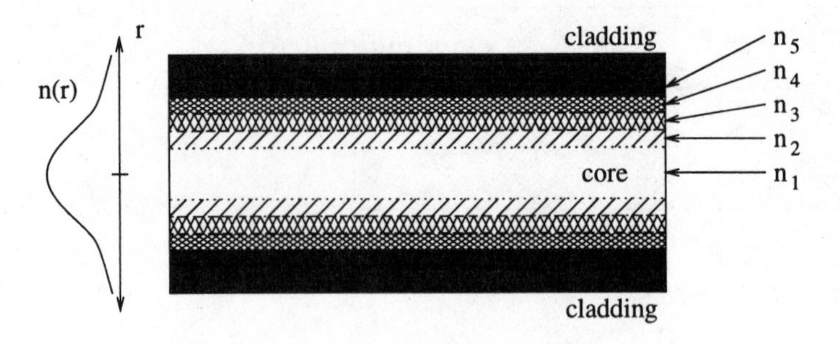

Figure 2.4 Graded-index fiber.

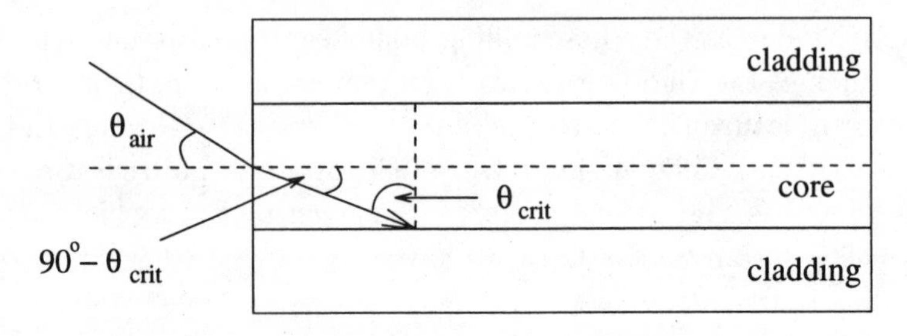

Figure 2.5 Numerical aperture of a fiber.

core-cladding boundary. As shown in Fig. 2.5, the maximum value of θ_{air} can be derived from

$$\begin{aligned} n_{\text{air}} \sin \theta_{\text{air}} &= n_{\text{core}} \sin(90^\circ - \theta_{\text{crit}}) \\ &= n_{\text{core}} \sqrt{1 - \sin^2 \theta_{\text{crit}}} \end{aligned} \tag{2.2}$$

From Eqn. (2.1), since $\sin \theta_{\text{crit}} = n_{\text{clad}} / n_{\text{core}}$, we can rewrite Eqn. (2.2) as

$$n_{\text{air}} \sin \theta_{\text{air}} = \sqrt{n_{\text{core}}^2 - n_{\text{clad}}^2} \tag{2.3}$$

The quantity $n_{\text{air}} \sin \theta_{\text{air}}$ is referred to as NA, the *numerical aperture* of the fiber, and θ_{air} is the maximum angle with respect to the normal at the air-core boundary, so that the incident light that enters the core will experience total internal reflection inside the fiber.

2.1.2 Single-Mode vs. Multimode Fiber

A mode in an optical fiber corresponds to one of possibly multiple ways in which a wave may propagate through the fiber. It can also be viewed as a standing wave in the transverse plane of the fiber. More formally, a mode corresponds to a solution of the wave equation which is derived from Maxwell's equations and subject to boundary conditions imposed by the optical fiber waveguide.

An electromagnetic wave propagating along an optical fiber consists of an electric field vector, **E**, and a magnetic field vector, **H**. Each field can be broken down into three components. In the cylindrical coordinate system, these components are $E_\rho, E_\phi, E_z, H_\rho, H_\phi$, and H_z, where ρ is the component of the field which is normal to the wall (core-cladding boundary) of the fiber, ϕ is the component of the field which is tangential to the wall of the fiber, and z is the component of the field which is in the direction of propagation. Fiber modes are typically referred to using the notation HE_{xy} (if $H_z > E_z$), or EH_{xy} (if $E_z > H_z$), where x and y are both integers. For the case $x = 0$, the modes are also referred to as transverse-electric (TE) in which case $E_z = 0$, or transverse-magnetic (TM) in which case $H_z = 0$.

Although total internal reflection may occur for any angle θ_{core} which is greater than θ_{crit}, light will not necessarily propagate for all of these angles. For some of these angles, light will not propagate due to destructive interference between the incident light and the reflected light at the core-cladding interface within the fiber. For other angles of incidence, the incident wave and the reflected wave at the core-cladding interface constructively interfere in order to maintain the propagation of the wave. The angles for which waves do propagate correspond to *modes* in a fiber. If more than one mode may propagate through a fiber, the fiber is called multimode. In general, a larger core diameter or higher operating frequencies allow a greater number of modes to propagate.

The number of modes supported by a multimode optical fiber is related to the normalized frequency V which is defined as

$$V = k_0 a \sqrt{n_{\text{core}}^2 - n_{\text{clad}}^2} \tag{2.4}$$

where $k_0 = 2\pi/\lambda$, a is the radius of the core, and λ is the wavelength of the propagating light in vacuum. In multimode fiber, the number of modes, m, is given approximately by

$$m \approx \frac{1}{2} V^2. \tag{2.5}$$

The advantage of multimode fiber is that its core diameter is relatively large; as a result, injection of light into the fiber with low coupling loss[2] can be accomplished by using inexpensive, large-area light sources, such as light-emitting diodes (LEDs).

The disadvantage of multimode fiber is that it introduces the phenomenon of *intermodal dispersion.* In multimode fiber, each mode propagates at a different velocity due to different angles of incidence at the core-cladding boundary. This effect causes different rays of light from the same source to arrive at the other end of the fiber at different times, resulting in a pulse which is spread out in the time domain. Intermodal dispersion increases with the propagation distance. The effect of intermodal dispersion may be reduced through the use of *graded-index* fiber, in which the region between the cladding and the core of the fiber consists of a series of gradual changes in the index of refraction (see Fig. 2.4). However, even with graded-index multimode fiber, intermodal dispersion may still limit the bit rate of the transmitted signal and may limit the distance that the signal can travel.

One way to limit intermodal dispersion is to reduce the number of modes. From Eqns. (2.4) and (2.5), we observe that this reduction in the number of modes can be accomplished by reducing the core diameter, by reducing the numerical aperture, or by increasing the wavelength of the light.

By reducing the fiber core to a sufficiently small diameter and by reducing the numerical aperture, it is possible to capture only a single mode in the fiber. This single mode is the HE_{11} mode, also known as the *fundamental mode.* Single-mode fiber usually has a core size of about 10 μm, while multimode fiber typically has a core size of 50 to 100 μm (see Fig. 2.2). A step-index fiber will support a single mode if V in Eqn. (2.4) is less than 2.4048 [Ishi91].

Thus, single-mode fiber eliminates intermodal dispersion, and can, hence, support transmission over much longer distances. However, it introduces the problem of concentrating enough power into a very small core. LEDs cannot couple enough light into a single-mode fiber to facilitate long distance communications. Such a high concentration of light energy may be provided by a semiconductor laser, which can generate a narrow beam of light.

2.1.3 Attenuation in Fiber

Attenuation in optical fiber leads to a reduction of the signal power as the signal propagates over some distance. When determining the maximum distance that a signal can propagate for a given transmitter power and receiver

[2] Coupling loss measures the power loss experienced when attempting to direct light into a fiber.

sensitivity, one must consider attenuation. Receiver sensitivity is the minimum power required by a receiver to detect the signal. Let $P(L)$ be the power of the optical pulse at distance L km from the transmitter and A be the attenuation constant of the fiber (in dB/km). Attenuation is characterized by [Henr85]

$$P(L) = 10^{-AL/10} P(0) \tag{2.6}$$

where $P(0)$ is the optical power at the transmitter. For a link length of L km, $P(L)$ must be greater than or equal to P_r, the receiver sensitivity. From Eqn. (2.6), we get

$$L_{\max} = \frac{10}{A} \log_{10} \frac{P(0)}{P_r} \tag{2.7}$$

The maximum distance between the transmitter and the receiver (or the distance between amplifiers[3]) depends more heavily on the constant A than on the optical power launched by the transmitter. Referring back to Fig. 2.1, we note that the lowest attenuation ($\approx$ 0.2 dB/km) occurs at approximately 1550 nm.

2.1.4 Dispersion in Fiber

Dispersion is the widening of a pulse duration as it travels through a fiber. As a pulse widens, it can broaden enough to interfere with neighboring pulses (bits) on the fiber, leading to intersymbol interference. Dispersion thus limits the bit spacing and the maximum transmission rate on a fiber-optic channel.

As mentioned earlier, one form of dispersion is *intermodal dispersion.* This is caused when multiple modes of the same signal propagate at different velocities along the fiber. Intermodal dispersion does not occur in a single-mode fiber.

Another form of dispersion is *material* or *chromatic dispersion.* In a dispersive medium, the index of refraction is a function of the wavelength. Thus, if the transmitted signal consists of more than one wavelength, certain wavelengths will propagate faster than other wavelengths. Since no laser can create a signal consisting of an exact single wavelength, or more precisely, since any information carrying signal will have a nonzero spectral width (range

[3]The amplifier sensitivity is usually equal to the receiver sensitivity, while the amplifier output is usually equal to optical power at a transmitter.

of wavelengths/frequencies in the signal), material dispersion will occur in most systems[4].

A third type of dispersion is *waveguide dispersion.* Waveguide dispersion is caused because the propagation of different wavelengths depends on waveguide characteristics such as the indices and shape of the fiber core and cladding.

At 1300 nm, material dispersion in a conventional single-mode fiber is near zero. Luckily, this is also a low-attenuation window (although loss is lower at 1550 nm). Through advanced techniques such as *dispersion shifting*, fibers with zero dispersion at a wavelength between 1300 nm and 1700 nm can be manufactured [Powe93]. In a dispersion-shifted fiber, the core and cladding are designed such that the waveguide dispersion is negative with respect to the material dispersion, thus canceling the total dispersion. However, the dispersion will only be zero for a single wavelength. If the signal bandwidth is low, that may be good enough. However, the installed fiber is the standard fiber, and we have to live with its dispersion properties.

2.1.5 Nonlinearities in Fiber

Nonlinear effects in fiber may potentially have a significant impact on the performance of WDM optical communication systems. Nonlinearities in fiber may lead to attenuation, distortion, and cross-channel interference. In a WDM system, these effects place constraints on the spacing between adjacent wavelength channels, limit the maximum power on any channel, and may also limit the maximum bit rate.

Nonlinear Refraction

In optical fiber, the index of refraction depends on the optical intensity of signals propagating through the fiber [Chra90]. Thus, the phase of the light at the receiver will depend on the phase of the light sent by the transmitter, the length of the fiber, and the optical intensity. Two types of nonlinear effects caused by this phenomenon are self-phase modulation (SPM) and cross-phase modulation (XPM).

SPM is caused by variations in the power of an optical signal and results in variations in the phase of the signal. The amount of phase shift introduced

[4]Even if an unmodulated source consisted of a single wavelength, the process of modulation would cause a spread of wavelengths.

by SPM is given by

$$\phi_{NL} = n_2 k_0 L |E|^2 \tag{2.8}$$

where n_2 is the nonlinear coefficient for the index of refraction, $k_0 = 2\pi/\lambda$, L is the length of the fiber, and $|E|^2$ is the optical intensity. In phase-shift-keying (PSK) systems, SPM may lead to a degradation of the system performance, since the receiver relies on the phase information. SPM also leads to the spectral broadening of pulses, as explained below. Instantaneous variations in a signal's phase caused by changes in the signal's intensity will result in instantaneous variations of frequency around the signal's central frequency. For very short pulses, the additional frequency components generated by SPM combined with the effects of material dispersion will also lead to spreading or compression of the pulse in the time domain, affecting the maximum bit rate and the bit error rate.

Cross-phase modulation (XPM) is a shift in the phase of a signal caused by the change in intensity of a signal propagating at a different wavelength. XPM can lead to asymmetric spectral broadening, and combined with SPM and dispersion, may also affect the pulse shape in the time domain.

Although XPM may limit the performance of fiber-optic systems, it may also have advantageous applications. XPM can be used to modulate a pump signal at one wavelength from a modulated signal on a different wavelength. Such techniques can be used in wavelength conversion devices and are discussed in Section 2.6.

Stimulated Raman Scattering

Stimulated Raman Scattering (SRS) is caused by the interaction of light with molecular vibrations. Light incident on the molecules creates scattered light at a longer wavelength than that of the incident light. A portion of the light traveling at each frequency in a Raman-active fiber is downshifted across a region of lower frequencies. The light generated at the lower frequencies is called the Stokes wave. The range of frequencies occupied by the Stokes wave is determined by the Raman gain spectrum which covers a range of around 40 THz below the frequency of the input light. In silica fiber, the Stokes wave has a maximum gain at a frequency of around 13.2 THz less than the input signal.

The fraction of power transferred to the Stokes wave grows rapidly as the power of the input signal is increased. Under very high input power, SRS will cause almost all of the power in the input signal to be transferred to the Stokes wave.

In multiwavelength systems, the shorter-wavelength channels will lose some power to each of the higher-wavelength channels within the Raman gain spectrum. To reduce the amount of loss, the power on each channel needs to be below a certain level. In [Chra84], it is shown that in a 10-channel system with 10 nm channel spacing, the power on each channel should be kept below 3 mW to minimize the effects of SRS.

Stimulated Brillouin Scattering

Stimulated Brillouin Scattering (SBS) is similar to SRS, except that the frequency shift is cause by sound waves rather than molecular vibrations [Chra90]. Other characteristics of SBS are that the Stokes wave propagates in the opposite direction of the input light, and SBS occurs at relatively low input powers for wide pulses (greater than 1 μs), but has negligible effect for short pulses (less than 10 ns) [Agra89]. The intensity of the scattered light is much greater in SBS than in SRS, but the frequency range of SBS, on the order of 10 GHz, is much lower than that of SRS. Also, the gain bandwidth of SBS is only on the order of 100 MHz.

To counter the effects of SBS, one must ensure that the input power is below a certain threshold. Also, in multiwavelength systems, SBS may induce crosstalk between channels. Crosstalk will occur when two counter-propagating channels differ in frequency by the Brillouin shift, which is around 11 GHz for wavelengths at 1550 nm. However, the narrow gain bandwidth of SBS makes SBS crosstalk fairly easy to avoid.

Four-Wave Mixing

Four-Wave Mixing (FWM) occurs when two wavelengths, operating at frequencies f_1 and f_2, respectively, mix to cause signals at $2f_1 - f_2$ and $2f_2 - f_1$. These extra signals, called sidebands, can cause interference if they overlap with frequencies used for data transmission. Likewise, mixing can occur between combinations of three or more wavelengths. The effect of FWM in WDM systems can be reduced by using unequally spaced channels [FTCM94].

Four-wave mixing can be used to provide wavelength conversion, as will be shown in Section 2.6.

Summary

Nonlinear effects in optical fibers may potentially limit the performance of WDM optical networks. Such nonlinearities may limit the optical power on each channel, limit the maximum number of channels, limit the maximum transmission rate, and constrain the spacing between different channels.

It is shown that, in a WDM system using channels spaced 10 GHz apart and a transmitter power of 0.1 mW per channel, a maximum of about 100 channels can be obtained in the 1550 nm low-attenuation region [Chra90].

The details of optical nonlinearities are very complex, and beyond the scope of this book. However, they are a major limiting factor in the available number channels in a WDM system, especially those operating over distances greater than 30 km [Chra90]. The existence of these nonlinearities suggests that WDM protocols which limit the number of nodes to the number of channels do not scale well. For further details on fiber nonlinearities, the reader is referred to [Agra89].

2.1.6 Couplers

Coupler is a general term that covers all devices that combine light into or split light out of a fiber. A splitter is a coupler that divides the optical signal on one fiber to two or more fibers. The most common splitter is a 1×2 splitter, as shown in Fig. 2.6(a). The *splitting ratio*, α, is the amount of

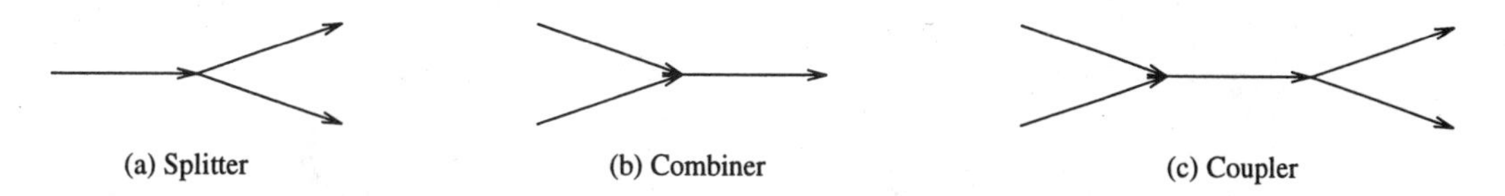

Figure 2.6 Splitter, combiner, and coupler.

power that goes to each output. For a two-port splitter, the most common splitting ratio is 50:50, though splitters with any ratio can be manufactured [Powe93]. Combiners (see Fig. 2.6(b)) are the reverse of splitters, and when turned around, a combiner can be used as a splitter. An input signal to the combiner suffers a power loss of about 3 dB. A 2×2 coupler (see Fig. 2.6(c)), in general, is a 2×1 combiner followed immediately by a 1×2 splitter, which has the effect of broadcasting the signals from two input fibers onto two output fibers. One implementation of a 2×2 coupler is the *fused biconical tapered coupler* which basically consists of two fibers fused together. In addition to the 50:50 power split incurred in a coupler, a signal also experiences *return loss.* If the signal enters an input of the coupler, roughly half of the signal's power goes to each output of the coupler. However, a small amount of power is reflected in the opposite direction and is directed back to the inputs of the coupler. Typically, the amount of power returned by a coupler is 40-50 dB below the input power. Another type of loss is *insertion loss.* One source of insertion loss is the loss incurred when directing the light from a fiber into the coupler device; ideally, the axes of the fiber core and the coupler input port must be perfectly aligned, but full perfection may not be achievable due to the very small dimensions.

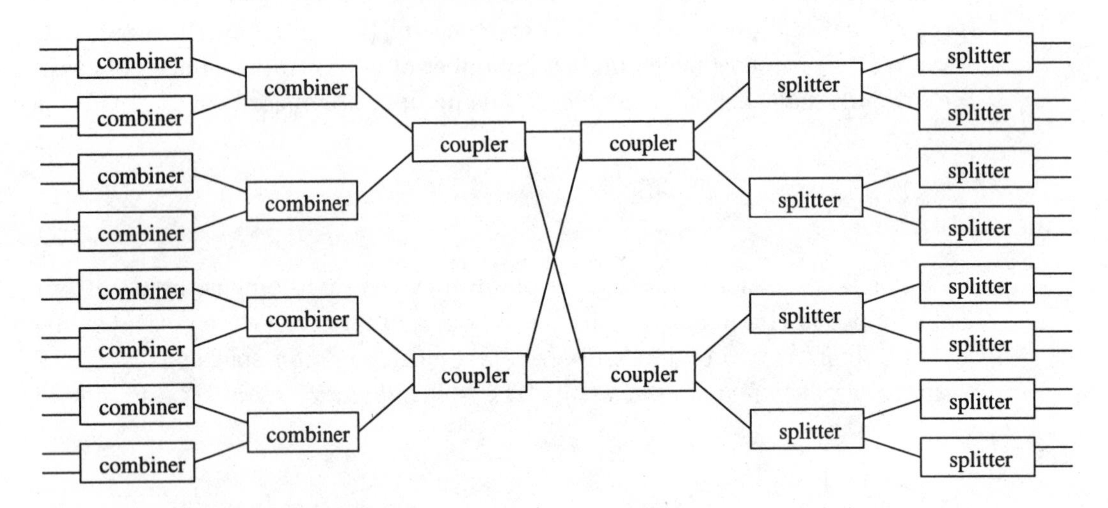

Figure 2.7 A 16×16 passive-star coupler.

The passive-star coupler (PSC) is a multiport device in which light coming into any input port is broadcast to every output port. The PSC is attractive because the optical power that each output receives P_{out} equals

$$P_{\text{out}} = \frac{P_{\text{in}}}{N} \tag{2.9}$$

where P_{in} is the optical power introduced into the star by a single node and N is the number of output ports of the star. Note that this expression ignores the *excess loss,* caused by flaws introduced in the manufacturing process, that the signal experiences when passing through each coupling element. One way to implement the PSC is to use a combination of splitters, combiners, and couplers as shown in Fig. 2.7. (However, this implementation of a PSC and the implementation of the 2×2 coupler in Fig. 2.6(c) are not the most power efficient. An alternative, power-efficient design of the PSC will be explored as an exercise in Chapter 3.) Another implementation of the star coupler is the integrated-optics planar star coupler in which the star coupler and waveguides are fabricated on a semiconductor, glass (silica), or polymer substrate. A 19×19 star coupler on silicon has been demonstrated with excess loss of around 3.5 dB at a wavelength of 1300 nm [DHKK89]. In [OkTa91], an 8×8 star coupler with an excess loss of 1.6 dB at a wavelength of 1550 nm was demonstrated.

2.2 Optical Transmitters

In order to understand how a tunable optical transmitter works, we must first understand some of the fundamental principles of lasers and how they work. Then, we will discuss various implementations of tunable lasers and their properties. Good references on tunable laser technology include [Gree93, Brac90, LeZa89].

2.2.1 How a Laser Works

The word *laser* is an acronym for *Light Amplification by Stimulated Emission of Radiation.* The key words are stimulated emission, which is what allows a laser to produce intense high-powered beams of coherent light (light which contains one or more distinct frequencies).

In order to understand stimulated emission , we must first acquaint ourselves with the energy levels of atoms. Atoms that are stable (in the ground state) have electrons that are in the lowest possible energy levels. In each atom, there are a number of discrete levels of energy that an electron can have; thus, we refer to them as states. In order to change the level of an electron in the ground state, the atom must absorb energy. When an atom absorbs energy, it becomes excited, and moves to a higher energy level. At this point, the electron is unstable, and usually moves quickly back to the ground state by releasing a *photon*, a particle of light.

However, there are certain materials whose states are *quasi-stable*, which means that the substances are likely to stay in the excited state for longer periods of time, without constant excitation. By applying enough energy (either in the form of an optical pump or in the form of an electrical current) to a substance with quasi-stable states for a long enough period of time, *population inversion* occurs, which means that there are more electrons in the excited state than in the ground state. As we shall see, this inversion allows the substance to emit more light than it absorbs.

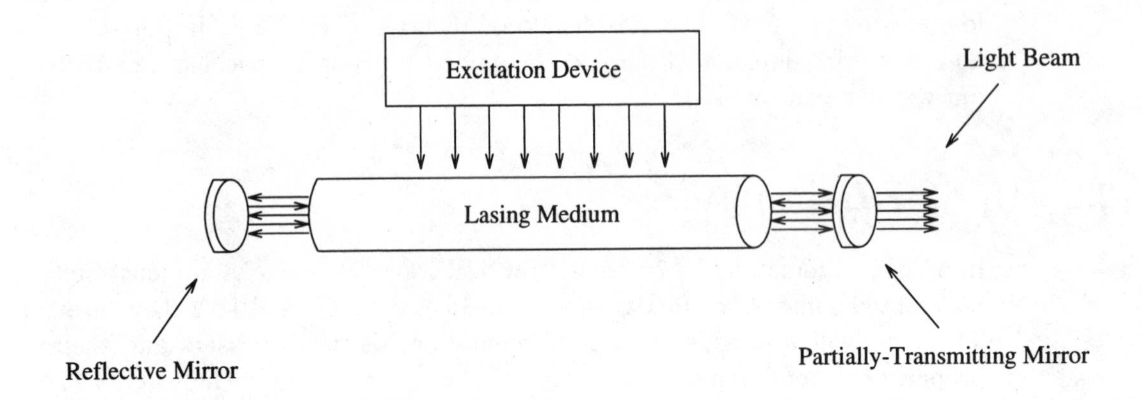

Figure 2.8 The general structure of a laser.

Figure 2.8 shows a general representation of the structure of a laser. The laser consists of two mirrors which form a cavity (the space between the mirrors), a lasing medium which occupies the cavity, and an excitation device. The excitation device applies current to the lasing medium, which is made of a quasi-stable substance. The applied current excites electrons in the lasing medium, and when an electron in the lasing medium drops back to the ground state, it emits a photon of light. The photon will reflect off the mirrors at each end of the cavity, and will pass through the medium again.

Stimulated emission occurs when a photon passes very closely to an excited electron. The photon may cause the electron to release its energy and return to the ground state. In the process of doing so, the electron releases another photon which will have the same direction and coherence as the stimulating photon. Photons for which the frequency is an integral fraction of the cavity length will coherently combine to build up light at the given frequency within the cavity. Between "normal" and stimulated emission, the light at the selected frequency builds in intensity until energy is being removed from the medium as fast as it is being inserted. The mirrors feed the photons back and forth, so further stimulated emission can occur and higher intensities of light can be produced. One of the mirrors is partially transmitting, so that some photons will escape the cavity in the form of a narrowly focused beam of light. By changing the length of the cavity, the frequency of the emitted light can be adjusted.

The frequency of the photon emitted depends on its change in energy levels. The frequency is determined by the equation

$$f = \frac{E_i - E_f}{h} \tag{2.10}$$

where f is the frequency of the photon, E_i is the initial (quasi-stable) state of the electron, E_f is the final (ground) state of the electron, and h is Planck's constant. In a gas laser, the distribution for $E_i - E_f$ is given by an exponential probability distribution, known as the *Boltzmann distribution*, which changes depending on the temperature of the gas. Although many frequencies are possible, only a single frequency, which is determined by the cavity length, is emitted from the laser.

Semiconductor Diode Lasers

The most useful type of laser for optical communications is the semiconductor diode laser. The simplest implementation of a semiconductor laser is the bulk laser diode, which is a p-n junction with mirrored edges perpendicular to the junction (see Fig. 2.9). To understand the operation of the semiconductor diode requires a brief diversion into semiconductor physics.

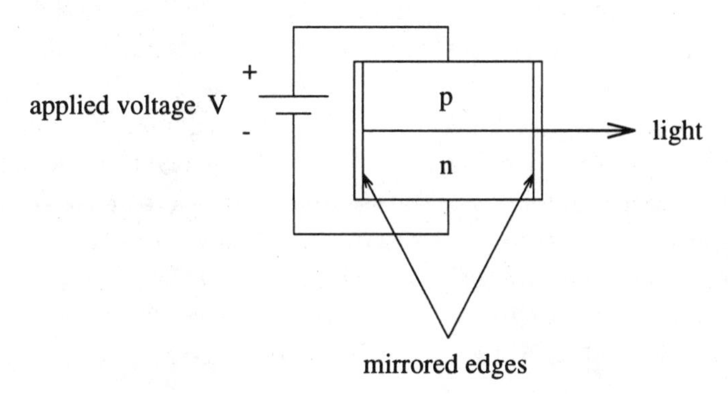

Figure 2.9 Structure of a semiconductor diode laser.

In semiconductor materials, electrons may occupy either the valence band or the conduction band. The valence band and conduction band are analogous to the ground state and excited state of an electron mentioned in the previous section. The valence band corresponds to an energy level at which an electron is not free from an atom. The conduction band corresponds to a level of energy at which an electron has become a free electron and may move freely to create current flow. The region of energy between the valence band and the conduction band is known as the *band gap*. An electron may not occupy any energy levels in the band gap region. When an electron moves from the valence band to the conduction band, it leaves a vacancy, or *hole*, in the valence band. When the electron moves from the conduction band to the valence band, it recombines with the hole and may produce the spontaneous emission of a photon. The frequency of the photon is given by Eqn. (2.10), where $E_i - E_f$ is the band gap energy. The distribution of the energy levels which electrons may occupy is given by the *Fermi-Dirac distribution*.

A semiconductor may be *doped* with impurities to increase either the number of electrons or the number of holes. An n-type semiconductor is doped with impurities which provide extra electrons. These electrons will remain in the conduction band. A p-type semiconductor is doped with impurities which increase the number of holes in the valence band. A p-n junction is formed by layering p-type semiconductor material over n-type semiconductor material.

In order to produce stimulated emission, a voltage is applied across the p-n junction to forward bias the device and cause electrons in the "n" region to combine with holes in the "p" region, resulting in light energy being released at a frequency related to the band gap of the device. By using different types of semiconductor materials, light with various ranges of frequencies may be released. The actual frequency of light emitted by the laser is determined by the length of the cavity formed by mirrored edges perpendicular to the p-n junction.

An improvement to the bulk laser diode is the multiple-quantum-well (MQW) laser. Quantum wells are thin alternating layers of semiconductor materials. The alternating layers create potential barriers in the semiconductors which confine the position of electrons and holes to a smaller number of energy states. The quantum wells are placed in the region of the p-n junction. By confining the possible states of the electrons and holes, it is possible to achieve higher-resolution, low-linewidth lasers (lasers which generate light with a very narrow frequency range).

2.2.2 Tunable and Fixed Lasers

The previous section provided an overview of a generic model of a laser, but the transmitters used in WDM networks often require the capability to tune to different wavelengths. This section briefly describes some of the more-popular, tunable and fixed, single-frequency laser designs.

Laser Characteristics

Some of the physical characteristics of lasers which may affect system performance are laser linewidth, frequency stability, and the number of longitudinal modes.

The laser linewidth is the spectral width of the light generated by the laser. The linewidth affects the spacing of channels and also affects the amount of dispersion that occurs when the light is propagating along a fiber. As was mentioned in Section 2.1.4, the spreading of a pulse due to dispersion will limit the maximum bit rate.

Frequency instabilities in lasers are variations in the laser frequency, three examples of which are mode hopping, mode shifts, and wavelength chirp [MoTo93]. Mode hopping occurs primarily in injection-current lasers and is a sudden jump in the laser frequency caused by a change in the injection current above a given threshold. Mode shifts are changes in frequency due to temperature changes. Wavelength chirp is a variation in the frequency due to variations in injection current. In WDM systems, frequency instabilities may limit the placement and spacing of channels. In order to avoid large shifts in frequency, methods must be utilized which compensate for variations in temperature or injection current. One approach for temperature compensation is to package with the laser a thermoelectric cooler element which produces cooling as a function of applied current. The current for the thermoelectric cooler may be provided through a thermistor, which is a temperature-dependent resistor.

The number of longitudinal modes in a laser is the number of wavelengths that it can amplify. In lasers consisting of a simple cavity, wavelengths for which an integer multiple of the wavelength is equal to twice the cavity length will be amplified (i.e., wavelengths λ for which $n\lambda = 2L$, where L is the length of the cavity, and n is an integer). The unwanted longitudinal modes produced by a laser may result in significant dispersion; therefore, it is desirable to implement lasers which produce only a single longitudinal mode.

Some primary characteristics of interest for tunable lasers are the *tuning range*, the *tuning time*, and whether the laser is continuously tunable (over its tuning range) or discretely tunable (only to selected wavelengths). The tuning range refers to the range of wavelengths over which the laser may be operated. The tuning time specifies the time required for the laser to tune from one wavelength to another.

Mechanically Tuned Lasers

Most mechanically tuned lasers use a Fabry-Perot cavity that is adjacent to the lasing medium (i.e., an *external cavity*) to filter out unwanted wavelengths. Tuning is accomplished by physically adjusting the distance between two mirrors on either end of the cavity such that only the desired wavelength constructively interferes with its multiple reflections in the cavity. This approach to tuning results in a tuning range that encompasses the entire useful gain spectrum of the semiconductor laser [Brac90], but tuning time is limited to the order of milliseconds due to the mechanical nature of the tuning and the length of the cavity. The length of the cavity may also limit transmission rates unless an external modulator is used. External cavity lasers tend to have very good frequency stability.

Acoustooptically and Electrooptically Tuned lasers

Other types of tunable lasers which use external tunable filters include acousto-optically and electrooptically tuned lasers. In an acoustooptic or electrooptic laser, the index of refraction in the external cavity is changed by using either sound waves or electrical current, respectively. The change in the index results in the transmission of light at different frequencies. In these types of tunable lasers, the tuning time is limited by the time required for light to build up in the cavity at the new frequency.

An acoustooptic laser combines a moderate tuning range with a moderate tuning time. While not quite fast enough for packet switching with multi-gigabit per second channels, the 10 μs tuning time is a vast improvement over that of mechanically tuned lasers (which have millisecond tuning times). Electrooptically tuned lasers are expected to tune on the order of some tens of nanoseconds. Neither of these approaches allow continuous tuning over a range of wavelengths. The tuning range is limited by the range of frequencies generated by the laser (the laser's gain spectrum) and the range of wavelengths resolvable by the filter [Brac90].

Injection-Current-Tuned Lasers

Injection-current-tuned lasers form a family of transmitters which allow wavelength selection via a diffraction grating. The Distributed Feedback (DFB) laser uses a diffraction grating placed in the lasing medium. In general, the grating consists of a waveguide in which the index of refraction alternates periodically between two values. Only wavelengths which match the period and indices of the grating will be constructively reinforced. All other wavelengths will destructively interfere, and will not propagate through the waveguide. The condition for propagation is given by:

$$D = \frac{\lambda}{2n}$$

where D is the period of the grating [MoTo93]. The laser is tuned by injecting a current which changes the index of the grating region.

If the grating is moved to the outside of the lasing medium, the laser is called a Distributed Bragg Reflector (DBR) laser. The tuning in a DBR laser is discrete rather than continuous, and tuning times of less than 10 ns have been measured [Brac90]. In [DSGP93], a tuning time of 0.5 ns is reported for a DBR laser with a tuning range of 4 nm, which is capable of supporting eight wavelengths. Because the refractive index range in the DBR laser is limited, the DBR laser has a low maximum tuning range (around 10 nm), which can provide up to 25 channels [Alfe93]. One of the drawbacks of the DBR laser is that it is susceptible to mode hopping. Typical linewidths for injection current semiconductor lasers range from about 1 MHz to 50 MHz. MQW lasers offer narrower linewidths which can be on the order of hundreds of kHz [MoTo93].

Laser Arrays

An alternative to tunable lasers is the laser array, which contains a set of fixed-tuned lasers and whose advantage/application is explained below. A laser array consists of a number of lasers which are integrated into a single component, with each laser operating at a different wavelength. The advantage of using a laser array is that, if each of the wavelengths in the array is modulated independently, then multiple transmissions may take place simultaneously. The drawback is that the number of available wavelengths in a laser array is fixed and is currently limited to about 20 wavelengths. Laser arrays with up to 21 wavelengths have been demonstrated in the laboratory [ZaFa92], while a laser array with four wavelengths has actually been deployed in a network prototype [LeCh96].

2.2.3 Optical Modulation

In order to transmit data across an optical fiber, the information must first be encoded, or modulated, onto the laser signal. Analog techniques include amplitude modulation (AM), frequency modulation (FM), and phase modulation

(PM). Digital techniques include amplitude-shift keying (ASK), frequency-shift keying (FSK), and phase-shift keying (PSK).

Of these techniques, binary ASK is currently the preferred method of digital modulation because of its simplicity. In binary ASK, also known as on-off keying (OOK), the signal is switched between two power levels. The lower power level represents a "0" bit, while the higher power level represents a "1" bit.

In systems employing OOK, modulation of the signal can be achieved by simply turning the laser on and off (direct modulation). In general, however, this can lead to chirp, or variations in the laser's amplitude and frequency, when the laser is turned on. A preferred approach for high bit rates (≥ 2 Gbps) is to have an external modulator which modulates the light coming out of the laser. The external modulator blocks or passes light depending on the current applied to it.

The Mach-Zehnder interferometer, described later in Section 2.3.2, can be used as a modulation device. A drive voltage is applied to one of two waveguides creating an electric field which causes the signals in the two waveguides to either be in phase or 180° out of phase, resulting in the light from the laser being either passed through the device or blocked. Mach-Zehnder amplitude modulators which offer bandwidths of up to 18 GHz are currently available [Dela94]. One of the advantages of using integrated-optics devices such as the Mach-Zehnder interferometer is that the laser and modulator can be integrated on a single structure, which may potentially be cost effective. Also, integrating the laser with the modulator eliminates the need for polarization control and results in low chirp.

2.2.4 Summary

Table 2.1 summarizes the characteristics of the different types of tunable transmitters. We observe that there is a trade-off between the tuning range of a transmitter and its tuning time.

2.3 Optical Receivers and Filters

Tunable optical filter technology is a key in making WDM networks realizable. Good sources of information on these devices include [Gree93, Brac90, KoCh89].

Tunable Transmitter	*Approx. Tuning Range (nm)*	*Tuning Time*
Mechanical (external cavity)	500	1–10 ms
Acoustooptic	83	~10 μs
Electrooptic	7	1–10 ns (estimated)
Injection-Current (DFB and DBR)	10	1–10 ns

Table 2.1 Tunable optical transmitters and their associated tuning ranges and times.

2.3.1 Photodetection

In receivers employing *direct detection*, a photodetector converts the incoming photonic stream into a stream of electrons. The electron stream (i.e., electrical current) is then amplified and passed through a threshold device. Whether a bit is a logical 0 or 1 depends on whether the stream is above or below a certain threshold for a bit duration. In other words, the decision is made based on whether or not light is present during the bit duration.

The basic detection devices for direct detection optical networks are the PN photodiode (a p-n junction) and the PIN photodiode (an intrinsic material is placed between "p" and "n" type material). In its simplest form, the photodiode is basically a reverse-biased p-n junction. Through the photoelectric effect, light incident on the junction will create electron-hole pairs in both the "n" and the "p" regions of the photodiode. The electrons released in the "p" region will cross over to the "n" region, and the holes created in the "n" region will cross over to the "p" region, thereby resulting in a current flow.

The alternative to direct detection is *coherent detection* in which phase information is used in the encoding and detection of signals. Coherent-detection-based receivers use a monochromatic laser as a local oscillator. The incoming optical stream, which is at a slightly different frequency from the oscillator, is combined with the signal from the oscillator, resulting in a signal at the difference frequency. This difference signal, which is in the microwave range, is amplified, and then photodetected. While coherent detection is more elaborate than direct detection, it allows the reception of weak signals from a noisy background. However, in optical systems, it is difficult to maintain the phase information required for coherent detection (see [Aziz91]). Since semiconductor lasers have nonzero linewidths, the transmitted signal consists of a number of frequencies with varying phases and amplitudes. The effect is that the phase of the transmitted signal experiences random but significant fluctuations around the desired phase. These phase fluctuations make it difficult to recover the original phase information from the transmitted signal, thus limiting the performance of coherent detection systems.

2.3.2 Tunable Optical Filters

This section discusses several types of tunable optical filters and the properties of each type, while Section 2.3.3 examines fixed-tuned optical filters. The feasibility of many local WDM networks is dependent upon the speed and range of tunable filters. Overviews of tunable filter technology can be found in [Gree93] and [Brac90].

Filter Characteristics

Tunable optical filters are characterized primarily by their tuning range and tuning time. The tuning range specifies the range of wavelengths which can be accessed by a filter. A wide tuning range allows systems to utilize a greater number of channels. The tuning time of a filter specifies the time required to tune from one wavelength to another. Fast tunable filters are required for many WDM local area networks (LANs) based on broadcast-and-select architectures.

Some filters, such as the etalon (described in the following section), are further characterized by two parameters: *free spectral range* and *finesse*. In some filters, the transfer function, or the shape of the filter passband, repeats itself after a certain period. The period of such devices is referred to as the *free spectral range* (FSR). In other words, the filter passes every frequency which is a distance of $n \times$ FSR from the selected frequency, where n is a positive integer. For example, in Fig. 2.10, if the filter is tuned to frequency f_1, then all frequencies labeled with a 1 will be passed by the filter; tuning the filter to the next frequency, f_2, will allow all frequencies labeled with a 2 to be passed by the filter; etc. The free spectral range usually depends on various physical parameters in the device, such as cavity lengths or waveguide lengths.

The *finesse* of a filter is a measure of the width of the transfer function. It is the ratio of free spectral range to channel bandwidth, where the channel bandwidth is defined to be the 3-dB bandwidth of a channel.

The number of channels in an optical filter is limited by the FSR and finesse. All of the channels must fit within one FSR. If the finesse is high, the

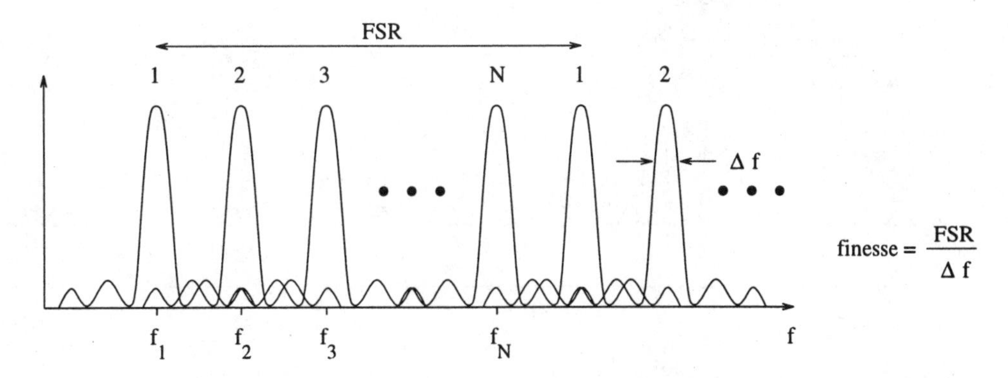

Figure 2.10 Free spectral range and finesse of a tunable filter capable of tuning to N different channels.

transfer functions (passband peaks) are narrower, resulting in more channels being able to fit into one FSR. With a low finesse, the channels would need to be spaced further apart to avoid crosstalk, resulting in fewer channels. One approach to increasing the number of channels is to cascade filters with different FSRs [Gree93]. Figure 2.11 shows the filter passbands for a high-resolution filter and a low-resolution filter, each with four channels within a FSR. By cascading these filters, up to 16 unique channels may be resolved.

The Etalon

The etalon consists of a single cavity formed by two parallel mirrors. Light from an input fiber enters the cavity and reflects a number of times between the mirrors. By adjusting the distance between the mirrors, a single wavelength can be chosen to propagate through the cavity, while the remaining wavelengths destructively interfere. The distance between the mirrors may be adjusted mechanically by physically moving the mirrors, or may be adjusted by changing the index of the material within the cavity. Many modifications (e.g., multicavity and multipass) to the etalon can be made to improve the number of resolvable channels [Brac90]. In a multipass filter , the light passes through the same cavity multiple times, while in a multicavity filter , multiple etalons of different FSRs are cascaded to effectively increase the finesse. An example of a mechanically tuned etalon is the Fabry-Perot filter . In [HuHa90], it was found that the maximum number of channels for a single-cavity Fabry-Perot filter is $0.65F$, where F is the finesse. A two-pass filter was found to have a maximum of $1.4F$ channels, and two-cavity filters up to $0.44F^2$ channels.

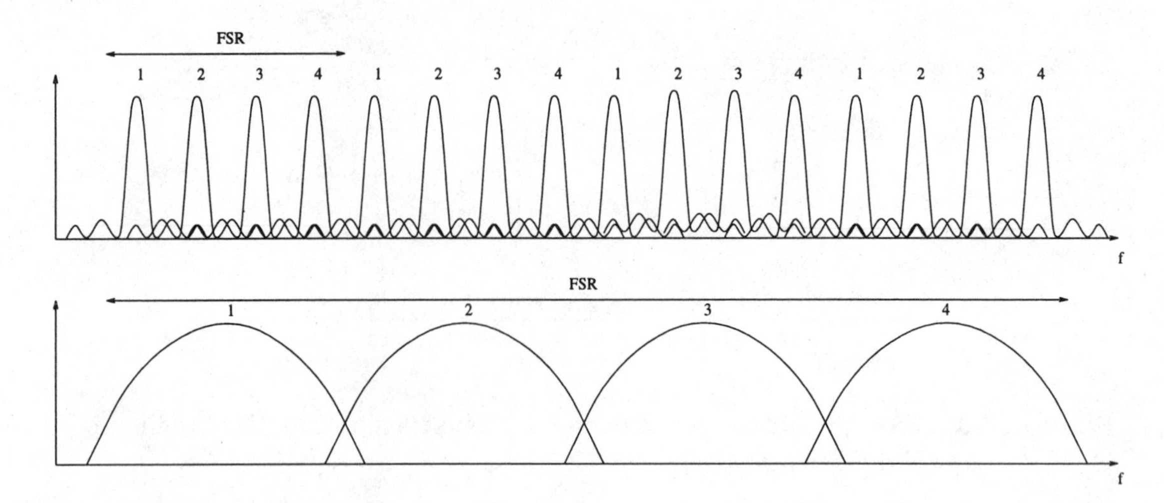

Figure 2.11 Cascading filters with different FSRs.

While the Fabry-Perot etalon can be made to access virtually the entire low-attenuation region of the fiber and can resolve very narrow passbands, it has a tuning time on the order of tens of milliseconds, due to its mechanical tuning. This makes it unsuitable for many packet-switched applications in which the packet duration is much smaller than the tuning time. The Fabry-Perot etalon is used as a tunable receiver in the Rainbow optical WDM local-area network prototype [DGLR90].

The Mach-Zehnder Chain

In a Mach-Zehnder (MZ) interferometer, a splitter splits the incoming wave into two waveguides, and a combiner recombines the signals at the outputs of the waveguides (see Fig. 2.12). An adjustable delay element controls the optical path length in one of the waveguides, resulting in a phase difference between the two signals when they are recombined. Wavelengths for which the phase difference is 180° are filtered out. By constructing a chain of these elements, a single desired optical wavelength can be selected.

While the MZ chain promises to be a low-cost device because it can be fabricated on semiconductor material, its tuning time is still on the order of milliseconds, and its tuning control is complex, requiring that the setting of the delay element in each stage of the MZ chain be based on the settings in previous stages of the chain [Gree93]. The high tuning time is due to thermal elements used in implementing the delay elements. Recent advances

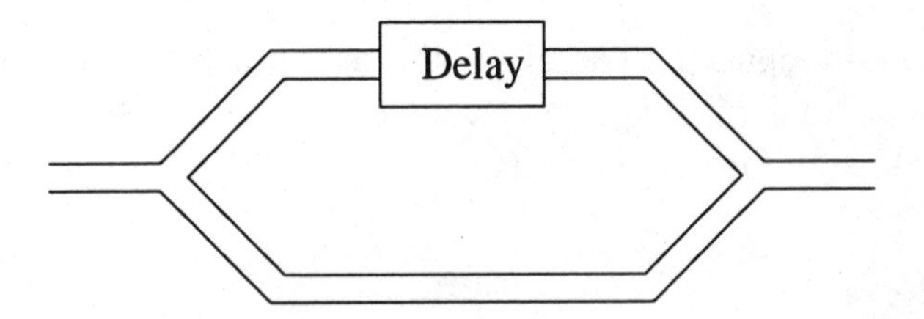

Figure 2.12 Structure of a Mach-Zehnder interferometer.

have produced a very fast (approximately 100 ns) tuning MZ filter, but the number of channels is limited (to 8) and the insertion loss is high [BMSW94].

Acoustooptic Filters

A fast tuning time is obtained when acoustooptic filters are used. Radio frequency (RF) waves are passed through a transducer. The transducer is a piezoelectric crystal that converts sound waves to mechanical movement. The sound waves change the crystal's index of refraction, which enables the crystal to act as a grating. Light incident upon the transducer will diffract at an angle that depends on the angle of incidence and the wavelength of the light [Gree93]. By changing the RF waves, a single optical wavelength can be chosen to pass through the material while the rest of the wavelengths destructively interfere.

The tuning time of the acoustooptic filter is limited by the flight time of the surface acoustic wave to about 10 μs [Brac90, NewF94b]. However, the tuning range for acoustooptic filters covers the entire 1300 nm to 1560 nm spectrum [Brac90]. This tuning range may potentially allow about 100 channels.

If more than one RF wave is passed through the grating simultaneously, more than one wavelength can be filtered out [CSBH89]. This allows the filter to be effectively tuned to several channels at the same time. However, the received signal is the superposition of all of the received wavelengths; therefore, if more than one of those channels is active, crosstalk will occur. The selection of up to five wavelengths was reported in [CSBH89]. One manufacturer has introduced an acoustooptic filter [NewF94a] which has an 80-nm tuning range, 10-μs tuning time, and the ability to simultaneously select up to 10 wavelengths [NewF94b].

One drawback of acoustooptic filters is that, because of their wide transfer function, they are unable to filter out crosstalk from adjacent channels if the channels are closely spaced. Therefore, the use of acoustooptic filters in a multiwavelength system places a constraint on the channel spacing, thus limiting the allowable number of channels.

Electrooptic Filters

Since the tuning time of the acoustooptic filter is limited by the speed of sound, crystals whose indices of refraction can be changed by electrical currents can be used. Electrodes, which rest in the crystal, are used to supply current to the crystal. The current changes the crystal's index of refraction, which allows some wavelengths to pass through while others destructively interfere [Gree93]. Since the tuning time is limited only by the speed of electronics, tuning time can be on the order of several nanoseconds, but the tuning range (and thus, the number of resolvable channels) remains quite small at 16 nm (on the order of 10 channels) [Brac90].

Liquid-Crystal Fabry-Perot Filters

Liquid-crystal filters appear to be a promising new filter technology. The design of a liquid-crystal filter is similar to the design of a Fabry-Perot filter, but the cavity consists of a liquid crystal (LC). The refractive index of the LC is modulated by an electrical current to filter out a desired wavelength, as in an electrooptic filter. Tuning time is on the order of microseconds (submicrosecond times are expected to be achievable), and tuning range is 30 to 40 nm [SnJo94]. These filters have low power requirements and are inexpensive to fabricate. The filter speed of LC filter technology promises to be high enough to handle high-speed packet switching in broadcast-and-select WDM networks.

Summary

Table 2.2 summarizes the state of the art in tunable receivers. As has been stated earlier, tuning range and tuning time seem inversely proportional, except for LC Fabry-Perot filters.

2.3.3 Fixed Filters

An alternative to tunable filters is to use fixed filters or grating devices. Grating devices typically filter out one or more different wavelength signals from

Tunable Receiver	*Approx. Tuning Range (nm)*	*Tuning Time*
Fabry-Perot	500	1–10 ms
Acoustooptic	250	$\sim$10 μs
Electrooptic	16	1–10 ns
LC Fabry-Perot	30	0.5–10 μs

Table 2.2 Tunable optical filters and their associated tuning ranges and times.

a single fiber. Such devices may be used to implement optical multiplexers and demultiplexers or receiver arrays.

Grating Filters

One implementation of a fixed filter is the diffraction grating. The diffraction grating is essentially a flat layer of transparent material (e.g., glass or plastic) with a row of parallel grooves cut into it [Hech92]. The grating separates light into its component wavelengths by reflecting light incident with the grooves at all angles. At certain angles, only one wavelength adds constructively; all others destructively interfere. This allows us to select the wavelength(s) we want by placing a filter tuned to the proper wavelength at the proper angle. Alternatively, some gratings are transmissive rather than reflective and are used in tunable lasers (see DFB lasers in Section 2.2.2).

An alternative implementation of a demultiplexer is the *waveguide grating router (WGR)* in which only one input is utilized. WGRs will be discussed in Section 2.5.2.

Fiber Bragg Gratings

In a fiber Bragg grating, a periodical variation of the index of refraction is directly photo-induced in the core of an optical fiber. A Bragg grating will reflect a given wavelength of light back to the source while passing the other wavelengths. Two primary characteristics of a Bragg grating are the reflectivity and the spectral bandwidth. Typical spectral bandwidths are on the order of 0.1 nm, while a reflectivity in excess of 99% is achievable [ISII95]. While inducing a grating directly into the core of a fiber leads to low insertion loss, a drawback of Bragg gratings is that the refractive index in the grating varies with temperature, with increases in temperature resulting in longer wavelengths being reflected. An approach for compensating for temperature variations is presented in [ASWC96]. Fiber Bragg gratings may be used in the implementation of multiplexers, demultiplexers, and tunable filters.

Thin-Film Interference Filters

Thin-film interference filters offer another approach for filtering out one or more wavelengths from a number of wavelengths. These filters are similar to fiber Bragg grating devices with the exception that they are fabricated by depositing alternating layers of low index and high index materials onto a substrate layer. Thin-film filter technology suffers from poor thermal stability, high insertion loss, and poor spectral profile. However, advances have been made which address some of these issues [ScSp96].

2.4 Optical Amplifiers

Although an optical signal can propagate a long distance before it needs amplification, both long-haul and local lightwave networks can benefit from optical amplifiers. All-optical amplification may differ from opto-electronic amplification in that it may act only to boost the power of a signal, not to restore the shape or timing of the signal. This type of amplification is known as 1R (regeneration), and it provides total data transparency (the amplification process is independent of the signal's modulation format). 1R-amplification is emerging as the choice for transparent all-optical networks of tomorrow. However, in today's digital networks (e.g., Synchronous Optical Network (SONET) and Synchronous Digital Hierarchy (SDH)), which use the optical fiber only as a transmission medium, the optical signals are amplified by first converting the information stream into an electronic data signal, and then retransmitting the signal optically. Such amplification is referred to as 3R (regeneration, reshaping, and reclocking). The reshaping of the signal reproduces the original pulse shape, eliminating much of the noise. Reshaping applies primarily to digitally-modulated signals, but in some cases may also be applied to analog signals. The reclocking of the signal synchronizes the signal to its original bit timing pattern and bit rate. Reclocking applies only to digitally-modulated signals. Another approach to amplification is 2R (regeneration and reshaping), in which the optical signal is converted to an electronic signal which is then used to directly modulate a laser. 3R and 2R techniques provide less transparency than the 1R technique; and in future optical networks, the aggregate bit rate of even just a few channels might make 3R and 2R techniques less practical.

Also, in a WDM system with electronic amplification, each wavelength would need to be separated before being amplified electronically, and then recombined before being retransmitted. Thus, in order to eliminate the need for optical multiplexers and demultiplexers in amplifiers, optical amplifiers must boost the strength of optical signals without first converting them to electrical signals. A drawback is that optical noise, as well as the signal, will be amplified. Also, the amplifier introduces spontaneous emission noise.

Optical amplification uses the principle of stimulated emission, similar to the approach used in a laser. The two basic types of optical amplifiers are semiconductor laser amplifiers and rare-earth-doped-fiber amplifiers, which will be discussed in the following sections. A general overview of optical amplifiers can be found in [Maho93].

2.4.1 Optical Amplifier Characteristics

Some basic parameters of interest in an optical amplifier are gain, gain bandwidth, gain saturation, polarization sensitivity, and amplifier noise.

Gain measures the ratio of the output power of a signal to its input power. Amplifiers are sometimes also characterized by *gain efficiency*, which measures the gain as a function of pump power in dB/mW.

The *gain bandwidth* of an amplifier refers to the range of frequencies or wavelengths over which the amplifier is effective. In a network, the gain bandwidth limits the number of wavelengths available for a given channel spacing.

The *gain saturation* point of an amplifier is the value of output power at which the output power no longer increases with an increase in the input power. When the input power is increased beyond a certain value, the carriers (electrons) in the amplifier are unable to output any additional light energy. The saturation power is typically defined as the ouput power at which there is a 3 dB reduction in the ratio of output power to input power (the small-signal gain).

Polarization sensitivity refers to the dependence of the gain on the polarization of the signal. The sensitivity is measured in dB and refers to the gain difference between the TE and TM polarizations.

In optical amplifiers, the dominant source of *noise* is *amplified spontaneous emission* (ASE), which arises from the spontaneous emission of photons in the active region of the amplifier (see Fig. 2.13). The amount of noise generated by the amplifier depends on factors such as the amplifier gain spectrum, the noise bandwidth, and the *population inversion parameter* which specifies the degree of population inversion that has been achieved between two energy levels. Amplifier noise is especially a problem when multiple amplifiers are *cascaded.* Each subsequent amplifier in the cascade amplifies the noise generated by previous amplifiers.

2.4.2 Semiconductor Laser Amplifier

A semiconductor laser amplifier (see Fig. 2.13) consists of a modified semiconductor laser. A weak signal is sent through the active region of the semiconductor, which, via stimulated emission, results in a stronger signal being emitted from the semiconductor.

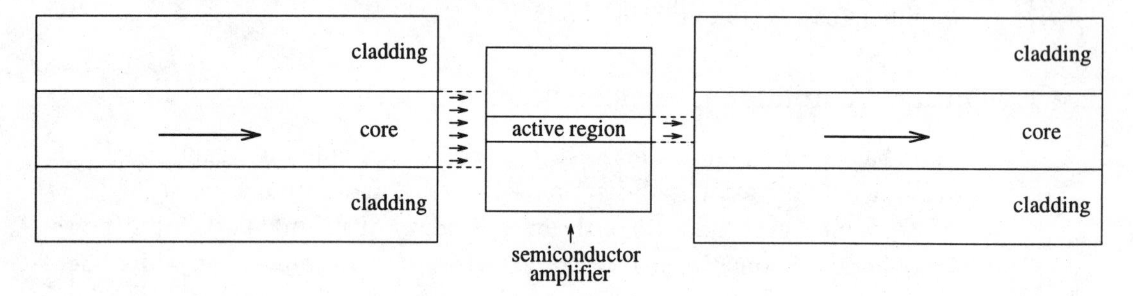

Figure 2.13 A semiconductor optical amplifier.

The two basic types of semiconductor laser amplifiers are the Fabry-Perot amplifier, which is basically a semiconductor laser, and the traveling-wave amplifier (TWA). The primary difference between the two is in the reflectivity of the end mirrors. Fabry-Perot amplifiers have a reflectivity of around 30%, while TWAs have a reflectivity of around 0.01% [Maho93]. In order to prevent lasing in the Fabry-Perot amplifier, the bias current is operated below the lasing threshold current. The higher reflections in the Fabry-Perot amplifier cause Fabry-Perot resonances in the amplifier, resulting in narrow passbands of around 5 GHz. This phenomenon is not very desirable for WDM systems; therefore, by reducing the reflectivity, the amplification is performed in a single pass and no resonances occur. Thus, TWAs are more appropriate than Fabry-Perot amplifiers for WDM networks.

Today's semiconductor amplifiers can achieve gains of 25 dB with a gain saturation of 10 dBm, polarization sensitivity of 1 dB, and bandwidth range of 40 nm [Maho93].

Semiconductor amplifiers based on multiple quantum wells (MQW) are currently being studied. These amplifiers have higher bandwidth and higher gain saturation than bulk devices. They also provide faster on-off switching times. The disadvantage is a higher polarization sensitivity.

An advantage of semiconductor amplifiers is the ability to integrate them with other components. For example, they can be used as gate elements in switches. By turning a drive current on and off, the amplifier basically acts like a gate, either blocking or amplifying the signal.

2.4.3 Doped-Fiber Amplifier

Optical doped-fiber amplifiers are lengths of fiber doped with an element (rare earth) which can amplify light (see Fig. 2.14). The most common doping element is erbium, which provides gain for wavelengths between 1525 nm and 1560 nm. At the end of the length of the fiber amplifier, a laser transmits a strong signal at a lower wavelength (referred to as the *pump wavelength*) back up the fiber. This pump signal excites the doped atoms into a higher energy level. This allows the data signal to stimulate the excited atoms to release photons. Most erbium-doped fiber amplifiers (EDFAs) are pumped by lasers with a wavelength of either 980 nm or 1480 nm. The 980-nm pump wavelength has shown gain efficiencies of around 10 dB/mW, while the 1480-nm pump wavelength provides efficiencies of around 5 dB/mW. Typical gains are on the order of 25 dB. Experimentally, EDFAs have been shown to achieve gains of up to 51 dB with the maximum gain limited by internal Rayleigh backscattering in which some of the light energy of the signal is scattered in the fiber and directed back toward the signal source [HaDL92]. The 3-dB gain bandwidth for the EDFA is around 35 nm (see Fig. 2.15), and the saturation power is around 10 dBm [Maho93].

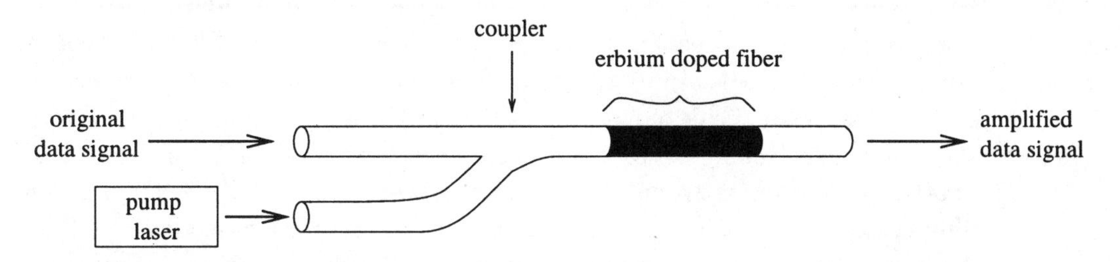

Figure 2.14 Erbium-doped fiber amplifier.

For the 1300-nm region, the praseodymium-doped fluoride fiber amplifier (PDFFA) has recently been receiving attention. These amplifiers have low crosstalk and noise characteristics, while attaining high gains. They are able to operate over a range of around 50 nm in the 1280 nm to 1330 nm range. In [YaSh95], a PDFFA was developed which had a 40.6 dB gain. Recent developments on PDFFAs are presented in [Whit95].

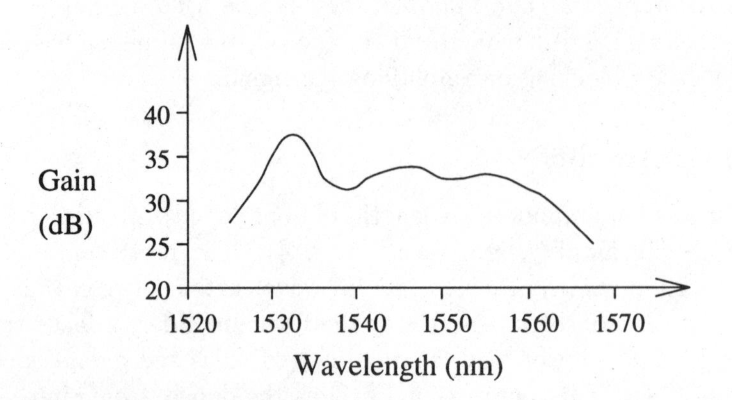

Figure 2.15 The gain spectrum of an erbium-doped fiber amplifier with input power = −40 dBm.

A limitation to optical amplification is the unequal gain spectrum of optical amplifiers. The EDFA gain spectrum is shown in Fig. 2.15 (from [Sams97]). While an optical amplifier may provide gain across a range of wavelengths, it will not necessarily amplify all wavelengths equally. This characteristic, accompanied by the fact that optical amplifiers amplify noise as well as signal, and the fact that the active region of the amplifier can spontaneously emit photons which also cause noise, limits the performance of optical amplifiers. Thus, a multiwavelength optical signal passing through a series of amplifiers will eventually result in the power of the wavelengths being uneven.

A number of approaches to equalizing the gain of an EDFA have been studied. In [TaLa91], a notch filter (a filter which attenuates the signal at a selected frequency) centered at around 1530 nm is used to suppress the peak in the EDFA gain (see Fig. 2.15). However, when multiple EDFAs are cascaded, another peak appears around the 1560 nm wavelength. In [WiHw93], a notch filter centered at 1560 nm is used to equalize the gain for a cascade of EDFAs. Another approach to flattening the gain is to adjust the input transmitter power such that the power on all received wavelengths at the destination is equal [ChNT92]. A third approach to gain equalization is to demultiplex the individual wavelengths and then attenuate selected wavelengths such that all wavelengths have equal power. In [EGZJ93], this approach is applied to a WDM interoffice ring network.

Amplifier Type	*Gain Region*	*Gain Bandwidth*	*Gain*
Semiconductor	Any	40 nm	25 dB
EDFA	1525–1560 nm	35 nm	25–51 dB
PDFFA	1280–1330 nm	50 nm	20–40 dB

Table 2.3 Amplifier characteristics.

2.5 Switching Elements

Most current networks employ electronic processing and use the optical fiber only as a transmission medium. Switching and processing of data are performed by converting an optical signal back to its "native" electronic form. Such a network relies on electronic switches. These switches provide a high degree of flexibility in terms of switching and routing functions; however, the speed of electronics is unable to match the high bandwidth of an optical fiber. Also, an electrooptic conversion at an intermediate node in the network introduces extra delay. These factors have motivated a push toward the development of all-optical networks in which optical switching components are able to switch high-bandwidth optical data streams without electrooptic conversion. In a class of switching devices currently being developed, the control of the switching function is performed electronically with the optical stream being transparently routed from a given input of the switch to a given output. Such transparent switching allows the switch to be independent of the data rate and format of the optical signals. For WDM systems, switches which are wavelength dependent are also being developed.

Switches can be divided into two classes. A switch in the first of these classes, called *relational devices*, establishes a relation between the inputs and the outputs. The relation is a function of the control signals applied to the device and is independent of the contents of the signal or data inputs. A property of this device is that the information entering and flowing through it cannot change or influence the current relation between the inputs and the outputs. An example of this type of device is the directional coupler as it is used in switching applications. Thus, the strength of a relational device, which allows signals at high bit rates to pass through it, is that it cannot sense the presence of individual bits that are flowing through itself. This characteristic is also known as *data transparency*. The same feature may sometimes also be a weakness since it causes loss of flexibility (i.e., individual portions of a data stream cannot be switched independently).

The second class of devices is referred to as *logic devices.* In such a device, the data, or the information-carrying signal that is incident on the device, controls the state of the device in such a way that some Boolean function, or combination of Boolean functions, is performed on the inputs. In a logic device, some of its components must be able to change states or *switch* as fast as or faster than the signal bit rate [Hint90]. This ability gives the device some added flexibility but limits the maximum bit rate that can be accommodated.

Hence, *relational devices* are needed for circuit switching, and *logic devices* are needed for packet switching.

In the following sections, we review a number of different optical switch elements and architectures.

2.5.1 Fiber Crossconnect Elements

A fiber crossconnect element switches optical signals from input ports to output ports. These type of elements are usually considered to be wavelength insensitive, i.e., incapable of demultiplexing different wavelength signals on a given input fiber.

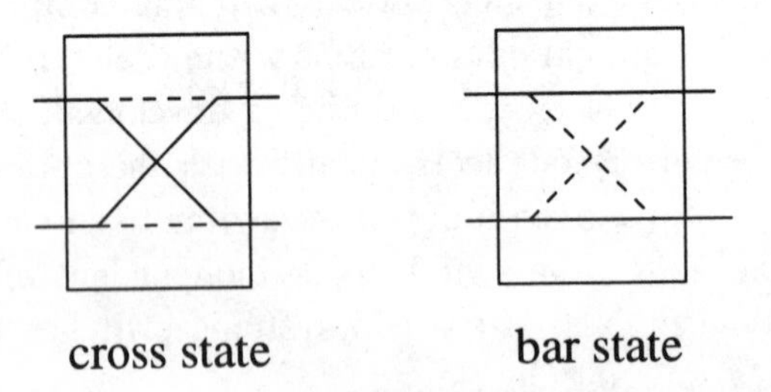

Figure 2.16 2×2 crossconnect elements in the cross state and bar state.

A basic crossconnect element is the 2×2 crosspoint element. A 2×2 crosspoint element routes optical signals from two input ports to two output ports and has two states: cross state and bar state (see Fig. 2.16). In the cross state, the signal from the upper input port is routed to the lower output port, and the signal from the lower input port is routed to the upper output port. In the bar state, the signal from the upper input port is routed to the upper output port, and the signal from the lower input port is routed to the lower output port.

Optical crosspoint elements have been demonstrated using two types of technologies: (1) the generic directive switch [Alfe88], in which light is physically directed to one of two different outputs, and (2) the gate switch, in which optical amplifier gates are used to select and filter input signals to specific output ports.

Directive Switches

The directional coupler (see Fig. 2.17(a) [Alfe88]) consists of a pair of optical channel waveguides that are parallel and in close proximity over some finite interaction length. Light input to one of the waveguides couples to the second waveguide via evanescent[5] coupling. The coupling strength corresponds to the interwaveguide separation and the waveguide mode size which in turn depends upon the optical wavelength and the confinement factor[6] of the waveguide. If the two waveguides are identical, complete coupling between the two waveguides occurs over a characteristic length which depends upon the coupling strength. However, by placing electrodes over the waveguides, the difference in the propagation constants in the two waveguides can be sufficiently increased so that no light couples between the two waveguides. Therefore, the cross state corresponds to zero applied voltage, and the bar state corresponds to a nonzero switching voltage. Unfortunately, the interaction length needs to be very accurate for good isolation, and these couplers are wavelength specific.

Switch fabrication tolerances, as well as the ability to achieve good switching for a relatively wide range of wavelengths, can be overcome by using the so-called reversed delta-beta coupler (see Fig. 2.17(b)). In this device, the electrode is split into at least two sections. The cross state is achieved by applying equal and opposite voltages to the two electrodes. This approach has been shown to be very successful [Alfe88].

The balanced bridge interferometric switch (see Fig. 2.17(c)) consists of a input 3-dB coupler, two waveguides sufficiently separated so that they do not couple, electrodes to allow changing the effective path length over the two arms, and a final 3-dB coupler. Light incident on the upper waveguide is split in half by the first coupler. With no voltage applied to the electrodes, the optical path length of the two arms enters the second coupler in phase. The second coupler acts like the continuation of the first, and all the light is crossed over to the second waveguide to provide the cross state. To achieve the bar state, voltage is applied to an electrode, placed over one of the interferometer arms to electrooptically produce a 180° phase difference between the two arms.

---

[5]An evanescent wave is the part of a propagating wave which travels along or outside of the waveguide boundary.

[6]The confinement factor determines the fraction of power that travels within the core of the waveguide.

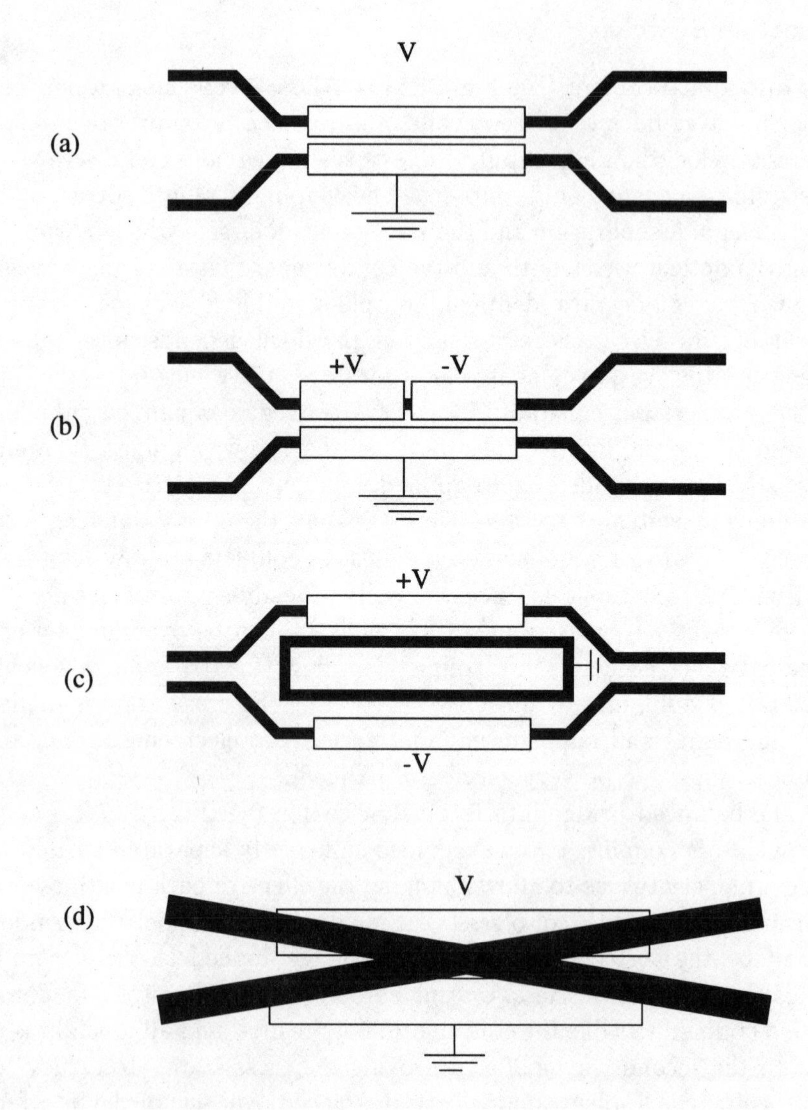

Figure 2.17 Schematic of optical crosspoint elements.

In this case, the two inputs from the arms of the interferometer combine at the second 3-dB coupler out of phase, with the result that light remains in the upper waveguide.

The intersecting waveguide switch is shown in Fig. 2.17(d). This device can be viewed as a directional coupler (see Fig. 2.17(a)) with no gap between the waveguides in the interaction region. When properly fabricated, both cross and bar states can be electrooptically achieved with good crosstalk performance.

Other types of switches include the mechanical fiber-optic switch and the thermo-optic switch. These devices offer slow switching (about milliseconds) and may be employed in circuit-switched networks. One mechanical switch, for example, consists of two ferrules, each with polished end faces that can rotate to switch the light appropriately [Ande95]. Thermo-optic waveguide switches, on the other hand, are fabricated on a glass substrate and are operated by the use of the thermo-optic effect. One such device uses a zero-gap directional-coupler configuration with a heater electrode to increase the waveguide index of refraction [LeSu94].

Gate Switches

In the $N \times N$ gate switch, each input signal first passes through a $1 \times N$ splitter. The signals then pass through an array of N^2 gate elements, and are then recombined in $N \times$ combiners and sent to the N outputs. The gate elements can be implemented using optical amplifiers which can either be turned on or off to pass only selected signals to the outputs. The amplifier gains can compensate for coupling losses and losses incurred at the splitters and combiners. A 2×2 amplifier gate switch is illustrated in Fig. 2.18. A disadvantage of the gate switch is that the splitting and combining losses limit the size of the switch.

Amplifier gate switches of size 8×8 are commercially available. In [Opti95], an 8×8 switch is described which uses semiconductor optical amplifiers to provide lossless switching. It operates around the 1300-nm region, has an optical bandwidth of 40 nm, has low polarization dependence (1 dB), and has a fairly low crosstalk (below −40 dB). The disadvantages are that the switch is bulky (weight of 50 lbs) and expensive.

A possible less-expensive alternative is the integrated-optics amplifier gate switch. Such switches of size 4×4 have been demonstrated experimentally which operate around the 1550-nm region and have a fiber-to-fiber gain of

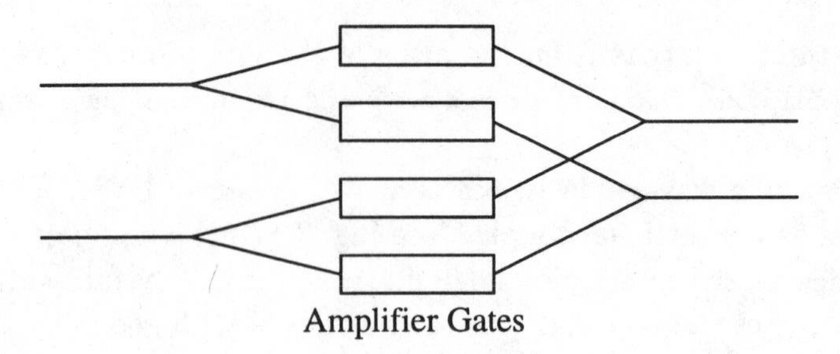

Figure 2.18 A 2 × 2 amplifier gate switch.

5 dB with crosstalk levels below −40 dB [GLTJ92]. However, the main disadvantage is the high polarization sensitivity (6 to 12 dB).

2.5.2 Nonreconfigurable Wavelength Router

A wavelength-routing device can route signals arriving at different input fibers (ports) of the device to different output fibers (ports) based on the wavelengths of the signals. Wavelength routing is accomplished by demultiplexing the different wavelengths from each input port, optionally switching each wavelength separately, and then multiplexing signals at each output port. The device can be either *nonreconfigurable*, in which case there is no switching stage between the demultiplexers and the multiplexers, and the routes for different signals arriving at any input port are fixed (these devices are referred to as routers rather than switches), or *reconfigurable*, in which case the routing function of the switch can be controlled electronically. In this section, we will discuss wavelength routers, while Section 2.5.3 will cover reconfigurable wavelength switches.

A nonreconfigurable wavelength router can be constructed with a stage of demultiplexers which separate each of the wavelengths on an incoming fiber, followed by a stage of multiplexers which recombine wavelengths from various inputs to a single output. The outputs of the demultiplexers are hardwired to the inputs of the multiplexers. Let this router have P incoming fibers, and P outgoing fibers. On each incoming fiber, there are M wavelength channels. A 4 × 4 nonreconfigurable wavelength router with $M = 4$ is illustrated in Fig. 2.19. The router is nonreconfigurable because the path of a given wavelength channel, after it enters the router on a particular input fiber, is fixed. The wavelengths on each incoming fiber are separated using a grating demultiplexer. And finally, information from multiple WDM channels are multiplexed before launching them back onto an output fiber. In between the demultiplexers and multiplexers, there are direct connections from each demultiplexer output to each multiplexer input. Which wavelength on which input port gets routed to which output port depends on a "routing matrix" characterizing the router; this matrix is determined by the internal "connections" between the demultiplexers and multiplexers.

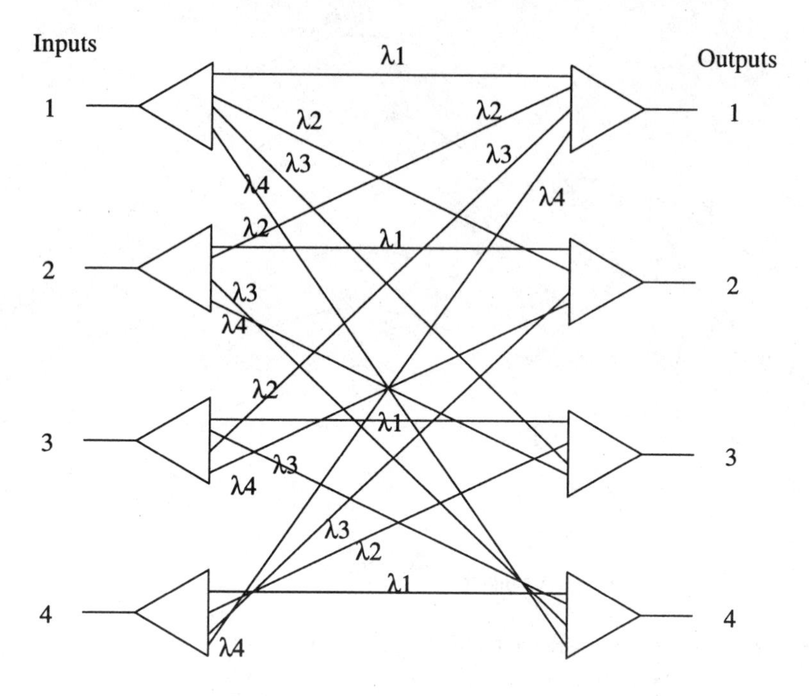

Figure 2.19 A 4 × 4 nonreconfigurable wavelength router.

Waveguide Grating Routers

One implementation of a wavelength router is the waveguide grating router (WGR), which is also referred to as an arrayed waveguide grating (AWG) multiplexer. A WGR provides a fixed routing of an optical signal from a given input port to a given output port based on the wavelength of the signal. Signals of different wavelengths coming into an input port will each be routed to a different output port. Also, different signals using the same wavelength can be input simultaneously to different input ports, and still not interfere with each other at the output ports. Compared to a passive-star coupler in which a given wavelength may only be used on a single input port, the WGR with N input and N output ports is capable of routing a maximum of N^2

connections, as opposed to a maximum of N connections in the passive-star coupler. Also, because the WGR is an integrated device, it can easily be fabricated at low cost. The disadvantage of the WGR is that it is a device with a fixed routing matrix which cannot be reconfigured.

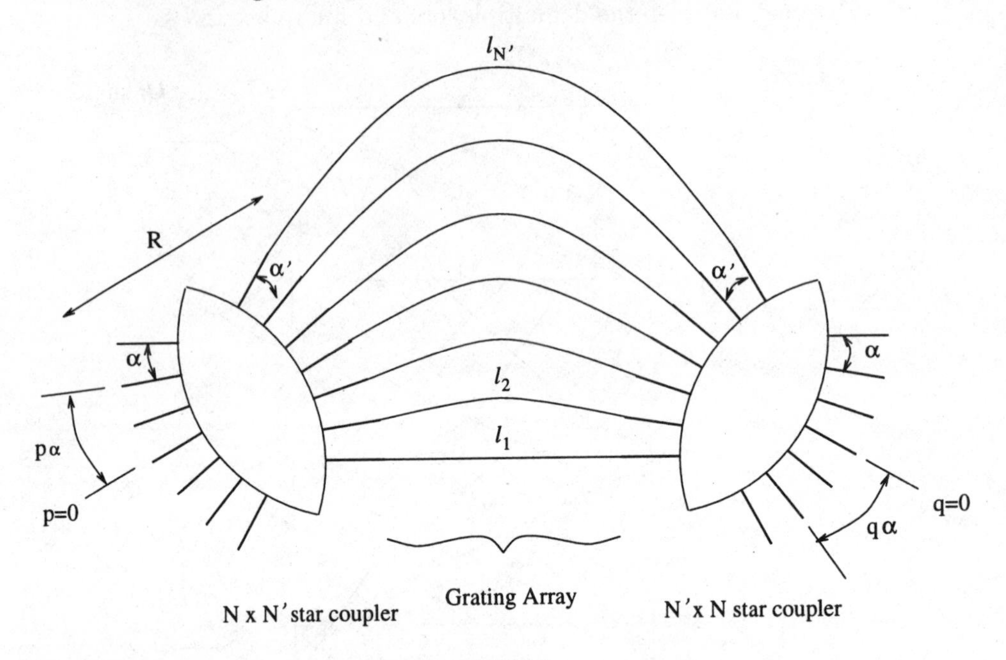

Figure 2.20 The waveguide grating router (WGR).

The WGR, shown in Fig. 2.20, can be used as a nonreconfigurable wavelength-router, or it can be used to build a tunable optical transmitter or a tunable optical receiver. It consists of two passive-star couplers connected by a grating array. The first star coupler has N inputs and N' outputs (where $N \ll N'$), while the second one has N' inputs and N outputs. The inputs to the first star are separated by an angular distance of α, and their outputs are separated by angular distance α'. The grating array consists of N' waveguides, with lengths $l_1, l_2, \ldots, l_{N'}$ where $l_1 < l_2 < \cdots < l_{N'}$. The difference in length between any two adjacent waveguides is a constant Δl.

In the first star coupler, a signal on a given wavelength entering from any of the input ports is split and transmitted to its N' outputs which are also the N' inputs of the grating array. The signal travels through the grating array, experiencing a different phase shift in each waveguide depending on the length of the waveguides and the wavelength of the signal. The constant difference in the lengths of the waveguides creates a phase difference of $\beta \times \Delta l$

in adjacent waveguides, where $\beta = 2\pi n_{\text{eff}}/\lambda$ is the propagation constant in the waveguide, n_{eff} is the effective refractive index of the waveguide, and λ is the wavelength of the light. At the input of the second star coupler, the phase difference in the signal will be such that the signal will constructively recombine only at a single output port.

Two signals of the same wavelength coming from two different input ports will not interfere with each other in the grating because there is an additional phase difference created by the distance between the two input ports. The two signals will be combined in the grating, but will be separated out again in the second star coupler and directed to different outputs. This phase difference is given by $kR(p - q)\alpha\alpha'$, where k is a propagation constant which doesn't depend on wavelength, R is the constant distance between the two foci of the optical star, p is the input port number of the router, and q is the output port number of the router. The total phase difference is:

$$\phi = \frac{2\pi \times \Delta l}{\lambda} + kR(p - q)\alpha\alpha' \tag{2.11}$$

The transmission power from a particular input port p to a particular output port q is maximized when ϕ is an integer multiple of 2π. Thus, only wavelengths λ for which ϕ is a multiple of 2π will be transmitted from input port p to output port q. Alternately, for a given input port and a given wavelength, the signal will only be transmitted to the output port which causes ϕ to be a multiple of 2π.

Prototype devices have been built on silicon with $N = 11$ and $N' = 11$, giving a channel spacing of 16.5 nm; and $N = 7$ and $N' = 15$, giving a channel spacing of 23.1 nm [DrEK91]. In [ZiDJ92], a 15×15 waveguide grating multiplexer on InP is demonstrated to have a free spectral range of 10.5 nm and channel spacing of 0.7 nm in the 1550-nm region. In [OkMS95], a 64×64 arrayed-waveguide multiplexer on silicon is demonstrated with a channel spacing of 0.4 nm. WGRs with flat passbands have also been developed [OkSu96, TrBe97]. Other applications of the WGR, such as tunable transmitters and tunable receivers, are presented in [GlKW94]. These tunable components are implemented by integrating the WGR with switched amplifier elements. An amplifier element may either be activated, in which case it amplifies the signal passing through it, or it may be turned off, in which case it prevents any signals from passing through it. By using only a single input port of the WGR, each wavelength on that input port will be routed to a different output port. By placing an amplifier element at each output port of the WGR, we may filter out selected wavelengths by activating or deactivating the appropriate amplifiers. The outputs of the amplifier elements may then be multiplexed into a signal containing only the desired wavelengths.

2.5.3 Reconfigurable Wavelength-Routing Switch

A reconfigurable wavelength-routing switch (WRS), also referred to as a wavelength selective crossconnect (WSXC), uses photonic switches inside the routing element. The functionality of the reconfigurable WRS, illustrated in Fig. 2.21, is as follows. The WRS has P incoming fibers and P outgoing fibers. On each incoming fiber, there are M wavelength channels. Similar to the nonreconfigurable router, the wavelengths on each incoming fiber are separated using a grating demultiplexer.

The outputs of the demultiplexers are directed to an array of M $P \times P$ optical switches between the demultiplexer and the multiplexer stages. All signals on a given wavelength are directed to the same switch. The switched signals are then directed to multiplexers which are associated with the output ports. Finally, information streams from multiple WDM channels are multiplexed before launching them back onto an output fiber.

Space-division optical-routing switches may be built from 2×2 optical crosspoint elements [ScAl90] arranged in a banyan-based fabric. The space-division switches (which may be one per wavelength [ShCS93]) can route a signal from any input to any output on a given wavelength. Such switches based on relational devices [Hint90] are capable of switching very high-capacity signals. The 2×2 crosspoint elements that are used to build the space-division switches may be slowly tunable and they may be reconfigured to adapt to changing traffic requirements. Switches of this type can be constructed from off-the-shelf components available today.

Networks built from such switches are more flexible than passive, nonreconfigurable, wavelength-routed networks, because they provide additional control in setting up connections. The routing is a function of both the wavelength chosen at the source node, as well as the configuration of the switches in the network nodes.

2.5.4 Photonic Packet Switches

Most of the switches discussed above are *relational devices* , i.e., they are useful in a circuit-switched environment where a connection may be set up over long periods of time. Here, we review optical packet switches that have been proposed in the literature. These switches are composed of *logic devices*,

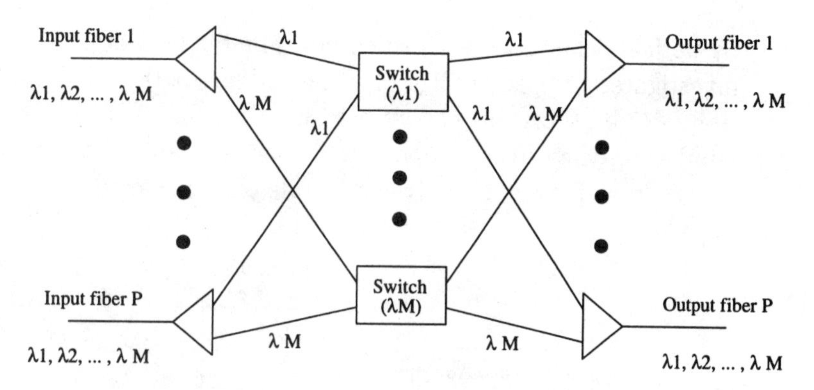

Figure 2.21 A P × P reconfigurable wavelength-routing switch with M wavelengths.

instead of *relational devices* used before, so that the switch configuration is a function of the data on the input signal.

In a packet-switched system, there exists the problem of resource contention when multiple packets contend for a common resource in the switch. In an electronic system, contention may be resolved through the use of buffering; however, in the optical domain, contention resolution is a more complex issue, since it is difficult to implement components which can store optical data. A number of switch architectures which use delay lines to implement optical buffering have been proposed. A delay line is simply a long length of fiber which introduces propagation delays that are on the order of packet transmission times.

The Staggering Switch

The *staggering switch*, which is an "almost-all-optical" packet switch has been proposed in [Haas93]. In an "almost-all-optical" network, the data path is fully optical, but the control of the switching operation is performed electronically. One of the advantages of such switching over its electronic counterpart is that it is transparent, i.e., except for the control information, the payload may be encoded in an arbitrary format or at an arbitrary bit rate. The main problem in the implementation of packet-switched optical networks is the lack of random-access optical memory.

The staggering switch architecture employs an output-collision-resolution scheme that is controlled by a set of delay lines with unequal delays. The architecture is based on two rearrangeably nonblocking stages interconnected by optical delay lines with different amounts of delay. The work in [Haas93] investigates the probability of packet loss and the switch latency as a function of link utilization and switch size. In general, with proper setting of the number of delay lines, the switch can achieve an arbitrary low probability of packet loss. Figure 2.22 gives a simple overview of the switch architecture.

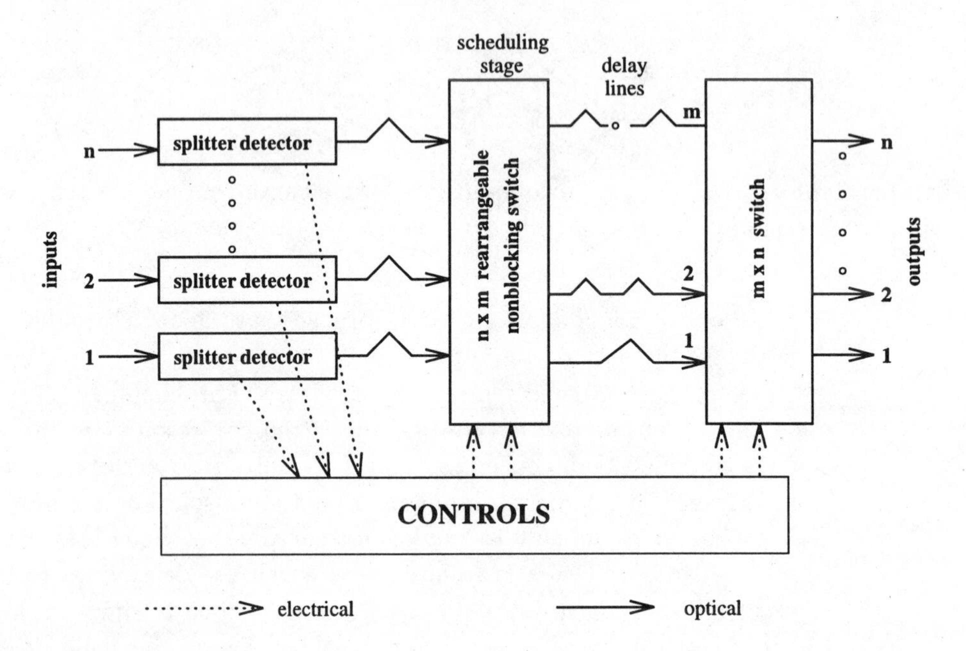

Figure 2.22 The staggering switch architecture.

Contention Resolution by Delay Lines (CORD)

Another architecture which deals with contention in a packet-switched optical network is the Contention Resolution by Delay Lines (CORD) architecture [CFKM96]. The CORD architecture consists of a number of 2×2 crossconnect elements and delay lines (see Fig. 2.23). Each delay line functions as a buffer for a single packet. If two packets contend for the same output port, one packet may be switched to a delay line while the other packet is switched to the proper output. The packet which was delayed can then be switched to the same output after the first packet has been transmitted.

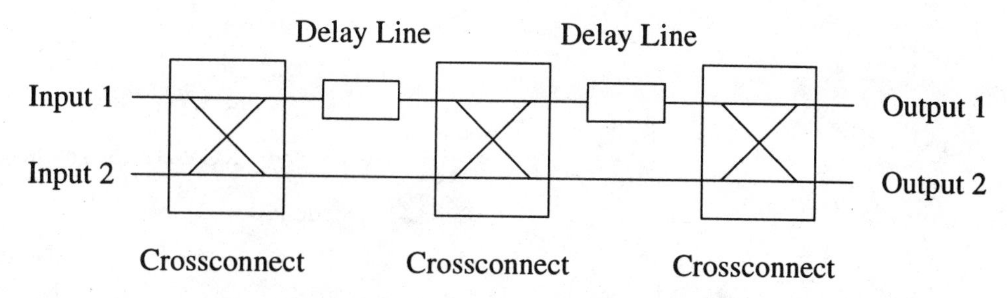

Figure 2.23 The CORD architecture.

The HLAN Architecture

The HLAN architecture, described in [BCHK96], avoids channel collisions by using a slotted system with empty slot markers that indicate when a node can write data into a slot. HLAN is implemented on a helical unidirectional bus in which the physical fiber wraps around twice to visit each node three times, and the fiber medium is considered to be divided into three nonoverlapping "segments" (see Fig. 2.24). A headend periodically generates equal-sized frames of empty slots and puts them on the bus. The Guaranteed-Bandwidth traffic is transmitted on the GBW segment, the Bandwidth-On-Demand traffic is transmitted on the BOD segment, and data is received on the RCV segment. All users receive traffic on the third segment, i.e., all receivers are downstream of the transmitters. Each node in the network is equipped with a header/slot processor, protocol logic units, clock recovery mechanisms, and optoelectronic buffers. Though the network has been designed to operate at 100 Gbps, it is scalable in principle to faster media rates.

2.6 Wavelength Conversion

Consider the network in Fig. 2.25. It shows a wavelength-routed network containing two WDM crossconnects (S1 and S2) and five access stations (A through E). Three lightpaths have been set up (C to A on wavelength λ_1, C to B on λ_2, and D to E on λ_1). To establish a lightpath, we require that the *same* wavelength be allocated on all the links in the path. This requirement is known as the *wavelength-continuity constraint* (e.g., see [BaMu96]). This constraint distinguishes the wavelength-routed network from a circuit-switched network which blocks calls only when there is no capacity along any of the links in the path assigned to the call. Consider the example in Fig. 2.26(a). Two

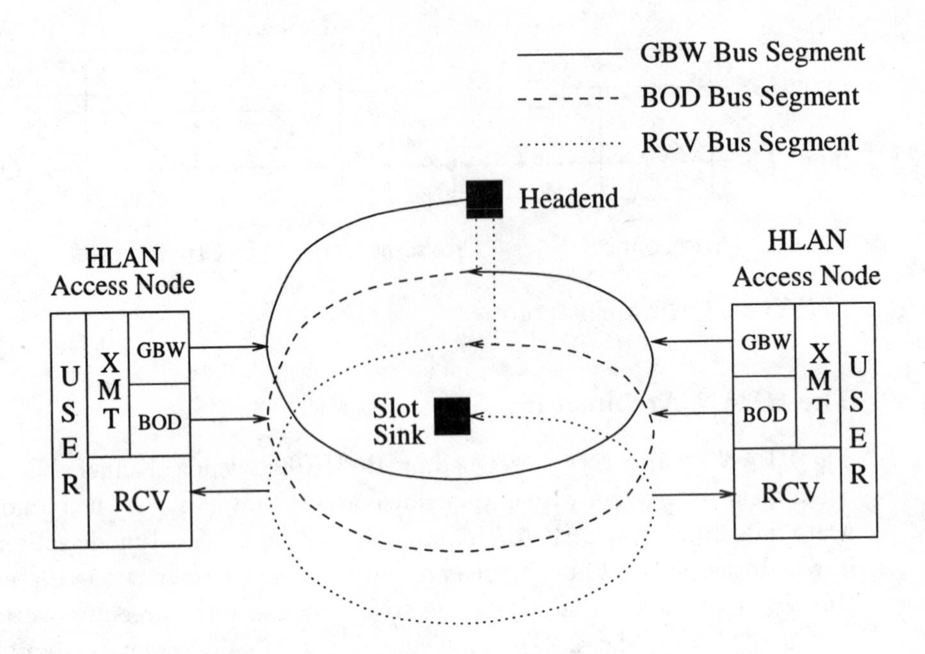

Figure 2.24 The HLAN architecture.

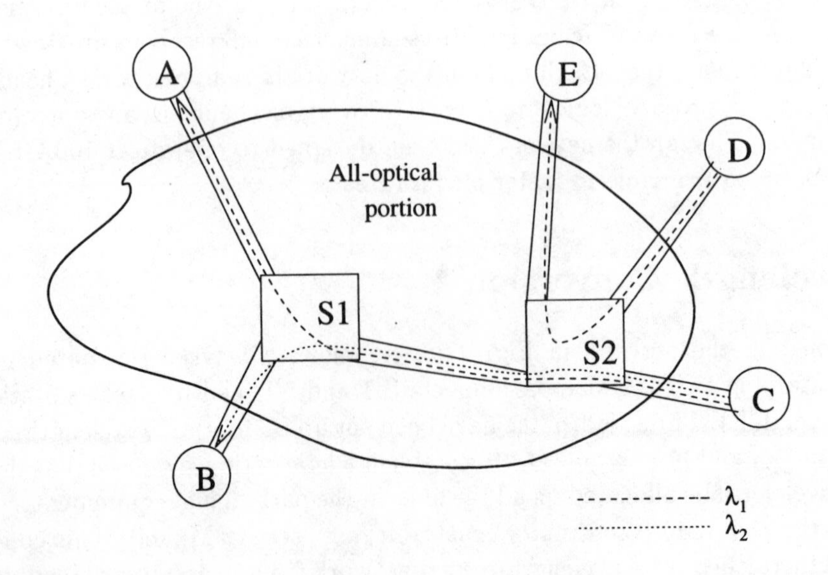

Figure 2.25 An all-optical wavelength-routed network.

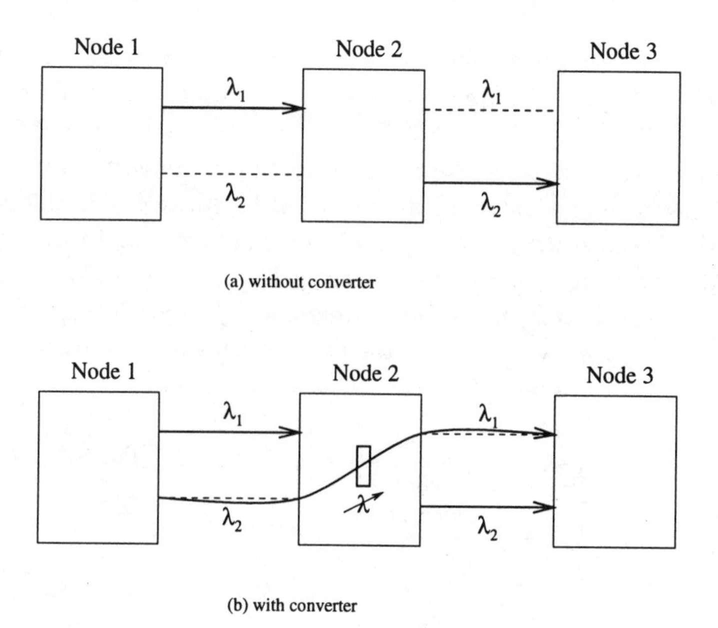

Figure 2.26 Wavelength-continuity constraint in a wavelength-routed network.

lightpaths have been established in the network: (i) between Node 1 and Node 2 on wavelength λ_1, and (ii) between Node 2 and Node 3 on wavelength λ_2. Now suppose a lightpath between Node 1 and Node 3 needs to be set up. Establishing such a lightpath is impossible even though there is a free wavelength on each of the links along the path from Node 1 to Node 3. This is because the available wavelengths on the two links are *different*. Thus, a wavelength-routed network may suffer from higher blocking as compared to a circuit-switched network.

It is easy to eliminate the wavelength-continuity constraint, if we were able to *convert* the data arriving on one wavelength along a link into another wavelength at an intermediate node and forward it along the next link. Such a technique is actually feasible and is referred to as *wavelength conversion.* In Fig. 2.26(b), a wavelength converter at Node 2 is employed to convert data from wavelength λ_2 to λ_1. The new lightpath between Node 1 and Node 3 can now be established by using the wavelength λ_2 on the link from Node 1 to Node 2, and then by using the wavelength λ_1 to reach Node 3 from Node 2. Notice that a single lightpath in such a *wavelength-convertible* network can use a different wavelength along each of the links in its path. Thus, wavelength conversion may improve the efficiency in the network by resolving the wavelength conflicts of the lightpaths. The impact of wavelength conversion on WDM wide-area network (WAN) design is further elaborated in Section 2.7.5.

The function of a wavelength converter is to convert data on an input wavelength onto a possibly different output wavelength among the N wavelengths in the system (see Fig. 2.27). In this figure and throughout this section, λ_s denotes the input signal wavelength; λ_c, the converted wavelength; λ_p, the pump wavelength; f_s, the input frequency; f_c, the converted frequency; f_p, the pump frequency; and CW, the continuous wave (unmodulated) generated as the pump signal.

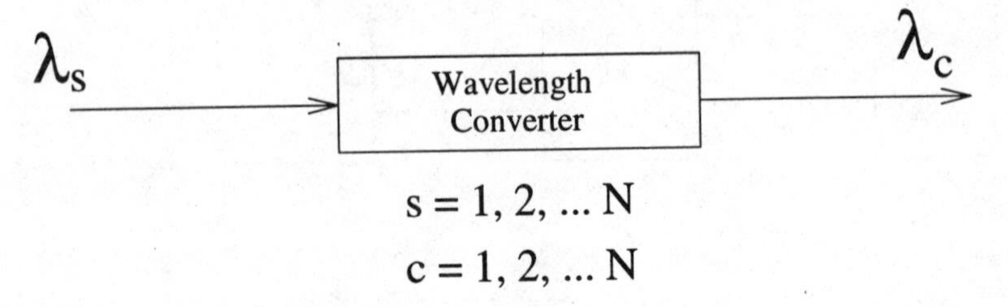

Figure 2.27 Functionality of a wavelength converter.

An ideal wavelength converter should possess the following characteristics [DMJD96]:

- transparency to bit rates and signal formats,
- fast setup time of output wavelength,
- conversion to both shorter and longer wavelengths,
- moderate input power levels,
- possibility for same input and output wavelengths (i.e., no conversion),
- insensitivity to input signal polarization,
- low-chirp output signal with high extinction ratio[7] and large signal-to-noise ratio, and
- simple implementation.

[7]The *extinction ratio* is defined as the ratio of the optical power transmitted for a bit "0" to the power transmitted for a bit "1."

2.6.1 Wavelength Conversion Technologies

Several researchers have attempted to classify and compare the several techniques available for wavelength conversion [DMJD96, MDJD96, SaIa96, Wies96, Yoo96]. The classification of these techniques presented in this section follows that in [Wies96]. Wavelength conversion techniques can be broadly classified into two types: *opto-electronic wavelength conversion*, in which the optical signal must first be converted into an electronic signal; and *all-optical wavelength conversion*, in which the signal remains in the optical domain. All-optical conversion techniques may be subdivided into techniques which employ *coherent effects* and techniques which use *cross modulation.*

Opto-Electronic Wavelength Conversion

In opto-electronic wavelength conversion [Fuji88], the optical signal to be converted is first translated into the electronic domain using a photodetector (labeled R in Fig. 2.28, from [Mest95]). The electronic bit stream is stored in the buffer (labeled FIFO for the First-In-First-Out queue mechanism). The electronic signal is then used to drive the input of a tunable laser (labeled T) tuned to the desired wavelength of the output (see Fig. 2.28). This method has been demonstrated for bit rates up to 10 Gbps [Yoo96]. However, this method is much more complex and consumes a lot more power than the other methods described below [DMJD96]. Moreover, the process of opto-electronic (O/E) conversion adversely affects the transparency of the signal, requiring the optical data to be in a specified modulation format and at a specific bit rate. All information in the form of phase, frequency, and analog amplitude of the optical signal is lost during the conversion process.

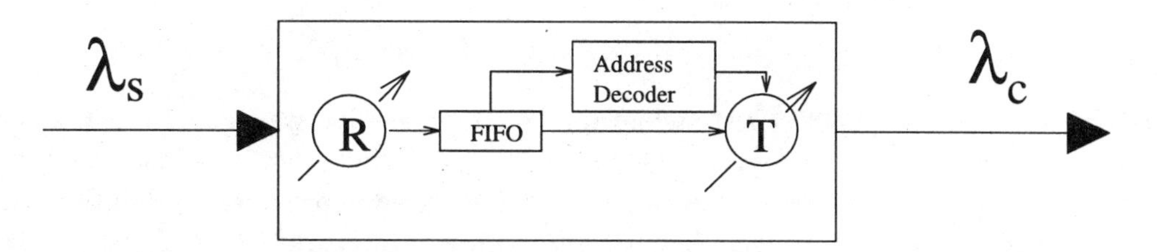

Figure 2.28 An opto-electronic wavelength converter.

Wavelength Conversion Using Coherent Effects

Wavelength conversion methods using coherent effects are typically based on wave-mixing effects (see Fig. 2.29). Wave-mixing arises from a nonlinear optical response of a medium when more than one wave is present. It results in the generation of another wave whose intensity is proportional to the product of the interacting wave intensities. Wave-mixing preserves both phase and amplitude information, offering strict transparency. It is also the only approach that allows simultaneous conversion of a set of multiple input wavelengths to another set of multiple output wavelengths and could potentially accommodate signal with bit rates exceeding 100 Gbps [Yoo96]. In Fig. 2.29, the value $n = 3$ corresponds to Four-Wave Mixing (FWM) and $n = 2$ corresponds to Difference Frequency Generation (DFG). These techniques are described below.

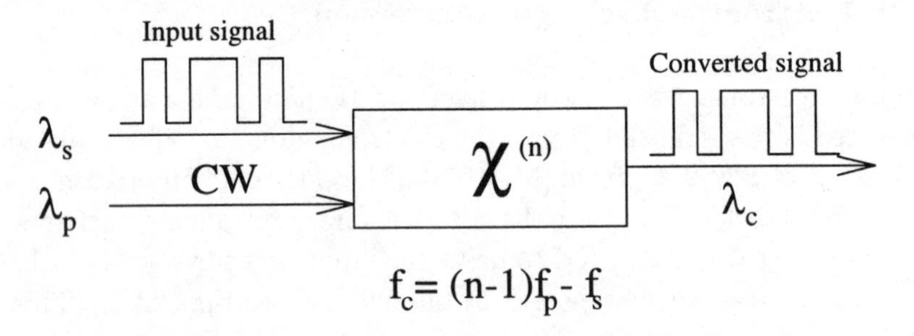

Figure 2.29 A wavelength converter based on nonlinear wave-mixing effects.

- *Four-Wave Mixing (FWM)*: FWM (also referred to as four-photon mixing) is a third-order nonlinearity in silica fibers, which causes three optical waves of frequencies f_i, f_j, and f_k ($k \neq i, j$) to interact in a multichannel WDM system [TCFG95] to generate a fourth wave of frequency given by

$$f_{ijk} = f_i + f_j - f_k$$

FWM is also achievable in other passive waveguides such as semiconductor waveguides and in an active medium such as a semiconductor optical amplifier (SOA). This technique provides modulation-format independence [Schn94] and high bit rate capabilities [LuRa94]. However, the conversion efficiency from pump energy to signal energy of this technique is not very high, and it decreases swiftly with increasing conversion span (shift between pump and output signal wavelengths) [ZPVN94].

- *Difference Frequency Generation (DFG)*: DFG is a consequence of a second-order nonlinear interaction of a medium with two optical waves: a pump wave and a signal wave [Yoo96]. DFG is free from satellite signals which appear in FWM-based techniques. This technique offers a full range of transparency without adding excess noise to the signal [YCBK95]. It is also bidirectional and fast, but it suffers from low efficiency and high polarization sensitivity. The main difficulties in implementing this technique lie in the phase-matching of interacting waves [YCBK96] and in fabricating a low-loss waveguide for high conversion efficiency [Yoo96].

Wavelength Conversion Using Cross Modulation

Cross-modulation wavelength conversion techniques utilize active semiconductor optical devices such as semiconductor optical amplifiers (SOAs) and lasers. These techniques belong to a class known as optical-gating wavelength conversion [Yoo96].

- *Semiconductor Optical Amplifiers (SOAs) in XGM and XPM mode*: The principle behind using an SOA in the cross-gain modulation (XGM) mode is shown in Fig. 2.30 (from [DMJD96]). The intensity-modulated input signal modulates the gain in the SOA due to gain saturation. A continuous-wave (CW) signal at the desired output wavelength (λ_c) is modulated by the gain variation so that it carries the same information as the original input signal. The CW signal can either be launched into the SOA in the same direction as the input signal (codirectional), or launched into the SOA in the opposite direction as the input signal (counterdirectional). The XGM scheme gives a wavelength-converted signal that is inverted compared to the input signal. While the XGM scheme is simple to realize and offers penalty-free conversion at 10 Gbps [DMJD96], it suffers from the drawbacks due to inversion of the converted bit stream and extinction ratio degradation for the converted signal.

  The operation of a wavelength converter using SOA in cross-phase modulation (XPM) mode is based on the fact that the refractive index of the SOA is dependent on the carrier density in its active region. An incoming signal that depletes the carrier density will modulate the refractive index and thereby result in phase modulation of a CW signal (wavelength λ_c) coupled into the converter [DMJD96, LaPT96]. The

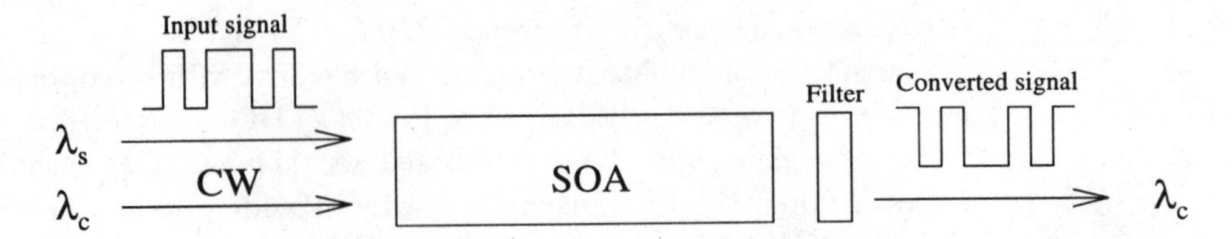

Figure 2.30 A wavelength converter using co-propagation based on XGM in an SOA.

SOA can be integrated into an interferometer so that an intensity-modulated signal format results at the output of the converter. Techniques involving SOAs in XPM mode have been proposed using nonlinear optical loop mirrors (NOLMs) [EiPW93], Mach-Zender interferometers (MZI) [DJMP94] and Michelson interferometers (MI) [MDJP94]. Figure 2.31 shows an asymmetric MZI wavelength converter based on SOA in XPM mode (from [DMJD96]). With the XPM scheme, the converted output signal can be either inverted or noninverted, unlike in the XGM scheme where the output is always inverted. The XPM scheme is also very power efficient compared to the XGM scheme [DMJD96].

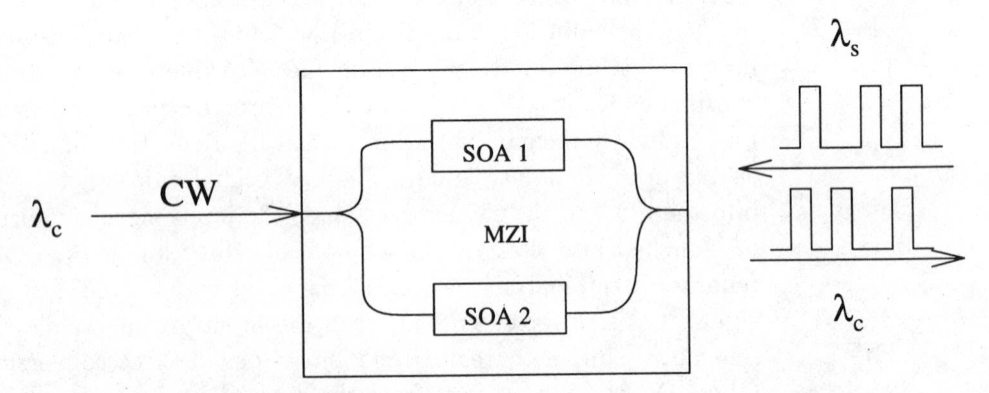

Figure 2.31 An interferometric wavelength converter based on XPM in SOAs.

- *Semiconductor Lasers*: Using single-mode semiconductor lasers, lasing mode intensity is modulated by input signal light through lasing mode gain saturation. The obtained output (converted) signal is inverted compared to the input signal. This gain suppression mechanism has been employed in a Distributed Bragg Reflector (DBR) laser to convert signals at 10 Gbps [YSIY96].

In the method using saturable absorption in lasers (e.g., [KOYM87]), the input signal saturates the absorption of carrier transitions near the band gap and allows the probe beam to transmit (see Fig. 2.32). This technique shows a bandwidth limit of 1 GHz due to carrier recombinations [Yoo96].

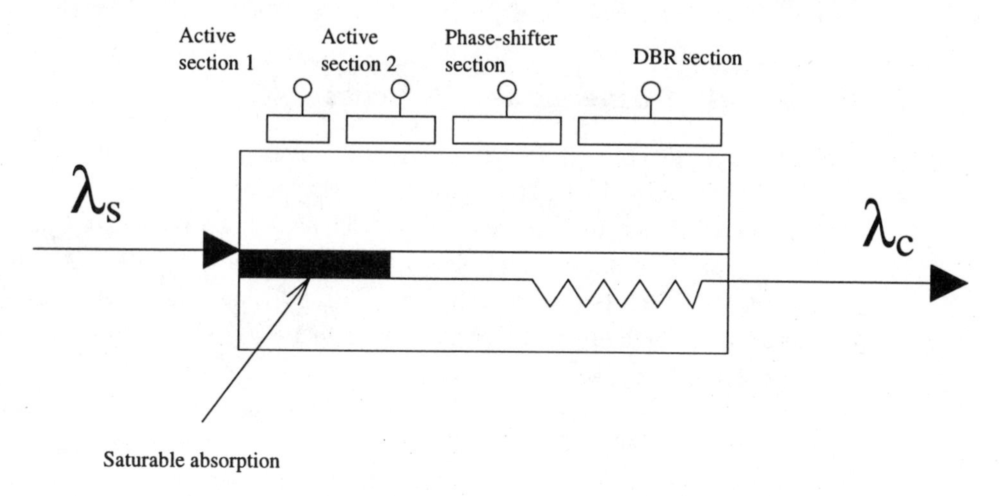

Figure 2.32 Conversion using saturable absorption in a laser.

Summary

In this subsection, we reviewed the various techniques and technologies used in the design of a wavelength converter. The actual choice of the technology to be employed for wavelength conversion in a network depends on the requirements of the particular system. However, it is clear that opto-electronic converters offer only limited digital transparency. Moreover, deploying multiple opto-electronic converters in a WDM crossconnect, e.g., in a WRS, requires sophisticated packaging to avoid crosstalk among channels. This leads to increased cost per converter, further making this technology less attractive than all-optical converters [Yoo96]. Other disadvantages of opto-electronic converters include complexity and large power consumption [DMJD96]. Among all-optical converters, converters based on SOAs using the XGM and the XPM conversion scheme presently seem well suited for system use. Converters based on FWM, though transparent to different modulation formats, perform inefficiently [DMJD96]. However, wave-mixing converters are the only category of wavelength converters that offer the full range of transparency while also allowing simultaneous conversion of a set of input wavelengths to another set of output wavelengths. In this respect, DFG-based methods offer great promise. Further details on comparison of various wavelength-conversion techniques can be found in [MDJD96, Wies96, Yoo96].

In the next subsection, we examine various switch architectures that may be employed in a wavelength-convertible network.

2.6.2 Wavelength Conversion in Switches

As wavelength converters become readily available, a vital question comes to mind: Where do we place them in the network? An obvious location is in the switches (crossconnects) in the network. A possible architecture of such a wavelength-convertible switching node is the dedicated wavelength-convertible switch (see Fig. 2.33, from [LeLi93]). In this architecture, referred to as a wavelength interchanging crossconnect (WIXC), each wavelength along each output link in a switch has a *dedicated* wavelength converter (shown as boxes labeled WC in Fig. 2.33), i.e., an $M \times M$ switch in an N-wavelength system requires $M \times N$ converters. The incoming optical signal from a link at the switch is first wavelength demultiplexed into separate wavelengths. Each wavelength is switched to the desired output port by the nonblocking optical switch. The output signal may have its wavelength changed by its wavelength converter. Finally, various wavelengths combine to form an aggregate signal coupled to an outbound fiber. The switch architecture shown in Fig. 2.33 is similar to that of the reconfigurable WRS (also WSXC) shown in Fig. 2.21 with additional wavelength converters added after the switching elements.

However, the dedicated wavelength-convertible switch is not very cost efficient since all of its converters may not be required all the time [InMu96]. An effective method to cut costs is to share the converters. Two architectures have been proposed for switches sharing converters [LeLi93]. In the share-per-node structure (see Fig. 2.34(a)), all the converters at the switching node are collected in a converter bank. A converter bank is a collection of a few wavelength converters (e.g., two in each of the boxes labeled WC in Fig. 2.34), each of which is assumed to have identical characteristics and can convert any input wavelength to any output wavelength. This bank can be accessed by any wavelength on any incoming fiber by appropriately configuring the larger optical switch in Fig. 2.34(a). In this architecture, only the wavelengths which require conversion are directed to the converter bank. The converted wavelengths are then switched to the appropriate outbound fiber link by the second optical switch. In the share-per-link structure (see Fig. 2.34(b)), each

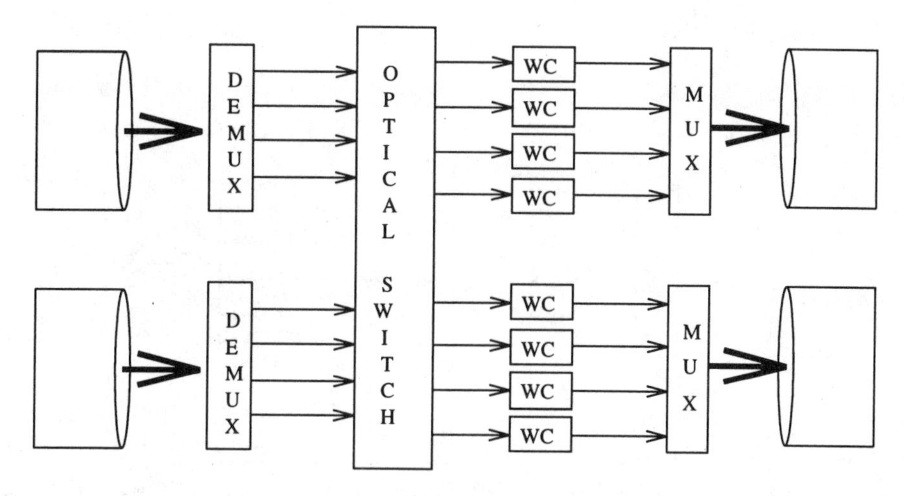

Figure 2.33 A switch which has dedicated converters at each output port for each wavelength.

outgoing fiber link is provided with a dedicated converter bank which can be accessed only by those lightpaths traveling on that particular outbound link. The optical switch can be configured appropriately to direct wavelengths toward a particular link, either with conversion or without conversion.

When opto-electronic wavelength conversion is used, the functionality of the wavelength converter can also be performed at the access stations instead of at the switches. The share-with-local switch architecture proposed in [LeLi93] (see Fig. 2.35) and the simplified network access station architecture proposed in [KoAc96b] (see Fig. 2.36) fall under this category. In the share-with-local switch architecture, selected incoming optical signals are converted to electrical signals by a receiver bank. A signal can then be either dropped locally or retransmitted on a different wavelength by a transmitter bank. In Fig. 2.36, an optical signal on wavelength W1 can be switched to a network access station where it is converted to an electronic signal. The signal can then be retransmitted by the network access station on a new wavelength W2.

2.7 Designing WDM Networks: Systems Considerations

In designing a WDM network, it is important to keep in mind not only the desired functionality of the network, but also the capabilities and limitations of available optical network components. In this section, we present some of

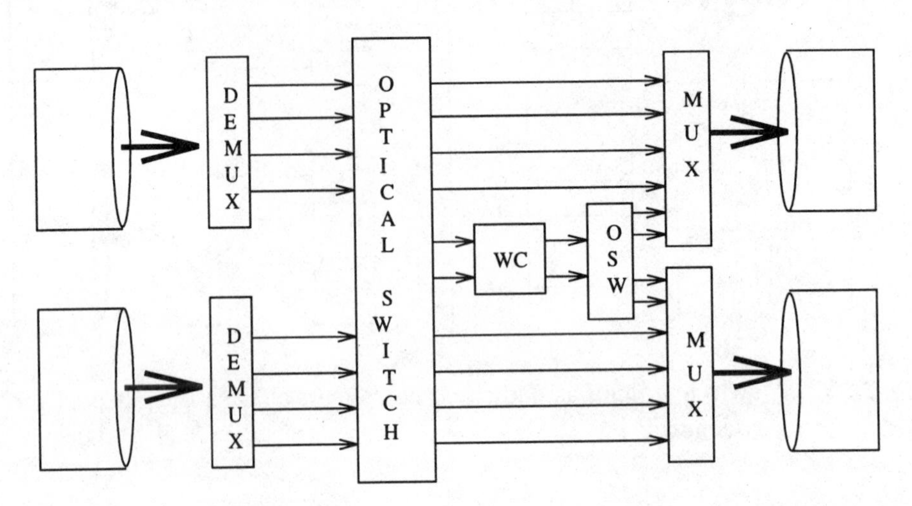

(a) Share-per-node wavelength-convertible switch architecture

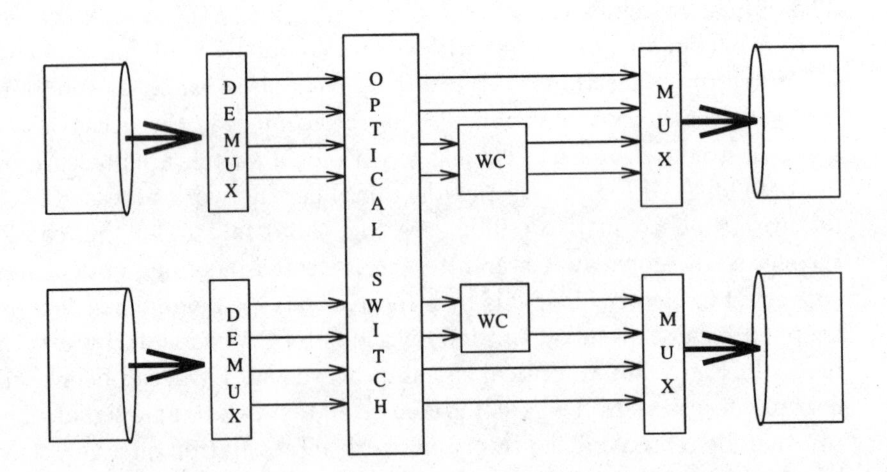

(b) Share-per-link wavelength-convertible switch architecture

Figure 2.34 Switches which allow sharing of converters.

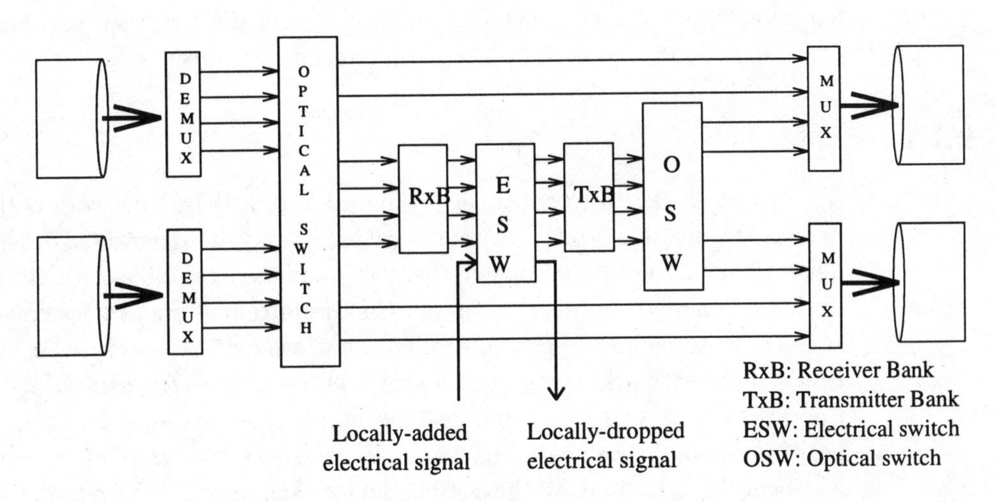

Figure 2.35 The share-with-local wavelength-convertible switch architecture.

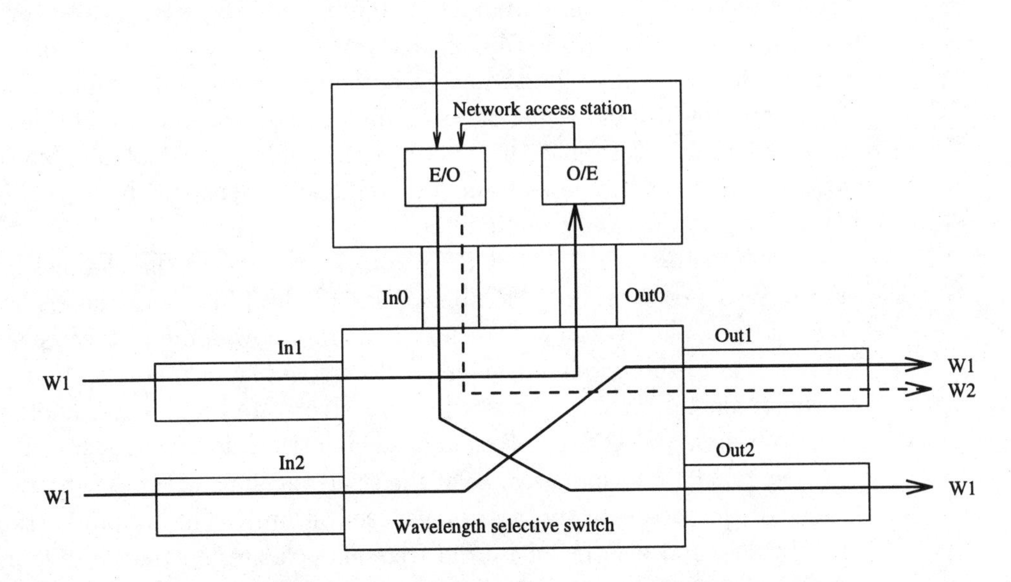

Figure 2.36 Architecture which supports electronic wavelength conversion.

the issues involved in designing optical networks, describe some of the physical constraints which must be considered, and discuss how various optical components may be used to satisfy networking requirements.

2.7.1 Channels

An important factor to consider in the design of a WDM network is the number of wavelengths to use. In some cases, it may be desirable to design the network with the maximum number of channels attainable with the current device technology, subject to tuning time requirements and cost constraints. Another approach is to assign a different wavelength to each node, although this type of network doesn't scale very well. In wide-area networks (WANs), the objective is often to minimize the number of wavelengths for a desired network topology or traffic pattern. In any case, the maximum number of wavelengths is limited by the optical device technology. The number of channels is affected primarily by the total available bandwidth or spectral range of the components and the channel spacing.

The bandwidth of the fiber medium, as mentioned in Section 2.1, is limited to the low-attenuation regions around 1300 nm and 1550 nm. These regions have bandwidths of approximately 200 nm (25 THz) each. However, optical networks may not necessarily be able to take advantage of this entire range due to the bandwidth limitations of optical components. Amplifiers have an optical bandwidth of around 35 to 40 nm, injection-current tunable lasers have a tuning range of around 10 nm, while the tuning range of tunable receivers varies from the entire low-attenuation region of fiber for Fabry-Perot filters to around 16 nm for electrooptic filters.

Some factors which affect the channel spacing are the channel bit rates, the optical power budget, nonlinearities in the fiber, and the resolution of transmitters and receivers. We now illustrate how some of these parameters may relate to the maximum number of channels in a WDM system. We will assume that tunable transmitters and receivers are being used, and we would like to design a WDM passive-star-based network for N nodes.

Let ΔT be the tuning range of the transmitters and let ΔR be the tuning range of the receivers (both are measured in nm). The available transceiver bandwidth, BW_T is given by the frequency range in which the transmitter tuning range intersects with the receiver tuning range.

Using the identity

$$\Delta f = \frac{c\Delta\lambda}{\lambda^2}$$

the frequency needed for BW_T is

$$\Delta f = \frac{cBW_T}{\lambda^2}$$

where λ is the "center" wavelength.

If we want each channel to have a bit rate of B Gbps, $2B$ GHz of bandwidth will be needed for encoding, assuming a modulation efficiency of 2 Hz/bps. According to [Brac90], a channel spacing of at least 6 times the channel bit rate is needed to minimize crosstalk on a WDM system. Thus, if we want w channels, we need

$$2B \cdot w + 6B(w-1) \text{ GHz}$$

Thus, the maximum number of resolvable channels for this network is

$$W = \frac{\Delta f + 6B}{8B}$$

Although a maximum of W channels may be accommodated, in some cases, it may be desirable to use fewer than W channels, e.g., in a shared-channel WDM optical LAN [TrMu96, TrMu97]. A higher number of channels may provide more network capacity, but it also results in higher network costs, and in some cases, may require more complex protocols.

2.7.2 Power Considerations

In any network, it is important to maintain adequate signal-to-noise ratio (SNR) in order to ensure reliable detection at the receiver. In a WDM network, signal power can degrade due to losses such as attenuation in the fiber, splitting losses, and coupling losses. Some of the losses may be countered through the use of optical amplifiers, and an important consideration in designing a WDM network is the design and appropriate placement of amplifiers.

There are three main applications for optical amplifiers in a lightwave network [Gree93]. The first application is as a *transmitter power booster*, which is placed immediately after the transmitter in order to provide a high power signal to the network. This allows the signal to undergo splitting at couplers or to travel longer distances. The second application is as a *receiver preamplifier*, which boosts the power of a signal before detection at a receiver photodetector. The third application is as an *in-line amplifier*, which is used within the network to boost degraded signals for further propagation. Each of these situations requires the amplifier to have different characteristics. A discussion of the requirements and design of multistage EDFAs for various applications is given in [DeNa95]. Table 2.4 summarizes some of these requirements.

EDFA Application	*Gain (dB)*	*Noise Figure (dB)*	*Saturated Output (dBm)*
Transmitter power booster	Moderate	Moderate	High
Preamplifier	High	Low	Moderate
In-line amplifier	High	Low	High

Table 2.4 Requirements for EDFA applications.

For in-line amplifier applications, there is the additional issue of amplifier placement. Amplifiers need to be placed strategically throughout the network in a way which guarantees that all signals are adequately amplified while minimizing the total number of amplifiers being used. A study of this problem in local access networks is reported in [RaIM97].

When utilizing cascades of in-line amplifiers, one must also consider issues such as ASE noise introduced by the amplifiers, and the unequal gain spectrum of the amplifiers. The accumulation of ASE noise in a cascade of amplifiers may seriously degrade the SNR. If the input signal power is too low, ASE noise may cause the SNR to fall below detectable levels; however, if the signal power is too high, the signal combined with ASE noise may saturate the amplifiers. The unequal gain spectrum of the EDFA places limitations on the usable bandwidth in WDM systems. When multiple EDFAs are cascaded, the resulting gain bandwidth may be significantly reduced from the gain bandwidth of a single EDFA. An initial bandwidth of 30 nm can potentially be reduced to less than 10 nm after a cascade of 50 EDFAs [Maho95].

Although recent developments in amplifier technology have solved many of the power loss and noise problems in optical networks, network designers should not rely solely on amplifiers for resolving power issues, but should also consider other options. For example, to avoid splitting losses in network interconnections, it might be worthwhile to consider using wavelength-routing devices, such as the wavelength-routing switch or the waveguide grating router, instead of wavelength-independent devices, such as the amplifier gate switch or the passive-star coupler.

2.7.3 Additional Considerations

Other device considerations in the design of WDM networks include crosstalk and dispersion.

Crosstalk may either be caused by signals on different wavelengths (interband crosstalk), or by signals on the same wavelength on another fiber (intraband crosstalk) [Maho95]. Interband crosstalk must be considered when determining channel spacing. In some cases, intraband crosstalk may be removed through the use of appropriate narrowband filters. Intraband crosstalk usually occurs in switching nodes where multiple signals on the same wavelength are being switched from different inputs to different outputs. This form of crosstalk is more of a concern than interband crosstalk because intraband crosstalk cannot be removed through filtering, and may accumulate over a number of nodes. The degree of intraband crosstalk depends in part on the switch architecture.

As was mentioned in Section 2.1.4, dispersion in an optical communication system causes a pulse to broaden as it propagates along the fiber. The pulse broadening limits the spacing between bits, and thus limits the maximum transmission rate for a given propagation distance. Alternatively, it limits the maximum fiber distance for a given bit rate.

Apart from the device considerations mentioned above, there are architectural considerations in designing a WDM network. The topology of the physical optical fiber buried in the ground may influence the choice of which transmitter-receiver pairs to operate on which wavelengths. The need for fault-tolerance and reliability affects the choice of the network architectures. Moreover, the emerging standards on optical wavelengths and channel spacing (e.g., ITU-T, MONET) will influence the design of the network components.

2.7.4 Elements of Local-Area WDM Network Design

A local area WDM network will typically consist of a number of nodes which are connected via two-way optical fibers either to some physical network medium or directly to other nodes. In this section, we will present some of the issues involved in selecting the hardware for both the network medium and the nodes.

The Network Medium

The simplest and most popular interconnection device for a local area WDM network is the passive-star coupler which provides a broadcast medium (see

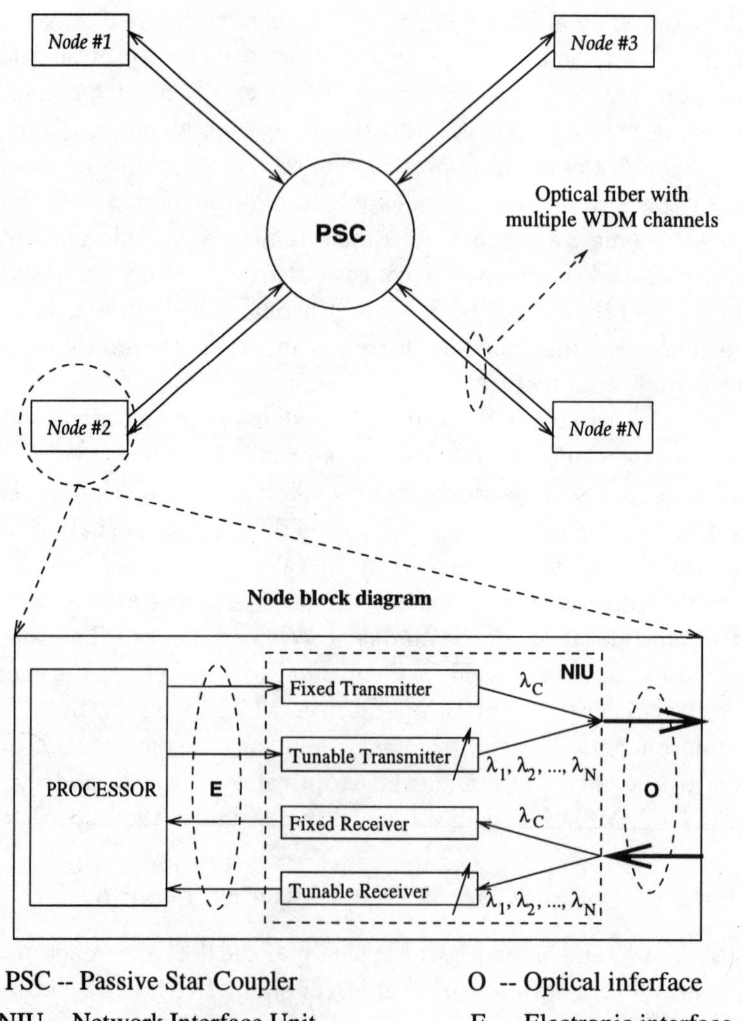

Figure 2.37 Broadcast-and-select WDM local optical network with a passive-star coupler network medium.

Fig. 2.37). The broadcast capability of the star coupler combined with multiple WDM channels allows for a wide range of possible media access protocols [Mukh92a, Mukh92b]. Also, since the star coupler is a passive device, it is fairly reliable. The drawback of having a passive network medium is that the network nodes may be required to handle additional processing and may require additional hardware in order to route and schedule transmissions. The broadcast capability of the star coupler also prevents the reuse of wavelengths to create more simultaneous connections.

Network Nodes

Another important consideration in the design of a WDM network is the hardware at each node. Each node in a network typically consists of a workstation connected to the network medium via optical fiber, and the node may potentially access any of the available wavelength channels on each fiber. In designing the network interface for each node, one must select the number of transmitters and receivers, as well as the type of transmitters and receivers – fixed-tuned or tunable – to place at each node. These decisions usually depend on the protocol, degree of access, and connectivity desired in the network, as well as on practicality and cost considerations.

A WDM network protocol may either be a single-hop protocol [Mukh92a], in which communication takes place directly between two nodes without being routed through intermediate nodes; or a multihop protocol [Mukh92b], in which information from a source node to a destination node may be routed through the electronics at intermediate nodes in the network. In general, multihop networks require less tuning than single-hop networks.

At a minimum, each node must be equipped with at least one transmitter and one receiver. When both the transmitters and the receivers are fixed tuned to certain wavelength channels, and there is more than one channel, then a static multihop topology must be established over the passive-star coupler. The topology is created by establishing connections between pairs of nodes on given wavelengths. An overview of multihop protocols and topologies is provided in [Mukh92b].

A more flexible approach would be to use either a tunable transmitter and/or a tunable receiver. The tunability allows the network to be dynamically reconfigured based on traffic patterns, and it also allows the implementation of single-hop protocols. A number of single-hop WDM protocols based on nodes with tunable transmitters and/or tunable receivers are presented in [Mukh92a]. Additional transmitters and receivers at each node may help to increase the connectivity of the network and may also be used to help coordinate transmissions. In some cases, the network may have a control channel which may be used for pretransmission coordination (pretransmission coordination allows a node to preannounce its transmission so that the receiving node may get ready for reception, e.g., by appropriately tuning its receiver). Each node may then be equipped with an additional fixed transmitter and an additional fixed receiver, each permanently tuned to the control channel.

The tuning latency of tunable transmitters and receivers may be an important factor in choosing components, depending on the type of network being implemented. A single-hop network generally requires tunable components to create connections on demand and requires some amount of coordination in order to have the source node's transmitter and the destination node's receiver tuned to the same channel for the duration of an information transfer. In this case, the tuning time of transmitters and receivers may have a significant impact on the performance of the network. On the other hand, most multihop networks require tunability only for infrequent reconfigurations of the network based on changing traffic patterns; thus, the tuning time of components in a multihop network is not as critical as in the case of a single-hop network.

Chronologically, the multihop concept came first precisely because fast device tuning was difficult to achieve in the past.

Node Separation in Passive-Star Coupler WDM Local Area Networks

Given the output power of the transmitters and the receiver sensitivity, we can compute the maximum distance allowable between network nodes. Assume that all nodes are D meters from the passive-star coupler (PSC), and that the input-to-output power splitting ratio of the PSC is given by Eqn. (2.9). Then, the maximum value of D such that the optical signals reaching each receiver are strong enough to be received (D_{max}) can be computed by combining Eqns. (2.9) and (2.7), so that

$$D_{max} = \frac{10}{A} \log_{10} \frac{P_t}{NP_r}$$

where P_t is the optical power of the transmitter and P_r is the minimum amount of power that the receiver needs to resolve the optical signal.

2.7.5 WDM Wide-Area Network Design Issues

Current wide-area networks are designed as electronic networks with fiber links. However, these networks may not be able to take full advantage of the bandwidth provided by optical fiber, because electronic switching components may be incapable of switching the high volume of data which can be transmitted on the fiber links. It is anticipated that the next generation of optical networks will make use of optical routers and switching elements to allow all-optical lightpaths to be set up from a source node to a destination node, thus bypassing electronic bottlenecks at intermediate switching nodes. Also, WDM will allow multiple lightpaths to share each fiber link. The concept of WDM lightpaths is analogous to a multilane highway which can be used to bypass stoplights on city roads. Another concept in WAN design is *wavelength reuse.* Since each wavelength may be used on each fiber link in the network, multiple lightpaths which do not share any links may use the same wavelength. For example, in Fig. 2.38, wavelength λ_1 is used to set up one lightpath from node A to node C, and another lightpath from node G to node H. (Such wavelength reuse is not possible in a passive-star-based WDM network.)

The issue of setting up lightpaths and routing the lightpaths over the physical fibers and switches in a wide-area WDM network is an optimization problem in which the overall network performance must be balanced against the consumption of network resources. The degree of freedom in designing the lightpaths depends in part on the type of switching elements or crossconnects used in the access nodes or switching nodes. If wavelength-insensitive crossconnect devices are used, then each signal on a given input fiber must be routed to the same output fiber. Wavelength-sensitive switching devices offer more flexibility, allowing different wavelengths arriving on a single input fiber to be directed independently to different output fibers. However, this approach may still result in conflicts at the nodes if two signals on the same wavelength arriving on different input ports need to be routed to the same output port. The conflict may be resolved by incorporating wavelength converters at each node, and converting one of the incoming signals to a different wavelength (see Section 2.6). If wavelength-conversion facilities are not available at switching nodes, then a lightpath must have the same wavelength on all of the fiber links through which it traverses; this is referred to as the *wavelength-continuity constraint* (see Fig. 2.26). Another approach for resolving conflicts is to find an alternate route in the network for one of the two conflicting lightpaths, and in some cases an alternate wavelength.

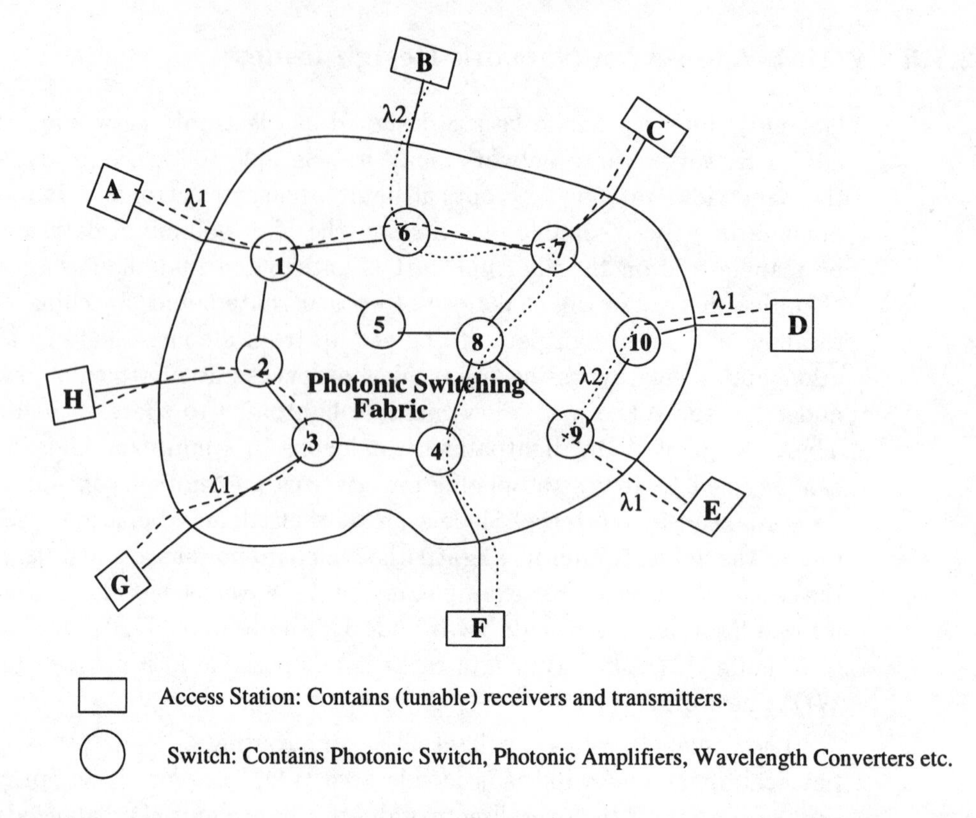

Figure 2.38 Lightpath routing in a WDM WAN.

In designing an optical network, it is important to recognize what can and cannot be accomplished by optical switching devices. Current optical crossconnects may be able to switch optical information based on input ports or wavelengths, but they cannot efficiently demultiplex time-division-multiplexed (TDM) data, i.e., time-based data packets, within an optical data stream. Also, in a reconfigurable optical switch, the time required to reconfigure the switch is often long when compared to the speed of data passing through the switch. Therefore, optical switches seem more appropriate for switching large streams of data or for setting up somewhat static routes in the network based on wavelengths rather than for switching individual packets.

2.8 Experimental WDM Lightwave Networks

While WDM is anticipated to be the technology for future networks, it is important to know what has been currently accomplished using WDM technology. This section describes a number of experimental WDM testbeds and projects in both the local area and wide-area domains.

2.8.1 Local Area Network Testbeds

Two examples of WDM local area network testbeds are Bellcore's LAMBDANET [GGKV90] and IBM's Rainbow [DGLR90]. Like in many practical networks, simplicity has been chosen over "smart" protocols. Both LAMBDANET and Rainbow are single-hop networks.

LAMBDANET

In LAMBDANET, each node is equipped with one fixed transmitter consisting of a DFB laser and N fixed receivers, where N is the number of nodes in the network. The incoming wavelengths are separated using a diffraction-grating wavelength demultiplexer, and each individual wavelength is sent to an optical receiver. Each node's transmitter is fixed on that node's *home wavelength.* This allows single-hop connectivity without requiring tunable components, as well as not requiring the design of smart protocols. In [GGKV90], an aggregate throughput of 36 Gbps was demonstrated for a 16-node system. While the advantages of LAMBDANET include simplicity of design and architectural support for multicasting, we don't consider the architecture scalable, because it requires N data wavelengths. The practicality of requiring N receivers per node is also a limitation, as for systems with large N, the cost per node might become large. However, recent advances in receiver array technology may help reduce the system's cost [GGKV90].

Rainbow

IBM's Rainbow-I testbed was designed to support up to 32 IBM PS/2s connected via thirtytwo 200-Mbps WDM channels (for an aggregate throughput of 6.4 Gbps) [JaRS93]. Each node is equipped with a fixed DFB laser transmitter and a tunable Fabry-Perot filter receiver. Each node's transmitter is fixed to a *home channel.* When a node has a packet to send, it tunes its receiver to the destination's home channel, and then transmits a connection setup request by repeatedly sending the destination node's address. When a receiver is idle, it scans all wavelengths in round-robin fashion, until it finds a channel with a setup request containing the receiver's own address. It then transmits an acknowledgment to the source node, and two-way communication can begin. Though a simple protocol, the setup time may be too long for packet switching. Like LAMBDANET, Rainbow does not scale well. Rainbow was the first WDM testbed to demonstrate tunable WDM components working in a real environment.

The Rainbow-II optical network [HaKR96] is an extension of the Rainbow-I network. It consists of similar optical hardware and the same media-access protocol as the Rainbow-I network, but also incorporates some higher-layer protocols. The Rainbow-II network supports 32 nodes, each transmitting at rates of up to 1 Gbps.

2.8.2 Wide-Area Network Testbeds

A number of government-funded programs have been established in Europe and in the U.S. to investigate WDM wide-area networks. Some of the more notable projects include the Research and development in Advanced Communications technologies in Europe (RACE) MultiWavelength Transport Network (MWTN) project and the DARPA-sponsored Multiwavelength Optical Networking (MONET), Optical Networks Technology Consortium (ONTC), and All-Optical Network (AON) projects. These programs focus on the design of WDM networks as well as the design of the optical components used in the networks.

RACE MultiWavelength Transport Network

The European RACE-MWTN program [Hill93, Joha96] involves a consortium of European companies and universities. Its objective is to design an all-optical transport network layer employing optical switches, crossconnects, transmitters, receivers, and amplifiers. The RACE-MWTN demonstration network spans 130 km and has been demonstrated with four wavelengths in the 1550 nm band. The project has developed two basic optical networking elements: the optical crossconnect (OXC), and the optical add/drop multiplexer (OADM).

The OXC is an optical switching element for switched networks. It is constructed using 8×8 digital switches and 4×4 laser amplifier gate switches, and uses EDFAs at its input and output ports. The OXC is able to perform wavelength-based switching through the use of tunable filters. Four types of tunable filters – acoustooptic filters, integrated multigrating filter, multilayer thin-film filters, and Fabry-Perot filters – have been used. Wavelength conversion takes place by converting a signal to electronics and retransmitting on a different wavelength.

The OADM adds or drops one or more wavelengths from a fiber while allowing the other optical signals to pass through. It makes use of 2 × 2 space switches and 2 × 2 acoustooptic switches. The OADM is intended primarily for ring or bus networks.

MONET

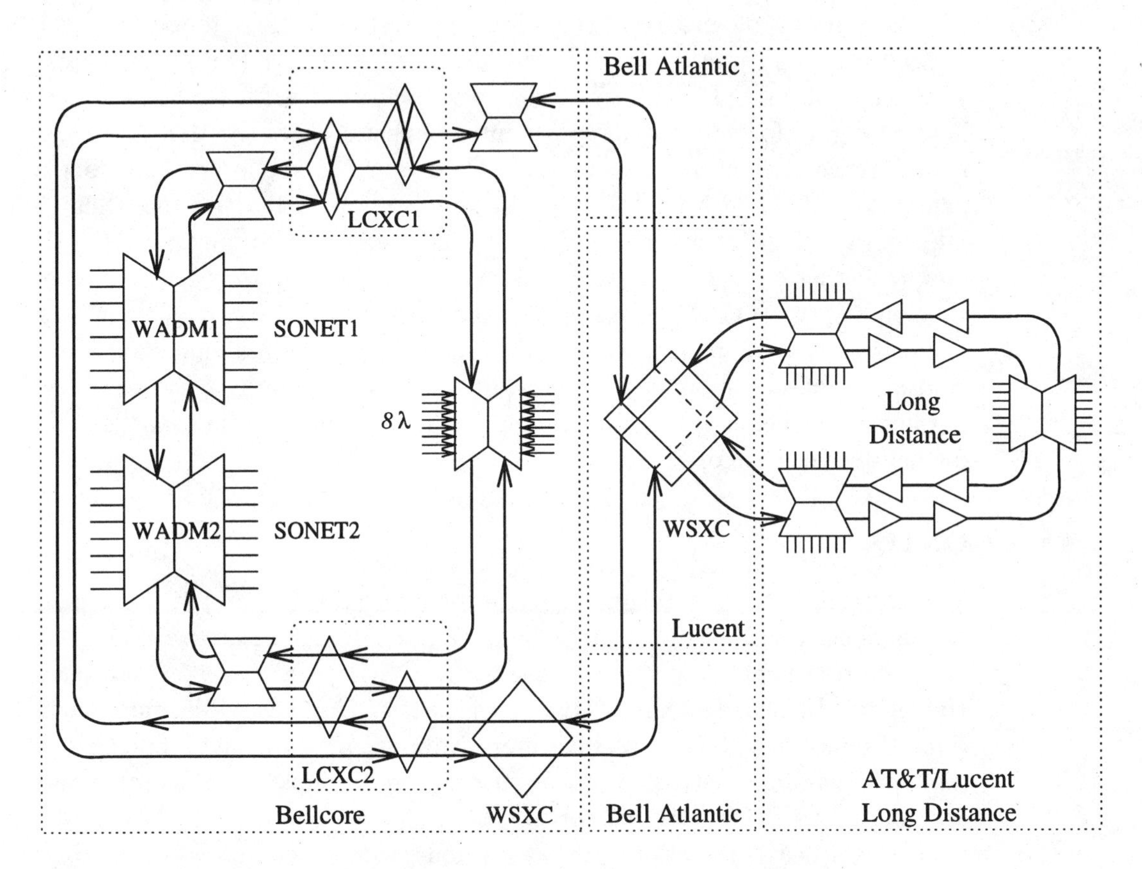

Figure 2.39 MONET New Jersey Area Network.

The Multiwavelength Optical Networking (MONET) consortium [WaAl96] includes the American Telephone and Telegraph Company (AT&T), Bellcore, Lucent Technologies, Bell Atlantic, BellSouth, Pacific Telesis, and Southwestern Bell Communications (SBC). The goal of the program is to develop a transparent multiwavelength optical network, define the required enabling technologies, and explore the potential of WDM networks.

Currently, three testbeds are being constructed in the New Jersey area – local exchange, crossconnect, and long distance (see Fig. 2.39, from [VoGa97]). The purpose of the long distance testbed is to study the transmission of various optical signals over distances of 2,000 km. The crossconnect testbed will connect the long distance testbed with the local exchange and will allow researchers to study various crossconnect devices and network management software. The local exchange testbed will demonstrate various local area network topologies and some of the properties of LAN topologies such as scalability and interoperability. As of February '97, MONET offers eight wavelengths ("MONET-compliant") in the 1550-nm region, supporting analog and digital formats and supports interconnection between the three testbeds. Transmission of 2.5-Gbps channels over 2,290 km is currently being demonstrated. The MONET program employs *laser arrays*, consisting of DFB arrays with integrated modulators and integrated coupler; *switch fabrics*, including 4×4 $LiNbO_3$ switches (WSXC), 2×2 liquid crystal switches with bulk optics (LCXC), and InP digital switch technologies; *receiver arrays*, comprising of eight photodetectors with integrated preamplifiers; *wavelength converters*, based on cross-phase modulation with integrated DFB laser and parametric conversion; and *optical amplifiers*, based on gain-flattened EDFAs for multiwavelength operation.

ONTC

The Optical Networks Technology Consortium [ChEl96] program includes members such as Bellcore, Colombia University, Hughes Research Laboratories, Northern Telecom, Rockwell Science Center, Case Western Reserve University, United Technology Research Center, Uniphase Telecommunications Product, and Lawrence Livermore National Laboratories. The goal of the project is to construct a testbed to explore various WDM components and to study ATM/WDM networks.

The testbed consists of four access nodes connected to two fiber rings via 2×2 WDM crossconnects (see Fig. 2.40). The two fiber rings are also connected to each other using a 2×2 WDM crossconnect.

Each access node is equipped with an ATM switch and SONET optical interfaces. The node may insert data into or remove data from the network, and may also provide packet forwarding for multihop communications.

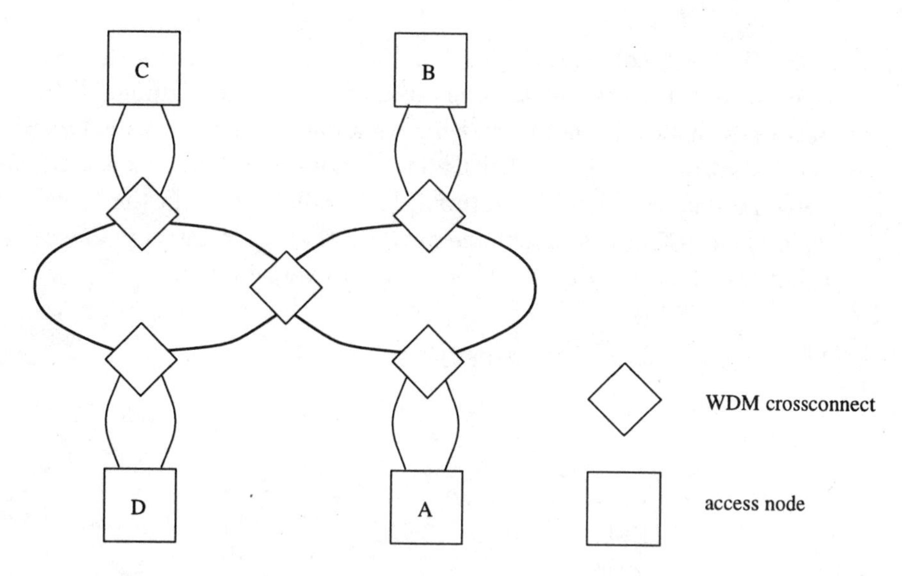

Figure 2.40 ONTC testbed.

Key network elements include EDFAs, acoustooptic tunable filters, hybrid wavelength selective crossconnect switches, wavelength add/drop multiplexers, multi-wavelength transmitters, and multi-wavelength receivers.

AON

AT&T, Digital Equipment Corporation (DEC), and Massachusetts Institute of Technology Lincoln Laboratory (MIT-LL) had formed a precompetitive consortium for ARPA to address the challenges of utilizing the evolving Terahertz capability of optical fiber technology to develop a national information infrastructure capable of providing a flexible transport layer [Alex93]. The main motivations for this architecture are the following.

- The architecture should scale gracefully to accommodate thousands of nodes, and provide for a nationwide communication infrastructure.
- The architecture should be "future-proof", i.e., modular and flexible to incorporate future developments in technology.

The architecture, shown in Fig. 2.41, is based on WDM, and provides scalability through wavelength reuse and TDM techniques. The architecture employs a three-level hierarchy. At the lowest level are Level-0 networks, each consisting of a collection of LANs. Each Level-0 network shares wavelengths internally, but there is extensive reuse of wavelengths among different Level-0 networks. A Level-1 network, which is a metropolitan area network, interconnects a set of Level-0 networks, and provides wavelength reuse among Level-0 networks via wavelength routers. The highest level is the Level-2 network, which is a nationwide backbone network that interconnects Level-1 networks, using wavelengths routers, wavelength converters, etc.

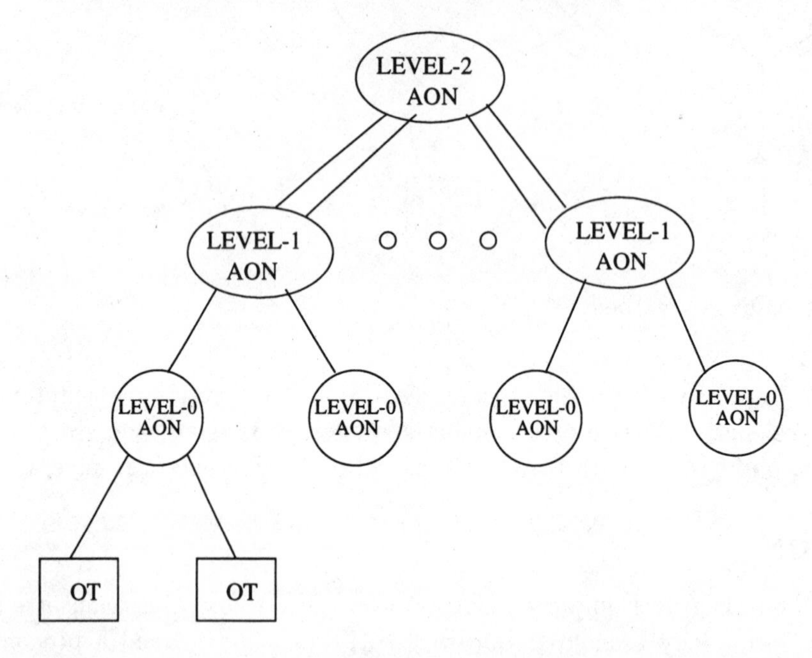

Figure 2.41 The AT&T/MIT-LL/DEC AON testbed architecture.

The services provided by the architecture are classified as follows.

- Type A service provides a dedicated optical path for point-to-point, or point-to-multipoint communication. This would provide for Gbps circuit-switched digital or analog sessions.

- Type B service uses TDM over an optical path to provide circuit-switched sessions, with bandwidths in the range from a few Mbps to the full channel rate of a few Gbps.

- Type C service is packet-switched and would be used internally for control, scheduling, network management, and for user datagram services.

2.9 Summary

Recent advances in the field of optics have opened the way for the practical implementation of WDM networks. In this chapter, we have provided a brief overview of some of the optical WDM devices currently available or under development, as well as some insight into the underlying technology. As optical device technology continues to improve, network designers need to be ready to take advantage of new device capabilities, while keeping in mind the limitations of such devices.

Exercises

2.1. An IP-based network application is to be built on top of a fiber-optic communication network. The application programmer considers two options for error correction. One option is to simply use the TCP/IP protocol. The other option is to use the UDP/IP protocol, which the programmer calculates will have a 10% higher bandwidth when transmissions are error free. The programmer has included a cyclic redundancy check in the UDP packets, but does not have a good scheme for retransmitting individual packets. Thus, when an error is detected, an average of 125 megabytes of data will have to be retransmitted. Which scheme will yield the higher average bandwidth if the fiber bit error rates is: (a) 10^{-9}? (b) 10^{-15}?

2.2. Consider a step-index fiber which has a core refractive index of 1.495. What is the maximum refractive index of the cladding in order for light entering the fiber at an angle of 60 degrees to propagate through the fiber? Air has refractive index of 1.0.

2.3. Find the numerical aperture in a graded-index fiber with two layers shown in Fig. 2.43. Compare the answer with the numerical aperture of the step-index fiber shown in Fig. 2.42. Can we use geometric optics to deal with situations where the wavelength and core diameter are of the same order of magnitude (e.g., single-mode fiber)?

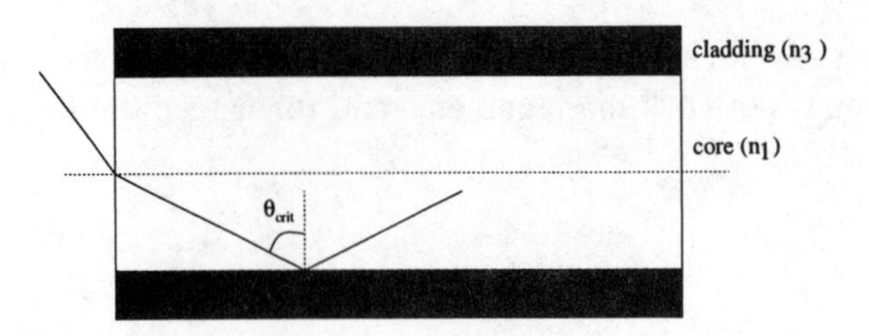

Figure 2.42 Critical angle in a step index fiber.

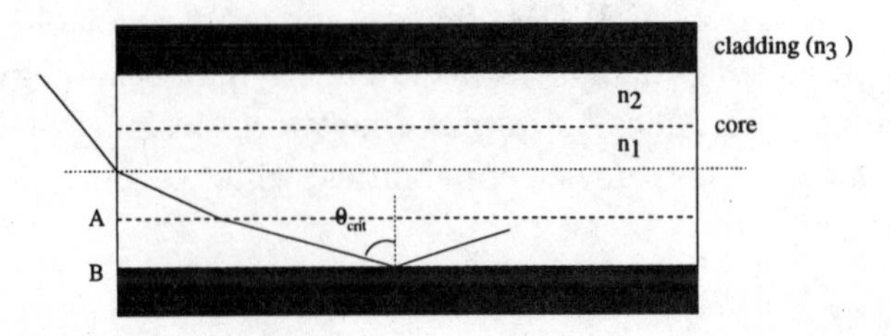

Figure 2.43 Critical angle in a graded index fiber.

2.4. Consider a step-index multimode fiber in which the refractive indices of the cladding and core are 1.35 and 1.4, respectively. The diameter of the core is 50 μm. Approximately how many modes are supported by the fiber for a signal at a wavelength of 1550 nm?

2.5. Find the approximate number of modes in a 100 μm core step-index multimode fiber at 850 nm. Assume the refractive index of glass to be 1.5 and that of the cladding to be 1.47.

2.6. Consider an optical link in which power at the transmitter is 0.1 mW and the minimum power required at the receiver is 0.08 mW. The attenuation constant for the fiber material is 0.033 dB/km. What is the maximum length of the optical link, assuming that there are no amplifiers?

2.7. Describe the various types of dispersion and explain how the effects of each type of dispersion can be reduced.

2.8. Consider a single mode optical fiber with an attenuation of 0.2 dB/km and a dispersion limit of 200 Gbps·km. The transmitter power is 1 mW and the receiver sensitivity is 10^{-5} mW. The link operates at a rate of 2.4 Gbps. Assume a 10 dB power margin for losses in connectors. Calculate the maximum length of the optical link.

2.9. Suppose we have a system which has 3 channels operating at 1549.32 nm, 1554.13 nm, and 1558.98 nm. At which frequencies will we have sidebands as a result of four-wave mixing? (Use $c = 2.998 \times 10^8$.)

2.10. Consider a broadcast star network with $N = 2^k$ nodes where k is an integer. The network is built out of 2×2 couplers with excess loss β and coupling coefficient α. Each transmitter has a laser with power P_t.

(a) Find the power levels received by the receivers when a single transmitter (say transmitter T_1) transmits. That is, determine how many different power levels are received by the N receivers, and how many receivers receive each of these levels.
Hint: Construct a tree with the transmitter at the root.

(b) Suppose your goal as the network designer is to maximize the minimum power received by any of the receivers from a transmitter. Find the optimal value of α for this design criterion. Explain your answer.

(c) Now suppose you have a different design criterion. Your new goal is to maximize the expected value of the power between a random transmitter/receiver pair. Assume that each such pair is equally likely. How would you select your couplers? Explain your answer.
Hint: You can exploit the symmetry to fix the transmitter.

2.11. Suppose a 1 mW, 1550 nm signal is transmitted across a 5 km fiber, through an 8×8 passive star coupler, and through another 15 km of fiber before reaching its destination. No amplifiers are used. What is the power of the signal at the destination?

2.12. A 16×16 passive-star hub has been constructed from combiners, couplers, and splitters as in Fig. 2.7. Each combiner, coupler, and splitter results in a 3 dB power loss. Each host is up to 10 km away from the star, with a signal attenuation of 0.2 dB/km. If each host must receive signal power of at least 0.01 mW to clearly recognize signals, how strong must each host's transmission signal be?

2.13. Consider a unidirectional fiber bus with N nodes. (Assume that N is even.) All the couplers have an excess loss β ($\ll 1$). (Let $\gamma = 1 - \beta$ and use γ in your solution instead of β.) The ith coupler has a coupling coefficient α_i for $1 \leq i \leq N$. The coupling coefficients can be independently selected. Optimize the coupling coefficients to maximize the worst-case power transfer between a transmitter and a receiver. Compare the resulting worst-case power with the case where all couplers are identical and optimized. Hint: First assume some transmitter k is the worst-off. Argue why, in the optimal solution, all the receivers to the right of k will see the same attenuation from transmitter k. Similarly all the transmitters to the left of k should be equally badly off. Obtain a recursion for α_i assuming k is known and find the value of k.

2.14. Consider the following simplified model of a direct detection binary FSK system. By using a pair of optical filters and a pair of photodetectors in the receiver, we observe two Poisson distributed photon counts: X_0 and X_1. When the data bit is 0, the parameter of X_0 is $\lambda + \lambda_d$ while that of X_1 is λ_d. (Here λ_d models the dark current.) Conversely, when the data bit is 1, X_0 has parameter λ_d and X_1 has $\lambda + \lambda_d$. X_0 and X_1 are statistically independent when conditioned on the value of the data bit.

(a) Obtain the Maximum Likelihood processing of X_0 and X_1 explicitly.

(b) Find the probability of a bit error as a function of λ and λ_d. You may leave your answer as a double series.

(c) Repeat part (b) when there is no dark current ($\lambda_d = 0$). Now your answer must have a simple form.

2.15. In this problem, you will investigate the relationship between the finesse F and the reflection coefficient R of a Fabry-Perot filter.

(a) Show that for $R \simeq 1$, the finesse can be approximated as

$$F \simeq -\frac{\pi}{\ln R}$$

(b) Find the exact and the approximate expression for R in terms of F. Evaluate the accuracy of the approximation for $F = 10$ and $F = 100$.

2.16. In this problem, you will consider the worst-case crosstalk in a WDM environment with Fabry-Perot (FP) filters. Assume that the filter is lossless ($A = 0$).

(a) Show that the power transfer function $T(f)$ of the FP can be written as

$$T(f) = \frac{1-R}{1+R} \sum_{m=-\infty}^{\infty} R^{|m|} e^{j2\pi m f/P}$$

where P is the free spectral range.

(b) Using the result in (a), show that the worst-case interference

$$C_{\max} = \sum_{i=1}^{M-1} T\left(\frac{iP}{M}\right)$$

is given by

$$C_{\max} = M \frac{1-R}{1+R} \frac{1+R^M}{1-R^M} - 1$$

(c) Using the approximation in Problem 2.15, show that when $F \gg 1$

$$C_{\max} \simeq \frac{\pi M}{2F} \coth\left(\frac{\pi M}{2F}\right) - 1$$

where $\coth(x) = (e^x + e^{-x})/(e^x - e^{-x})$. Find, numerically, the maximum value of M/F such that $C_{\max} \leq 1$. Comment on your result.

2.17. Consider a Mach-Zehnder filter chain of K cascaded filters with

$$\Delta L_i = 2^{i-1} \Delta L \quad i = 1, 2, \ldots, K$$

Show that the power transfer function of this chain is given by

$$T(f) = \frac{\sin^2(\pi M f/P)}{M^2 \sin^2(\pi f/P)}$$

where $M = 2^K$ and $P = c/\Delta L$.

2.18. Optical amplifiers saturate at high levels of output power. Suppose the saturation power of an erbium-doped fiber amplifier is 20 mW, and the amplifier gain is 5 dB/mW of pump power. The pump power is set to 5 mW. What is the largest amount of input power that can be amplified without driving the amplifier into saturation.

2.19. An optical amplifier delivers an output power P_{out} in response to an input power P_{in} as described by the following equation

$$P_{out} = a\,(1 - \exp(-bP_{in}))$$

where a and b are constants.

(a) What is the saturation power P_{sat} of this amplifier?

(b) Find the power gain of the amplifier in the linear operating region (i.e., small input power).

(c) Suppose this amplifier is to be placed in a transmission link of length L. The fiber has an attenuation factor of α per unit length, i.e., after a distance l, a factor $e^{-\alpha l}$ of the original power remains. The transmitter has a laser with power P_t. The goal is to maximize the received power P_r. Find the optimal position x (measured from the transmitter) of the amplifier. Comment on your result.
Hint: The only root of the equation $u = e^u - 1$ is at $u = 0$.

2.20. (a) Use four 2×2 optical crosspoint elements to construct a 4×4 Banyan interconnect.

(b) How many rows and how many columns of 2×2 crossbars would be required for an $N \times N$ Banyan interconnect.

(c) Suppose we have an 8×8 space-division Banyan optical routing switch. Label the inputs from 0 to 7 and label the outputs from 0 to 7. Suppose we need to simultaneously route Input 5 to Output 2 and Input 7 to Output 0. Can this routing be accomplished? If yes, give the routes through the switch, otherwise explain why the routing isn't possible.

2.21. What are the uses of wavelength conversion in a WDM network? Consider the two wavelength-convertible switch (WCS) architectures shown in Fig. 2.44. Construct a set of connection requests that can be routed by the share-per-node WCS and cannot be routed by the share-per-link WCS, and vice-versa.

2.22. Suppose we want to design a system with 16 channels, each channel with a rate of 1 Gbps. How much bandwidth is required for the system?

2.23. Suppose we have a fiber medium with a bandwidth of about 20 nm. The center wavelength is 0.82 μm. How many 10 GHz channels can

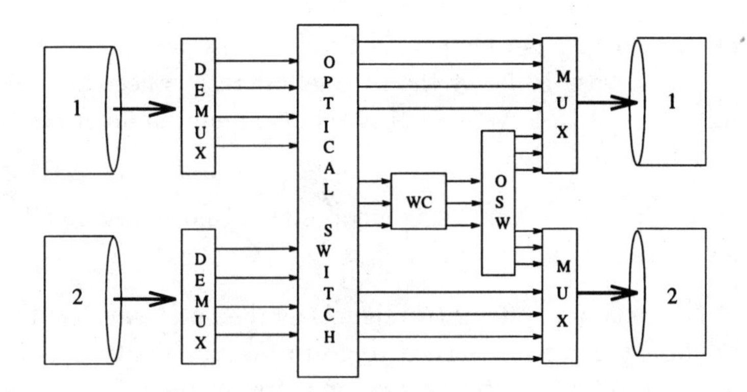

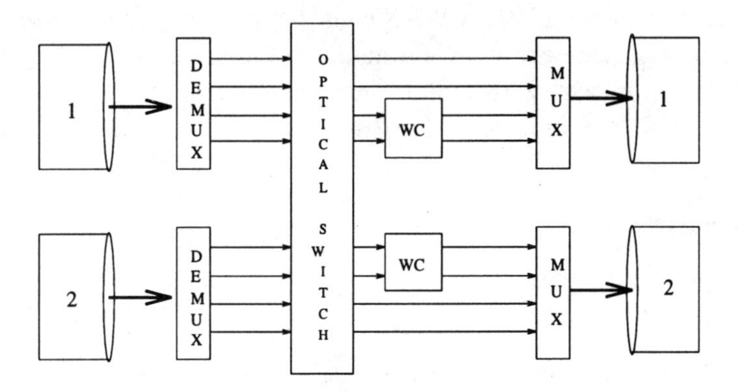

Figure 2.44 Two architectures for wavelength convertible routers: (a) share-per-node, (b) share-per-link.

be accommodated by the fiber? Calculate the maximum number of channels for a center wavelength of 1.5 μm.

2.24. Consider an optical communication system in which the transmitter tuning range is from 1450 nm to 1600 nm, and the receiver tuning range is from 1500 nm to 1650 nm. How many 1 Gbps channels can be supported in the system?

2.25. Consider a WDM passive-star-based network for N nodes. Let the tuning range of the transmitters be 1550 nm to 1560 nm, and let the tuning range of the receivers be 1555 nm to 1570 nm. Assume that the desired bit rate per channel is 1 Gbps. Also assume that a channel spacing of at least 10 times the channel bit rate is needed to minimize crosstalk on a WDM system. Find the maximum number of resolvable channels for this system.

2.26. In which type of network, single-hop or multihop, is a smaller tuning latency more critical? Why?

2.27. In a WDM network node, if two signals on the same wavelength arriving from different input ports need to go to the same output port, then a conflict may occur. Describe two or more methods for resolving this conflict. Discuss the advantages and disadvantages of each solution.

2.28. Figure 2.45 shows a WDM WAN constructed using WRSs at each node. Assume that there are sufficient number of fibers between the node pairs (not shown in Fig. 2.45). Show how the following connection requests can be satisfied (you may have to write a program which tries out various possibilities).

- *node* 3 $\rightarrow$ *node* 1
- *node* 3 $\rightarrow$ *node* 1
- *node* 1 $\rightarrow$ *node* 2
- *node* 1 $\rightarrow$ *node* 2
- *node* 2 $\rightarrow$ *node* 1
- *node* 2 $\rightarrow$ *node* 3

2.29. Suppose we are given the network in Fig. 2.38 and have two wavelengths available. We wish to set up the following connections:

i. H-2-3-4-8-9-E
ii. C-7-8-4-F
iii. B-6-7-8-9-E
iv. D-10-7-C

At which nodes are wavelength converters required, and how many conversions are required at these nodes? Explain.

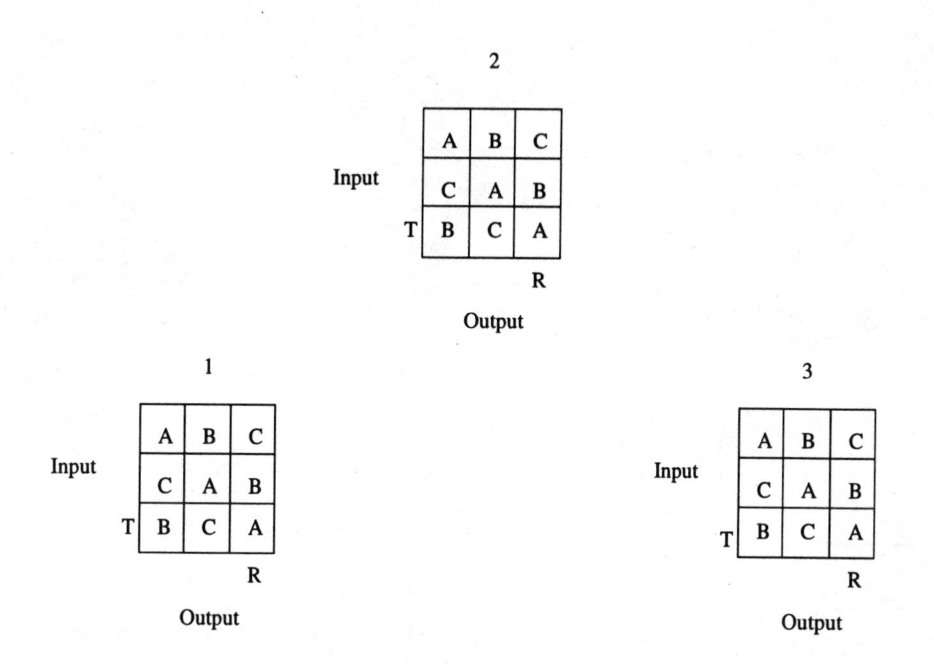

Figure 2.45 T=Transmitter, R=Receiver. All connections begin at transmitters and end at receivers

Part II

Broadcast (Local) Optical Networks

Chapter 3

Single-Hop Networks

This chapter will start with a discussion on how a local lightwave network employing WDM may be constructed based on a pasive-star coupler. Such networks can be broadly categorized into one of two types – single-hop and multihop. Various approaches to single-hop network design will be discussed in the remainder of this chapter. While this chapter provides *breadth* of discussion on single-hop networks, a particular single-hop protocol – the IBM Rainbow protocol – will be discussed and analyzed in *depth* in Chapter 4.

The corresponding *breadth* and *depth* treatment of multihop networks will be provided in Chapters 5 and 6, respectively. The multihop network to be analyzed in depth in Chapter 6 is called GEMNET.

3.1 A Passive-Star-Based Local Lightwave Network

A local lightwave network can be constructed by exploiting the capabilities of optical technology, viz., WDM and tunable optical transceivers (transmitters or receivers), as follows. The vast optical bandwidth of a fiber is carved up into smaller-capacity channels, each of which can operate at peak electronic processing speeds (viz., over a small wavelength range) of, say, a few Gbps. By tuning its transmitter(s) to one or more wavelength channels, a node can transmit into those channel(s); similarly, a node can tune its receiver(s) to receive from the appropriate channels. The system can be configured as a broadcast-and-select network in which all of the inputs from various nodes are combined in a WDM passive-star coupler and the mixed optical information is broadcast to all outputs (see Fig. 3.1).

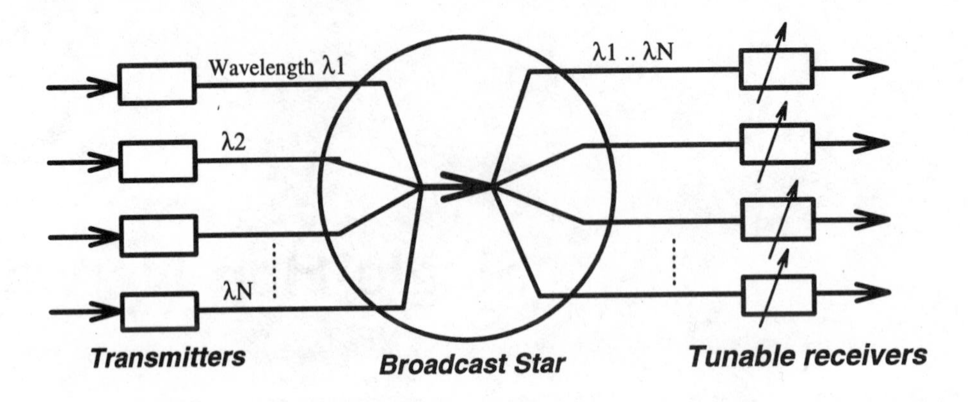

Figure 3.1 A broadcast-and-select WDM network.

An $N \times N$ star coupler can be considered to consist of an $N \times 1$ combiner followed by a $1 \times N$ splitter[1]; thus, the signal strength incident from any input can be (approximately) equally divided among all of the N outputs. The passive-star topology is attractive (1) because of its logarithmic splitting loss in the coupler (since the splitter portion of the coupler circuit is essentially a binary tree type structure) and (2) because there is no tapping or insertion loss (as in a linear bus). In addition, the passive property of the optical star coupler is important for network reliability since no power is needed to operate the coupler; also it allows information relaying without the bottleneck of electrooptic conversion.

In general, the physical topology – instead of being a star – can be a linear bus or a tree (see Fig. 3.2); also, the topology of the data-collection part of the network and that of the data-distribution portion need not be identical. (The bus we show here is the unidirectional folded kind, but note that there is a dual-bus topology also with two unidirectional buses carrying information in opposite directions.) From a network protocol perspective, all three structures – star, bus, and tree – can be considered *equivalent* since,

[1] If the splitters and combiners are made out of couplers, then this arrangement may not be very power efficient. An alternate arrangement would be to have a butterfly arrangement (also called a multistage interconnection network) which has $\log_2 N$ stages of 2×2 couplers with $N/2$ couplers per stage. See Problem 3.2.

in all of them, information from a sender to a receipient must flow through a *central funneling point.* However, the bus has an additional *attempt-and-defer capability* (to be discussed in Section 3.6) under which a node, before or during its transmission, can "sense" activity on the bus from *upstream transmissions.*

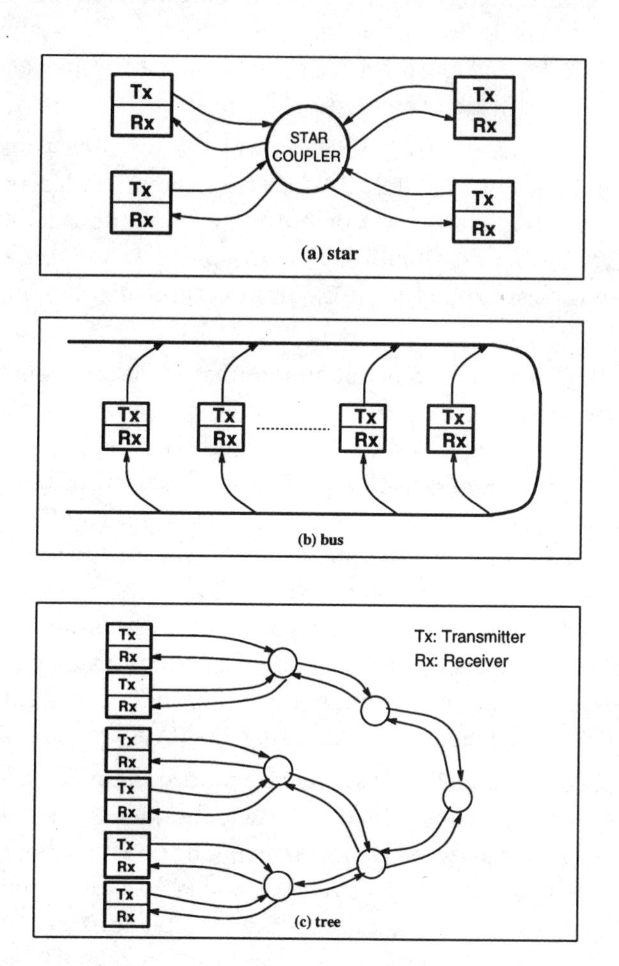

Figure 3.2 Alternative physical topologies for a WDM local lightwave network.

The passive-star network typically can support a larger number of users than a linear-bus topology because power loss and tapping loss in linear buses limit the number of users that can be attached without adding broadband optical amplifiers. However, interest in the linear and the tree structures has been revitalized due to the recent development of the erbium-doped broadband fiber amplifier (EDFA), and such networks are being examined for deployment as metropolitan area networks (MANs) (also referred to as *access networks* and *passive optical networks* in the literature [TaWB95]).

The optimal physical topology design problem may be referred to as the cable plant design problem to determine that the necessary power budget is satisfied [BaFG90]). A related problem on optimal amplifier placements in designing such a network will be examined in Chapter 15.

From an architectural perspective, given any of the broadcast-and-select physical network topologies of Fig. 3.2, the fact that the input lasers (transmitters) or the output filters (receivers) or both can be made tunable opens up a multitude of networking possibilities. The tunable transceivers are used differently depending on the type of network architecture chosen – single-hop or multihop.

In a *single-hop* network, a significant amount of dynamic coordination between nodes is required. For a packet transmission to occur, one of the transmitters of the sending node and one of the receivers of the destination node must be tuned to the same wavelength for the duration of the packet's transmission. In the single-hop environment, it is also important that transmitters and receivers be able to tune to different channels quickly so that packets may be sent or received in quick succession. Currently, the tuning time for transceivers is relatively long compared to packet transmission times, and the tunable range of these tranceivers (the number of channels they can scan) is small. So, the key challenge in designing single-hop network architectures is to develop protocols for efficiently coordinating the data transmissions.

In a *multihop* network, a node is assigned one or more channels to which its transmitters and receivers are to be tuned. These assignments are only rarely changed, usually to improve network performance. Connectivity between any arbitrary pair of nodes is achieved by having all nodes also act as intermediate routing nodes. The intermediate nodes are responsible for routing packets among the WDM channels such that a packet sent out on one of the sender's transmit channels finally gets to the destination on one of the destination's receive channels, possibly after multihopping through a number of intermediate nodes. A number of different multihop architectures are possible, with a range of operational properties (e.g., ease of routing) and performance characteristics (e.g., average packet delay, number of hops that must be traversed, and efficient use of links). As can be inferred, in a single-hop network, there are no intermediate nodes in the optical path.

Both single-hop and multihop networks have their own strengths (and weaknesses) based on, for example, optical transceiver tuning capabilities. Accordingly, we study both approaches. For both single-hop and multihop networks, it is important to keep in mind that any design must be not only simple to implement (i.e., based on realistic assumptions about the properties of optical components), but also scalable to accommodate large user populations. Our attention will focus on such realizable single-hop approaches in this chapter, and will shift to multihop approaches in Chapter 5.

3.2 Characteristics of a Single-Hop System

For a single-hop system to be efficient, the bandwidth allocation among the contending nodes must be managed dynamically. Such systems can be classified into two categories – those employing pretransmission coordination vs. those not requiring any pretransmission coordination.

Pretransmission coordination systems employ a *single*[2], *shared control channel* through which nodes arbitrate their transmission requirements, and the actual data transfers take place through a number of data channels. Idle nodes may be required to monitor the control channel. Before data packet transmission or data packet reception, a node tune its transmitter or its receiver, respectively, to the proper data channel.

Generally, no such control channel exists in systems that do not require any pretransmission coordination, and arbitration of transmission rights is performed either in a preassigned fashion or through contention-based data transmissions on the regular data channels (e.g., requiring nodes to either transmit on or receive from predetermined channels). As a result, *for a large user population whose size may be time-varying*, deterministic scheduling approaches fall out of favor so that pretransmission coordination may be the preferred choice.

An alternative classification of WDM systems can be developed based on whether the nodal transceivers are tunable or not. A node's network interface unit (NIU) can employ one of the following four structures:

1. Fixed Transmitter(s) and Fixed Receiver(s) — $(FT - FR)$

2. Tunable Transmitter(s) and Fixed Receiver(s) — $(TT - FR)$

---

[2]While at least one channel is required for control signalling, some approaches employ as many as N.

3. Fixed Transmitter(s) and Tunable Receiver(s) — $(FT - TR)$
4. Tunable Transmitter(s) and Tunable Receiver(s) — $(TT - TR)$

Fixed transceivers, which can only access some predetermined channels, are readily available in the market, but cost considerations often restrict the installation of a large number of such transceivers at each node. The $FT-FR$ structure is generally suitable for constructing multihop systems in which no dynamic system reconfiguration may be necessary, although a single-hop $FT - FR$ system (LAMBDANET) with a small number of nodes has been demonstrated [GGKV90] (see Section 3.3). $FT - FR$ and $TT - FR$ systems, because they employ fixed receivers, may not require any coordination in control channel selection between two communicating parties, while such coordination is usually necessary in systems employing $FT - TR$ and $TT - TR$ structures. If each node (actually each nodal transmitter) is assigned a different channel under the $FT - FR$ or $FT - TR$ structures, then no channel collisions will occur and simple medium access protocols can be employed, but the maximum number of nodes will be limited by the number of available channels. Systems based on the $TT - TR$ structure are the most flexible in accommodating a scalable user population, but they also have to deal with the channel switching overhead of the transceivers.

In addition, some systems require that a node be equipped with multiple transmitters or receivers. Accordingly, the following general classification for single-hop systems can be developed:

$$\begin{cases} FT^iTT^j - FR^mTR^n & \text{no pretransmission coordination} \\ CC - FT^iTT^j - FR^mTR^n & \text{control-channel (CC) based system} \end{cases}$$

where a node has i fixed transmitters, j tunable transmitters, m fixed receivers, and n tunable receivers. In this classification, the default values of i, j, m, and n, if not specified, will be unity. Also, wherever possible, the number of network nodes, if finite, will be denoted by M. Thus, Bellcore's LAMBDANET [GGKV90] is a $FT - FR^M$ system since each of the M nodes in the system requires one fixed transmitter and an array of M fixed receivers. The TT and TR portions of the classification are suppressed since the system requires no tunable transmitter or tunable receiver.

Most experimental WDM network prototypes belong to the single-hop category, and they do not employ any control channel for pretransmission coordination. Below, such experimental WDM systems are first outlined in Section 3.3, followed by other non-pretransmission-coordination systems in

Section 3.4 and pretransmission-coordination protocols in Section 3.5. While the above mechanisms can generally be employed on any physical topology, Section 3.6 is devoted to the specific case of a linear-bus physical topology, and protocols which can employ an attempt-and-defer mechanism are outlined.

3.3 Experimental WDM Systems

New experimental WDM systems are rapidly being developed. While [Brac90] documented the state of experimental systems in 1990, several new systems such as IBM's Rainbow [JaRS93], Columbia's TeraNet [GiTe91], and Stanford's STARNET [KaPo93] have since been developed. Initial work in this field was done by the British Telecom Research Lab (BTRL), whose experiment introduced the concept of a multiwavelength network, operating in the broadcast mode and using mechanically tunable filters at each receiver [PaSt86]. The AT&T Bell Labs experiment was the first to demonstrate channel spacings on the order of 1 nm [OHLJ85]. The Heinrich Hertz Institute (HHI) reported the first broadcast star demonstration of video distribution using coherent lightwave technology [BBCG86]. Other recent demonstrations include those by AT&T [Kami90, GlSc90] and by NTT [TONT90]. Results from a number of demonstrations are reported in [Kami90] – initially with two 45-Mbps channels and employing tunable receivers, later with two 600-Mbps channels, and followed by two 1.2-Gbps channels and employing tunable transmitters as well. An experimental system employing six 200-Mbps channels, spaced by 2.2 GHz, is reported in [GlSc90]. The work in [TONT90] demonstrates a system operating with a 5-GHz-spaced (equivalent wavelength spacing ≈ 0.04 nm), 16-channel system based on tunable receivers, where each channel can carry 622 Mbps (enough to accommodate a high-definition television channel).

Experimental demonstrations of subcarrier-multiplexing-based systems have also been recently reported [MWWP92, BESO92]. In [MWWP92], for example, a system with one control node and two data nodes is demonstrated. Fixed-wavelength transmitters are used at each node, and a 1-Mbps control information is subcarrier-multiplexed at frequencies $f_c = 710$ MHz, $f_1 = 720$ MHz, and $f_2 = 730$ MHz at the central controller, node 1, and node 2 (which are operated at three distinct wavelengths, viz., 1526 nm, 1543 nm, and 1553 nm, respectively). A receiving node employs a tunable filter. It has a passband in the 1526-nm wavelength region (to continuously receive from the control channel) and another tunable passband to receive from either of the data nodes (depending on the control information in the f_c subcarrier).

The following subsections elaborate on some experimental demonstrations and prototypes of single-hop WDM networks.

3.3.1 LAMBDANET

In Bellcore's LAMBDANET demonstration [GGKV90], a $FT - FR^M$ system with M nodes, each transmitter was equipped with a laser transmitting at a fixed wavelength. Via a broadcast star at the center of the network, each of the wavelengths in the network was broadcast to every receiving node. The experiment demonstrated the use of an array of M receivers at each node in the network, employing a grating demultiplexer to separate the different optical channels. Experiments report that 18 wavelengths were successfully transmitted at 2 Gbps over 57.5 km. Although each node requires M receivers, advances in opto-electronic integrated circuits may reduce the impact of this limitation [GGKV90].

3.3.2 Rainbow

In IBM's Rainbow project [DGLR90], the experimental prototype takes the form of a direct-detection, circuit-switched metropolitan-area network (MAN) backbone consisting of 32 IBM PS/2s as network nodes, communicating with one another at 200-Mbps data rates. The network structure is a broadcast-star, but the lasers and filters are housed centrally near the star coupler. The lasers are tuned to fixed wavelengths, but the Fabry-Perot etalon filters are tunable in submillisecond switching times, i.e., this is a $FT - TR$ system. An in-band receiver polling mechanism is employed under which each idle receiver is required to continuously scan the various channels to determine if a transmitter wants to communicate with it. The transmitting node continuously transmits a setup request (a packet containing the destination node's address), and has its own receiver tuned to the intended destination's transmitting channel to listen for an acknowledgement from the destination for circuit establishment. The destination node, after reading the setup request, will send such an acknowledgement on its transmitter channel, thereby establishing the circuit. Because of its long setup-acknowledgement delay, this mechanism may not be very suitable for packet-switched traffic, although it would work well for circuit-switched applications with long holding times. Under the in-band polling protocol, nodes also need to employ a *timeout* mechanism after issuing a setup request; otherwise there exists the possibility of a *deadlock*. Details of the Rainbow protocol and its analysis are provided in Chapter 4.

The Rainbow-I prototype was demonstrated at Telecom '91 in Geneva. The lessons learned from the corresponding prototype development and demonstration are provided in [Gree92].

Rainbow-II, a follow-on to Rainbow-I, is an optical MAN that supports 32 nodes, each operating at 1 Gbps, over a distance of 10 km to 20 km [HaKR96]. It uses the same optical hardware and medium access control protocol as Rainbow-I, viz., a broadcast-star architecture with each node having a fixed transmitter and a tunable receiver that follows the in-band polling protocol. The goals of Rainbow-II are: (1) to provide connectivity to host computers using standard interfaces, e.g., to interconnect supercomputers via the standard high-performance parallel interface (HIPPI) while overcoming distance limitations, (2) to deliver a throughput of 1 Gbps to the application layer, and (3) to demonstrate real computing applications requiring Gbps bandwidth. The Rainbow-II prototype is deployed as an experimental testbed at the Los Alamos National Laboratory (LANL), where performance measurements and experimentation with gigabit applications are currently being conducted [HaKR96].

3.3.3 Fiber-Optic Crossconnect

The goal of the Fiber-Optic Crossconnect (FOX) demonstration [ACGK86] was to investigate the potential of using fast tunable lasers in a parallel processing environment (with fixed receivers), i.e., this is a $TT - FR$ system. The architecture employed two star networks, one for signals traveling from the processors to the memory banks, and the other for information flowing in the reverse direction. Since the utilization of the memory accesses is relatively slow, a binary exponential backoff algorithm was used for resolving contentions, and it was shown to achieve sufficiently good performance. Since the transmitters are tunable, for applications in which data packet transmission times are in the range 100 ns to 1 mus, transmitter tuning times less than a few tens of nanoseconds will ensure reasonable efficiency.

3.3.4 STARNET

STARNET is a WDM LAN, based on the passive-star topology [CAMS96]. It supports – and allows all of its nodes to be on – two virtual subnetworks: a high-speed reconfigurable packet-switched data subnetwork, and a moderate-speed fixed-tuned packet-switched control subnetwork. Each STARNET node is equipped with a single fixed-wavelength transmitter, which employs a combined modulation technique to simultaneously send data on both subnetworks on the same transmitter carrier wavelength. Each node also has two receivers, referred to as a main receiver (which opeartes at high speed, viz., 2.5 Gbps) and an auxliary receiver (which operates at a moderate speed of 125 Mbps, viz., at the rate of a fiber distributed data interface (FDDI) network). The auxiliary receiver at a node is tuned to the "previous node's transmitting wavelength" so that the moderate-speed subnetwork is a logical ring that carries control packet and is also FDDI-compatible. The main (2.5 Gbps) receiver can be tuned to any node, based on prevailing traffic conditions. The corresponding high-speed subnetwork may be operated as a multihop network (see Chapter 5) that allows electronic multihopping whenever required.

3.3.5 Other Experimental WDM Systems

Thunder and Lightning is a 30-Gbps ATM network using optical transmission and electronic switching [MeBo96]. Electronic switching, using 7.5-GHz Galium Arsenide (GaAs) HBT circuits fabricated by Rockwell, was chosen to simplify clock recovery, synchronization, routing, and packet buffering; and to facilitate the transition to manufacture.

In HYPASS [AGKV88], an extension of FOX, the receivers were made tunable as well (i.e., a $TT-TR$ system), resulting in vastly improved throughputs. Other reported experiments include BHYPASS, STAR-TRACK, passive photonic loop (PLL), and broadcast video distribution systems. Characteristics of these systems are discussed in [Brac90], and are not repeated here to conserve space. See also [GiTe91] for a report on Columbia's TeraNet prototype.

3.4 Other Non-Pretransmission Coordination Protocols

Among protocols that do not require any pretransmission coordination, some are based on fixed assignment of the channel bandwidth, whereas others are based on demand assignment. These protocols are categorized accordingly in the following subsections.

3.4.1 Fixed Assignment

A simple technique that allows one-hop communication is based on a fixed assignment technique, viz., time-division multiplexing (TDM) extended over a multichannel environment [ChGa88a]. Each node is equipped with one tunable transmitter and one tunable receiver; hence these systems are classified as $TT - TR$ systems. The tuning times are assumed zero and the transceiver tuning ranges are the entire set of N available channels. Time is divided into cycles, and it is predetermined at what point in a cycle and over what channel a pair of nodes is allowed to communicate.

For example, for the case of three nodes (numbered 1, 2, 3) and two channels (numbered 0 and 1), one can formulate the following channel allocation matrix which indicates a periodic assignment of the channel bandwidth, and, in which, $t = 3n$ where $n = 0, 1, 2, 3, \ldots$.

Channel No.	t	$t+1$	$t+2$
0	(1,2)	(1,3)	(2,1)
1	(2,3)	(3,1)	(3,2)

An entry (i, j) for channel k in slot l means that node i has exclusive permission to transmit a packet to node j over channel k during slot l. The allocation matrix can be generalized for an arbitrary number of nodes M and an arbitrary number of channels N [ChGa88a]. The allocation matrix can be further generalized to accommodate tuning times, (in integral number of slots), via a "staggerred" approach [BoSD93].

This scheme has the usual limitations of any fixed assignment technique, i.e., it is insensitive to the dynamic bandwidth requirements of the network nodes and it is not easily scalable in terms of the number of nodes. Also, the packet delay at light loads can be high.

The above approach can be extended to a versatile time-wavelength assignment algorithm in which node i is equipped with t_i transmitters and r_i receivers, all of which are tunable over all available channels [GaGa92a]. The scheduling algorithm is mindful of the fact the transceiver tuning times are nonnegligible. Specifically, the algorithm is designed such that, given a traffic demand matrix, it will minimize the tuning times in the schedule, while also attempting to reduce the packet delay.

Arbitrary switching times and nonuniform traffic loads can also be accommodated. One approach requires the establishment of a periodic TDM frame structure consisting of a transmission subframe followed by a switching subframe, so that all of the necessary switching functions of nodal transmitters/receivers are performed during the switching subframe [GaGa92a]. Another approach distributes the nodal switching requirements all over the frame, and is hence less restrictive and more efficient [PiSa94, AzBM96, BoMu96].

A brief examination of the characteristics of these "optimal" scheduling approaches is appropriate, and we will do so by reviewing the findings in one of these studies [AzBM96]. This study considers the scheduling of an arbitrary traffic matrix with a tunable transmitter and a fixed-tuned receiver at each node. Specifically, it quantifies the degradation due to nonzero tuning time on several performance criteria such as schedule completion time and average packet delay. For off-line scheduling, where the traffic matrix is known a priori, the effect of tuning delay is found to be small even if tuning time is as large as the packet transmission duration, and a lower bound on the expected schedule completion time is obtained. (Similar results have been reported in [PiSa94, BoMu96], with [BoMu96] allowing tuning time larger than a packet duration also.) The average packet delay is found to be insensitive to the tuning time under a near-optimal schedule where an idle transmitter tunes *just-in-time* to the appropriate channel just before its packet transmission. This study examines extensions of the approach to accommodate real-time and connection-oriented traffic.

The approach in [GaGa92a] is further extended in [GaGa92b] so that users, based on their traffic flow patterns with other users, can be grouped into separate communities. Users within a community are connected by their own local WDM star, but users can communicate with users in other communities (also in a single-hop fashion) via a remote WDM star. Again, given a traffic demand matrix, the algorithm determines the proper time-wavelength schedule.

3.4.2 Partial Fixed Assignment Protocols

The above fixed assignment protocol is too pessimistic because its main goal is to avoid both channel collision and *receiver collisions.* (A *receiver collision* occurs when a collision-free data packet transmission cannot be picked up by the intended destination since the destination's receiver may be tuned to some other channel for receiving data from some other source.) However, alternative protocols can be defined in which the channel allocation procedures are less restrictive [ChGa88a].

In the Destination Allocation (DA) protocol, the number of node pairs which can communicate over a slot is increased from the earlier value of

N (the number of channels) to M (the number of nodes). During a slot, a destination is still required to receive from a fixed channel, but more than one source can transmit to it in this slot. Thus, even though receiver collisions are avoided, the possibility of channel collision is introduced. For the 3-node, 2-channel case, an example slot allocation may be the following.

Channel No.	t	$t+1$
0	(1,2) (3,2)	(1,3)
1	(2,3)	(2,1) (3,1)

A Source Allocation (SA) protocol can also be defined in which the control of access to the channels is further reduced. Now, over a slot duration, N ($N \leq M$) source nodes are allowed to transmit, each over a different channel. Since a node can transmit to each of the remaining $(M-1)$ nodes, the possibility of receiver collisions is introduced. An example periodic slot allocation policy for the 3-node, 2-channel case may now be the following.

Channel No.	t	$t+1$	$t+2$
0	(1,2) (1,3)	(1,2) (1,3)	(2,1) (2,3)
1	(2,1) (2,3)	(3,1) (3,2)	(3,1) (3,2)

Finally, an Allocation Free (AF) protocol can be defined in which all source-destination pairs have full rights to transmit on any channel over any slot duration. Due to the possibility of receiver collisions, the latter two protocols (SA and AF) may not have much practical significance.

3.4.3 Random Access Protocols I

One can design random access protocols that require each node to be equipped with one tunable transmitter and one fixed receiver (i.e., it is a $TT-FR$ system). The channel on which a node will receive is directly determined by the node's address, e.g., based on the low-order bits of the node's address. The channel a receiver receives from is referred to as that node's *home channel.*

Two slotted-ALOHA protocols were proposed in [Dowd91], and both were shown to out-perform the control-channel-based slotted-ALOHA/ALOHA protocol [HaKS87] and its improved version [Mehr90] (the latter two protocols will be discussed later in Section 3.5). Under one of the protocols in [Dowd91], time is slotted on all the channels, and these slots are synchronized across all channels. A slot length equals a packet's transmission time. In the second protocol, each packet is considered to be of L minislots, and time across all channels is synchronized over minislots. Throughput calculations were performed for these two schemes, and slotting across the entire packet length was found to perform better than minislotting since the latter scheme increases the vulnerability period of a data packet transmission (just like pure ALOHA has poorer performance than slotted ALOHA). Also, not surprisingly, the maximum throughput on each data channel is found to be $1/e$, the value for the single-channel case.

3.4.4 Random Access Protocols II

A slotted-ALOHA and a random TDM protocol have been investigated in [GaKo91]. Unlike previous work, these protocols assume limited tuning range, but zero tuning time. Both of these protocols are based on slotted architectures. Any node, say node i, is equipped with a *single tunable transmitter* and *a number of fixed receivers* (i.e., this is a $TT - FR^x$ system where x is a system parameter). Let $T(i)$ and $R(i)$ be the set of wavelengths over which node i can transmit and receive, respectively. The assignment of transmitters and receivers to various nodes is performed such that the intersection of $T(i)$ and $R(j)$ is always non-null for all i and j, i.e., any two nodes can communicate with one another via one hop. The optimal node/transceiver assignment task is a challenging but open problem.

Under the slotted-ALOHA scheme, if node i wants to transmit to node j, it arbitrarily selects a channel from the set $T(i) \cap R(j)$, and transmits its packet on the selected channel with probability $p(i)$.

The random time-division multiple access (TDMA) scheme operates under the presumption that all network nodes, even though they are distributed, are capable of generating the same random number to perform the arbitration decision in a slot. It is indicated in [GaKo91] that this can be done by equipping all nodes with the same random number generator starting with the same seed. Thus, for every slot, and for each channel at a time, the distributed nodes generate the same random number, which indicates the identity of the node with the corresponding transmission right. Analytical Markov chain models for the slotted ALOHA and random TDMA schemes are formulated to determine the systems' delay and throughput performances.

3.4.5 The PAC Optical Network

In a $TT-FR$ system, packet collisions can be avoided by employing Protection-Against-Collision (PAC) switches at each node's interface with the network's star coupler. These collisions are avoided by allowing a node's transmitter access to a channel (through the PAC switch) only if the channel is available. Also, packets simultaneously accessing the same channel are denied access. The concept is similar to that in collision-avoidance stars [Alba83, SuMo89], except that collision-avoidance is now extended to a multichannel environment.

The PAC circuit probes the state of the selected channel (i.e., it performs carrier sensing) by using a n-bit burst which precedes the packet. The carrier burst is switched through a second $N \times N$ "control" star coupler, where it is combined with a fraction of all the packets coming out of the "main" star plus all carrier bursts trying to gain access to the "control" star (see Fig. 3.3). The resulting electrical signal controls the optical switch which connects the input to the network. The switch is closed only if no energy is detected on the selected channel from other nodes. When two or more nodes try to access the channel simultaneously, all of them detect the "carrier" and their access to the network is blocked. Blocked packets are reflected back to the sender. Because of its "carrier-sensing" mechanism, this approach is sensitive to propagation delays.

Note that the length n of the carrier burst will influence the characteristics of this mechanism. Also, it would be preferable to co-locate the individual PAC circuits as close to the passive star (also referred to as the *hub*) as possible.

3.5 Pretransmission Coordination Protocols

3.5.1 Partial Random Access Protocols

The simplest requirement for single-hop communication is that each node be equipped with a single tunable transmitter and a single tunable receiver, and the system employ a control channel, i.e., a $CC - TT - TR$ system. Such a class of systems was first studied in [HaKS87]. The tuning times are assumed zero and transceivers are assumed tunable over the entire wavelength range under consideration. A number of protocols and their performance capabilities are outlined in this work. Access to the control channel is provided via three random access protocols — ALOHA, slotted-ALOHA, and carrier sense

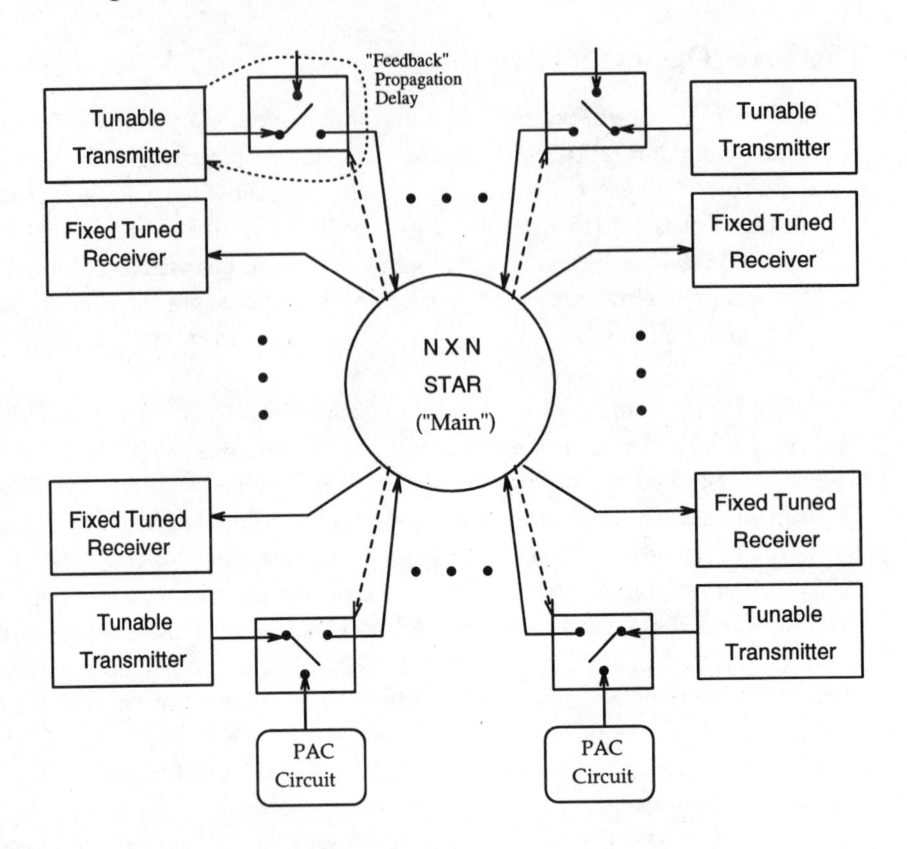

Figure 3.3 Architecture of the PAC optical packet network (the dashed lines are used to detect energy on the various channels from the "main" star).

multiple access (CSMA). ALOHA, CSMA, and an N-server switch mechanism are studied as subprotocols for the data channels. The specific protocols are discussed below.

Assume that time is normalized to the duration of a control packet transmission (which is fixed and is of size one unit). There are N data channels and data packets are of fixed length, L units [HaKS87]. A control packet contains three pieces of information – the source address, the destination address, and a data channel wavelength number (which may be chosen at random by the source and on which the corresponding data packet is to be transmitted).

Under the ALOHA/ALOHA protocol, a node transmits a control packet over the control channel at a randomly selected time, after which it immediately transmits the data packet on data channel i, $1 < i < N$, which was specified in its control packet (see Fig. 3.4). Note that the "vulnerable period" of the control packet equals two time units, extending from $t_0 - 1$ to $t_0 + 1$ where t_0 is the instant at which the control packet's transmission is started. That is, any other control packet transmitted during the tagged packet's "vulnerable period" would "collide" with (and destroy) the tagged packet. (Since different nodes can be at different distances from the hub, these times are specified relative to the activity seen at the hub.) However, even if the control packet transmission is successful, the corresponding data packet may still encounter a collision. This may happen if there is another successful control packet transmission over the period $t_0 - L$ to $t_0 + L$, and the data channel chosen by that control packet is also i. Using such arguments, the throughput performance of this protocol can be analytically obtained. However, what this and the other protocols in [HaKS87] ignore is the possibility of "receiver collisions." Even if the control and data packet transmissions occur without collision, the intended receiver of the destination node might not always be able to read either the control packet or the data packet if it is tuned to some other data channel for receiving data from some other source. For a large or infinite population system, the effect of receiver collisions on the system's performance is negligible [HaKS87, JiMu92b].

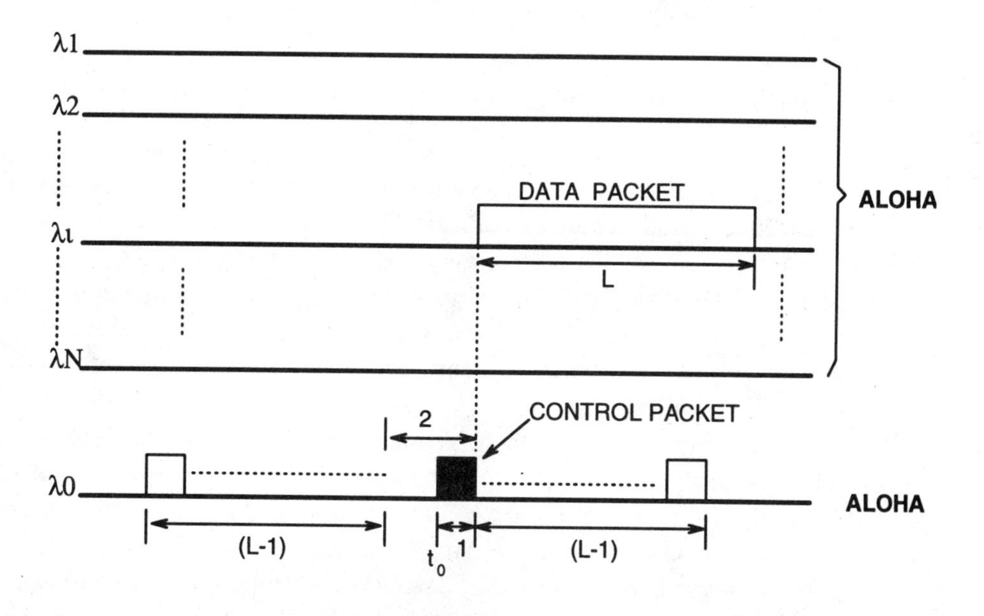

Figure 3.4 The ALOHA/ALOHA protocol.

The slotted-ALOHA/ALOHA protocol is similar, except that access to the control channel is via the slotted-ALOHA protocol. Other schemes outlined in [HaKS87] include ALOHA/CSMA, CSMA/ALOHA, and CSMA/N-server protocols. However, the main limitation of the CSMA-based schemes is that carrier sensing is based on near-immediate feedback, which may not be a practical feature of high-speed systems even for short distances in the range of a kilometer or so.

3.5.2 Improved Random Access Protocols

The approaches in Section 3.5.1 have been extended to obtain improved protocols and performance predictions [Mehr90]. The focus in is on realistic protocols which do not require any carrier sensing since the channel propagation delay in a high-speed environment may exceed the packet transmission time. Hence, slotted-ALOHA for the control channel and ALOHA and the N-server mechanism for the data channels are examined. Another improvement in the protocols is also studied. Specifically, it is required that a node delay its access to a data channel until after it learns that its transmission on the control channel has been successful. As a result, better throughput performance can be achieved, and such results for the improved protocols has been analytically demonstrated [Mehr90].

Bimodal Throughput, Nonmonotonic Delay, and Receiver Collisions

Both the original set of protocols in [HaKS87] and the improvements in [Mehr90] ignored "receiver collisions," stating (1) that the probability of receiver collisions is small for large population systems and (2) that they would be taken care of by higher-level protocols. A receiver collision occurs when a source transmits to a destination without any channel collision; however, the destination may be tuned to some other channel receiving information from some other source.

The study in [JiMu92b] first shows that the slotted-ALOHA/delayed-ALOHA protocol in [Mehr90] can have a *bimodal throughput characteristic.* Basically, *if the number of data channels is small, the data channel bandwidth is underdimensioned* and the data channels are the bottleneck. *If there is a large number of data channels, then the control channel's bandwidth is underdimensioned* and it is the bottleneck. See Fig. 3.5 for some representative throughput results.

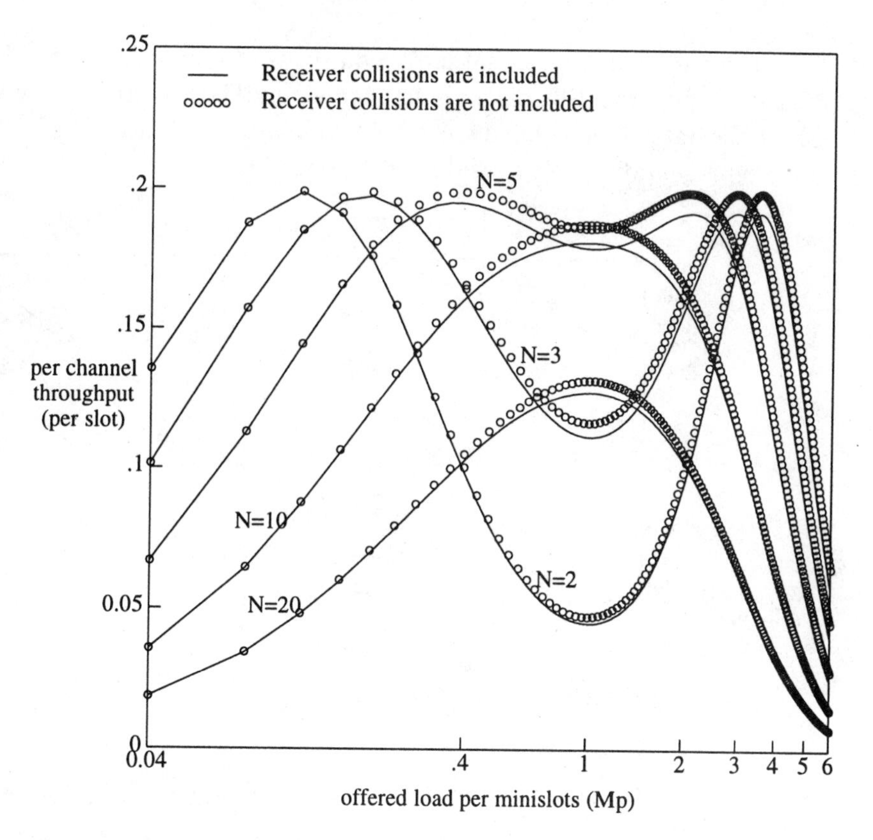

Figure 3.5 Bimodal throughput characteristics of the slotted-ALOHA/delayed-ALOHA protocol for $L = 10$ slots per data packet and $N =$ number of data channels.

The study in [JiMu92b] also finds a useful relationship for *optimally dimensioning the available bandwidth* (viz., properly selecting the number of data channels) so that neither is the bottleneck. Specifically, it is required that, under the slotted-ALOHA/delayed-ALOHA protocol with L-slot data packets, the number of data channels should be given by

$$N = \left\lfloor \frac{2L - 1}{e} \right\rfloor$$

Additional investigations in [JiMu92b] reveal that the system has an interesting delay characteristic, viz., that the mean packet delay is not necessarily monotonically increasing with increase in offered load or throughput. For example, for some sets of system parameters, such as short backoff period, the mean packet delay can actually reduce even though the offered load is increased. Of course, this can only happen when the throughput also decreases due to the data channel bandwidth being underdimensioned. Figure 3.6 shows some representative delay results from [JiMu92b].

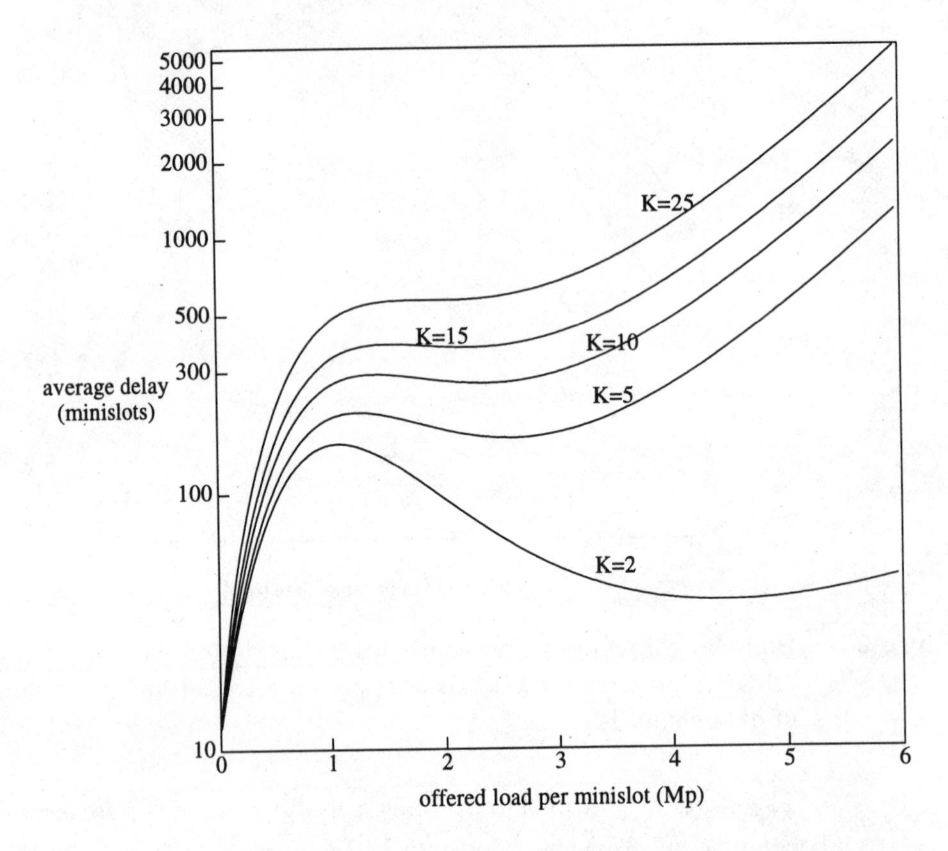

Figure 3.6 Nonmonotonic delay characteristics of the slotted-ALOHA/delayed-ALOHA protocol for L=10 slots per data packet, N=3 data channels, zero propagation delay, and different values of the backoff parameter K.

The work in [JiMu92b] also studies the slotted-ALOHA/delayed-ALOHA protocol's performance degradation due to receiver collisions for finite population systems, and the corresponding throughput results are also shown in Fig. 3.5. As expected, the throughput reduction due to receiver collisions is more prominent when the system population is smaller [JiMu92b].

3.5.3 Extended Slotted-ALOHA and Reservation-ALOHA Protocols

Two sets of slotted-ALOHA protocols and a set of Reservation-ALOHA protocols are defined in [SuGK91a], for the same setting as before, viz., one tunable transmitter and one tunable receiver per node. The first set of slotted ALOHA protocols parallel those in [HaKS87] in the sense that a data packet is transmitted after a control packet transmission, independent of whether the control packet transmission is successful or not.

One of the protocols can be understood by considering Fig. 3.7(a). A *cycle* is defined to be a contiguous set of $N + L$ minislots, where N is the number of data channels (and there is an additional control channel) and L is the data packet length. A node which has a data packet to send will arbitrarily choose one of the N control minislots and will transmit a control packet in it. If it chose the ith control minislot in a cycle, it will transmit its data packet in the ith data channel during the same cycle. This fixed assignment of a control minislot to each data channel ensures that, if a control packet is successful, then the corresponding data packet will also be successful.

The first set of slotted-ALOHA-based protocols in [SuGK91a] also includes five variations of the above access mechanism. Note that the mechanism in Fig. 3.7(a) is quite wasteful since, during a cycle, the data channels are unused during the first N minislots, while the control channel is idle during the last L minislots of the cycle. To improve the protocol's efficiency, the variation in Fig. 3.7(b) can be employed. Now, each cycle consists of L minislots, where $L > N$, but the control channel preassignment mechanism spans consecutive cycles. That is, a node transmitting a control packet in the ith control minislot of the Kth cycle will transmit its corresponding data packet in the ith data channel of the $(K + 1)^{\text{th}}$ cycle. Note that, now, the wastage is reduced to only the last $(L - N)$ minislots on the control channel in each cycle.

Variations of the above two schemes are possible, e.g., one need not preassign a control minislot to a data channel, i.e., a node can select a control minislot and a data channel via two different random choices. The final protocol in the first set of slotted-ALOHA-based protocols in [SuGK91a] employs asynchronous cycles on the different data channels. The control channel still consists of periods of N minislots followed by $(L - N)$ idle minislots and data channels are preassigned to the control minislots, but, now, a data transmission on channel i starts immediately after the control packet transmission in minislot i.

The second set of slotted-ALOHA protocols in [SuGK91a] parallel the

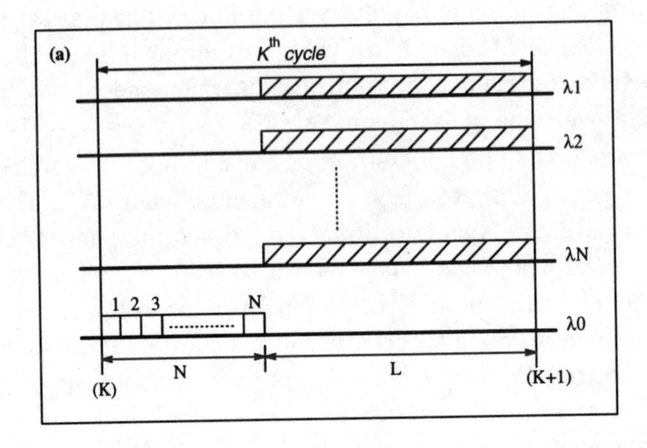

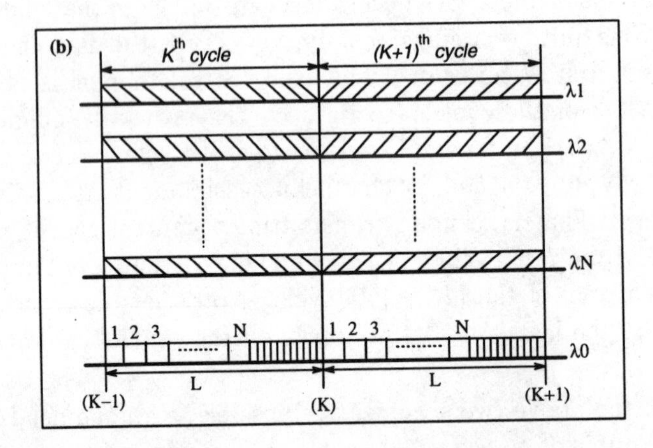

Figure 3.7 The extended slotted-ALOHA protocols: (a) simple case, (b) higher concurrency to reduce channel wastage. Note: λ_0 is the control channel and $\lambda_1, \lambda_2, \ldots, \lambda_N$ are data channels.

improved approaches in [Mehr90] (employing delayed feedback), viz., that a data packet transmission is initiated only after the node learns that its control packet transmission is successful.

In order to accommodate circuit-switched traffic or traffic with long holding times, e.g., file transfers, two Reservation-ALOHA-based protocols are also outlined in [SuGK91a]. The first builds up on the improved slotted ALOHA technique in Fig. 3.7(a). The data channels are preassigned to control minislots, and a node transmits its data packet (which is the first data packet of a multipacket message) in the same cycle only if its control packet is successful (as in [Mehr90]; feedback from the channel is assumed to be immediate). If both the control and data packet transmissions succeed, then the node essentially reserves the same data channel in all subsequent cycles until its use of the data channel is completed. It does so by transmitting a *jam* signal in the corresponding minislot of subsequent cycles, while transmitting its data packets continuously on the corresponding data channel. The second Reservation ALOHA protocol builds up on the improved protocol with asynchronous cycles on the different data channels.

Some further improvements to the above architectures are also possible [SuGK92a, SuGK91b]. Specifically, for the architecture in Fig. 3.7(a), note that, instead of wasting the data channels during the first N minislots of a cycle, one can make use of these minislots for control operation as well. Thus, the multi-control-channel protocol in [SuGK91b] extends the control operation over all channels. Now, one can define communities of interest, and all nodes belonging to a community must tune their receiver during the control part of a cycle to a particular predetermined channel. Any node wishing to send a data packet must therefore transmit its control packet to the destination on the above predefined channel. Data transmission can occur over any of the available data channels (randomly selected by the source before its control packet transmission and so indicated in the control packet). Also, data channel collisions may occur.

Data channel collisions are avoided in the protocols in [SuGK92a] by requiring that a cycle consist of a control slot (which is further divided into M minislots, one preassigned to each node), an information slot (of X minislots) through which a receiver can notify the senders of whether their control packet transmissions are successful or not, followed by a data slot through which collision-free data transmissions occur. This scheme will have value only for small propagation delay systems, and that too when fast tunable devices become available (because of the multiple channel tuning times involved over a cycle for a data packet transmission).

A multiple-channel-based reservation scheme is studied in [SuGK92b], where, in a cycle, all channels have a contention slot, followed by a control slot, followed by data slots. For related work on a multi-control-channel-based protocol, please see [HuRS93].

3.5.4 Receiver Collision Avoidance (RCA) Protocol

In the previous protocols, the main difficulty in detecting receiver collisions arose due to the simplicity of the systems, viz., the availability of only one tunable receiver per node to track both the control channel and the data channel activities. However, even for such a simple system, it turns out that, by adding some intelligence to the receivers, receiver collisions can be avoided and resolved at the data link (medium access control) layer. Thus, the Receiver Collision Avoidance (RCA) protocol [JiMu93a] operates under the same basic system parameters as before, viz., one tunable transmitter and one tunable receiver per node and a contention-based control channel. In addition, the protocol accommodates the fact that transceiver tuning times can be nonzero (of duration T slots, say). For simplicity of presentation, all nodes are assumed to be D slots away from the hub and $N = L$, but these conditions can be generalized [JiMu92a]. The protocol is briefly outlined below.

Channel Selection

Before a control packet is sent, the sender should decide which channel will be used to transmit the corresponding data packet. In order to avoid data channel collision, the RCA protocol proposes a simple and fixed data channel assignment policy. For the case $N = L$, each control slot is numbered 1 through N, periodically, as in a TDM system. Specifically, each control slot is assigned a fixed wavelength which will be the channel number on which a data packet will be transmitted if the corresponding control packet is successfully sent in that slot. Not only is this assignment scheme simple, but also it guarantees that the corresponding data channel transmission will be collision-free. For the cases $N \neq L$, please see [JiMu93a].

Node Activity List (NAL)

Each node maintains a Node Activity List (NAL) which contains information on the control channel history during the most recent $2T+L$ slots. Each entry contains the slot number and a status (Active or Quiet). If the status is Active (which means that a successful control packet is received), the corresponding NAL entry will also contain the source address, the destination address, and the wavelength selected, which are copied from the corresponding control packet. NAL may not be available (or its information is outdated) if the local receiver has been receiving on some data channel.

Packet Transmission

Consider a packet generated at transmitter i and destined for receiver j. Transmitter i will send out a control packet only if the following condition holds: node i's NAL does not contain any entry with either node i or node j as a packet destination. The control packet thus transmitted will be received back at node i after $2D$ slots, during which time node i's receiver must also be on the control channel. Based on the NAL updated by node i's receiver, if a successful control packet to node i (without receiver collision) is received during the $2T+L$ slots prior to the return of the control packet, then a receiver collision is detected and the current transmission procedure has to be aborted and restarted. Otherwise, transmitter i starts to tune its transmitter to the selected channel at time $t+2D+1$, and the tuning takes T slots, after which L slots are used for data packet transmission, which is followed by another T-slot duration during which the transmitter tunes back to the control channel.

Packet Reception

The packet reception procedure is quite straightforward and is left as an exercise for the reader.

Performance

The maximum throughput achievable (over all data channels) under the RCA protocol is $1/e \approx 0.368$. This maximum is affected very slightly even when the transceiver tuning T is increased to several times the packet transmission time. For additional related work, see [JiMu93b, JiMu92a, Jia93].

3.5.5 Reservation Protocols

A single transmitter and a single receiver per node are the minimal requirements for a single-hop system, but the protocol and the system's performance can be improved by equipping nodes with multiple transceivers. The dynamic time-wavelength division multiple access (DT-WDMA) protocol [ChDR90] requires that each node be equipped with two transmitters and two receivers – one transmitter and one receiver at each node are always tuned to the control channel, each node has exclusive transmission rights on a data channel on which its other transmitter is always tuned to, and the second receiver at each node is tunable over the entire wavelength range, i.e., this is a $CC - FT^2 - FRTR$ system.

If there are N nodes, the system requires $N + 1$ channels – N for data transmission and the $(N + 1)$-th for control. Access to the control channel is TDM-based. The system is slotted with slots synchronized over all channels at the passive star (hub). A slot on the control channel consists of N minislots, one for each of the N nodes. Each minislot contains a source address field, a destination field, and an additional field by which the source node can signal the priority of the packet it has queued up for transmission, e.g., the priority information could be the delay the packet would experience from its arrival instant until the time it would reach the hub when it is transmitted. Note that control information is transmitted collision-free, and after transmitting in a control minislot, the node transmits the data packet in the following slot over its own dedicated data channel. By monitoring the control channel over a slot, a node determines if it is to receive any data over the following data slot. If a receiver finds that more than one node is transmitting data to it over the next data slot, it checks the priority fields of the corresponding minislots, and selects the one with highest priority. To receive the data packet, the node simply tunes its receiver to the source node's dedicated transmission channel. Figure 3.8 elaborates on the protocol's operation.

A novelty of this mechanism is that, even though there may be a "collision" in the sense that two or more nodes might have transmitted data packets to the same destination over a data slot duration, *exactly one of these transmissions will always be successfully received.* Also, this mechanism has an *embedded acknowledgement feature* since all other nodes can learn about successful data packet transmissions by following the same distributed arbitration protocol. In addition, the mechanism supports arbitrary propagation delays between the various nodes and the passive hub. The main limitation of the system is its scalability property since it requires that each node's transmitter have its own dedicated data channel. An additional issue is that this mechanism requires infinitely fast receivers or requires that the receiver tuning time be part of the slot duration (which may lead to a reduction of the protocol's efficiency). Without this limitation, for a large user population, the peak throughput of

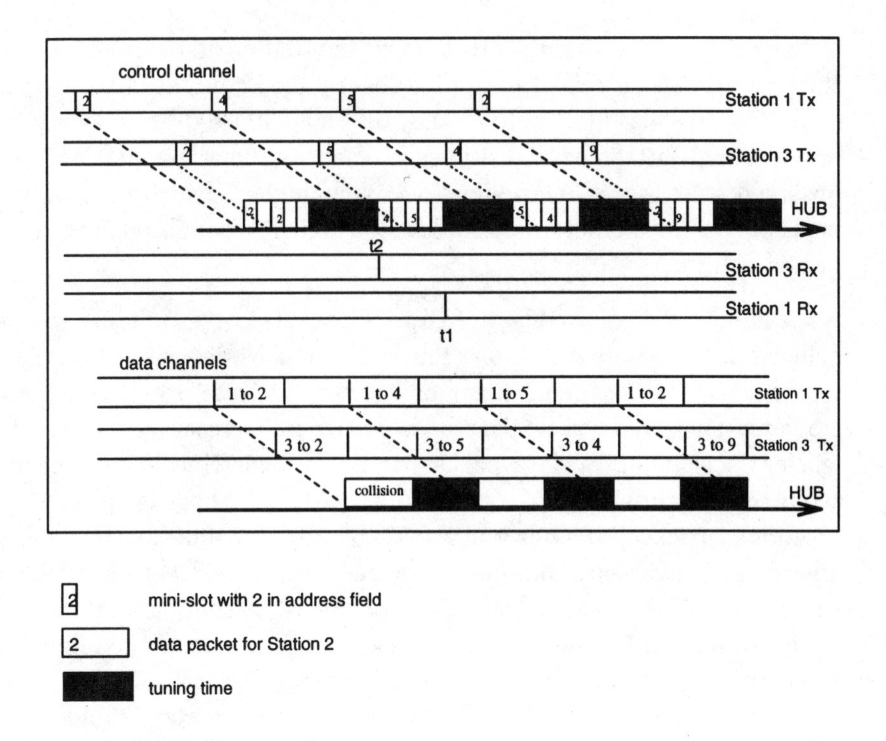

Note: t_1 and t_2 are instants when Stations 1 and 3, respectively, learn about the status of their first control packet transmissions (to station 2) in this figure.

Figure 3.8 The dynamic time-wavelength division multiple access (DT-WDMA) protocol.

the system is $1 - 1/e \approx 0.632$ [ChDR90].

A number of extensions to the DT-WDMA protocol have been reported. Note that in DT-WDMA, when two or more nodes transmit data to the same destination, the destination will read one of the data packets while ignoring the others. One proposal is to use an optical delay line to essentially buffer one or more of the collided packets which would have otherwise been lost [ChFu91]. If the destination is not going to receive packets from any source over the next data slot, then it can read a previously buffered packet. The larger the capacity of the optical delay line, the lower will be the fraction of lost packets. Note that packet loss can still occur if a large number of successive slots have collisions for the same destination. Different policies (FIFO vs. LIFO) exist depending on the order in which previously collided packets vs. new packets are presented to the destination by the delay line buffer. Simulation results in [ChFu91] indicate that, with a delay line buffer of 10 or so, the network's throughput can be raised to approximately 0.95 compared to approximately 0.632 for no delay-line buffer (the case in [ChDR90]). The above results are obtained for the asymptotic case of a large user population. Also, it is observed that the FIFO policy (of giving higher priority to older collided packets) results in better performance (higher throughput and lower mean packet delay).

Avoiding the rebroadcast of control packets corresponding to previously collided data packets is also possible [ChYu91]. Nodes are now made more intelligent so that they can remember information from previous control packet transmissions, and participate in a distributed (reservation) algorithm for efficient scheduling of data packet transmissions. Specifically, each node is required to maintain an $N \times N$ backlog matrix B whose element b_{ij} indicates how many packets at node i are available for transmission to node j. Since all packet arrivals are announced through the broadcast channel, all nodes can maintain identical copies of B locally by assuming that all nodes are equidistant from the hub. All nodes use B and the same scheduling algorithm to compute the same transmission schedule T, which is an $N \times N$ matrix of binary entries such that $t_{ij} = 1$ indicates that node i should transmit a packet to node j over the next data slot (on node i's dedicated data channel), and $t_{ij} = 0$ otherwise. A proper T matrix must have at most one nonzero element per row, and one nonzero element per column. Two algorithms to compute the best possible T exist [ChYu91]. One of them is the Maximum Remaining Sum (MRS) algorithm which can find a suboptimal T in a small number of operations [$O(N^2)$], while the other is the System of Distinct Representative (SDR) algorithm which can find the optimal schedule T, but it is compute-intensive [and requires $O(N^4)$ operations]. Typical numerical examples indicate that the loss of scheduling efficiency of MRS (as compared with SDR) is not very significant [ChYu91]. Numerical examples also indicate that the maximum utilization of a data channel under the improved scheme can approach unity (as compared with 0.632 for the original DT-WDMA protocol). However, the scheduling algorithms in [ChYu91] also assume that all nodes are equidistant from the WDM star coupler, but an extension which eliminates this restriction will be very desirable.

Two additional protocols are outlined in [ChZA92], which are also aimed at improving the DT-WDMA protocol. Just like in [ChYu91], each node maintains a backlog matrix [ChZA92]). The first algorithm, called Dynamic Allocation Scheme (DAS), requires that each node execute an identical algorithm based on a common random number generator with the same seed at all nodes (as in [GaKo91]). Thus, all transmitters mutually arrive at the same conclusion. First, a transmitter is selected randomly, and among the destinations for which it has queued packets, one receiver is chosen randomly (the same at all nodes). The chosen transmitter will transmit to the chosen receiver over the next slot. The process is repeated to select the other transmitters and the corresponding receivers they will be transmitting to in a similar fashion, except that transmitters and receivers which have already been scheduled are excluded (since each node has one data transmitter and one data receiver). The second protocol in [ChZA92], called Hybrid TDM (HDTM), requires that time on the data channels be divided into frames consisting of $M + X$ slots, where M is the number of nodes and X is a positive integer. After every $\lfloor M/X \rfloor$ slots, one slot is left "open" into which a node may transmit to any receiver. Who gets to transmit into such open slots is again determined by the same random-number-based mechanism as in the DAS protocol. Obviously, since HDTM has these open slots, it can provide lower delays than TDM, especially for nonuniform traffic. It is found that HDTM has lower signaling needs than DAS, but DAS also performs close to optimal when the channel propagation delay is zero [ChZA92]. Finally, we note that these schemes can become useful if mechanisms which can synchronize random number generators at distributed network locations can be developed.

A reservation protocol that can accommodate variable-length messages has been recently proposed [JiMI95].

3.5.6 The TDMA/N-Server Protocol

This protocol [BoDo91, DoBo92] employs a TDM control channel, but it improves upon the DT-WDMA protocol [ChDR90] by allowing (1) a user population larger than the number of data channels, (2) variable-length data packets, and (3) possibly faster access to the data channel (by not requiring that a node delay its data packet transmission until the beginning of the next control cycle). Each node has its own control slot on the control channel, allocated in TDM fashion, but the number of data channels, M, is limited (possibly less than the number of nodes). Also, each node is equipped with only one tunable transmitter for both control and data packet transmissions. However, as in [ChDR90], each node has two receivers – one fixed on the control channel and another tunable over all data channels. Thus, this system can be classified as $CC-TT-FRTR$. A control packet consists of the source address (possibly redundant due to the control channel being of TDM type), destination address, selected channel number, and length of the data packet (variable). After a control packet transmission, a data packet transmission is initiated on the previously-selected data channel α slots later, where $\alpha = \max(t_s, t_r)$, t_s is the transmitter's tuning time, and t_r is the tuning time of the intended receiver.

Collisionless data transmissions are ensured by requiring that each node maintain two status tables. One of them keeps track of the status of the data channels (to eliminate data channel collisions), while the other keeps track of the status of the tunable receiver at each node (in order to avoid receiver collisions on the data channel). The status of these two tables (which are identical at all nodes) is maintained by having the nodes monitor the control packet transmissions on the control channel and update the tables accordingly.

3.6 Special Case: Linear Bus with Attempt-and-Defer Nodes

The previous protocols were based on a centralized hub architecture, independent of the network's physical topology. For a linear-topology physical network (see Fig. 3.2(b)), however, nodes can also employ a multiple access mechanism for sharing their transmissions on a common data channel. In particular, a node on the linear multichannel bus can be equipped with a sense tap with which it can access a channel via an *attempt-and-defer* policy. By tuning a sense tap to a channel, a node can determine if there is any activity on that channel from its upstream nodes. If there is activity, the node will defer until the activity has ended. Otherwise, the node can transmit on the channel until its transmission is completed or until it senses activity at its sense tap, at which time it aborts its packet transmission, defering to the upstream transmission. This is therefore a preemptive priority system.

Until recently, protocols based on the above attempt-and-defer mechanism were not considered serious contenders for implementation in a local lightwave network because each tap on the linear bus results in a 3-dB power loss (due to halving of the power at each splitter) so that, with a power margin of a few tens of dBs, only a small number of nodes could be accommodated (without amplification of the optical signal). However, the emergence of the erbium-doped fiber amplifier (EDFA) should rekindle interest in the linear-bus-based lightwave networks because optical signal amplification is no longer a problem. Protocols based on the linear topology, viz., AMTRAC [ChGa88b] and multichannel p(i)-persistent [BaMu93c], are outlined below.

3.6.1 AMTRAC

The AMTRAC approach simultaneously exploits the combined performance advantages of both multichannel and train-oriented protocols. Under this mechanism, a node is equipped with a single receiver tuned to a fixed channel, while a node's transmitter is tunable, so this is a $TT - FR$ system.

Let there be N nodes and M parallel channels (see Fig. 3.9(a)), and let the internode distance be D. Time is slotted with slot duration D, and a cycle is defined to be of length $2(N-1)$ slots (see Fig. 3.9(b)). Transceiver tuning times are assumed negligible.

A cyclic structure is imposed on the system. A node is assigned a scheduling point on each channel in each cycle. If node i wishes to transmit a packet to node j which happens to receive packets only from channel c (since its receiver is always tuned to this channel), then node i will check to see if channel c is free at the beginning of slot $i-1$ in a cycle. If it is free, node i will transmit its packet; otherwise it will wait for the next cycle to repeat the same procedure.

The protocol has been refined to incorporate multihopping when it is known that single-hopping would result in longer delay [ChGa90].

3.6.2 Multichannel Probabilistic Scheduling

The network structure is the same as before, viz., a single tunable transmitter and a fixed-tuned receiver at each node. All of the M parallel channels are slotted and synchronized, and the slot length equals the packet transmission time. At the beginning of every slot, out of the M channels, node i chooses a channel c with probability p_{ic}. After probabilistically selecting a channel (say channel c), node i transmits the packet (if any) at the head of its buffer for channel c if the slot arriving on that channel is sensed to be empty.

In [BaMu93c], the proper transmission probabilities (p_{ic}) for various fairness criteria (e.g., equal average packet delay at each node) have been analytically obtained. Also, to keep pace with the fluctuations (if any) of the offered traffic at the various nodes, the transmission probabilities can be dynamically adjusted.

3.7 Summary

A large sampling of contributions made to date on the research and development of wavelength-division multiplexing (WDM) based local lightwave

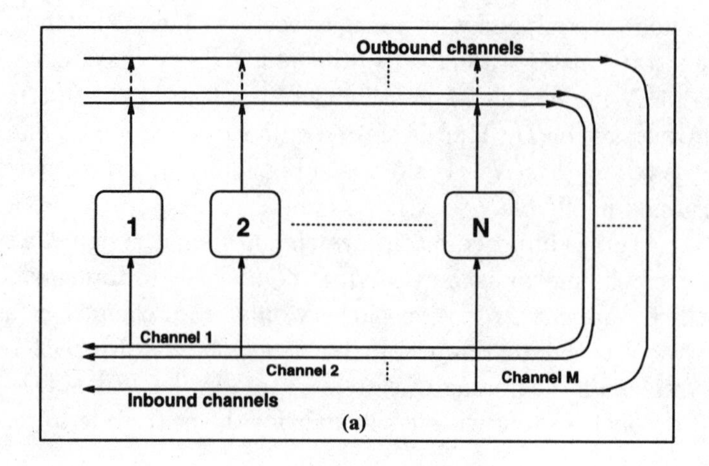

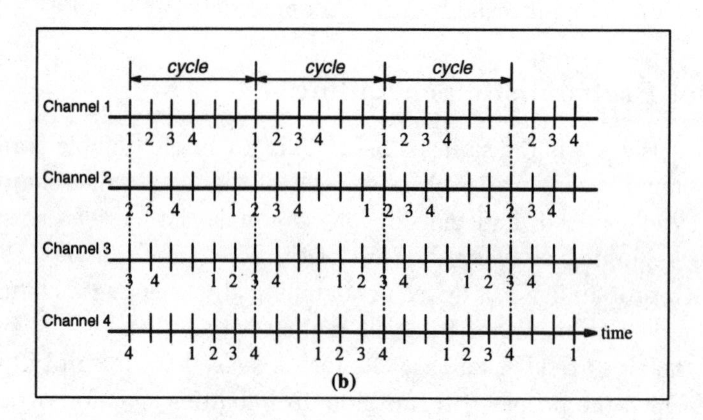

Figure 3.9 The multichannel bus network: (a) network structure with tunable transmitters and fixed receivers, and (b) a cycle in AMTRAC for N=4 and M=4.

networks employing the single-hop approach was reviewed in this chapter. Various architectures, protocols, and experimental prototypes belonging to single-hop networks were examined. Figure 3.10 provides a summarized classification of single-hop systems.

Much more exciting research and development on single-hop systems is expected. Single-hop systems which can accommodate a large and time-varying user population, and which properly utilize the available channel capacity based on wavelength-agile transceivers (with nonzero tuning times and limited tuning ranges) are candidates for further study. It is expected that reservation-based protocols as well as protocols that can simultaneously accommodate circuit-switched traffic, narrowband packet traffic (short packets), and bulk data transfers (wideband packet traffic) will be developed. Protocols that can accommodate multicast traffic are expected to be very useful as well.

Although most of the single-hop architectures and protocols discussed in this chapter are meant for LANs/MANs, some of them can also be applied to situations where the propagation between the end users is negligible, e.g., the end-users may be co-located, e.g., as photonic switching fabrics or as multiprocessor interconnects.

The following chapter will examine in depth a specific single-hop protocol that has been implemented, viz., the IBM Rainbow protocol.

Exercises

3.1. Compare the physical topologies: star, bus, and tree with respect to: (a) number of simultaneous connections, (b) scalability, and (c) delay.

3.2. Consider the following two designs of a passive-star coupler using 2×2 couplers.

(a) One approach is shown in Fig. 3.1. Determine the number of 2×2 couplers needed for this design, and prove that each output gets $1/N^2$ of the input power.

(b) The second approach is the butterfly arrangement (also called a multistage interconnection network) mentioned in the text. Determine the number of 2×2 couplers needed for this design, and prove that each output gets $1/N$ of the input power.

(c) What are the trade-offs between the two designs.

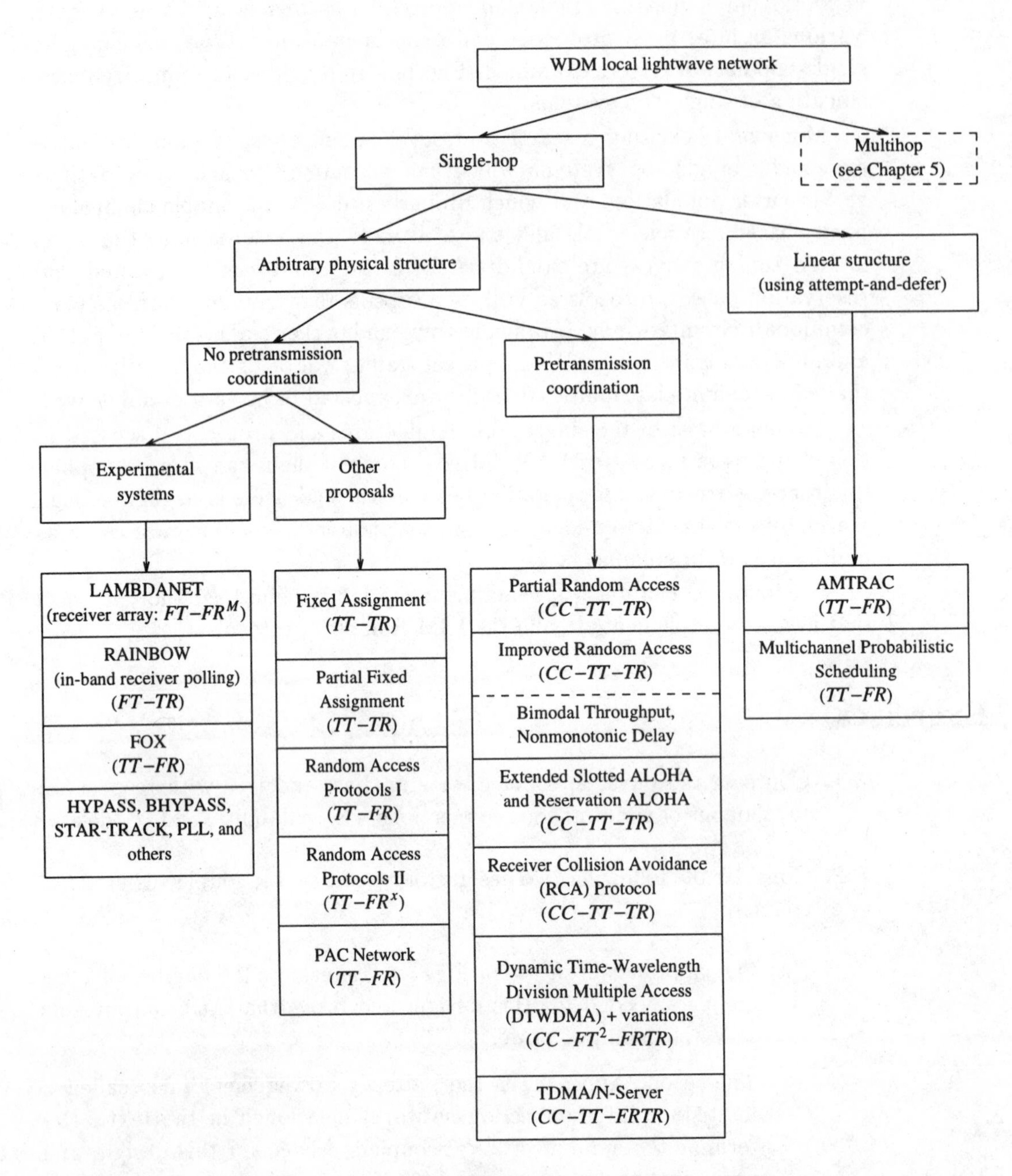

Figure 3.10 Classification of single-hop network architectures.

3.3. Compare the maximum power loss in an N node passive-star coupler to the maximum power loss in an N node linear bus. Assume that the transmitted power is P. Consider only losses from splitting and coupling.

3.4. Compare protocols with pretransmission coordination to those protocols without pretransmission coordination.

3.5. Compare single-hop WDM systems which employ fixed transceivers to single-hop WDM systems which employ tunable transceivers.

3.6. Consider a single-hop network with four nodes and three channels.
(a) Give a schedule for the fixed channel assignment technique.
(b) Give a schedule for the Destination Allocation protocol.

3.7. Consider an optical LAN consisting of three nodes connected to a PSC. Assume two channels. Consider the following three scheduling protocols:

(a) fixed assignment,

(b) destination allocation, and

(c) source allocation.

Show the schedules for three time slots. Assume uniform traffic conditions. Calculate the maximum throughput per channel that can be achieved by employing the protocols. Now assume that a packet corresponding to every (*source*, *destination*) pair arrives with probability p every three time slots. Calculate the throughput by taking account of channel and receiver collisions. Calculate the maximum expected throughput for the protocols.

When can source/destination assignment protocols perform better than fixed assignment?

3.8. A single-hop $TT - TR$ network with 10 nodes connected via a passive star coupler is to use a fixed channel assignment. Each transmitter and receiver is capable of tuning to three noninterfering channels. How many TDM slots are required? Show that there is no interference. Assume negligible tuning times.

3.9. A partial fixed-assignment protocol is used in a passive star network. There are 10 nodes in the network and three channels available. What is the shortest time slot scheme (i.e., the scheme with the shortest period)?

Avoid receiver collision, but not necessarily channel collision. When might such a scheme be desirable, and when might it not?

3.10. A fixed-assignment protocol is designed for a system with N nodes and W channels. The time slot duration is T. Find a lower bound on the expected packet delay in the system. Compare this with the lower bound on a partial fixed-assignment protocol. By what factor do these values differ?

3.11. Consider a two-channel, four-node broadcast-and-select WDM network. Each node is equipped with a single fixed transmitter and a single tunable receiver. Design a fixed assignment schedule for the following traffic matrix. Assume that the tuning time is equal to one time slot. Be sure to indicate which node is tuning its receiver during which slot.

$$A = \begin{bmatrix} 0 & 1 & 4 & 3 \\ 2 & 0 & 1 & 2 \\ 3 & 2 & 0 & 2 \\ 1 & 1 & 2 & 0 \end{bmatrix}$$

3.12. Consider a system using a fixed assignment protocol. Each node is equipped with a fixed transmitter and a tunable receiver. For arbitrary values of N (number of nodes), W (number of channels), L, (tuning time for a receiver), and traffic matrix a_{ij} where $i, j = 1, 2, \ldots, N$, find a lower bound on the optimal schedule length.

3.13. Suppose we have a system with four nodes. Each node is equipped with a single tunable transmitter and two fixed receivers. Assign channels to each of the transmitters and receivers ($T(i)$ and $R(i)$) for the random TDM protocol. What must we consider when assigning channels?

3.14. Consider a packet network in which the packet arrival process is a Poisson process with rate λ. In this network, users can start packet transmissions at times $T, 2T, 3T, \ldots$, where T is the length of a packet. All the packets that arrive during the time interval $[(n-1)T, nT)$ are transmitted during the time "slot" $[nT, (n+1)T)$. When a single packet is transmitted in a time slot, that packet is successfully received by the receiver. When two or more packets are transmitted in the same slot, a collision occurs and all the packets are lost.

(a) Find the probability P_k that k packets are transmitted in a time slot.

(b) The *throughput* S of this network is defined as the fraction of time slots which result in successful packet transfer. Find S as a function of λ and T.

(c) Find the value of λ that maximizes the throughput and the resulting throughput.

3.15. Consider the following WDM protocol. The network has a large number of nodes and W channels. Each node has W fixed-tuned receivers, one per channel, and a tunable transmitter. The time is slotted, and packets arrive according to a Poisson process with rate G packets/slot. When a node has a packet to transmit, it selects a wavelength randomly and transmits the packet in the first time slot on that wavelength.

(a) Find the probability that a wavelength is successfully utilized in a given time slot.

(b) Find the traffic rate that maximizes the throughput per channel and the resulting throughput.

(c) Find the average number of packets successfully carried by this network per time slot.

3.16. Consider a random-access $TT - FR$ system in which the home channel for node j is specified by $\lambda_{j\,\mathrm{mod}\,8}$. There are 64 nodes in this slotted ALOHA network, and the capacity of each channel is 2.5 Gbps. What is the maximum bandwidth of the network assuming uniform traffic, i.e., if every node has packets to send to every other node, what is the total of all the links bandwidth. Also, under these circumstances, what is the maximum average bandwidth of a single communication link?

3.17. In the ALOHA/ALOHA protocol, calculate the throughput as a function of the normalized offered load and calculate the maximum throughput. Assume that there are N stations, and packets arrive to each station according to a Poisson process with rate λ. Traffic is uniformly distributed across all stations. (Normalized offered load is defined as the rate of arrivals per packet transmission time.)

3.18. In slotted-ALOHA/ALOHA, what is the length of the vulnerable period for the control packet? For the data packet? Assume control packets to be one time unit in length, and data packets to be L time units in length.

3.19. In the ALOHA/ALOHA partial random-access protocol, a node transmits its data packet immediately after it transmits its control packet, regardless of whether the control packet succeeds or not. Suppose a node can transmit its data packet only when the corresponding control packet is successful. Compute the throughput per channel for the improved protocol. Assume N data channels in the system.

3.20. Explain the reason behind bimodal throughput seen in a slotted-ALOHA / delayed-ALOHA protocol. What is the optimal number of channels for a system with data packet length equal to 15 slots.

3.21. Why are CSMA protocols usually not very attractive for a single-hop WDM optical network?

3.22. Describe the procedure for receiving a packet in the Receiver Collision Avoidance (RCA) protocol.

3.23. In the Receiver Collision Avoidance (RCA) protocol, we set the size of the Node Activity List (NAL) to $2T + L$ slots. Explain why.

3.24. Describe possible scenarios in which receiver collisions may occur in the RCA protocol. Assume all nodes are D slots away from the hub, T is the tuning time, and L is the data packet length.

3.25. Consider the DT-WDMA protocol with N stations. Suppose that a packet arrives to a station in a given time slot with probability p, and no packets arrive to the station in the time slot with probability $(1-p)$. Calculate the probability that a transmitted packet will encounter a receiver collision. Assume tuning time is included in the time slot.

3.26. Show that under uniform traffic conditions and a large user population, the peak throughput in a DT-WDMA system is given by $1 - 1/e$.

3.27. In DT-WDMA, nodes can be made more intelligent by storing an $N \times N$ matrix B, called the "backlog" matrix at each node. Element b_{ij} denotes the number of packets at node i destined for node j. We can find an optimal algorithm which constructs a transmission schedule T, where $t_{ij} = 1$ denotes that node i should transmit a data packet to node j in the next slot. An optimal algorithm maximizes the number of transmission in a slot. What is the maximum throughput that can be achieved using this algorithm? (A qualitative justification is sufficient.)

3.28. Given a STARNET network of 100 nodes, and control packets of length 100 bytes, with an average distance between nodes of 100 m and negligible processing time, how long will a simple broadcast take, given that the information in the broadcast signal is embedded in the control packets?

3.29. For LAMBDANET, approximately how much of the low-loss region of bandwidth was utilized?

3.30. Professor W. D. Myer receives a small grant to set up an experimental WDM testbed. He decides to have four nodes in his network, each node with a fixed transmitter and tunable receiver.
(a) Should Professor Myer go for pretransmission control? Why or why not?
(b) Suddenly Professor Myer gets a lot of money and decides to construct a large network with pretransmission coordination, with each node having two tunable transceivers. How would this network be represented using the notation introduced in this chapter?

3.31. Consider the linear bus with attempt and defer nodes. Suppose the bus has five nodes over 5 km, and we require the received power at each node to be at least 30% of the transmitted power. How many amplifiers are required? Assume that amplifiers provide a gain of 25 dB.

3.32. An engineer is asked to come up with a transmission protocol for a single-hop network in which each node has a tunable transmitter and a tunable receiver. There are N nodes in the network and W channels. His solution employs a control channel. A frame consists of $N+L$ slots, where the length of a data packet is L. The first N slots in the frame correspond to control slots. Host i puts the number of the destination host to which it wants to transmit in slot i, $1 \leq i \leq N$. When more than one host wants to transmit to the same destination, the station with the lowest index wins. Further, the choice of channel is implicit – the first winner transmits on Channel 1, the second winner transmits on Channel 2, etc. Note that some winners may not be able to transmit because of the constraint on the number of channels. Comment on the characteristics of the protocol and suggest ways of improving it.

Chapter 4

Single-Hop Case Study: IBM Rainbow Protocol

A number of architectures and protocols for single-hop WDM networks have been proposed, and some prototypes have been built as well, as was reviewed in Chapter 3. The current chapter is devoted to an in-depth case study on the performance characteristics of a specific single-hop system, viz., the Rainbow prototype built at IBM which is a WDM local optical network testbed [DGLR90, JaRS93, HaKR96].

4.1 Introduction

The original Rainbow-I WDM local optical network prototype could support up to 32 IBM PS/2 stations (or nodes) connected in a star topology (Fig. 4.1) over a range of 25 km [DGLR90]. [Because it provides coverage larger than that provided by a local area network (typically a few km), this network is referred to as a metropolitan-area network (MAN).] The system requires as many WDM channels as there are stations. Data can be transmitted on each WDM channel at a rate of up to 300 Mbps. The Rainbow-II network, which is a follow-on to Rainbow-I, also supports 32 nodes, and employs the same optical hardware and multiple access protocol as Rainbow-I; however, the data rates on Rainbow-II have been raised to 1 Gbps [HaKR96]. Thus, our model of the network protocol is equally applicable to both Rainbow-I and Rainbow-II.

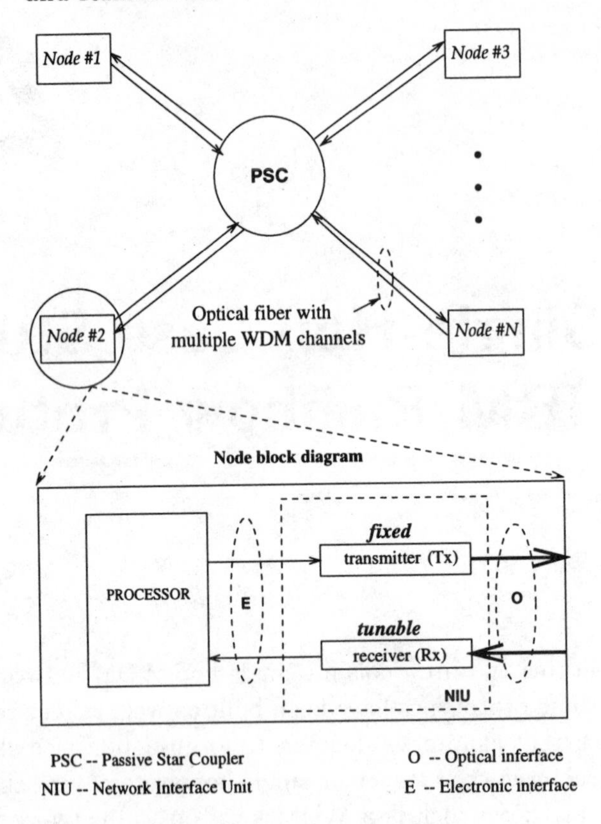

Figure 4.1 Passive-star topology for Rainbow.

In the Rainbow architecture, each station is equipped with a single fixed transmitter, which is tuned to its own unique wavelength channel, and a single tunable Fabry-Perot filter, which can be tuned to any wavelength ($FT-TR$). Tuning to any particular channel may take up to 25 ms. The tunable receiver scans across all the channels, looking for connection requests or acknowledgements from other stations. Rainbow's protocol is also referred to as an *in-band polling protocol.*

The Rainbow system is intended primarily as a circuit-switched network. The large filter tuning time results in a high connection setup time. This makes the system impractical for packet switching.

This chapter provides an analytical model for the Rainbow medium-access protocol and presents an analysis of the system using the equilibrium point analysis (EPA) technique. The EPA technique is a means of analyzing complex systems by assuming that the system is always at an equilibrium point [FuTa83]. This technique has been successfully used to analyze a number of communication systems, e.g., satellite systems, and has been found to provide accurate results.

The signaling protocol for Rainbow is described in Section 4.2. Section 4.3 presents a model of the system, which is then analyzed using the EPA technique in Section 4.4. Section 4.5 provides some representative numerical examples which show the effects that various system parameters – such as the timeout duration, offered load, and message size – have on the throughput and delay performance of the system.

4.2 Rainbow Protocol

The signaling protocol for Rainbow is as follows. Each station is assigned its own unique channel on which its transmitter is fixed tuned. Upon the arrival of a message at Station A and destined for Station B, Station A first tunes its receiver to Channel λ_B so that it will be able to pick up Station B's acknowledgement signal. Station A then begins to send a continuous request signal on Channel λ_A. This request signal consists of a periodically repeated message which contains the identities of both the requesting station and the intended destination. If Station B's receiver, which is continuously scanning across all channels, comes across the request on Channel λ_A, the receiver will stop on that channel, and Station B's transmitter will send out an acknowledgement on Channel λ_B. Station A's receiver, which is tuned to Channel λ_B, will receive the acknowledgement and will now know that Station B's receiver is tuned to Channel λ_A. Station A's transmitter will then begin transmitting the message on Channel λ_A. This establishes a full duplex connection.[1] Upon completion of the transmission, both stations re-

[1] In the actual prototype system [JaRS93], the filters used do not allow round-robin-type scanning. Instead, the filter must scan back and forth across all wavelengths (the so-called elevator-type scanning). Also, the filters are not able to tune directly to a particular wavelength. They must perform "elevator scanning" to find the appropriate channel before stopping. Here, we examine the original Rainbow system presented in [DGLR90]. We do not model the system in [JaRS93] because it is difficult to model elevator-type scanning, and also because we expect that newer filters, which do not have these limitations, may be employed in the future. In the remainder of this chapter, the term "scan" will be used to mean round-robin scanning, instead of elevator-type scanning.

sume scanning for requests. Note that all stations perform their operations asynchronously and independently.

With this protocol, there is the possibility of *deadlock*. If two stations begin sending connection setup requests to each other nearly simultaneously, they will both have to wait until the other sends an acknowledgement, but since both stations are waiting for each other, acknowledgements will never be sent. To avoid this problem, the Rainbow protocol also includes a *timeout mechanism*. If an acknowledgement is not received within a certain timeout period measured from the message arrival instant, the connection attempt is aborted (i.e., the connection is blocked), and the station returns to scanning mode.

4.3 Model

Round-robin (or polling) systems have been modeled and analyzed extensively in the literature (see, for example, [Taka86]). In our current setting of the Rainbow protocol, although a station's receiver is performing a round-robin operation, that operation may be interrupted by the station's transmitter. "Vacation models" do not appear to be applicable here because the system's performance characteristics are now determined by the transmission ("vacation") process. The modeling challenge is to relate the transmitter and receiver operations at a station in a simple manner.

Our model makes the following assumptions:

- There are N stations.
- There is no queueing. Each station has a single buffer to store a message, and any arrival to a nonempty buffer is blocked. A message departs from the buffer after it is completely transmitted.[2]
- The sending station, upon a message arrival, tunes its receiver to the channel of the target station prior to sending the connection setup request.
- Stations monitor the channels in a round robin fashion, in the sequence: $1, 2, \ldots, N, 1, 2, \ldots$.

[2]The delay performance of this protocol with buffers has also been analyzed in [MoAz95a, MoAz95b] in the more general context of hybrid multiaccess, e.g., WDMA/CDMA. Our current setting on pure WDMA is a special case of the approach in [MoAz95a, MoAz95b].

- Time is slotted with slot length equal to 1 μs. This was chosen to provide a fine level of granularity in the system's model.
- It takes a fixed amount of time, τ slots, to tune a receiver to any particular channel.
- Messages arrive at each station according to a Bernoulli process with parameter σ, i.e., in any slot, a station with an empty slot can have a message arrival with probability σ.
- Message lengths are geometrically distributed with the average message length being $1/\rho$ slots. Message lengths are used to model connection holding times of circuits in the Rainbow prototype.
- The propagation delay between each station and the passive-star coupler is R slots. Given that signal propagation delay in fiber is approximately 5 μs/km, the value of R can be quite large, e.g., $R = 50$ slots for a station-to-star distance of 10 km.
- The timeout duration is denoted by ϕ (in time slots).
- Transmission times for request and acknowledgement messages are negligible.

The state diagram for the model is given in Fig. 4.2. A station can be in any state and remains in that state for a geometrically distributed amount of time if it is in the transmission state, TR, or a fixed amount of time (one slot) if it is in any other state. A station departs from state TR with probability ρ at the end of a time slot and remains in state TR with probability $1 - \rho$. The states are defined as follows:

- $TU_1, TU_2, \ldots, TU_\tau$: These are states during which a station's receiver is scanning across the channels for requests. It takes τ time slots to tune to a particular station. From each of these states, an arrival can occur with probability σ. From the state TU_τ, if there is no arrival, the station either finds a request on the channel to which it has just completed tuning with some probability M and proceeds to send an acknowledgement, or it doesn't find a request with probability $1 - M$ and proceeds to tune its receiver to the next channel.
- $TU'_1, TU'_2, \ldots, TU'_\tau$: After an arrival occurs, the station immediately begins tuning its receiver to the channel of the destination. This process takes τ time slots. After tuning, the station begins to transmit a request.

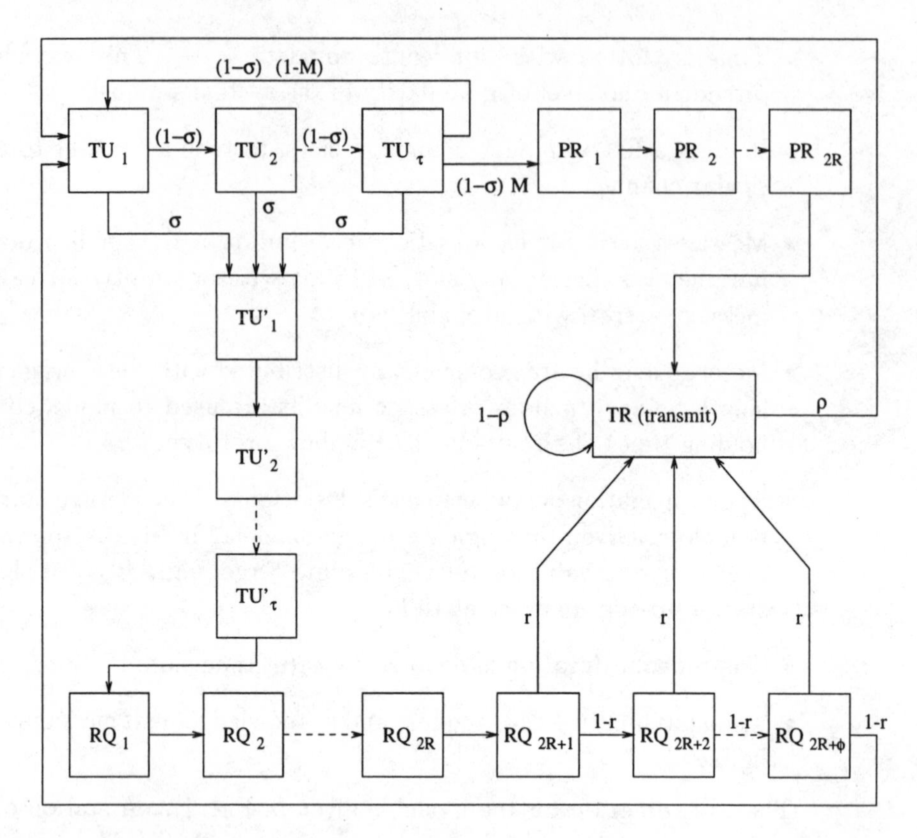

Figure 4.2 State diagram for the Rainbow model.

- $RQ_1, RQ_2, \ldots, RQ_{2R+\phi}$: Upon sending a request, it takes R time slots of propagation delay for the request to reach the destination (see Fig. 4.3). At the earliest, an acknowledgement will be received after a propagation delay of $2R$ time slots. The station continues to send the request signal for a duration of ϕ time slots or until it receives an acknowledgement, whichever occurs first, where ϕ is the timeout duration. After sending a request for a duration of ϕ slots, the station must wait an additional $2R$ time slots of propagation delay for an acknowledgement. This ensures that all acknowledgements will result in a connection. If no acknowledgement is received, the current message is "timed out" and considered "lost," and the station returns to scanning mode. The probability of getting an acknowledgement is denoted by r, which is the same for each of the states RQ_{2R+1} to $RQ_{2R+\phi}$ since the system is memoryless, and an acknowledgement can be sent at any time by the acknowledging station. (The parameter r will be related to the probability M later in this analysis.) When an acknowledgement is received, the station immediately begins transmission of its message and goes into the transmission state TR.

- $PR_1, PR_2, \ldots, PR_{2R}$: The station enters these states if it finds a request while scanning. Upon identifying the request, the station sends an acknowledgement to the requesting station. The acknowledgement takes R time slots of propagation delay to reach the station requesting the connection, after which the requesting station begins its transmission. It takes another R slots of propagation delay for the message to arrive at the destination station, after which the station goes into the transmission state, TR, to receive the message. A connection will always be established if an acknowledgement has been sent.

- TR(transmission): In this state, a station is either transmitting or receiving a message. A station may stay in this state for a duration of more than one time unit and depart with probability ρ at the end of a slot. Upon completion of message transmission or reception, the station returns to the scanning operation.

We define N_{TU_i} to be the expected number of stations in state TU_i, $N_{TU'_i}$ is the expected number of stations in state TU'_i, N_{RQ_i} is the expected number of stations in state RQ_i, N_{PR_i} is the expected number of stations in state PR_i, and N_{TR} is the expected number of stations in the transmission state.

The system can be modeled as a Markov chain with state space vector:

$$\begin{aligned} \mathrm{N} = & \Big\{N_{TU_1}, N_{TU_2}, \ldots, N_{TU_\tau}, N_{TU'_1}, N_{TU'_2}, \ldots, N_{TU'_\tau}, N_{RQ_1}, N_{RQ_2}, \ldots, \\ & N_{RQ_{2R+\phi}}, N_{PR_1}, N_{PR_2}, \ldots, N_{PR_{2R}}, N_{TR}\Big\} \end{aligned}$$

It is difficult to analyze this system using Markov analysis techniques because of the very large state space. Therefore, we analyze the system at an equilibrium point using equilibrium point analysis (EPA).

4.4 Analysis

In EPA, the system is assumed to be always operating at an equilibrium point. This is an approximation since the system actually moves around

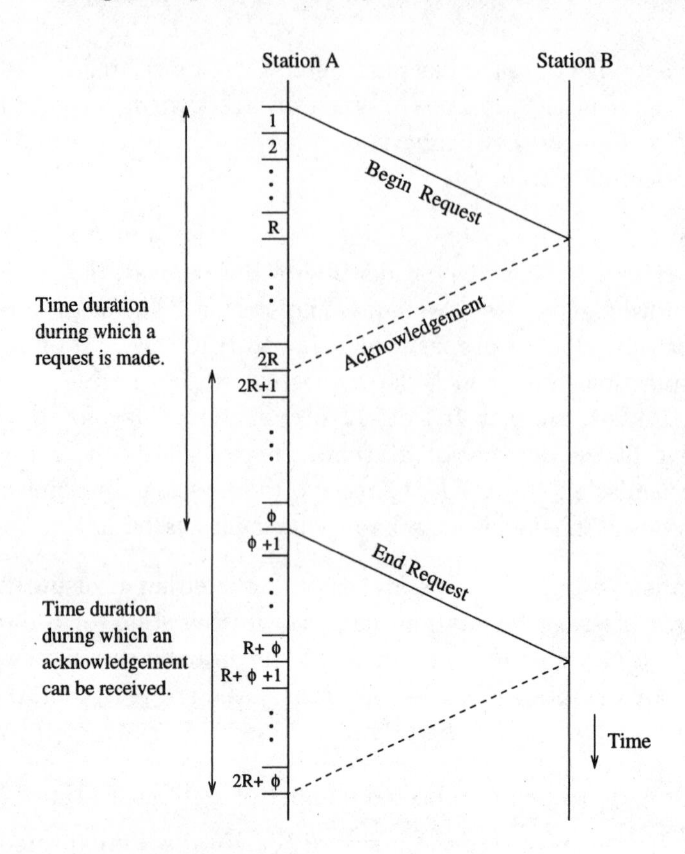

Figure 4.3 Timing for connection setup.

the equilibrium point. At an equilibrium point of the system, the expected increase in the number of stations in any state is zero. Thus, the expected number of stations entering each state must be equal to the number of stations departing from each state in each time slot.

By writing the flow equation for each state, we obtain a set of K equations with K unknowns, where K is the number of states. Also, the flow equations can be written such that the expected number of stations in each state is expressed in terms of the expected number of stations in state TU_1. Thus, if we are able to solve for N_{TU_1}, we will have the solution for the entire system.

The flow equations can be written as follows. (The N's, which previously represented the random variables, now take on the corresponding average values.)

$$N_{TU_i} = (1 - \sigma)^{i-1} N_{TU_1} \text{ for } i = 2, 3, \ldots, \tau \tag{4.1}$$

$$N_{PR_1} = N_{PR_2} = \cdots = N_{PR_{2R}} = (1-\sigma)^{\tau} M \times N_{TU_1} \tag{4.2}$$

$$\begin{aligned} N_{TU'_1} = N_{TU'_2} &= \cdots = N_{TU'_{\tau}} \\ &= N_{RQ_1} = N_{RQ_2} = \cdots \\ &= N_{RQ_{2R}} = [1-(1-\sigma)^{\tau}]\, N_{TU_1} \end{aligned} \tag{4.3}$$

$$N_{RQ_{2R+j}} = (1-r)^{j-1}\,(1-(1-\sigma)^{\tau})\, N_{TU_1} \quad \text{for } j = 1, 2, \ldots, \phi \tag{4.4}$$

$$\begin{aligned} \rho \times N_{TR} &= N_{PR_{2R}} + \sum_{j=1}^{\phi} r \times N_{RQ_{2R+j}} \\ &= \left[(1-\sigma)^{\tau} M + \left\{1-(1-r)^{\phi}\right\} \times \{1-(1-\sigma)^{\tau}\}\right] \times N_{TU_1} \end{aligned} \tag{4.5}$$

We now need to solve for the unknown variables N_{TU_1}, r, and M. M is the probability that a request will be found by a scanning station. This is equal to the probability that another station is in states RQ_{R+1} to $RQ_{R+\phi}$, and that the request is intended for the scanning station. That is,

$$M = \frac{1}{N-1} \times \frac{1}{N} \left(\sum_{i=R+1}^{R+\phi} N_{RQ_i} \right) \tag{4.6}$$

Substituting Eqns. (4.3) and (4.4) into Eqn. (4.6), and simplifying yields:

$$\begin{aligned} M &= \frac{1}{N-1} \times \frac{1}{N} [1-(1-\sigma)^{\tau}] \\ &\quad \times \left\{R + \frac{1}{r}\left(1-(1-r)^{\phi-R}\right)\right\} N_{TU_1} \end{aligned} \tag{4.7}$$

The rate of flow into the active state from the request states must equal the rate of flow into the active state from the state PR_{2R}. This is because a station can only begin transmission if there is another station that will be receiving the transmission. This leads to the equation:

$$N_{PR_{2R}} = \sum_{i=1}^{\phi} r \times N_{RQ_{2R+i}} \tag{4.8}$$

which, upon substitutions from Eqns. (4.2) and (4.4), yields:

$$(1-\sigma)^{\tau} M = [1-(1-\sigma)^{\tau}]\left[1-(1-r)^{\phi}\right] \tag{4.9}$$

In steady state, the sum of the stations in each state is equal to the total number of stations in the system, i.e.,

$$N = \sum_{i=1}^{\tau} N_{TU_i} + \sum_{i=1}^{\tau} N_{TU'_i} + \sum_{i=1}^{2R+\phi} N_{RQ_i} + \sum_{i=1}^{2R} N_{PR_i} + N_{TR} \tag{4.10}$$

or

$$\begin{aligned} N &= [1-(1-\sigma)^{\tau}] \times \left\{\frac{1}{\sigma} + \tau + 2R + \left(\frac{1}{r}+\frac{1}{\rho}\right)\left[1-(1-r)^{\phi}\right]\right\} N_{TU_1} \\ &\quad + \left[\left(2R+\frac{1}{\rho}\right)(1-\sigma)^{\tau} M\right] N_{TU_1} \end{aligned} \tag{4.11}$$

Equations (4.8), (4.9), and (4.11) can be solved simultaneously for the variables r, M, and TU_1, which can then be used to provide the steady state solution to the entire system.

The primary measures of interest are throughput, delay, and timeout probability. The normalized throughput is defined as the expected fraction of stations in the active state (which is also the fraction of a channel's bandwidth that is utilized):

$$S = \frac{N_{TR}}{N} \tag{4.12}$$

Delay is defined to be the time from a message's arrival to the system until the time that the message completes its transmission. This consists of the time required to tune to the destination station's channel, the propagation delay for the request and acknowledgement signals, the time until an acknowledgement is received, and the message transmission time. Delay is measured in slots, and is given by:

$$D = \tau + 2R + \sum_{k=1}^{\phi} k \times (1-r)^{k-1} \times r + \frac{1}{\rho} \tag{4.13}$$

The timeout probability is defined as the probability that a station will time out after entering the request mode, i.e.,

$$p_{TO} = (1 - r)^{\phi} \tag{4.14}$$

Also of interest is the blocking probability, which is the probability of an arrival being blocked. This is equal to the probability that a station is not in scanning mode. The equation for blocking probability is:

$$p_{BL} = 1 - \frac{\sum_{i=1}^{\tau} N_{TU_i}}{N} \tag{4.15}$$

4.5 Illustrative Examples

For illustration purposes, a system with the following default parameters is considered:

- $N = 32$ stations,
- slot length $= 1\ \mu$s,
- $R = 50$ slots (corresponding to a 10 km distance between each station and the star coupler),
- $\tau = 1000$ slots (corresponding to a 1 ms receiver tuning time),
- $\rho = 10^{-5}$ (corresponding to a mean message length of 100 ms),
- $\sigma = 10^{-4}$ (corresponding to a message arrival rate of 100 msg/s at each station), and
- $\phi = 10^{4}$ slots (corresponding to a timeout of 10 ms),

In our numerical examples presented below, we shall study the effect of some of these parameters on the system performance by varying them around the default values.

Figure 4.4 shows the normalized throughput versus message arrival rate for different values of timeout duration. As the arrival rate increases, the throughput will first increase as more messages become available for transmission, and then will eventually begin to decrease as the number of stations in request mode begin to outnumber the stations that are available to acknowledge requests. Note that, for a given arrival rate, there is an optimum timeout

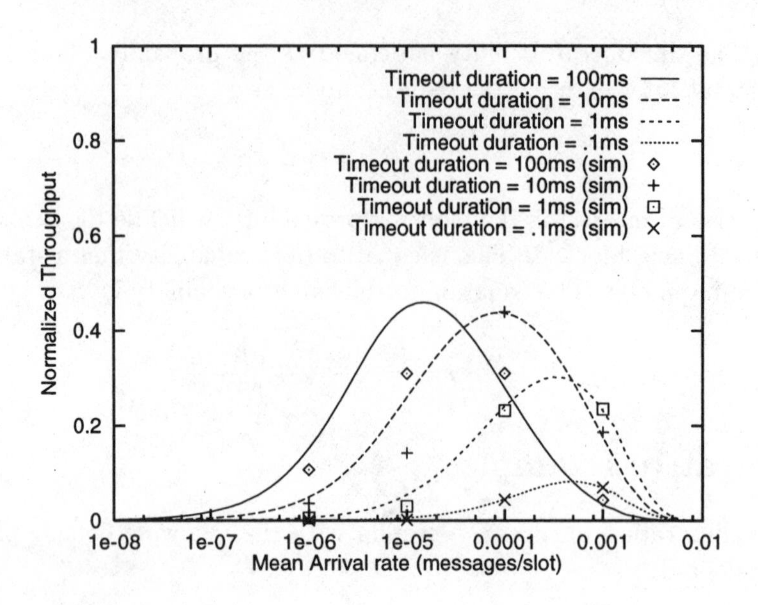

Figure 4.4 Throughput vs. arrival rate. Slot = 1 μs, $N = 32$, $R = 50$ μs, $\tau = 1$ ms, $1/\rho = 100$ ms.

duration at which the system achieves its maximum throughput. Also, with timeout duration equaling 10 ms or higher, and other parameters as chosen, the peak system throughput is approximately 0.45, which means that, if each channel is operating at 300 Mbps (as in the Rainbow-I prototype), the effective information rate on each channel equals 135 Mbps. To verify the accuracy of the analytical model, results from a simulation of the Rainbow protocol are also included in Fig. 4.4. We find that, for higher values of message arrival rate, there is excellent agreement between the analysis and simulation, but the agreement is not so good for low message arrival rates. An explanation on the cause of this discrepancy is provided later.

Normalized throughput versus mean message size ($1/\rho$) is plotted in Fig. 4.5. As expected, the throughput approaches unity as the message size is increased to a thousand seconds or higher.

Figure 4.6 plots the normalized throughput versus timeout duration for various arrival rates. For the higher arrival rates ($\sigma = 10^{-4}, \sigma = 10^{-3}$), as the timeout duration is increased, the throughput first increases and then decreases. For low timeout durations, requests are timing out too quickly, before they can be acknowledged. As the timeout duration increases, more requests will be acknowledged, resulting in higher throughput. As the timeout

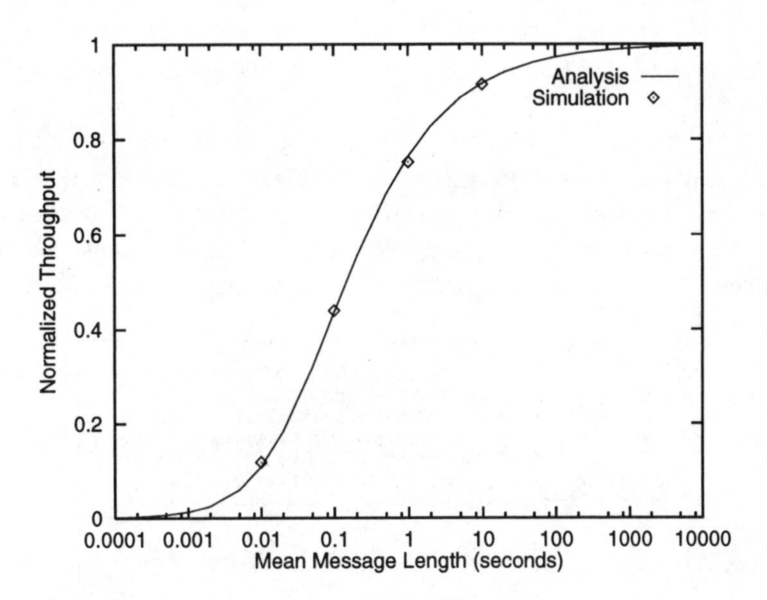

Figure 4.5 Throughput vs. message size. Slot = 1 μs, $N = 32$, $R = 50$ μs, $\tau = 1$ ms, $\sigma = 10^{-4}$ msg/slot, $\phi = 10$ ms.

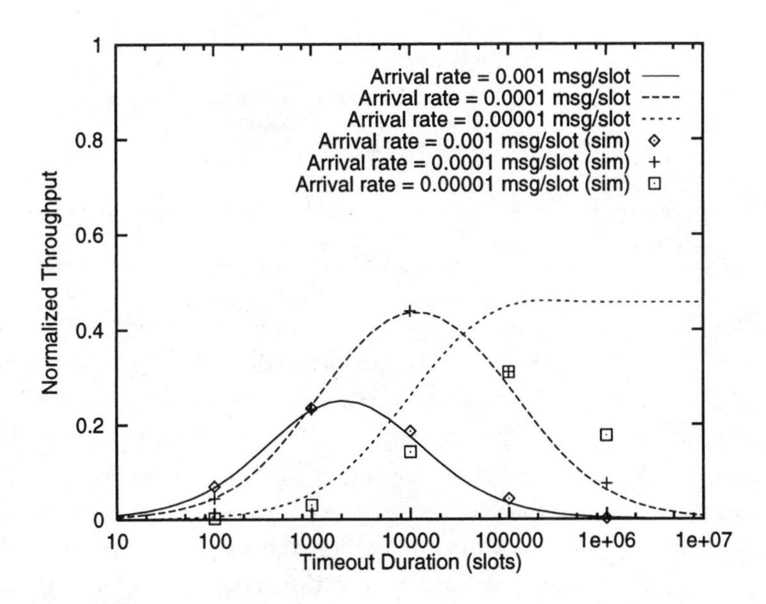

Figure 4.6 Throughput vs. timeout duration. Slot = 1 μs, $N = 32$, $R = 50$ μs, $\tau = 1$ ms, $1/\rho = 100$ ms.

duration increases even further, stations are spending longer amounts of time in the request mode, resulting in fewer stations being available to acknowledge requests.

This phenomenon is confirmed in Fig. 4.7, which examines the timeout probability as a function of the timeout duration. For low timeout durations, all of the requests are timing out, but as the timeout duration is increased, fewer requests time out. Eventually, the timeout probability levels off at some value.

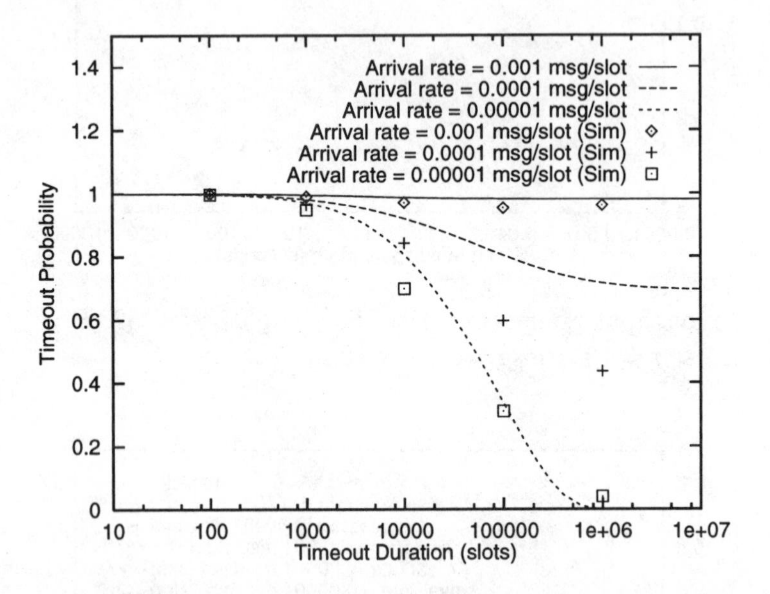

Figure 4.7 Timeout probability vs. timeout duration. Slot $= 1\ \mu$s, $N = 32$, $R = 50\ \mu$s, $\tau = 1$ ms, $1/\rho = 100$ ms.

For the lower arrival rates, the analytical and simulation results do not match very well. The analysis indicates that the throughput reaches some constant value as the timeout duration goes to infinity, while the simulation indicates that the throughput continues to decrease. This discrepancy is caused by the way deadlocks are modeled in the analysis.

In the analytical model, the deadlock probability is included in the timeout probability. A request times out when the target of the request is either in request mode or engaged in a connection (transmitting or receiving a message). For low arrival rates, most of the stations will be scanning for requests, and only a few stations will be requesting connections. The stations that are requesting connections will have a high probability of being acknowledged,

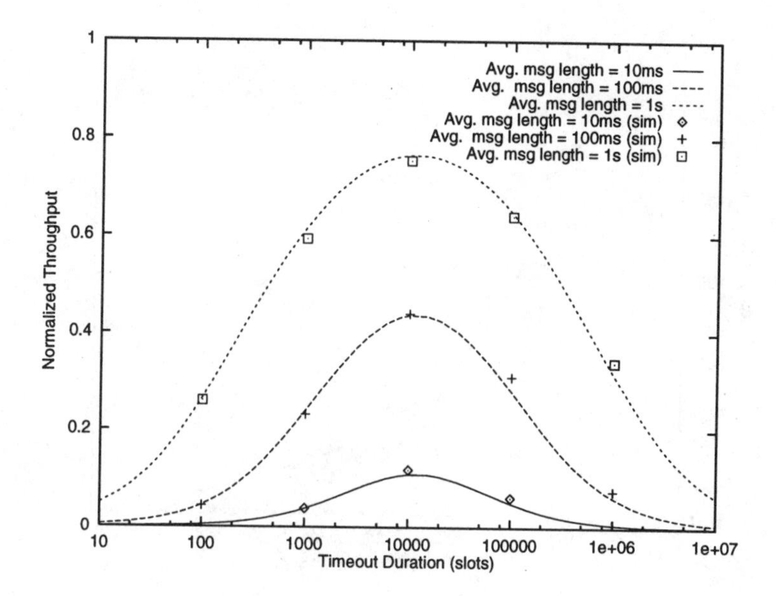

Figure 4.8 Throughput vs. timeout duration for different message lengths. Slot $= 1\ \mu s$, $N = 32$, $R = 50\ \mu s$, $\tau = 1$ ms, $\sigma = 10^{-4}$ msg/slot.

especially as the timeout duration increases. However, in the simulation, a deadlock can occur if any station being requested is also requesting a connection. This results in a higher probability of deadlock than in the analytical model for low arrival rates.

A significant result is that, for a given arrival rate, there is a timeout duration which optimizes the throughput. It is further shown in Fig. 4.8 that this optimal timeout duration is not significantly affected by the message lengths. This leads to the conclusion that the protocol can be improved by having a dynamic timeout duration, i.e., by varying the timeout duration based on the arrival rates at each station, it may be possible to maximize the throughput for specific arrival rates.

Figure 4.9 plots the average delay versus normalized throughput behavior of the system using the timeout duration ϕ as a parameter. As the timeout duration increases beyond a certain value, the throughput decreases while the delay continues to increase. For lower timeout durations, requests will either be serviced quickly or will time out quickly. The station will spend only a small amount of time in the request mode. Examining Eqn. (4.13), the bulk of the delay is equal to $\tau + 2R + 1/\rho$. In this example, the tuning time and propagation delays are small compared to the message length, so the

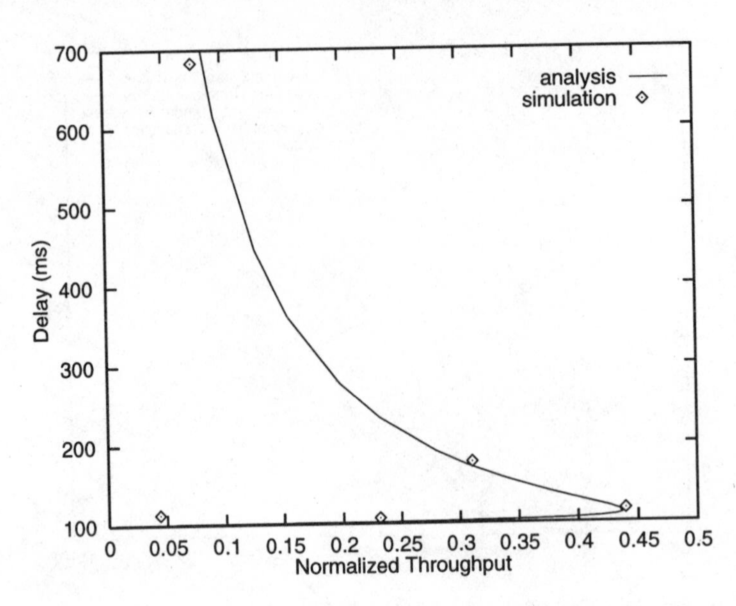

Figure 4.9 Delay vs. throughput with parameter ϕ. Slot $= 1$ μs, $N = 32$, $R = 50$ μs, $\tau = 1$ ms, $\sigma = 10^{-4}$ msg/slot, $1/\rho = 100$ ms.

delay is approximately $1/\rho$ which is equal to 100 ms. As the timeout duration increases beyond the point of optimal throughput, stations will spend longer periods in request mode before the requests are acknowledged. Thus, the summation term in Eqn. (4.13) contributes to a larger portion of the delay.

Figure 4.10 shows the average delay versus normalized throughput characteristics with increasing arrival rate, σ. Analytical results indicate that the arrival rate does not have a significant effect on the average delay of the system. This is because the only delay element that is changing is the amount of time spent in request mode. For a given timeout value ϕ, this duration will vary from 1 to ϕ. For higher values of ϕ relative to $1/\rho$, however, we expect that variations in arrival rate will have a greater effect on delays.

Figure 4.11 shows the average delay versus normalized throughput for increasing message size. As expected, the delay increases for larger message sizes with the $1/\rho$ term dominating Eqn. (4.13).

In Fig. 4.12, normalized throughput is plotted versus timeout duration for 16, 32, and 64 stations. This graph shows that the optimal timeout duration is not affected by the number of stations. This is because the optimal timeout duration depends on the ratio of stations that are scanning to the stations that are in request mode, rather than the total number of stations in the system.

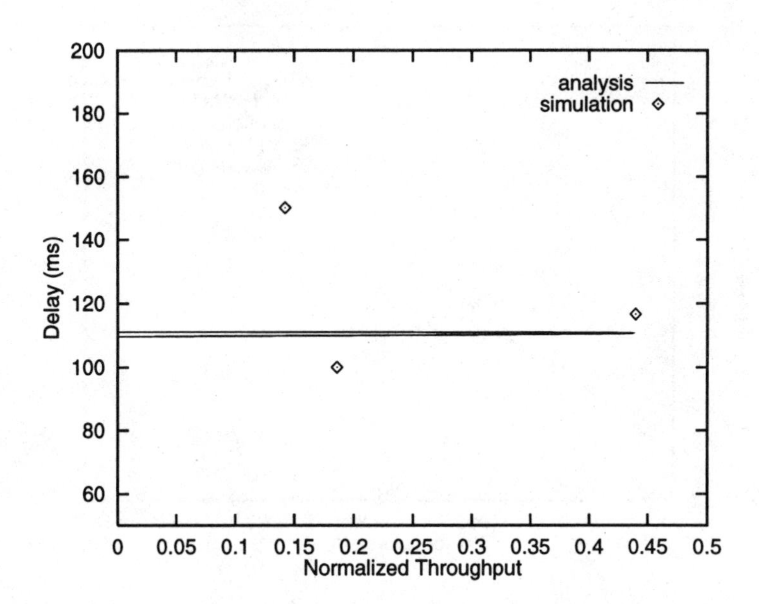

Figure 4.10 Delay vs. throughput with parameter σ. Slot $= 1\ \mu$s, $N = 32$, $R = 50\ \mu$s, $\tau = 1$ ms, $1/\rho = 100$ ms, $\phi = 10$ ms.

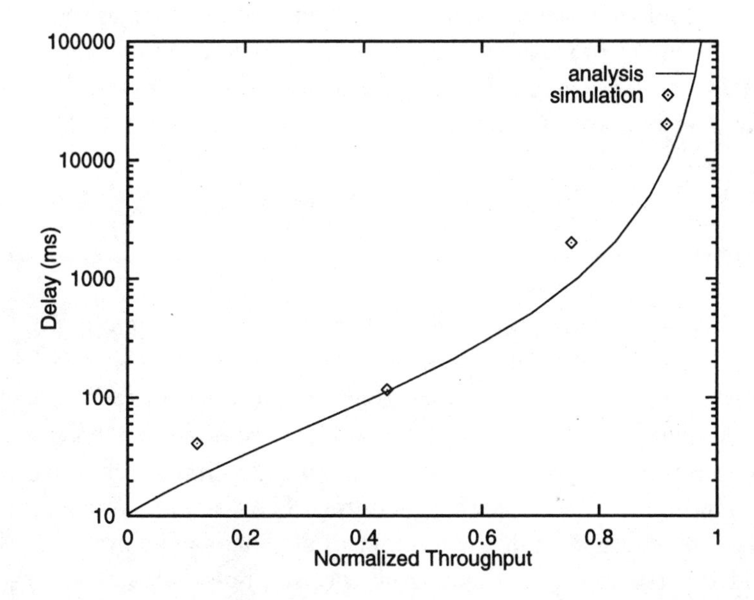

Figure 4.11 Delay vs. throughput for varying message size. Slot $= 1\ \mu$s, $N = 32$, $R = 50\ \mu$s, $\tau = 1$ ms, $\sigma = 10^{-4}$ msg/slot, $\phi = 10$ ms.

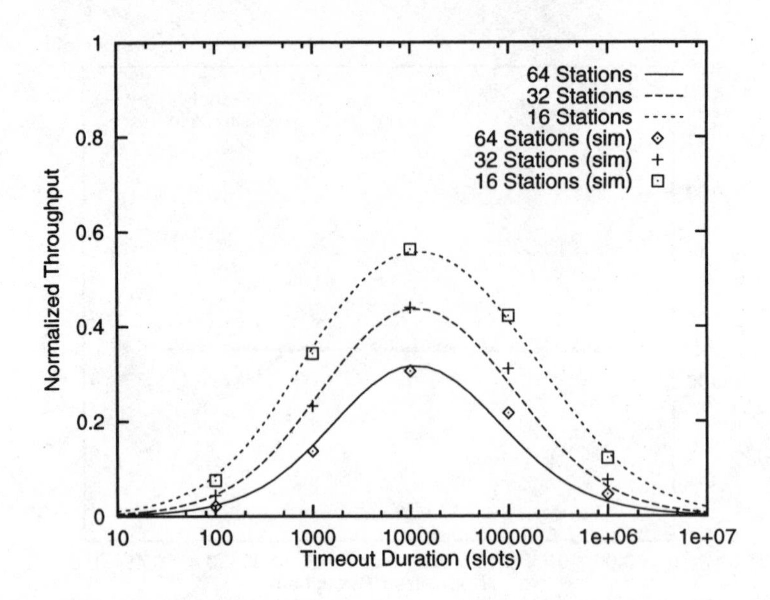

Figure 4.12 Throughput vs. timeout duration for different number of stations. Slot = 1 μs, $R = 50\ \mu$s, $\tau = 1$ ms, $\sigma = 10^{-4}$ msg/slot, $1/\rho = 100$ ms.

Also, we find that the per-channel (or per-station) throughput decreases for a larger number of stations. The receivers must tune across more channels to find a request. This indicates that the protocol is not very scalable. Increasing the number of stations reduces the per-channel throughput and requires more wavelengths.

4.6 Summary

This chapter provided an in-depth case study of a single-hop protocol used in WDM local lightwave network that has been built, viz., the IBM Rainbow network. Specifically, the study focused on a quantitative analysis of the performance characteristics of Rainbow's *in-band polling protocol.*

The Rainbow system is difficult to analyze directly, because each station can be in any of a large number of states. The size of the state space therefore grows exponentially with the number of stations in the system. This chapter provided a framework for analyzing the Rainbow protocol using the equilibrium point analysis (EPA) technique. By assuming that the system remains at its equilibrium point, we are able to reduce the complexity of the problem and obtain analytical results for an otherwise intractable system. Our analysis investigated the effect of system parameters, such as message arrival rate and timeout duration, on the performance of the system.

The EPA technique was found to provide good results when the arrival rates are high, but it was not very accurate for lower arrival rates. This is because the model doesn't properly capture the deadlock phenomenon for low arrival rates.

It was also found that, for a given set of system parameters, there exists an *optimal timeout duration* which will maximize the throughput of the system. In general, this optimal timeout duration seems to be independent of the message length and the number of stations, but depends on the arrival rate of messages.

Exercises

4.1. The Rainbow testbed utilized Fabry-Perot filters for the tunable receivers. In regards to the Rainbow medium-access control (MAC) protocol, how would you justify this choice of receiver?

4.2. Which kind of traffic is more well-suited to the Rainbow protocol – packet-switched traffic or circuit-switched traffic? Why?

4.3. Suppose station A is trying to set up a connection with station B. Draw simple state diagrams for stations A and B that illustrate this process.

4.4. Why is the timeout mechanism necessary in the Rainbow protocol? Show by example how, in the absence of a timeout mechanism, the system can become deadlocked. Give an alternate method of avoiding deadlocks in the Rainbow protocol.

4.5. Consider a Rainbow network with 32 stations. Each station is 10 m from the star coupler (i.e., propagation time is negligible), and the receiver tuning time is 1 ms. Two stations, A and B, wish to send a message to each other at the same time. What will happen assuming the protocol is as indicated in the text? Now suppose that station A transmits its message first. Station B will then transmit its message at some random time chosen from an uniform random distribution of duration over the range 0 to 100 ms. What is the probability that station B's message will be successfully transmitted? Assume that the timeout is sufficiently large to allow the scanning of all channels.

4.6. In this exercise, we will simulate the Rainbow protocol. Consider a Rainbow system with four nodes, four channels numbered 1 through 4, and a tuning time of 10 slots. Propagation delay is 5 slots. At time = 0, all receivers are parked on channel 1. Connection hold times are long (assume infinity for this exercise), and the timeout duration is finite. The connection requests are:

- Connections $(1 \rightarrow 2)$ and $(2 \rightarrow 3)$ at time = 0.
- Connections $(1 \rightarrow 3)$ and $(4 \rightarrow 3)$ at time = 10.

Which connection(s) will succeed?

4.7. Consider a Rainbow network with 100 nodes, an average scan time of 100 μs for each channel (including tuning time as well as signal detection time), and a round-robin channel scanning algorithm on each receiver. What is the expected time required to broadcast a packet of 1000 bytes from one node to all of the other nodes? Propagation time, as well as acknowledgement transmission time, is negligible. Also assume that the network has no traffic when the broadcast occurs. Bandwidth per channel is 250 Mbps.

4.8. If we allow synchronization among the receivers and senders in the Rainbow network with 100 nodes, can you find a scheme to significantly reduce the time of a broadcast from one node to all other nodes, given the same network and assumptions of the previous problem? What is the total broadcast time required by your scheme?

4.9. What is EPA? How does the EPA technique simplify the analysis? What is lost in the simplification?

4.10. Model the following system. There are N jobs and one server. A job can be in two states only: either it is being serviced (or queued), or it is idle. The service time is exponentially distributed with parameter μ, while the "idle" time (of the jobs) is exponentially distributed with parameter λ. Develop and solve the Markov chain for the system. Using EPA, find the average number of idle nodes.

Let $N = 10, \mu = 1$ and vary λ from 0.05 to 1 in steps of 0.05. Compare the average number of idle jobs calculated from EPA analysis with those computed from the Markov chain. Explain your results.

4.11. Why is the analysis of the Rainbow protocol inaccurate for large timeout values?

4.12. Prove that if N connections are attempted in an N node Rainbow system, *all* nodes will be "locked up", i.e., either a node will be involved in a deadlock or will "wait" for a node involved in a deadlock.

4.13. Through simulation, find out the average number of nodes "locked up" due to deadlock when E simultaneous connections (chosen randomly) are attempted in a system with N nodes.

4.14. Let $\Lambda_N(e)$ equal the average number of nodes that are "locked up" when E random connections are attempted simultaneously. Show how we can use $\Lambda_N(e)$ to improve the analytical model for the Rainbow protocol.

4.15. Using the analytical model of Rainbow, find the ratio of stations which are scanning to the stations which are requesting connections.

4.16. Explain the relationship between the timeout duration and the normalized throughput in the Rainbow system. What happens to the system throughput when the timeout duration is made too large? Too small? Why?

4.17. In the Rainbow state diagram, no state information is maintained regarding specific node identities. The state diagram applies to the case in which round-robin scanning is performed (i.e., scanning order is $1, 2, \ldots, N, 1, 2, \ldots, N$, etc.). Why can't this state diagram also apply to the protocol in which elevator-type scanning ($1, 2, \ldots, N-1, N, N-1, \ldots, 2, 1, 2$, etc.) is performed?

4.18. Consider the Rainbow protocol. In order to avoid deadlock, suppose the source node, upon tuning its receiver to the destination node's channel, doesn't proceed with its connection request if it finds that the destination node is busy transmitting its own connection request. It instead resumes scanning. How does this affect the performance of the protocol? How can this change be modeled in the state diagram shown in Fig. 4.2?

4.19. Suppose we modify the Rainbow protocol such that if a source node, upon tuning its receiver to the destination node's channel (state TU'_τ in the state diagram in Fig. 4.2), finds that the destination node is requesting a connection with the source node; and then, instead of sending a request, it sends an acknowledgement to the destination node's request.

In Fig. 4.2, this transition could be modeled as a link from state TU'_{τ} to state PR_1. What would be the probability associated with this link? How would the flow equations change?

Chapter 5

Multihop Networks

This chapter will provide a *breadth* treatment of multihop networks, covering various multihop topologies, their properties, and some related optimization problems. An in-depth case study of a specific multihop network called GEMNET will be presented in Chapter 6.

5.1 Characteristics of a Multihop System

In a multihop system, unlike the case in a single-hop system, the channel to which a node's transmitter or receiver is tuned (or is to be tuned) is relatively static, and this assignment is normally not expected to change except when a new global reassignment of all transceivers is deemed to be beneficial. It is unlikely that there will be a direct path between every node pair (in which case each node in a N-node network must be equipped with $N-1$ fixed-tuned tranceivers) so that, in general, a packet from a source to a destination may have to hop through some intermediate nodes, possibly zero. Different virtual structures will have different operational features (e.g., ease of routing) and different performance characteristics (e.g., minimal average packet delay, minimal number of hops, balanced link flows, etc.)

An example multihop architecture is shown in Fig. 5.1. The physical topology is a star (Fig. 5.1(a)) while the embedded virtual topology is a 2×2

torus (Fig. 5.1(b)). Note that, in this example, node 1 can communicate with nodes 2 and 3 directly via wavelength channels w_1 and w_2, but in order to reach node 4, information from node 1 should "multihop" either through node 2 or node 3.

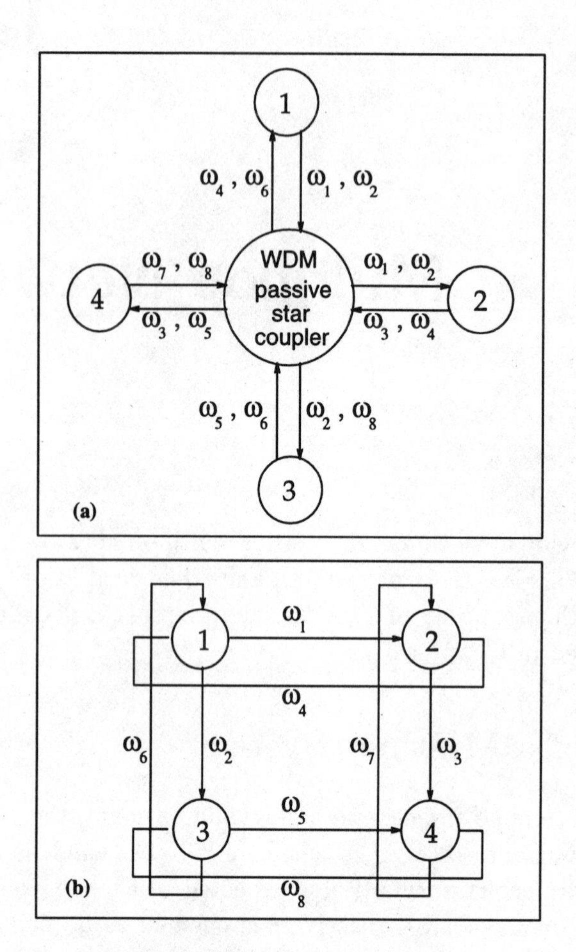

Figure 5.1 An example four-node multihop network: (a) physical topology, (b) logical topology.

While the transceiver tuning times play a vital role in determining the performance and characteristics of single-hop systems, they have little impact on multihop systems since the multihop virtual topology is essentially a static one. However, in designing a "good" multihop system, there are two other important issues which the system architect must address.

First, the virtual structure chosen must be close to "optimal" in some sense, e.g., the structure's *average (hop) distance* between nodes must be small, the average packet delay must be minimal, or the maximum flow on any link in the virtual structure must be minimal. Two nodes are at a *hop distance* of h if the *shortest* path between them requires h hops. In a multihop structure, each such hop means "travel to the star and back." The *maximum hop distance* between any two nodes is referred to as the structure's *diameter.* Multihop networks with small $\overline{h}$ and small diameter are desirable.

Second, the nodal processing complexity must also be small because the high-speed environment allows very little processing time; consequently, simple routing mechanisms must be employed. A routing-related subproblem is the buffering strategies at the intermediate nodes. Some approaches propose the use of "deflection routing" under which a packet, instead of being buffered at an intermediate node, may be intentionally misrouted but still reach its destination over a slightly longer path.

Multihop structures can be either irregular or regular. Irregular multihop structures generally address the optimality criterion directly, but the routing complexity can be large since they lack any structural connectivity pattern. Topological optimization of multihop architectures can be performed. Regular structures, because of their structured node-connectivity pattern, have simplified routing schemes; however, their regularity also constrains the set of solutions in addressing the optimality problem, and the number of nodes in a complete regular structure usually forms a special set of integers, rather than an arbitrary integer. Regular structures which have been studied include perfect shuffle (called ShuffleNet), de Bruijn graph, toroid (Manhattan Street Network, MSN), hypercube, linear dual bus, and a virtual tree. Characteristics of alternative routing strategies, including deflection routing, in ShuffleNet have also been studied quite extensively.

Attention must be paid to another piece of input to these designs, viz., the fact that the offered loads by the various nodes may not necessarily be symmetric, which is more pronounced with the proliferation of special-purpose networking equipment such as servers and gateways. Regular structures are generally amenable to uniform loading patterns, while irregular structures can generally be optimized for arbitrary workloads. The performance effect of nonuniform traffic and corresponding adaptive routing schemes to control congestion are important topics.

Finally, another pertinent issue which a multihop network architect must be mindful of is whether to employ "dedicated channels" or "shared channels." Under the case of dedicated channels, each virtual link employs a dedicated wavelength channel. However, since internode traffic may be bursty, the traffic on an arbitrary link is expected to be bursty as well, as a result of which some of the links' utilizations may be low. Consequently, the shared channel mechanism advocates the use of two or more virtual links to share the same channel in order to improve the channel utilization. However, this also introduces the need for a multiple access protocol on the channel, viz., an arbitration mechanism that governs access rights to the channel. Issues related to shared-channel strategies are discussed in Chapter 7.

5.2 Topological Optimization Studies

This section will first review the construction of optimal structures based on minimizing the maximum link flow [LaAc91], followed by optimizations based on minimization of the mean network-wide packet delay [BaFG90].

5.2.1 Flow-Based Optimization

Consider a network containing an arbitrary number of nodes N, which are indexed 1, 2, ... , N. Each node has T transmitters and T receivers. The capacity of each WDM channel is C units (say bps). The traffic matrix is given by λ_{sd}, where λ_{sd} is the traffic flow from source node s to destination node d for $s, d = 1, 2, \ldots, N$. The flow in link ij is denoted by f_{ij}, while the fraction of the λ_{sd} traffic flowing through link ij is denoted by $f_{ij}^{s}d$. Let Z_{ij} be the number of directed channels from node i to node j. Then, the capacity of link ij equals $C_{ij} = Z_{ij} \times C$. The fraction of the (i, j)-link capacity which is utilized equals f_{ij}/C_{ij}. An arbitrary topology will have a link with maximum utilization given by

$$max_{(i,j)} \left\{ \frac{f_{ij}}{C_{ij}} \right\}$$

Among various alternative topologies that are possible, the one that minimizes the above quantity is chosen to be the optimal interconnection pattern.

Formally, the above flow and wavelength assignment (FWA) problem can be set up as a mixed integer optimization problem with a min-max objective function subject to a set of linear constraints [LaAc91]. The main characteristic of this problem formulation is that it allows the traffic matrix to scale up by the maximum amount before its most heavily loaded link saturates. Another important characteristic is that only the node-to-node traffic intensities need to be known, and the solution is independent of the traffic type, which could be either circuit-switched or packet-switched (which, in turn, could be datagram-based or virtual-circuit-based).

Unfortunately, the search space for the connectivity diagram grows rapidly with increasing N. Hence, there exists a suboptimal and iterative algorithm which first determines a heuristic initial solution and then applies branch-exchange operations iteratively to improve the solution [LaAc91]. The initial solution, in turn, consists of a connectivity problem, which heuristically tries to maximize the one-hop path traffic (i.e., it attempts to connect nodes with more traffic between them in one hop), and this can be solved by a special version of the simplex algorithm. The second part of the initial solution is the routing problem, which can be formulated as a multicommodity flow problem with a nonlinear, nondifferentiable, convex objective function, and it can be solved by using the flow-deviation method [FrGK73]. Iterative improvement is performed by considering a number of least-utilized branches (say K) two at a time. A branch-exchange operation is performed by (1) swapping the transmitters (or receivers) of the two least-utilized branches, (2) re-solving the routing problem on the new connectivity diagram, and (3) accepting the swap if the new topology leads to a lower network-wide maximum link utilization. This procedure is repeated until no improvement is obtained.

Results obtained via the above algorithm for $N = 8$ and $T = 2$ are quite encouraging [LaAc91]. The connectivity diagrams and the corresponding link flows for uniform traffic, ring-type traffic, disconnected-type traffic, and centralized traffic do match with intuition. An improvement of the problem formulation can also accommodate the finite tuning range of the transceivers [LaAc90b].

5.2.2 Delay-Based Optimization

In designing an optimal virtual topology, an alternative objective may be to minimize the mean network-wide packet delay. The packet delay has two components – the first is due to the propagation delays encountered by the packet as it hops from the source through intermediate nodes to the final destination, and the second is due to queueing at the intermediate nodes. In a high-speed environment where the channel capacity C is quite large and the link utilizations are expected to be in the light-to-moderate range, the queueing delay component can be ignorable compared to the propagation delay component which is directly dependent on the "glass distance" between the nodes [BaFG90]. Thus, this optimization also requires knowledge of the distance matrix d_{ij}, where d_{ij} is the glass distance from node i to node j per the underlying physical topology. The mean network-wide packet delay can therefore be written as

$$\overline{D} = \sum_{i=1}^{N} \sum_{j=1}^{N} \frac{f_{ij} d_{ij}}{\nu \gamma} + \Delta$$

where ν = velocity of light in fiber, f_{ij} is the flow through link ij, $\gamma = \sum_{s=1}^{N} \sum_{d=1}^{N} \lambda_{sd}$ = total offered load to the network, and Δ is the nodal queueing delay component (see [BaFG90]).

Formally, the optimization can be stated as follows:

Given:	Traffic Matrix
	Distance Matrix
Objective:	Minimize Mean Packet Delay
Design Variables:	Virtual Topology
	Link Flows
Constraints:	Flow Conservation
	Nodal Connectivity (including the number of transmitters and receivers per node)

In [BaFG90], algorithms based on simulated annealing have been employed to solve both the dedicated-channel and the shared-channel cases, where time-division multiple access (TDMA) is employed for channel-sharing. A faster solution to the shared-channel topology has also been obtained in [BaFG90] by using genetic algorithms.

5.3 Regular Structures

Regular topologies which have been studied as candidates for multihop lightwave networks include perfect shuffle, de Bruijn graph, toroid, and hypercube. Their characteristics are outlined in the following subsections. Some general results and bound on regular multihop structures can be found in [HlKa91].

5.3.1 ShuffleNet

A (p, k) ShuffleNet can be constructed out of $N = kp^k$ nodes which are arranged in k columns of p^k nodes each (where $p, k = 1, 2, 3, \ldots$), and the kth column is wrapped around to the first in a cylindrical fashion. The nodal connectivity between adjacent columns is a p-shuffle, which is analogous to the shuffling of p decks of cards. This interconnection pattern can be defined more precisely as follows: (1) number the nodes in a column from top to bottom as 0 through $p^k - 1$, and (2) direct p arcs from node i to nodes $j, j+1, \ldots, j+p-1$ in the next column where $j = (i \bmod p^{k-1}) \cdot p$. A $(2,2)$ ShuffleNet is shown in Fig. 5.2.

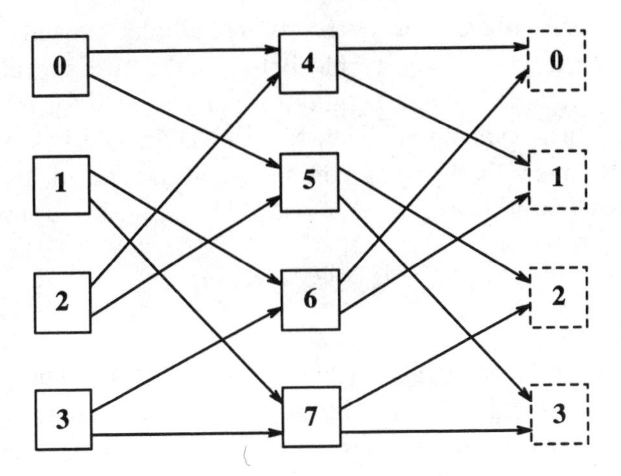

Figure 5.2 A $(2,2)$ ShuffleNet.

An important performance metric of this structure is the mean hop distance between any two randomly chosen nodes. From any "tagged" node in any column (say the first column), note that p nodes can be reached in one hop, another p^2 nodes in two hops, and so on, until all remaining $p^k - 1$ nodes in the first column are visited. Call this the first pass. In the second pass, all nodes which were not visited in the first pass can now be visited [although there can be multiple (shortest-path) routes for doing so]. For example, in the $(2,2)$ ShuffleNet in Fig. 5.2, node 6 can be reached from node 0 either via the path 0-5-3-6 or the path 0-4-1-6, both of which are "shortest paths." A preferred routing algorithm will be outlined later.

Thus, the number of nodes which are h hops away from a "tagged" node can be written as

$$n_h = \begin{cases} p^h & h = 1, 2, \ldots, k-1 \\ p^k - p^{h-k} & h = k, k+1, k+2, \ldots, 2k-1 \end{cases}$$

Then, the average number of hops between any two randomly selected nodes, given by $\sum_{h=1}^{2k-1} h n_h / (N-1)$, can be obtained as

$$\overline{h} = \frac{kp^k(p-1)(3k-1) - 2k(p^k-1)}{2(p-1)(kp^k-1)} \quad . \tag{5.1}$$

Note that the ShuffleNet structure's *diameter*, viz., the maximum hop distance between any two nodes, equals $2k - 1$.

Due to multihopping, note that only a fraction of a link's capacity is being actually utilized for carrying direct traffic between the two specific nodes connected by a link, while the remaining link capacity is used for forwarding of multihop traffic. In a *symmetric* (p, k) ShuffleNet in which the routing algorithm uniformly loads all the links, the above utilization of any link is given by $1/\overline{h}$. Since the network has $Np = kp^{k+1}$ links, the total network capacity equals

$$C = \frac{kp^{k+1}}{\overline{h}}$$

while the per-user throughput equals $C/N = p/\overline{h}$ [AcKa89]. Thus, different (p, k) combinations can yield different throughputs. Note that the per-user throughput may be increased by choosing a small k and a large p, so that the mean hop distance between nodes is reduced.

The following paragraphs examine alternative routing strategies that can be employed in ShuffleNet.

Simple Routing in ShuffleNet

There exist a number of approaches that deal with the routing problem in ShuffleNet. A simple addressing and fixed routing scheme is outlined first. A node in a (p, k) ShuffleNet is assigned the address (c, r) where $c \in 0, 1, \cdots, k\text{-}1$ is the node's column coordinate [labeled 0 through $k - 1$ from left to right (see Fig. 5.2)], and $r \in 0, 1, 2, \ldots, p^k - 1$ is the node's row coordinate (labeled 0 through $p^k - 1$ from top to bottom, using base-p digits). Thus, one may write $r = r_{k-1}r_{k-2}\ldots r_2r_1r_0$. This addressing scheme, along with the p-shuffle interconnection pattern, has the property that, from any node (c, r) where $r = r_{k-1}r_{k-2}\cdots r_2r_1r_0$, the row addresses of all the nodes reachable in the next column have the same first $k - 1$ p-ary digits (given by $r_{k-2}r_{k-3}\cdots r_2r_1r_0$), and they differ in only the least-significant digit. For routing purposes, it is required that the destination address (c^d, r^d) be included in every packet. When such a packet arrives at an arbitrary node $(\hat{c}, \hat{r})$, then, it is removed from the network if $(c^d, r^d) = (\hat{c}, \hat{r})$ (i.e., the packet has reached its destination). Otherwise, node $(\hat{c}, \hat{r})$ determines the column distance X between itself and the packet's destination (c^d, r^d) to be

$$X = \begin{cases} k + c^d - \hat{c} & c^d \neq \hat{c} \\ k & c^d = \hat{c} \end{cases}$$

Out of the p nodes in the next column to which node $(\hat{c}, \hat{r})$ may forward the current packet, it chooses the one whose least-significant digit is given by r^d_{X-1} (which is part of the destination node's address obtainable from the packet header). In particular, the packet is routed to the node with the identity $[(\hat{c}+1) \bmod k, \hat{r}_{k-2}\hat{r}_{k-3}\cdots\hat{r}_2\hat{r}_1\hat{r}_0 r^d_{X-1}]$. Note that this routing scheme follows the single shortest path between nodes $(\hat{c}, \hat{r})$ and (c^d, r^d) if the number of hops between them equals k or less; otherwise, it chooses one among several possible shortest paths. Also, note that the routing decision made at node $(\hat{c}, \hat{r})$ is independent of the packet's original source.

Effect of Nonuniform Traffic on ShuffleNet Performance

If the traffic between all node pairs is the same, then all of the ShuffleNet links may be uniformly loaded. In reality, however, the offered loading is expected to fluctuate and be nonuniform; so, the effect of nonuniform traffic on the network's traffic handling capability is of paramount importance.

There exist two alternative approaches to study the effect of traffic imbalance [EiMe88]. The first approach, referred to as extreme-value analysis, is based on the assumption that, even though the load distributions are uniform, the intensity of traffic sourced by individual nodes follows a normal distribution. Then, given the average and the variance of the aggregate per-channel load intensities, the probability that a channel is overloaded can be determined. The second approach, referred to as random load generation method, assumes that, even though the intensities of loads generated by all nodes are the same, the pattern of destination reference is random. In particular, all of the traffic generated at a node is assumed to be destined to exactly one of the other network nodes, determined randomly. It is found that, for realistic load patterns, the uniform load model predicts that the nodal throughput is reduced by a deloading factor of approximately 0.5 for small user populations to approximately 0.3 for large user populations [EiMe88].

Adaptive and Deflection Routing Strategies in ShuffleNet

An adaptive routing scheme for ShuffleNet (in order to deal with nonuniform traffic) has been developed [KaSh91]. The objective of this scheme is to ensure that packets avoid congestion or hot spots in the network. Basically, when a packet is more than k hops away from its destination in a (p, k) ShuffleNet, the packet is routed on the outgoing link with the least number of queued packets. If more than one such link exists, one is chosen at random. In addition, sometimes, even if a packet is less than k hops away from its destination (i.e., a single shortest path to the destination exists), the packet may be routed to one of the remaining and least-congested $p - 1$ outgoing links if the number of packets queued for the preferred link exceeds a certain threshold, while the queue size on the least-congested link is below a different and much-smaller threshold. Thus, although the packet is now "bumped" and has to take a longer path to its destination, it may still reach its destination faster since it can avoid the congested link(s) in the network.

In a multihop network, when two packets arrive at an intermediate node and contend for the same preferred outgoing link, one of them is usually allowed access to the link (possibly based on some priority mechanism). The other packet may be buffered at the node (i.e., the normal store-and-forward mechanism may be employed). To avoid this buffering, the intermediate node may choose an alternate strategy, viz., it can "deflect" or intentionally misroute the packet(s) which have just lost their contention(s) along its other (free) outgoing paths with the hope that the packet will eventually find its way back to its destination (over a slightly longer path while avoiding congested parts or hot spots in the network).

The advantage of employing deflection routing in a (p, k) ShuffleNet is that, if a packet at an intermediate node is more than p hops away from its destination, then multiple shortest paths exist between the intermediate node and the destination. The intermediate node can therefore be considered a "don't care" node, and all p outgoing links are equally suitable [AcSh91]. A large network (with a large p) has a large number of "don't care" nodes for an arbitrary path, and hence there are fewer contentions for preferred paths.

In [AcSh91], deflections at intermediate nodes are treated on a probabilistic basis, and it is found that deflection routing can reduce the mean number of hops, but it can also result in lower aggregate capacity, as compared to store-and-forward routing.

Deflections occur due to contention, and contentions may be resolved based on priority mechanisms. There are two metrics that may be used to resolve contentions: (1) the (remaining) *distance to the destination*, and (2) the *age* (the number of deflections already suffered by the contending packets). Under age-distance priority, packets which have suffered more deflections are given higher priority, and if there is a tie, the packet which is closest to its destination wins [KrHa90]. This approach can be generalized to consider both age-distance and distance-age priorities [ZhAc91]. In addition, an upper bound on the packet's age (i.e., on its number of deflections) can also be employed. Via analytical models which use $p = 2$ and employ independence assumptions (e.g., on the occupancy statuses on successive slots on an outgoing link), it is found that distance is a better discriminator than age since the distance-age-priority mechanism can provide lower delay, lower packet loss (for finite packet buffers at the nodes' external inputs), and higher saturation throughput [ZhAc91].

5.3.2 de Bruijn Graph

While ShuffleNet has been actively studied as a candidate for multihop lightwave networks, some other structures have been investigated as well. One such structure is the de Bruijn graph [SiRa94]. A (Δ,D) de Bruijn graph ($\Delta \geq 2$, $D \geq 2$) is a directed graph with the set of nodes $\{0, 1, 2, \ldots, \Delta - 1\}^D$ with an edge from node $a_1 a_2 \cdots a_D$ to node $b_1 b_2 \cdots b_D$ if and only if the following condition is satisfied:

$$b_i = a_{i+1}$$

where $a_i, b_i \in \{0, 1, 2, \ldots, \Delta - 1\}$ and $1 \leq i \leq D - 1$. Each node has in-degree and out-degree Δ, some of the nodes may have "self-loops," and the total number of nodes in the graph equals $N = \Delta^D$. An example $(2, 3)$ de Bruijn graph is shown in Fig. 5.3.

Routing in the de Bruijn graph is simple. A link from node A to node B can be represented by $(D + 1)$ Δ-ary digits, the first D of which represent node A, and the last D digits represent node B. In a similar fashion, any path of length k can be expressed by $D + k$ digits. In determining the shortest path from node $A = (a_1 a_2 \cdots a_D)$ to node $B = (b_1 b_2 \cdots b_D)$, one needs to consider the last several digits of A and the first several digits of B to obtain a perfect match over the largest possible number of digits. If this match is of size k digits, i.e., $(b_1 b_2 \cdots b_{D-k}) = (a_{k+1} a_{k+2} \cdots a_D)$, then the k-hop shortest path from node A to node B is given by $(a_1 a_2 \cdots a_D b_{D-k+1} b_{D-k+2} \cdots b_D)$.

An upper bound on the average hop distance between two arbitrary nodes

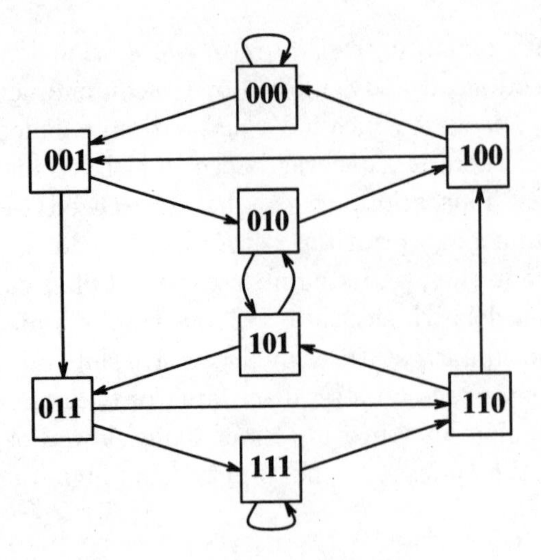

Figure 5.3 A (2, 3) de Bruijn graph.

in a de Bruijn graph follows [SiRa94]:

$$\overline{h}_{deBr} \leq D\frac{N}{N-1} - \frac{1}{\Delta - 1}$$

For a large Δ, this bound is close to the theoretical lower bound on the mean hop distance in an arbitrary directed graph with N nodes and maximum out-degree $\Delta \geq 2$ [SiRa94], thereby implying that the de Bruijn graph can be considered to be asymptotically optimal. The mean hop distances in (Δ,D) de Bruijn graphs and (p, k) ShuffleNets have been compared in [SiRa94], and it is found that, in general, for the same average number of hops, topologies based on de Bruijn graphs can support a larger number of nodes than can ShuffleNets. This is mainly due to the fact that the diameter (the maximum hop distance) in a ShuffleNet can be very large (it equals $2k - 1$ in a (p, k) ShuffleNet). Hence, ShuffleNet performs well when its diameter and consequently the number of nodes is small. ShuffleNet and de Bruijn graph are related in the sense that, if the second and subsequent columns in ShuffleNet are the same as the first, then one obtains a de Bruijn graph with degree p, diameter k, and p^k nodes [SiRa94].

An undesirable characteristic of the de Bruijn graph is that, even if the offered traffic to the network is fully symmetric, the link loadings can be unbalanced. This is due to the inherent asymmetry in the structure, e.g., in the (2,3) de Bruijn graph in Fig. 5.3, the self-loops on nodes "000" and "111" carry no traffic (and hence are wasted), and the link "1000" only carries traffic destined to node "000" while nink "1001" carries all remaining traffic generated by or forwarded through node "100." Expressions for the average link loading and the maximum link loading for uniform offered traffic have been obtained in [SiRa94]. As a result of the link-load asymmetry, the maximum throughput supportable by a de Bruijn graph is lower than that supportable by an equivalent ShuffleNet structure with the same number of nodes and the same nodal degree.

A simplified delay analysis based on M/M/1 queueing models for links and independence assumptions indicates that, for uniform loading, the average packet delay in a de Bruijn graph can be slightly lower than that in an equivalent ShuffleNet for low to moderate loads [SiRa94].

A longest-path routing scheme by which one can achieve load-balanced routing in the de Bruijn graph and get throughputs higher than ShuffleNets has also been developed [SiRa94].

5.3.3 Torus (Manhattan Street Network)

An $N \times M$ Manhattan Street Network (MSN) is a regular mesh structure of degree 2 with its opposite sides connected to form a torus. Unidirectional communication links connect its nodes into N rows and M columns, with adjacent row links and column links alternating in direction. An example 4×6 MSN is shown in Fig. 5.4. The MSN structure was originally proposed as a metropolitan-area network in [Maxe85, Maxe87], but recently its applicability as a virtual topology for a multihop lightwave network has been examined in [Ayan89]. A locally implementable, adaptive deflection routing algorithm was proposed for the MSN in [Maxe85], and since then considerable additional research has been conducted on this structure. Another advantage of the MSN is that it is highly modular and easily growable. In addition, architectures for optical deflection switches applicable in MSNs have been investigated [ChFu91]. Extension of the torus' hop-distance characteristics to higher dimensions has been recently reported in [BaMS94b].

Comparison of MSN and ShuffleNet

The work in [Ayan89] compares the characteristics of ShuffleNet and MSN, both with the same number of nodes and the same node degree (viz., 2). In particular, this work compares either network with 64 nodes, and it employs a technique based on *signal flow graphs* to enumerate the number of paths with

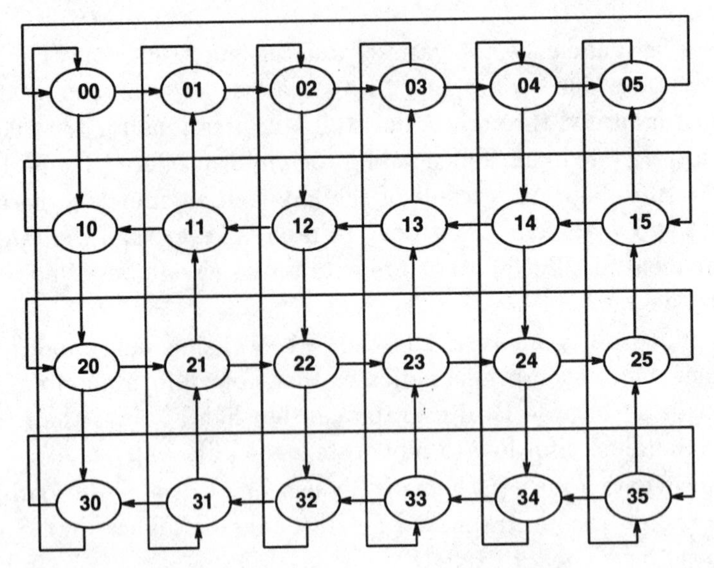

Figure 5.4 A 4×6 Manhattan Street Network (MSN) with unidirectional links.

different hop distances between nodes (when deflections are also considered). Such results provide insights into the number of alternate paths available to reach a node, thereby providing increased fault tolerance. The method requires that (1) the links of the network graph be labeled by dummy variables, (2) the transfer function of the resulting signal flow graph be solved, and (3) the transfer function coefficients be expanded into a Taylor series. These coefficients then correspond to the number of paths of a given length between a source node and a destination node.

For the example 64-node (2, 4) ShuffleNet in Fig. 5.5, consider the transfer function from node 0 to node 16 or node 17. Both of these transfer functions are given by [Ayan89]

$$T(D) = D + D^5 + 15D^9 + 225D^{13} + 3375D^{17} + \cdots$$

where D is a dummy variable representing one hop. The above transfer function indicates that, in going from node 0 to node 16, there is one 1-hop path, one 5-hop path, fifteen 9-hop paths, 225 13-hop paths, and so on.

Using such path enumeration methods, the 64-node MSN and 64-node ShuffleNet have been compared in [Ayan89]. It is found that, for the 64-node case,

1. in ShuffleNet, there are more paths to reach a node from any other node in a given number of hops,

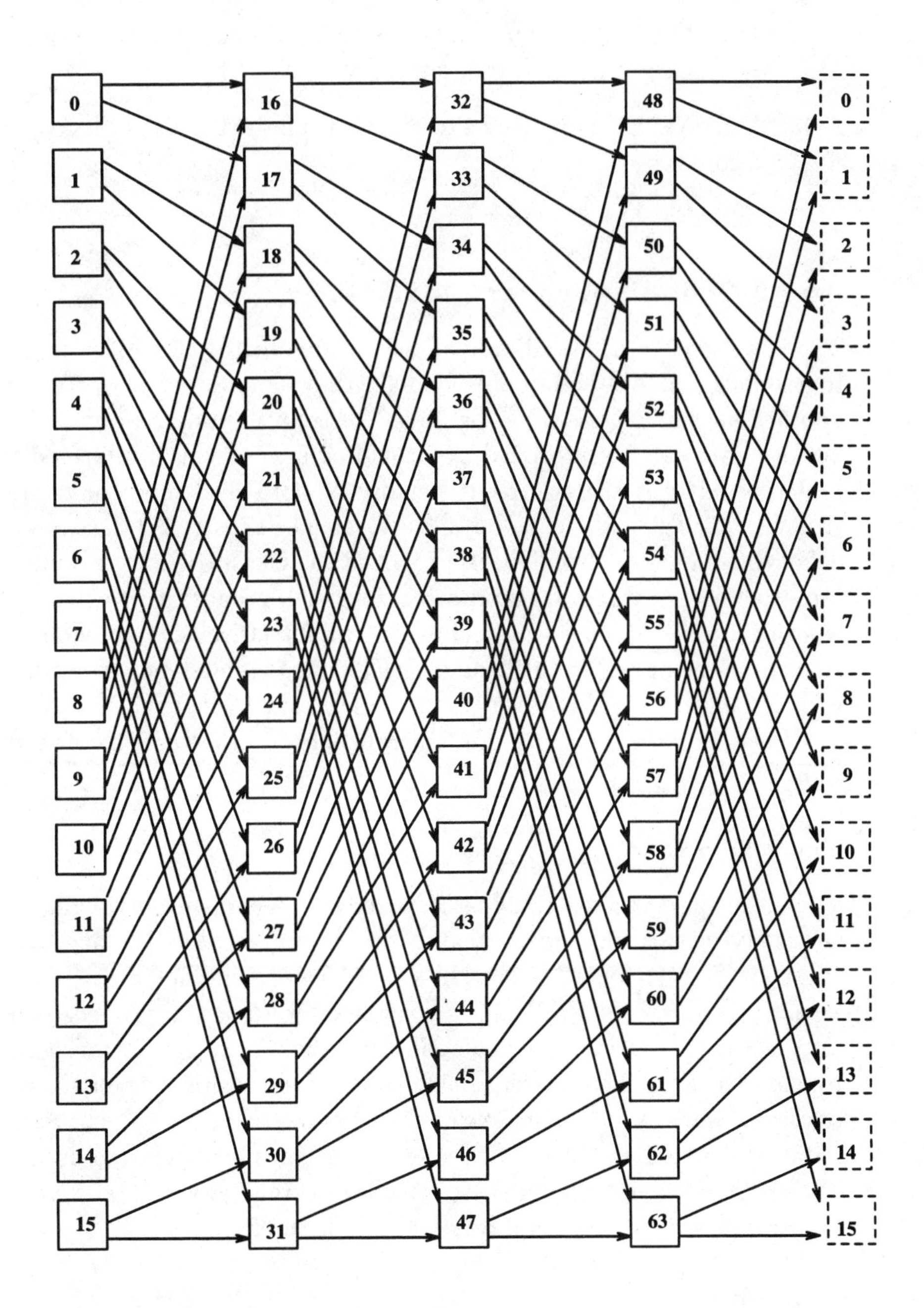

Figure 5.5 A (2, 4) ShuffleNet.

2. ShuffleNet has a larger number of distinct nodes reachable from a given node in a given number of hops,

3. ShuffleNet has more paths to reach the distinct nodes in a given number of hops, and

4. ShuffleNet has a smaller hop distance.

An All-Optical Switch for a MSN Node

The investigation of an all-optical switch architecture for MSN is motivated by the fact that, instead of simply deflecting a packet when it contends with another packet for the same preferred outgoing path, the optical switching performance can be improved by providing some form of intermediate node buffering without the electronic penalty of opto-electronic conversion, electronic buffering, and electrooptic conversion. Specifically, the use of optical delay lines as optical buffers is proposed in [ChFu91], and a number of alternative architectures incorporating such delay lines is investigated. It is shown that the proposed optical switch architectures can provide switching efficiencies comparable to those of electronic switches, but at optical transmission speeds. Finally, even though these architectures are studied in the MSN context, the concepts are generalizable to other multihop networks as well.

The Token Grid

The token grid can be considered to be a mechanism employing the MSN connectivity pattern, except that nodes communicate with one another by employing a number of tokens circulating around the network [Todd92]. Alternatively, the token grid can be considered to be a multidimensional extension of the token ring, so that nodes can share access to a mesh-connected topology by a number of overlapping rings. Nodes achieve transmission rights on the network by acquiring one of the several circulating tokens. Simple, distributed mechanisms are employed to enable rings to couple and decouple with one another in a dynamic fashion. Additional related work, referred to as the MultiMesh architecture, can be found in [ToHa92].

5.3.4 Hypercube

The hypercube interconnection pattern has been actively investigated for multiprocessor architectures, and it is starting to receive attention as a virtual topology for multihop lightwave networks as well [Dowd91, Dowd92, LiGa92]. Below, we first discuss the binary hypercube, followed by the generalized hypercube.

Binary Hypercube

The simplest form of the hypercube interconnection pattern is the binary hypercube [LiGa92]. A p-dimensional binary hypercube has $N = 2^p$ nodes, each of which have p neighbors. A node requires p transmitters and p receivers, and it employs one transmitter-receiver pair to communicate directly and bidirectionally with each of its p neighbors. Any node i with an arbitrary binary address will have as its neighbors those nodes whose binary address differs from node i's address in exactly one bit position. An 8-node binary hypercube is shown in Fig. 5.6.

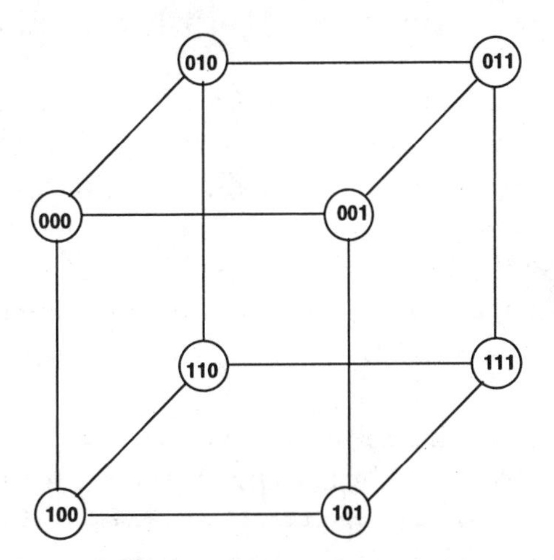

Figure 5.6 An eight-node binary hypercube.

The merits of this structure are its small diameter ($\log_2 N$) and short average hop distance [$(N \log_2 N)/(2(N-1))$]. Its disadvantage is that the nodal degree increases logarithmically with N.

Generalized Hypercube

The radix in the nodal address notation can be generalized to arbitrary integers, thereby resulting in the generalized hypercube structure. This structure can employ a mixed radix system to represent the node addresses. Let the number of nodes be given by $N = \prod_{i=1}^{p} n_i$, where the n_i are positive integers. A node's address P $(0 \leq P \leq N-1)$ is represented by the p-tuple $(m_p m_{p-1} \cdots m_1)$ where $0 \leq m_i \leq n_i - 1$. Thus, we have $P = \sum_{i=1}^{p} m_i w_i$ where $w_i = \prod_{j=1}^{i-1} n_j$.

The generalized hypercube shares similar merits and demerits as its binary version, except that it is more flexible in accommodating different numbers of nodes and their interconnection patterns.

5.3.5 Other Regular Multihop Topologies

Several other regular multihop topologies with nice properties have been examined in the literature, such as the Kautz graph [PaSe95] and CayleyNet [Tang94]. Please see the references for the descriptions and properties of these structures.

5.4 Near-Optimal Node Placement on Regular Structures

Given the flexibility of nodal interconnection patterns, one can construct an optimal regular structure which not only preserves a regular structure's simplified routing property, but also satisfies an optimality criterion such as minimum network-wide mean packet delay. Such studies have been reported for the linear dual-bus structure [TKBS91, BaMS94a], and other structures such as ring and ShuffleNet [Bane92, BaMu93a, BaMu93b]. In this section, various algorithms for placing nodes in a near-optimal fashion on a linear dual bus are reviewed.

Motivation for optimally structuring a linear bus is partly due to the standardization of the distributed queue dual bus (DQDB) as the medium access control protocol for the IEEE 802.6 metropolitan-area network (MAN). This network structure consists of two linear unidirectional buses. The incorporation of "slot reuse" techniques, which enable spatial reuse of the channel bandwidth, have also been proposed for DQDB. Thus, the network nodes may be considered to be connected via direct point-to-point links to form a linear multihop network, as shown in Fig. 5.7. The specific optimization problem may be stated as follows: Given that the network nodes must be connected linearly and that the node positions in the linear network may be adjusted by properly tuning their (optical) transmitters and receivers, what is the best pattern for interconnecting them?

In general, there are $(N!)/2$ different ways in which N nodes may be arranged in a linear fashion. From among these $(N!)/2$ structures, identifying the optimal structure(s) is a computationally intensive problem. Therefore, we

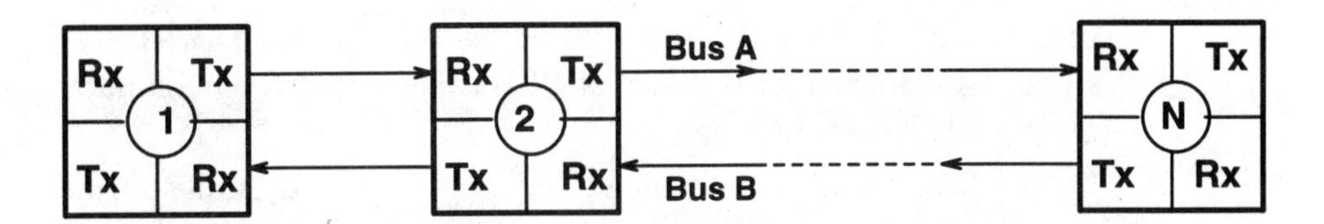

Figure 5.7 A linear dual-bus network.

investigate *fast* heuristic algorithms for constructing *near-optimal* structures. These algorithms can be classified into two categories – *flow-based heuristics* and *delay-based heuristics.* The flow-based heuristics are concerned with minimizing the maximum flow in any link, given that the network's traffic matrix is known. The delay-based heuristics require the knowledge of not only the traffic matrix but also the distance matrix, viz., the vector of distances between nodes and the hub. The goal of these algorithms is to find the node order which will minimize the network-wide mean packet delay. Below, our treatment of the heuristics will first consider static traffic patterns, but dynamic algorithms allowing traffic changes can also be developed.

5.4.1 Flow-Based Heuristics

These heuristics require that, for a given traffic matrix, the flow through the most heavily congested link in the network be minimized [TKBS91, BaMS94a]. The nodes are connected via full-duplex links and active interfaces, and traffic from the source node is relayed by intermediate nodes toward its destination where it is absorbed (see Fig. 5.7). The average traffic between nodes in such an N-node network may be represented by an $N \times N$ matrix $\mathbf{F}$, where f_{ij} represents the average traffic from node i to node j.

In the photonic bus network (PBNet) study [TKBS91], the algorithms build up the network around a partially formed PBNet by adding nodes to it one by one. Among the various heuristic algorithms studied in [TKBS91], the Min-Max algorithm is found to perform the best, and it is outlined below.

Min-Max

The optimality criteria of this algorithm is to locally minimize the maximum flow in any link in the PBNet as it is built up, after starting with a single network fragment (viz., a partial PBNet). New (unadded) nodes are selected on the basis of the maximum traffic flow between them and the "super node" consisting of all nodes currently belonging to the partial PBNet. The new node is added to the side of the partial PBNet such that the maximum utilization (on any link) is minimized.

In [BaMS94a], instead of adding nodes to a partially formed network fragment as in [TKBS91], the algorithms search for parallel techniques as well. Specifically, the algorithms build up the bits and pieces of the network in multiple fragments instead of around a single fragment.

SORTed First-Fit

In this algorithm, first the elements of the traffic matrix are sorted in non-decreasing order. Then, the algorithm steps through this sorted list to select candidate chains (of connected nodes) to be joined. Let u_{ij} be the next highest element in the sorted list. Then, if both nodes i and j are end nodes of two chains, a larger chain is formed by joining these two ends; otherwise the next highest element is considered. The time complexity of this algorithm is $O(N^2 \log(N))$.

First-Fit SUPERnodes

This algorithm is operated in $(N-1)$ steps, and at each step, two chains of nodes are connected together and the size of the effective traffic matrix is reduced by 1. If the two chains 'ik' and 'jl' are to be connected to form a longer chain, then the two end nodes to be connected are selected from the four possibilities (i, l), (i, j), (k, l), and (k, j) such that the maximum link-flow in the new chain is minimized. Then, the traffic matrix between chains is updated.

First-Fit on Binary TREE

This algorithm is based on a bottom-up design technique. Initially, the algorithm constructs chains of length two. Then, pairs of these chains are linked to obtain chains of length four (and possibly of length three as well, if N is odd). This process of "doubling" the length of the chains and "halving" their number is continued until a single chain is formed. The time complexities of this, the previous, and the Min-Max algorithms are $O(N^3)$.

Divide and Minimize Link Flow (DMF)

In this algorithm, first the N nodes are partitioned into two groups G_1 and G_2 consisting of $\lfloor N/2 \rfloor$ and $\lceil N/2 \rceil$ nodes, respectively. This partitioning attempts to minimize the flow through the link connecting the two groups G_1 and G_2 and is carried out as follows. Initially, one of the two nodes i and j with the minimum u_{ij} is placed in G_1 and the other one is placed in G_2. Then, from among the remaining nodes, node k is chosen such that the differential in traffic between node k and all nodes in G_1 and between node k and all nodes in G_2 is the maximum, and it is added to G_2. This process is repeated alternately for the two groups until all nodes are placed. Then, the nodes within each of these two groups are ordered as follows. First, a node from G_2 is chosen such that if this node were removed from G_2 and added to G_1, then the flow from the new G_1 to the new G_2 would be minimum (over all possible choices of k). This node is placed at the $(\lfloor N/2 \rfloor + 1)^{\text{th}}$ position in the linear topology being constructed. Using a similar approach, the other nodes in G_2 as well as the nodes in G_1 are arranged. Performance of this algorithm is generally superior to those of the previous algorithms, and its time complexity can be shown to be $O(N^2)$.

Iterative Approach

This approach is based on finding a Hamiltonian chain that optimizes a certain "cost function." Consider an N-dimensional surface, composed of points representing the values of the cost function for all different permutations of [1, 2, ... , N]. Then, this surface will have several local minima and one or few global minima. The iterative algorithm starts by picking randomly one point (σ) on this surface. Then, at each iteration, node σ_k is inserted at the place of node σ_r ($r < k$; $r = 1, 2, \ldots, N-1$; $k = r+1, r+2, \ldots, N$) if the maximum link-flow in the new sequence is lower than that in the previous one. If the maximum link-flow remains the same, then the new sequence is still retained if the total link-flow is reduced. By successive execution of this operation, a point on the surface is reached when no further minimization is possible. This point is either one of the local minima or the global one. By starting from different initial points, chances of hitting the global minimum can be arbitrarily increased. However, for each iteration, this algorithm takes $O(N^4)$ time.

5.4.2 Delay-Based Heuristics

Load over Distance

A set of algorithms for minimizing the average delay can be obtained by applying some of the flow-based algorithms to a transformed traffic matrix [BaMS94a]. In general, one would like to place two nodes close to each other if their combined distance from the hub is small. Also, two nodes which have a lot of traffic between them should be placed close to each other, so that this "heavy" traffic may travel through no or few intermediate nodes and thus may encounter a lower delay. Following these general guidelines, the traffic matrix is transformed by dividing each of its elements by the sum of the distances of the two corresponding nodes, and flow-based algorithms are applied to the transformed matrix.

Divide and Minimize Delay and Iterative Approach

These approaches are similar to their flow-based counterparts [BaMS94a].

5.4.3 Dynamic Load Balancing

The case where the prevailing traffic conditions may change is treated in [BaMS94a] so that the node sequence may need to be readjusted in order to maintain optimality. Specifically, the characteristics of a mechanism by which nodes can dynamically perform load balancing, and thereby reduce the maximum link-flow as well as the total link-flow of the system are investigated. The operations to achieve a better network configuration are performed by the nodes in a localized and distributed fashion using only local information available to them. See [BaMS94a, Bane92] for additional details.

5.5 Shared-Channel Multihop Systems

Channel-sharing was introduced in [Acam87] with the goal that the utilization of a multihop link can be improved if more than one transmitter-receiver pair is allowed to access the same wavelength channel. Generally, channel-sharing advocates the use of time-division multiplexing (TDM) as the multiple access mechanism for sharing a common channel. However, other channel arbitration strategies are studied in [Dowd91, Dowd92] in connection with a shared-channel hypercube architecture. Below, we discuss channel-sharing strategies proposed for ShuffleNet, MSN, and hypercube, followed by another channel-sharing concept called subcarrier multiplexing.

5.5.1 Channel-Sharing in ShuffleNet

In the original ShuffleNet work in [Acam87], (p, k) ShuffleNets with $p = 2$ are considered, and the following channel-sharing strategy is used. Recall that there are k columns of nodes and p^k nodes per column in a (p, k) ShuffleNet. All nodes in the same row share their transmissions on p (=2) channels. That is, although node i $(0 \leq i \leq p^k - 1)$ in an arbitrary column transmits to nodes $j, j+1, \ldots, j+p-1$ in the next column [where $j = (i \bmod p^{k-1})p$] via p distinct channels, all k of the nth transmitters $(n = 0, 1, \ldots, p)$ from the ith nodes in all columns must share the same channel.

Due to channel-sharing, a packet can reach its destination in fewer hops, on the average, i.e., *channel-sharing reduces the hop distance.* For $p = 2$, an upper bound on the expected number of hops is obtained to be [Acam87]

$$\overline{h'} = \frac{1}{2^k}[2 + (k-1)2^k] \tag{5.2}$$

The above result is not directly comparable to the mean hop distance for the dedicated-channel case in Eqn. (5.1) since Eqn. (5.2) includes self-hopping also (which was not included in Eqn. (5.1)).

Channel-sharing in generalized ShuffleNets is treated more rigorously in [HlKa91], but a different shared-channel allocation mechanism is employed. Specifically, it is required that, for $i = 0, 1, \ldots, p$, nodes i, $i+p^{k-1}$, $i+2p^{k-1}$, $\ldots$, and $i + (k-1)p^{k-1}$ in a column transmit on a shared channel which, in turn, is received by nodes $j, j+1, j+2, \ldots, j+p-1$ in the next column where $j = (i \bmod p^{k-1})p$. For other properties of shared-channel ShuffleNets, see [HlKa91].

5.5.2 Channel-Sharing in the Manhattan Street Network

The performance of the MSN and two of its shared-channel variants are compared in [GaLi92]. Note that a node in the MSN can be considered to be flanked by two imaginary rows and two imaginary columns. Each such imaginary row and imaginary column can be treated as a separate wavelength channel. In shared-channel variant 1 in [GaLi92], called Broadcast-Wavelength Structure 1 (BW1), each node is equipped with four transmitters and four receivers, and it can transmit on and receive from all four wavelengths flanking itself. In the other variant, called Broadcast-Wavelength Structure 2 (BW2), each node is equipped with two transmitters and two receivers, and it can transmit on and receive from one row channel and one column channel. For channel-sharing, TDM is used. Channel-sharing reduces the network diameter, the number of wavelengths required, and the average hop distance between nodes. It is also found that, for the uniform traffic case, both BW1 and BW2 can support much higher aggregate network throughput than the original MSN, and that BW1 can support more traffic than BW2 because nodes in BW1 have twice the number of transmitters and receivers than nodes in BW2 do.

5.5.3 Channel-Sharing in the Generalized Hypercube

Recall that the number of nodes in a generalized hypercube is given by $N = \prod_{i=1}^{p} n_i$ where p is the number of dimensions as well as the nodal degree and n_i is the number of nodes in the ith dimension. Channel-sharing is performed as follows [Dowd91, Dowd92]. There are $\prod_{j=1,j\neq i}^{p} n_j$ channels spanning the ith dimension, and the identity of these channels is given by $c_p c_{p-1} \cdots c_{i+1} X c_{i-1} \cdots c_1$ where $0 \leq c_i \leq n_i - 1$ and the X in position i denotes that the channel spans dimension i.

A node $(m_p m_{p-1} \cdots m_1)$ is connected to an i-dimensional channel $c_p c_{p-1} \cdots c_{i+1} c_i c_i - 1 \cdots c_1$ if $p_j = c_j$, $j \neq i$. Note that any node is connected to p such channels, all spanning different dimensions. Also, each i-dimensional channel has n_i nodes attached to it, where the node addresses in the mixed radix system are identical except in position i.

For channel arbitration, three control strategies are used. The first is random access with a control subchannel – in particular, a slotted-ALOHA control channel and ALOHA-based data channels are employed. The second approach is a random access protocol without a control subchannel; specifically, a slotted ALOHA approach is used to access the data channels. The third approach is a statically allocated media access protocol, viz., a multichannel TDMA mechanism [ChGa88a]. As expected, simulation results in [Dowd91] show that the random access approaches perform better at light loads, resulting in lower delay, but the TDMA approach can support higher throughputs. Additional details can be found in [Dowd92].

5.5.4 Multihop Systems Based on Subcarrier Multiplexing

The concept of subcarrier multiplexing was introduced in [Darc87]. It is based on the observation that, although rapid tuning between wavelength channels may not be feasible today, systems can be built which employ rapid tuning between microwave subcarriers within the same wavelength. A multihop architecture exploiting the above concept has been examined in [RaSi94].

Under this scheme, each wavelength channel is partitioned into a number of nonoverlapping subcarrier frequencies (subchannels). A node transmits on a fixed wavelength channel, but it can transmit on any subcarrier within the channel. Note that two or more nodes can transmit conflict-free information on the same channel if they do so on different subcarrier subchannels. Since a node can tune rapidly between subcarriers on its transmit wavelength, it can transmit packets in quick succession to other nodes which receive on various subcarriers within the transmitting node's wavelength channel. Information destined for nodes which do not receive on the wavelength of the transmitting node must be routed through intermediate nodes. The work in [RaSi94] demonstrates how multihop topologies such as ShuffleNet and de Bruijn graph can be realized by using the above subcarrier multiplexing approach.

Finally, TDM-based channel-sharing is also briefly discussed in the minimum-mean-packet-delay-based topological optimization work in [BaFG90]. A genetic algorithm was employed to solve the shared-channel optimization problem.

5.6 Summary

Figure 5.8 provides a summarized classification of multihop systems.

A significant amount of further research on multihop systems is expected. New architectures for multihop systems are also expected to be investigated, e.g., variants of the basic structures studied here in order to reduce the structure's mean hop distance, e.g., see Chapter 6 which examines a generalization of the perfect shuffle and de Bruijn graph that can be regular as well as grown by one node at a time. The modularity of regular structures, viz., how can one add nodes to or delete nodes from existing regular structures, is an important issue. One can refer to such structures as "injured" regular structures. How these multihop structures can reorganize their nodal connectivity pattern to restore optimality when the traffic pattern changes is another question which needs to answered.

Although this chapter's focus was on local lightwave networks employing

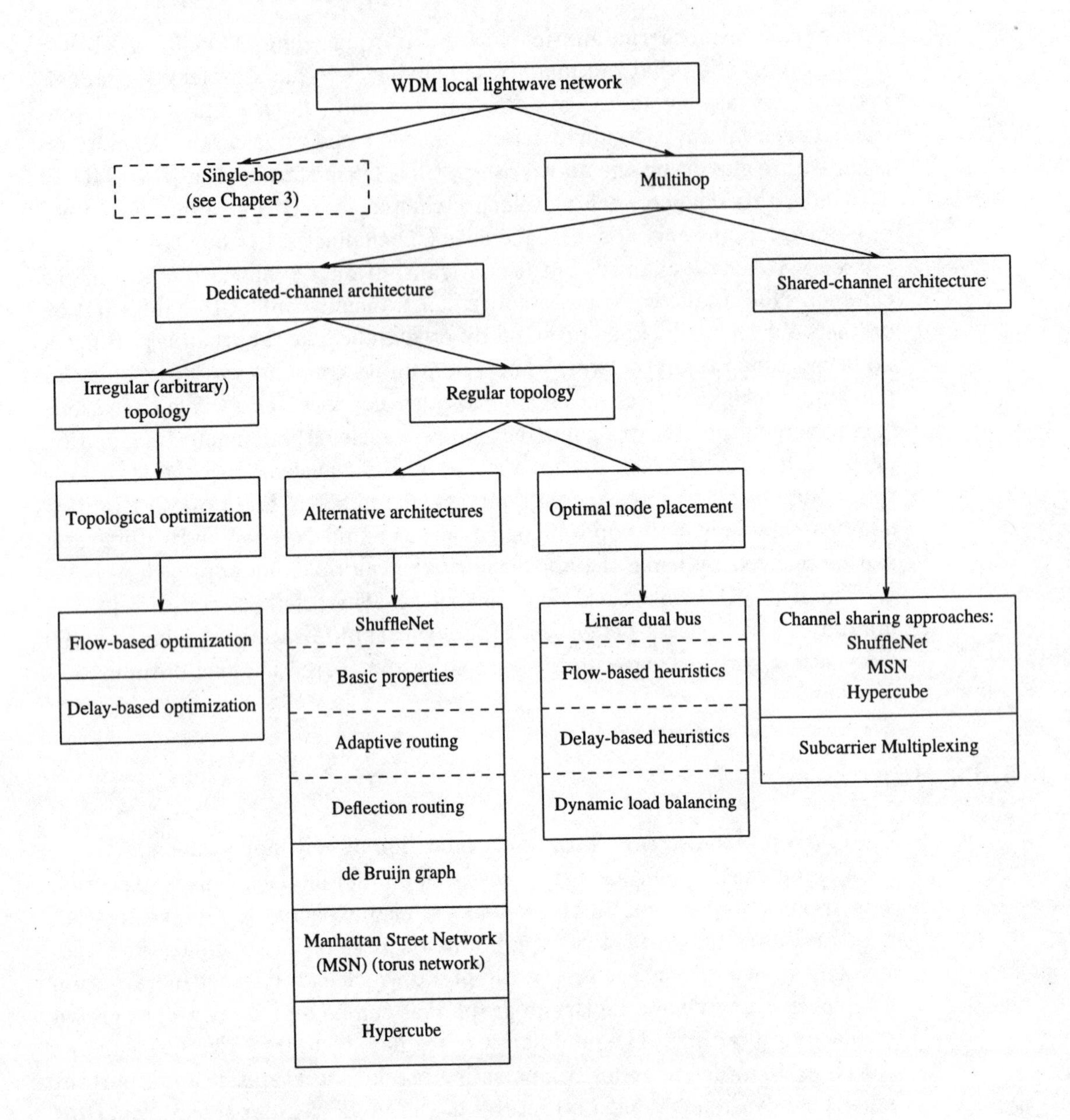

Figure 5.8 Classification of multihop network architectures.

the broadcast-and-select mechanism, work on optical WDM wide-area networks (WANs) has also been initiated based on wavelength-routing switches at intermediate nodes, so that an arbitrary multihop virtual topology may be embedded on an arbitrary physical fiber network (see Chapter 8). This category of networks does not need a centralized hub as local lightwave networks do. Hence, WDM WANs can employ spatial reuse of wavelengths in different parts of the network, thereby increasing the amount of concurrency in the network.

The following chapter will examine in depth a specific multihop virtual topology, viz., GEMNET.

Exercises

5.1. What are some of the advantages and disadvantages of passive-star coupler multihop networks compared to single-hop networks?

5.2. Consider an eight-node (2,2) ShuffleNet. Calculate the average delay for a uniform traffic matrix (the arrival rate for each source-destination pair is λ packets per second). Assume that the service time for a packet at a node is exponentially distributed with mean $1/\mu$ seconds. Assume shortest-path routing.

5.3. Describe how a packet gets routed from node 0 to node 7 in a $(2, 2)$ ShuffleNet.

5.4. Show how a packet gets routed from node 0 to node 6 in a $(2, 3)$ de Bruijn graph. What is the maximum hop distance in this graph?

5.5. Draw a (3,2) de Bruijn graph. Compute the average hop distance.

5.6. Compare and contrast the following topologies and calculate the average hop distance for each:

(a) (2,2) ShuffleNet

(b) (2,3) de Bruijn graph

(c) 8-node binary hypercube

(d) 8-node 4×2 Manhattan Street Network

5.7. Find the average hop distance for a 4×4 Manhattan Street Network.

5.8. Derive the average hop distance and diameter for a p-dimensional binary hypercube.

5.9. Derive the average hop distance of ShuffleNet [Eqn. (5.1)].

5.10. Suppose we wish to build a multihop network in which each node has a nodal degree of two, and the diameter of the network is three. We consider only ShuffleNet and Manhattan Street Network. Give the possible logical topologies and compare (i) the maximum number of nodes that can be supported, and (ii) the average hop distance.

5.11. A de Buijn graph topology is chosen by a major internet network service provider. The network must contain at least one million nodes.

(a) If a binary ($\Delta = 2$) de Bruijn graph is used, what is the minimum number of bits required to represent the label of a node?

(b) How many self-loops does the graph contain?

(c) Find a bound for $\overline{h}$.

(d) A routing algorithm finds the next node to send a packet to so that this next node is on a shortest path from node a to b. The computer used to implement the algorithm can perform operations on machine words of size 32 bits in 1 nanosecond. Operations include bit shift, bitwise and, bitwise or, add, subtract, equality test, and inequality test. Outline, or code, a fast routing algorithm or program that runs on node a where $a = (a_1, a_2, \ldots, a_D)$, and outputs the bit which is appended to a, after shifting, i.e., output is a_{D+1}, where $(a_2, \ldots, a_D, a_{D+1})$ is the label of the next node in the shortest path toward $b = (b_1, b_2, \ldots, b_D)$. Attempt to make use of low-level (bit) parallelism in your code, possibly packing a and b as unsigned integers, into machine words, as much as possible. Explain how the bits are stored in the words, including the order.

(e) Assuming that branches also take 1 nanosecond, find a bound on the running time of your routine. Do not consider compiler optimization.

5.12. In the following topologies, the presence of an undirected link is understood to be bidirectional, i.e., it may be replaced by two links going in opposite directions. Compute the diameter and average hop distance for the following logical topologies (graphs). Also, compute the total number of wavelengths (channels) required to implement these logical multihop topologies with a passive-star coupler as the physical topology.

(a) a 10-node unidirectional ring

(b) a 10-node bidirectional ring

(c) a 10-node complete graph

(d) the (10-node) Peterson graph (see Fig. 5.9 below)

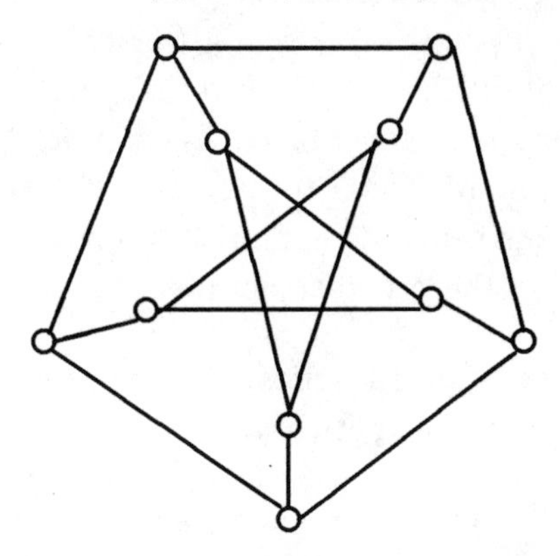

Figure 5.9 The (10-node) Peterson graph.

5.13. Consider an 8-node (2,3) de Bruijn graph. Determine the total number of paths from node 0 to node 5 that have at most 8 hops.

5.14. Find the transfer function between points A and B in the following graph. How many paths of distance 10 hops are there from A to B? How many paths of distance 15 hops are there from A to B?

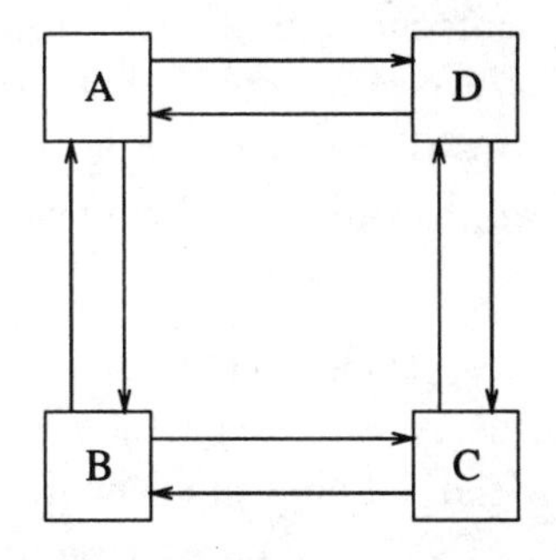

Figure 5.10 A bidirectional ring network.

5.15. Draw a two-dimensional radix-3 hypercube. Is the resulting graph isomorphic to a 3×3 torus? Will a two-dimensional radix-4 hypercube be isomorphic to a 4×4 torus? What is the number of edges in a d-dimensional radix-r hypercube?

5.16. Consider a multidimensional torus, with all links being bidirectional, and with all dimensions (M) being of the same size, i.e., an $N \times N \times \ldots \times N$ structure, of size N^M. Now consider the generalized hypercube with bidirectional links. Is it possible for a graph (logical topology) to be both? Does one class contain the other? In other words, can you find two graphs such that one is a generalized hypercube, but not a multidimensional torus, and the other is a torus, but not a generalized hypercube? Give a simple, one-sentence explanation of the difference the essential difference between these two structures.

5.17. Draw a generalized hypercube with 12 nodes. Explain how channel-sharing can be used in this hypercube network.

5.18. One of the disadvantages of a binary hypercube is that the degree of any node is unbounded (also true for generalized hypercubes). That is, for a hypercube of size $2^N = M$, the degree of a node is $N = \log_2 M$ for a binary hypercube. One solution to keep the nodal degree in check so that the structure can be scalable is to replace each node by a ring of N nodes. Each of the new nodes can then handle one of the connections required by the hypercube node. Such a graph is called a *cube-connected cycle*. It is also important to mention that the cycle must be ordered, i.e., hypercube nodes are connected when their labels differ by only one bit. The cycle must occur in bit order, so that connected cycles (hypercube nodes) are connected by corresponding cycle nodes. More precisely, if hypercube nodes a and b are connected, then their labels differ in one bit position, say the jth position. Thus, cycle node a_j and cycle node b_j must be connected. Also, for all cycles a, cycle nodes a_j and $a_{j+1 \bmod N}$ are connected.

When routing within the cube-connected cycle, a pair of nodes within a cycle must follow the cycle to form a shortest path. Outside of a cycle, routing can be accomplished without being much worse than the hypercube distance. Explain how. In other words, describe an efficient routing algorithm. Also, find a bound on the diameter of the graph. *Hint:* Consider the routing algorithm of a hypercube.

5.19. Consider a six-dimensional binary hypercube.

(a) How would you represent the root of this hypercube?

(b) Consider node (010110_2). List the members of the 3-cube to which this node belongs.

(c) Consider node (101011_2). Which node is its "partner" in the 6-cube? 4-cube?

5.20. Given three nodes and traffic matrix A, use the Min-Max flow-based heuristic to find a near-optimal node placement on a linear dual bus.

$$A = \begin{bmatrix} 0 & 2 & 3 \\ 5 & 0 & 7 \\ 4 & 3 & 0 \end{bmatrix}$$

5.21. Consider a four-node linear dual-bus network with the following traffic matrix B. Using the Min-Max heuristic, determine the ordering of the nodes. Start with a partial PBNet consisting only of node 2.

$$B = \begin{bmatrix} 0 & 2 & 3 & 1 \\ 5 & 0 & 2 & 4 \\ 1 & 4 & 0 & 3 \\ 1 & 4 & 1 & 0 \end{bmatrix}$$

5.22. List the advantages of channel-sharing.

Chapter 6

Multihop Case Study: GEMNET

While Chapter 5 reviewed a variety of multihop networks and related problems, this chapter is devoted to an in-depth case study of a specific multihop network, called GEneralized shuffle-exchange Multihop NETwork (GEMNET).

GEMNET is a generalization of shuffle-exchange networks and it can represent a family of network structures (including ShuffleNet and de Bruijn graph) for an arbitrary number of nodes. GEMNET employs a regular interconnection graph with highly desirable properties such as small nodal degree, simple routing, small diameter, and growth capability (viz., scalability). GEMNET can serve as a logical (virtual), packet-switched, multihop topology which can be employed for constructing the next generation of lightwave networks using WDM. Various properties of GEMNET are studied in this chapter.

6.1 Introduction

An attractive approach to interconnect computing equipment (nodes) in a high-speed, packet-switched network is to employ a regular interconnection graph. It is highly desirable that the graph have (1) small nodal degree (for low network cost), (2) simple routing (to allow fast packet processing), (3) small diameter (for short message delays), and (4) growth capability, viz., the graph should be scalable (so that nodes can be added to it at all times) with a modularity of unity [i.e., it should always be possible to add one node to (or delete one node from) an existing (regular) graph while maintaining regularity]. We examine such a new network structure, called GEMNET.

GEMNET can serve as a logical (virtual), multihop topology for constructing the next generation of lightwave networks using WDM. GEMNET is a regular multihop network architecture, it is a generalization of shuffle exchange networks, it represents a family of network structures, and it includes the well-studied ShuffleNet [HlKa91] and de Bruijn graph [SiRa94] as members of its family. Figure 6.1(b) shows a multihop (logical) network (a 10-node GEMNET) embedded on the physical star topology network of Fig. 6.1(a). In general, by using wavelength-routing switches, one can construct wide-area, multihop optical networks as well. (See Chapter 8.)

In a (K, M, P) GEMNET, $K \times M$ nodes – each of degree P – are arranged in a cylinder of K columns and M nodes per column so that nodes in adjacent columns are arranged according to a generalization of the shuffle-exchange connectivity pattern using directed links [RaSi94]. The generalization allows any number of nodes in a column as opposed to the constraint of P^K nodes in a column [HlKa91]. The logical topology in Fig. 6.1(b) is a (2,5,2) GEMNET.

In GEMNET, there is no restriction on the number of nodes as opposed to the cases in ShuffleNet and de Bruijn graph which can support only KP^K and P^D nodes, respectively, where $K, D = 1, 2, 3, \ldots$ and $P = 2, 3, 4, \ldots$. That is, GEMNET can represent arbitrary-sized networks in a regular graph; conversely, for any network size, at least two GEMNET configurations exist – one with $K = 1$, and the other with $M = 1$. GEMNET is also scalable in units of K if the nodes are equipped with either tunable transmitters or tunable receivers.

After describing the GEMNET architecture, we study its construction, routing, algorithms for balancing its link loads, mean as well as bounds on its hop distance, and algorithms to add nodes to (and delete nodes from) an existing GEMNET. We conclude with a discussion on which configuration of GEMNET is best based on different network parameters.

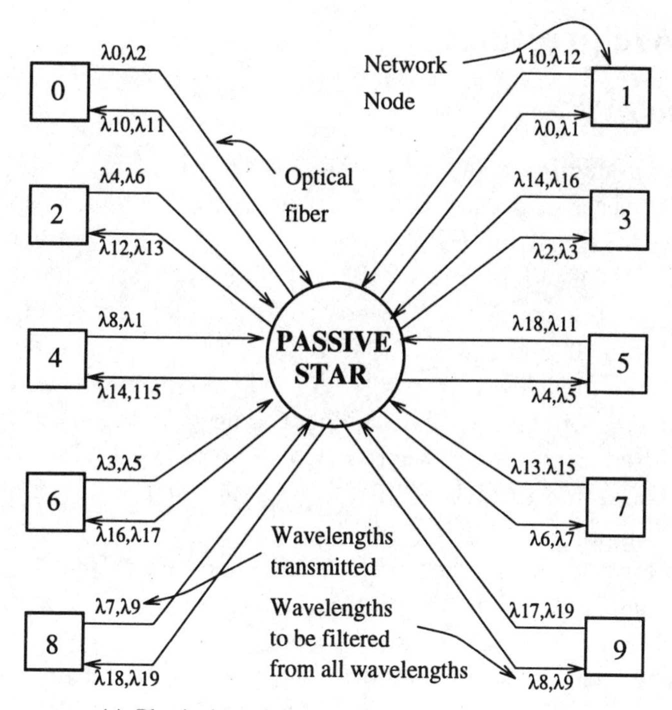

(a) Physical topology and transciever tuning pattern.

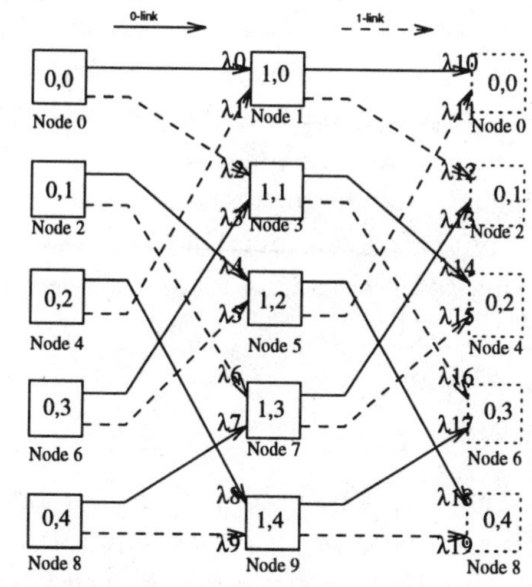

i-link from Node at (c,r) connects to Nodes at ((c+1) mod 2 , (2r+i) mod 5), i=0,1.

(b) Logical structure (virtual topology) corresponding to the transceiver assignments in (a).

Figure 6.1 A 10-node (2,5,2) GEMNET.

6.2 GEMNET Architecture

6.2.1 Interconnection Pattern

Let N denote the number of nodes in the network. If N is evenly divisible by an integer y, there exists a GEMNET with $K = y$ columns. In the corresponding (K, M, P) GEMNET, the $N = K \times M$ nodes are arranged in K columns $(K \geq 1)$ and M rows $(M \geq 1)$ with each node having degree P. Node a $(a = 0, 1, 2, \ldots, N-1)$ is located at the intersection of column c $(c = 0, 1, 2, \ldots, K-1)$ and row r $(r = 0, 1, 2, \ldots, M-1)$, or simply location (c, r), where $c = (a \bmod K)$ and $r = \lfloor a/K \rfloor$, where $\lfloor \cdot \rfloor$ represents the largest integer smaller than or equal to the argument. The P links emanating out of a node are referred to as i-links, where $i = 0, 1, 2, \ldots, P-1$. The i-link from node (c, r) is connected to node $(\hat{c}, [\hat{r} + i]_{\bmod M})$, for $i = 0, 1, 2, \ldots, P-1$ where $\hat{c} = [c+1]_{\bmod K}$ and $\hat{r} = [r \times P]$.

Note that, for a given number of nodes N, there are as many (K, M, P) GEMNETs as there are divisors for N. Specifically, when $K = 1$ or $M = 1$, we can accommodate any-sized network. However, $M = 1$ results in a ring with P parallel paths between consecutive nodes. This case is not considered further due to its large $(O(N))$ hop distance. Moreover, a (K, M, P) GEMNET reduces to a *ShuffleNet* when $M = P^K$; also, when $M = P^D$ and $K = 1$, it reduces to a *de Bruijn graph* of diameter[1] D, where $D = 2, 3, 4, \ldots$.

GEMNET's diameter is obtained as follows. Starting at any node, note that each and every node in a particular column can be reached for the first time on the $\lceil \log_P M \rceil^{\text{th}}$ hop, where $\lceil \cdot \rceil$ represents the smallest integer larger than or equal to the argument. This means that there were one or more nodes not covered in the previously visited column. Due to the cylindrical nature of GEMNET, the nodes in this column will be finally covered in an additional $K-1$ hops. Thus, $D = \lceil \log_P M \rceil + K - 1$.

6.2.2 Routing

Let (c_s, r_s) and (c_d, r_d) be the source node and the destination node, respectively. We define the "column distance" δ as the minimum hop distance in which the source node touches (covers) a node (not necessarily the destination) in the destination node's column. When $c_d \geq c_s$, we have $\delta = c_d - c_s$

[1]The diameter of a network is defined as the longest distance it takes to get from an arbitrary node to another arbitrary node while taking the shortest possible path between them.

because $(c_d - c_s)$ forward hops from any node in column c_s will cover a node in column c_d. When $c_d < c_s$, δ is given by $\delta = (c_d + K) - c_s$ because, after "sliding" c_d forward by K (i.e., $c_d + K$), due to wraparound, the situation becomes the same as when $c_d \geq c_s$. Thus, δ can be generalized as: $\delta = [(K + c_d) - c_s]_{\text{mod } K}$.

The hop distance from source node (c_s, r_s) to destination node (c_d, r_d) is given by the smallest integer h of the form $(\delta + jK)$, $j = 0, 1, 2, \ldots$, satisfying the following expression:

$$R = \left[M + r_d - \left(r_s \times P^h\right)_{\text{mod } M}\right]_{\text{mod } M} < P^h \tag{6.1}$$

R, called the *route code*, specifies a shortest route from the source node to the destination node when it is expressed as a sequence of h base-P digits. For example, if $R = (11)_{\text{base}10}$, $P = 3$, and $h = 4$, then R can be represented as $(0102)_{\text{base}3}$. In general, if $R = [\alpha_1\ \alpha_2\ \cdots\ \alpha_h]_{\text{base}P}$, then the node about to send the packet on its jth hop will route the packet to its α_j^{th} outgoing link. The maximum number of iterations needed to solve for R is just $\lceil D/K \rceil$, where D is the diameter of the network. When $K = 1$, the number of iterations is maximum and the complexity in computing R is $O(\log_P N)$. Such a simplified routing scheme is one of the main advantages of a regular structure.

To explain Eqn. (6.1), we first note that δ is the minimum number of hops required to reach a node in column c_d from a node in column c_s. However, multiple passes around the GEMNET may be required to reach the destination node, i.e., h will be of the form $\delta + jK$, where $j = 0, 1, 2, \ldots$. Define an all-0-link path to be the path traced, from a particular source node, by taking the 0-link out of every intermediate node (including the source node) for an arbitrary number of hops. Now, note that $[r_s \times P^h]_{\text{mod } M}$ is the row index of the node in column c_d reachable from the source node in h hops, by following the all-0-link path. Then, $(h-1)$ 0-links followed by a 1-link leads to the node with row index $[r_s \times P^h + 1]_{\text{mod } M}$ in column c_d, and so on. However, on the hth hop, a maximum of P^h nodes can be *covered*. The node reached on the hth hop from the source node by following the all-$(P-1)$-link path (defined similar to the all-0-link path) will be $[r_s \times P^h + (P^h - 1)]_{\text{mod } M}$. Thus, if R is less than P^h [which means that the destination node falls somewhere between the all-0-link path and the all-$(P-1)$-link path], then the destination node is reachable in h hops and its route code is given by R. In Eqn. (6.1), the addition of M and the *mod* operations are required to accommodate the wraparound of row indices.

Often, the P^h nodes covered on the hth hop could be *greater* than the number of nodes in that column. This means that multiple shortest paths may exist to some nodes in that column. Having calculated R, if $(R + jM) < P^h$ for $j = 1, 2, 3, \ldots$, then $(R + jM)$ is also a routing code with path-length h for any j that satisfies this inequality. Thus, if the shortest path from node a to node b is h hops, the number of shortest paths is given by

$$Y = \left\lceil \frac{P^h - R}{M} \right\rceil \tag{6.2}$$

Hence, for a given N, the number of alternate shortest paths increases as M decreases. The larger the number of shortest paths, the more opportunity there is to route a packet along a less-congested path and the greater is the network's ability to route a packet along a minimum-length path when a link or node failure occurs. The trade-off is that decreasing M will increase K, which, in general, will cause the average hop distance to increase.

6.3 GEMNET Properties

This section examines different shortest-path routing algorithms to balance GEMNET's link loads, analyze GEMNET's hop distance properties, and study how different GEMNET configurations perform relative to one another.

6.3.1 Routing Algorithms for Balancing Link Loading

In GEMNET, as in any other multihop network, an important goal is to balance the traffic on different links as much as possible. If a link's utilization is heavy, the corresponding link queue could become long and the delay for packets traversing that link could become significantly high. We study link loading properties of GEMNET by considering a uniform load which requires one unit of traffic to be moved between every source-destination pair.

So far, only shortest-length paths between source-destination pairs have been chosen for routing. However, if multiple shortest paths exist, then traffic can be routed over that shortest path which balances link flows as much as possible. The routing algorithm of Section 6.2.2 calculates R. If we choose to always route with the base-R route code of Eqn. (6.1) (henceforth referred to as the "unbalanced" routing scheme), the link loads tend to become unbalanced because, whenever there are multiple shortest paths, the "base-R" value is used.

Two alternative routing schemes, called "partially balanced" and "random," are found to perform better than the "unbalanced" scheme. Both of these schemes first calculate the base-R code as under the "unbalanced" scheme. Then, if only one shortest path to the destination exists, the base-R value is used to route the message. However, if multiple shortest paths exist, the "partially balanced" scheme will choose the route code R' such that, if $R + [(c_d \bmod P) \times M] < P^h$, then

$$R' = R + [(c_d \bmod P) \times M] \tag{6.3}$$

else set $R' = R$. When multiple paths exist, this approach spreads the traffic across different links, based on the destination node number. However, if the number of shortest paths exceeds P, this approach limits its selection to the first P such paths. The "random" scheme simply computes the number of alternate shortest paths from Eqn. (6.2) and assigns the route code

$$R'' = R + (M \times Z) \tag{6.4}$$

where Z is a uniformly distributed random integer in the range $[0, Y - 1]$ and Y is the number of shortest paths given by Eqn. (6.2). Some representative link loading statistics will be discussed later.

6.3.2 Bounds on the Average Hop Distance

A closed form solution for the average hop distance has not been found for all GEMNETs. However, based on the nature of the interconnection pattern, some fairly tight bounds can be placed on its average hop distance. In a GEMNET, beginning at a node, all of the nodes reachable on a certain hop count belong to a specific column, and are contiguous within that column (since row indices wraparound). This property can lead to some fairly tight bounds on the minimum and the maximum average hop distance. Below, these bounds and their maximum difference are analyzed. The interested reader is referred to [InBB94] for details of some of the derivations.

To determine the maximum average hop distance, note that if the set of nodes reached on hop i were to overlap all of the previously visited nodes, then the number of new nodes reached would be minimized. This pattern of visiting nodes would place an upper bound on the average hop distance for the network. This observation implies that, except for hops $D - K + 1$ through D, the number of new nodes covered on hop i equals P^i when $i < K$, and $P^i - P^{(i-K)}$ when $i \geq K$. For hop $D - K + 1$, we might cover fewer than P^i new nodes, and for hops $D - K + 2$ through D, we are guaranteed to cover fewer than P^i new nodes. Then, C, the last column in which we are guaranteed to cover P^i new nodes on hop i (as outlined above), equals

$$C = \begin{cases} D-K & \text{if } D-K \leq K-1 \\ K-1 & \text{otherwise} \end{cases} \tag{6.5}$$

Define the function $\delta(x,y) = 1$ if $x \leq y$, and $\delta(x,y) = 0$ otherwise. Then, it is a straightforward matter to verify that the maximum average hop distance bound for GEMNET equals

$$\begin{aligned} \overline{h}_{\max} &= \frac{1}{M \times K - 1} \left\{ \sum_{i=0}^{C} iP^i + \delta(C+1, D-K) \sum_{i=C+1}^{D-K} i(P^i - P^{i-K}) \right. \\ &\left. + \sum_{i=D-K+1}^{D} i[M - \delta(K,i)P^{i-K}] \right\} \end{aligned} \tag{6.6}$$

which, upon simplification, reduces to (see [InBB94] for details)

$$\overline{h}_{\max} = \frac{MK(D + \frac{1}{2} - \frac{1}{2}K) - K(\frac{P^{D-K+1}-1}{P-1})}{M \times K - 1} \tag{6.7}$$

When $C = D-K$, all nodes in the network are symmetric, so Eqn. (6.7) gives the *actual* average hop distance. Equation (6.7) also matches the average hop distance formula for ShuffleNet [HlKa91] and the upper bound for de Bruijn graph [SiRa91] when appropriate values for P, N, and K are used.

To determine the minimum average hop distance, we only want to visit new nodes on hop i (i.e., nodes which have not been visited before) until the contiguous property of the network forces some of the nodes reached to be previously visited nodes. This would place a lower bound on the average hop distance. So, P^i nodes, the maximum number of nodes one can possibly cover on hop i, will be covered until either hop $D-K$ or $D-K-1$. Let L be the last column in which P^i *new* nodes could be covered (as stated above). As described previously, the two possible values for L are $D-K$ and $D-K-1$. If the total number of nodes reached in a column on the $(D-K)^{\text{th}}$ hop, assuming P^i nodes covered on each hop is $\leq M$, then $L = D-K$, otherwise $L = D-K-1$. This is encompassed in the following definition of L:

$$L = \begin{cases} D-K & \text{if } P^{(D-K) \bmod K} \times \frac{(P^K)^{F+1}-1}{P^K-1} \leq M \\ D-K-1 & \text{otherwise} \end{cases} \tag{6.8}$$

where $F = \lfloor (D-K)/K \rfloor$. Now, the minimum average hop distance can be written as

$$\overline{h}_{\min} = \sum_{i=0}^{L} iP^i + \sum_{i=L+1}^{L+K} i \left(M - P^{i \bmod K} \sum_{j=0}^{\lfloor \frac{i}{K} \rfloor - 1} P^{Kj} \right) \tag{6.9}$$

whose closed form is

$$\begin{aligned}
\overline{h}_{\min} &= \frac{1}{N-1} \left(P \left[\frac{(L+1)P^L(P-1) - (P^{L+1}-1)}{(P-1)^2} \right] \right. \\
&+ \frac{MK(2L+K+1)}{2} - \delta\left(0, \left\lfloor \frac{L+1}{K} \right\rfloor - 1\right) \frac{(P^K)^G - 1}{P^K - 1} \\
&\times \left\{ (L+1)\frac{P^K-1}{P-1} + P^{Q+1}\frac{(K-Q)P^{K-Q-1}(P-1) - (P^{K-Q}-1)}{(P-1)^2} \right. \\
&+ \left. \frac{P^Q \delta(K-Q, K-1)}{(P-1)^2} \left[(P-1)(K-(K-Q)P^{-Q}) - P(1-P^{-Q}) \right] \right\} \\
&- P^{G \times K} \delta\left(\left\lfloor \frac{L+1}{K} \right\rfloor \times K + K, L+K \right) \left[(G \times K + K)\frac{P^{L-G\times K+1}-1}{P-1} \right. \\
&+ \left. \left. P\frac{(L-G\times K+1)P^{L-G\times K}(P-1) - (P^{L-G\times K+1}-1)}{(P-1)^2} \right] \right)
\end{aligned} \tag{6.10}$$

where $Q = (L+1) \bmod K$ and $G = \lfloor (L+1)/K \rfloor$. Equation (6.10) matches the formula for ShuffleNet [HlKa91] and the lower bound for de Bruijn graph [SiRa91] when appropriate values for P, N, and K are used [InBB94].

The largest potential difference between the maximum and the minimum average hop distance occurs when $P = 2$, $K = 1$, and $N (= M \times K)$ is large. When $P = 2$ and $K = 1$, the difference between the max and the min hop bounds equals

$$\Delta = \frac{ND - 2^D - 1}{N-1} - \frac{[(L+1)P^{L+1} - P^{L+2} + 2] + N(L+1) - (2^{L+1}-1)(L+1)}{N-1} \tag{6.11}$$

where the first and the second terms correspond to the max and the min bounds, respectively. It can be verified that, for both cases of L, $\Delta < 1$.

6.3.3 Which Configuration of GEMNET Is the Best?

Multiple GEMNET configurations exist for a given number of nodes, so the question of which one has the lowest average hop distance or optimum link loading naturally arises.

Figure 6.2 demonstrates how the average hop distance in different GEMNET configurations compare with one another. This figure considers a 64-node GEMNET with nodal degree of 2. In general, the larger the number of columns (i.e., the "fatter" the GEMNET) is beyond an "optimal" configuration, the higher is the mean hop distance. In this example, the two-column GEMNET exhibits the lowest hop distance, and it is superior, in this sense, to the corresponding one-column GEMNET (a de Bruijn graph) because the latter has two nodes (top and bottom nodes) with links that transmit back to themselves (self-loops), and these two nodes have larger individual mean hop distances (at most 1 greater) than the other nodes.

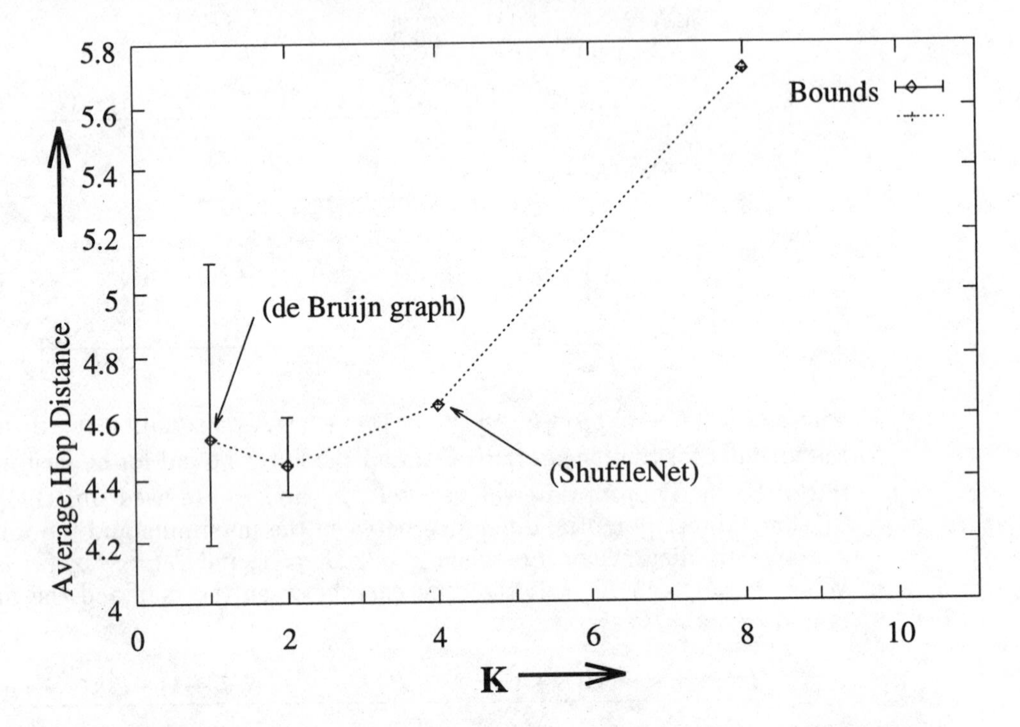

Figure 6.2 Bounds and average hop distance for a $P = 2$, 64-node GEMNET with different values of K.

In general, for a GEMNET with $P = 2$ and an even number of nodes, $K = 2$ gives the minimum average hop distance. For a GEMNET with $P = 2$, the minimum hop distance for a given N always has $K \leq 3$ columns. For $P = 2$ and an odd number of nodes, in general, the lowest average hop distance occurs for GEMNETs with $K = 1$. Also, for the GEMNETs with $P \geq 3$, in general, the best average hop distance is achieved by a GEMNET with $K = 1$. Most of these results can be seen in Table 6.1.

In terms of link loading, we observe from Table 6.1 that, in general, the "random" routing scheme performs better than the "unbalanced" and "partially balanced" schemes. Under the "unbalanced" scheme, the 0-links out of a node would be utilized more often when multiple paths exists. For the "partially balanced" scheme, the higher-numbered routes would not be chosen due to the mod-by-P operation (e.g., if there were four shortest paths and $P = 2$, only the two lowest-numbered routes would be considered). "Random" routing eliminates this bias toward the lowest-numbered routes; hence, it performs better. The "random" scheme also has the important advantage of being able to balance nonuniform traffic unlike fixed-path-routing schemes (like "unbalanced" and "partially balanced"). Henceforth, we limit our discussion to the "random" scheme only, but for additional results, please see Table 6.1.

Two opposing factors compete when we try to minimize the maximum link load. One is that, as a GEMNET is widened (by increasing K), a larger number of multiple shortest paths exist, allowing traffic to be better balanced, thereby decreasing the load on the most-congested link. On the other hand, as a GEMNET is widened, its average hop distance increases which will proportionally increase its average link load $\{= [(N-1) \times \overline{h}]/P\}$. For the "random" routing scheme, Table 6.1 reveals that an increase in the average link loading tends to increase the load on the most-congested link. For small N, the value of K that minimizes $\overline{h}$ also results in the minimum max-link load. [The optimum values of $\overline{h}$, $\overline{l_i}$ and $(l_i)\text{max}$ (for "random" routing), for a given (N, P) combination are italicized in Table 6.1.] However, as N gets larger, this trend does not hold. Due to various factors on which link loading depends, we can only state that, for large N, the value of K that minimizes the maximum link loading is generally slightly larger than the K which minimizes $\overline{h}$.

6.4 Scalability – Adapting the Size of a GEMNET

This section investigates how to scale the size of a GEMNET. Specifically, it examines different ways to grow a GEMNET. Approaches to decrease the size of a GEMNET are similar; see [InBB94] for more information. In addition, our approaches to scaling a GEMNET apply to its implementation based on a single passive star, because a topology reconfiguration under this implementation can be easily performed by retuning some of the transmitters and/or receivers at the various nodes. Reconfiguring a multistar or a switched

P	N	K	Network			Avg. Hop	D	Std. dev.	Avg. link	Fixed Routing		Part. Bal. Rounting		**Random Routing**	
			GN	SN	DB	dist. ($\overline{h}$)		of $\overline{h_i}$	load ($\overline{l}$)	Std. dev. of l_i	$(l_i)_{max}$: $(l_i)_{min}$	Std. dev. of l_i	$(l_i)_{max}$: $(l_i)_{min}$	Std. dev. of l_i	$(l_i)_{max}$: $(l_i)_{min}$
2	8	1	X		X	2.107	3	0.211	7.375	3.08	11:7	3.08	11:7	3.08	11:7
		2	X	X		*2.0*	3	0	*7.0*	2.83	11:3	1.0	8:6	2.09	*10*:4
		4	X			2.286	4	0	8.0	6.67	19:1	4.12	15:5	2.18	12:4
2	24	1	X			3.406	5	0.257	39.167	17.23	67:15	17.23	67:15	17.23	67:15
		2	X			3.348	5	0.043	38.5	15.45	65:15	15.26	63:21	15.43	65:16
		3	X	X		*3.261*	5	0	*37.5*	16.13	67:14	4.50	44:31	6.67	*50*:26
		4	X			3.478	6	0	40.0	18.49	64:12	9.0	49:25	6.27	58:31
2	64	1	X		X	4.532	6	0.224	142.766	38.78	208:63	38.78	208:63	38.78	208:63
		2	X			*4.448*	6	0.090	*140.125*	40.53	227:31	34.92	191:39	34.28	189:44
		4	X	X		4.635	7	0	146.0	78.34	331:45	33.88	220:100	15.42	*188*:116
		8	X			5.714	10	0	180.0	236.16	1023:17	168.22	772:70	11.09	216:152
2	256	1	X		X	6.417	8	0.221	818.164	192.73	1151:255	192.73	1151:255	192.73	1151:255
		2	X			*6.336*	8	0.093	*807.844*	203.73	1259:127	177.44	1126:159	174.71	1111:183
		4	X			6.514	9	0.020	830.5	384.11	1755:245	201.94	1285:433	99.98	*980*:562
		8	X			7.561	12	0	964.0	1241.58	8535:129	845.93	5676:370	24.08	1028:888
2	1024	1	X		X	8.377	10	0.220	4284.929	907.59	6309:1000	907.59	6309:1000	907.59	6309:1000
		2	X			*8.297*	10	0.093	*4243.875*	961.10	6469:511	842.87	5797:639	835.53	5701:749
		4	X			8.474	11	0.022	4334.625	1790.35	10593:1000	948.28	6840:1000	525.89	*5065*:1000
		8	X			9.517	14	0	4868.0	5972.13	55953:769	4001.38	33755:1000	55.43	5101:1000
3	81	1	X		X	*3.386*	4	0.066	*90.296*	26.46	138:40	26.46	138:40	26.46	138:40
	82	2	X			3.523	5	0.039	95.130	28.59	153:32	17.29	128:45	14.08	127:49
	81	3	X	X		3.562	5	0	95.0	62.48	265:31	21.26	147:60	11.78	*121*:69
	81	9	X			5.625	10	0	150.0	372.79	2000:7	284.11	1552:13	11.39	180:125
4	256	1	X		X	*3.599*	4	0.031	*229.406*	58.02	313:85	58.02	313:85	58.02	*313*:85
		2	X			3.828	5	0.017	244.062	162.53	917:31	90.29	533:43	45.31	333:92
		4	X			4.188	6	0	267.0	362.81	3165:57	166.04	1185:103	13.29	*313*:218
		8	X			5.867	10	0	374.0	1344.05	14407:9	941.53	9882:17	18.33	432:314

Table 6.1 Comparison of GEMNET (GN), ShuffleNet (SN) and de Bruijn (DB) graph. (Link loads are computed under the assumption of one unit of flow between every source-destination pair and different routing schemes. Also in this table, $\overline{h}$ is averaged over all the individual nodes' average hop distance ($\overline{h_i}$).)

(wavelength-routed) implementation of GEMNET would be more involved.

Since a one-column GEMNET can accommodate any number of nodes, there is always at least one interesting GEMNET (besides the one with P parallel rings corresponding to $M = 1$). Section 6.3.3 revealed that, for $P = 2$, the one-column and the two-column GEMNETs in general had the best average hop distance for an odd and even number of nodes, respectively. This property is very desirable for scalability since the one-column GEMNET can easily be grown by one node at a time (i.e., by adding an extra row), while the two-column GEMNET can be grown by two nodes at a time; also one node can be added to a two-column GEMNET with $2j$ nodes ($j = 1, 2, 3, \ldots$) by setting up the new structure as a one-column GEMNET with $2j + 1$ nodes. Since, in general, the best average hop distance for $P \geq 3$ is achieved by a one-column GEMNET, there is a dual benefit in that *the most scalable structure also possesses the best average hop distance.*

The easiest way to grow a GEMNET is to add one node at the bottom of each of its columns. Thus, a GEMNET can grow by K nodes at a time (i.e., with modularity K). The best location to add K nodes turns out to be the bottom row because the interconnection pattern is interrupted at the farthest point from the top-most row. Adding nodes closer to the top would have caused the interconnection pattern to be interrupted earlier. Thus, to add one row of nodes, each node must determine what are the row numbers of the nodes that it needs to connect to in the new *larger* network and then retune its transmitters or receivers accordingly. An example of growing a (1,6,2) GEMNET by one node to a (1,7,2) GEMNET in this fashion is shown in Fig. 6.3.

By examining the structure of a GEMNET, observe that, when adding a row of nodes, approximately the first M/P nodes in each column need not retune. Approximately the next M/P nodes perform one retuning, the next M/P nodes perform two retunings, and so on. To be exact, when adding K nodes to an N-node GEMNET in the fashion described above, we obtain

$$\text{number of retunings} = \sum_{i=2}^{P} \left\lceil \frac{N}{P} \right\rceil \times (i-1) \geq \frac{N \times (P-1)}{2} \qquad (6.12)$$

with the equality holding when N is divisible by P. Since the total number of links (and hence transmitters/receivers) equals NP, for a $P = 2$ GEMNET, this means that approximately one-fourth of the total number of transmitters or receivers in the network need to be retuned, while for a $P = 3$ GEMNET, approximately one-third of the transmitters or receivers need to be retuned.

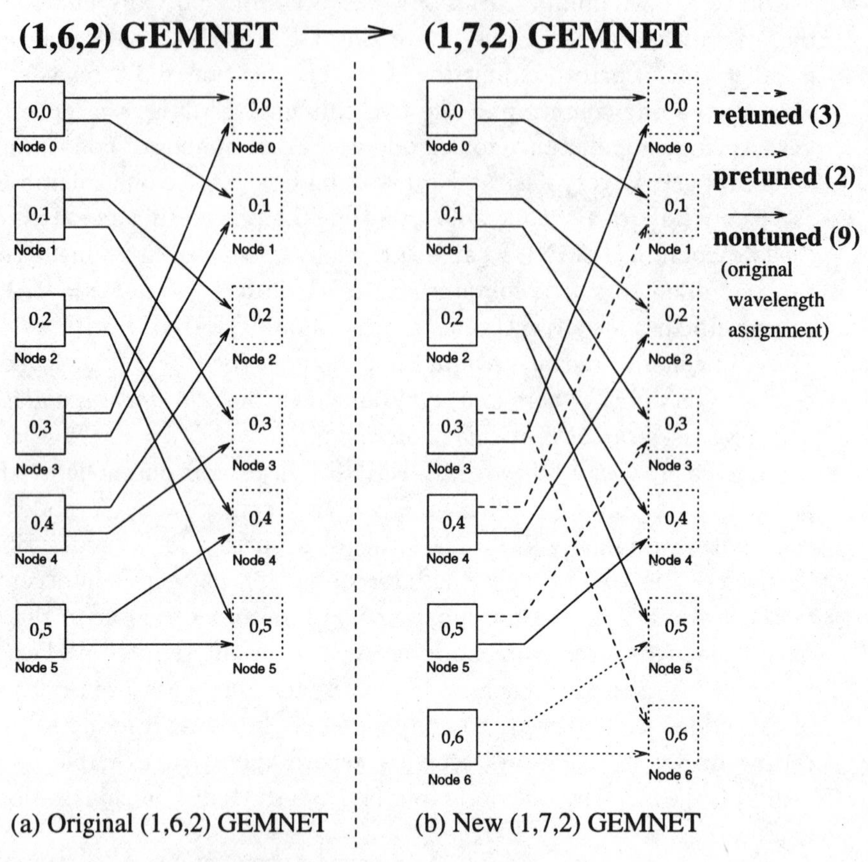

Figure 6.3 Growing a (1,6,2) GEMNET by one node.

Adding more than one row to a GEMNET can be performed in stages by adding one row at a time until all rows are added. Alternatively, the network can be scaled from the original setup to the final setup in one massive retuning operation, by "optimally" "renumbering" the individual nodes in order to minimize the number of retunings. Deletion of a row of nodes from GEMNET can also be performed similarly. The interested reader is referred to [InBB94] for more information.

6.5 Summary

Since multiple configurations of a GEMNET exist, one must consider its following properties before determining which configuration to choose: (1) scalability, (2) low average hop distance, and (3) balanced link loading. Our recommended approach is to use a one-column GEMNET since it provides the maximum flexibility in network sizes (i.e., good scalability property). For the $P = 2$ case, a one-column GEMNET will not give the best average hop distance for certain network sizes, but the average hop distance for a one-column GEMNET is very close to the best. The small benefit in average hop distance for the other configurations is probably not worth the loss in flexibility of network sizes. However, for the specific case of $P = 2$, a network with $K = 2$ has properties that may be beneficial. It always has a lower average hop distance than a one-column GEMNET for an even number of nodes.

For certain networks, the capacity of a link would be the determining factor in deciding which GEMNET to use. If, say, a one-column GEMNET has a maximum link load that is higher than the maximum carrying capacity of a link, then a wider GEMNET (with "random" routing) could be used since this will reduce the maximum link flow. However, if link capacity is not a bottleneck, then scalability in network sizes is a much more important property and therefore a one-column GEMNET is recommended.

Exercises

6.1. Discuss some of the salient features of GEMNET.

6.2. Show that a $(1, 2^N, 2)$ GEMNET is equivalent (having a one-to-one correspondence between all nodes and links) to a $(2, N)$ de Bruijn graph.

6.3. Show that a $(1, \Delta^N, \Delta)$ GEMNET is equivalent (having a one-to-one correspondence between all nodes and links) to a (Δ, N) de Bruijn graph.

6.4. Show that a (K, P^K, P) GEMNET is equivalent (having a one-to-one correspondence between all nodes and links) to a ShuffleNet.

6.5. Given nine nodes, find a GEMNET topology for which $K > 1$ and $M > 1$.

6.6. Under what conditions do we obtain the maximum and minimum hop distances in a GEMNET?

6.7. Explain Eqn. (6.1).

6.8. Find the average hop distance for a (2,5,2) GEMNET (see Fig. 6.1). Assume shortest path routing. Show how a packet is routed from node 0 to node 9.

6.9. Consider a (3,7,2) GEMNET. Find the route code R from source node (0,4) to destination node (2,1).

6.10. Finde the number of shortest paths in a (3,27,3) GEMNET between nodes $s = (1, 25)$ and $d = (0, 10)$.

6.11. Consider a lightly but uniformly loaded (K, M, P) GEMNET. Suppose one link fails. How will hop distance be affected? If shortest path routing is being used, how many connections will have to be rerouted?

6.12. Compute the diameter and average hop distance for a 3×3 GEMNET. Compute the base route code from node (1,0) to node (1,2). Find all shortest paths from node (1,0) to node (1,2).

6.13. We wish to build a 10-node network with a GEMNET topology, trying to achieve the best performance. Come up with a GEMNET topology and explain your choice. Assume each node has a degree of two.

6.14. Draw all possible GEMNETs having six nodes and a nodal degree of two. Which of these will have the shortest average hop distance? The longest?

6.15. If we are trying to minimize the maximum link load in the network, which GEMNET design is better: one with a higher number of columns or one with a higher number of rows?

6.16. A (4,4,2) GEMNET is to be upgraded into a (4,5,2) GEMNET by adding a row of nodes. What is the number of retunings required?

Chapter 7

Channel-Sharing and Multicasting

Consider a local lightwave network based on a passive-star physical topology. The number of channels, w, in such a WDM network is limited by technology and is usually less than the number of nodes, N, in the network. This chapter provides a general method using *channel-sharing* to construct practical multihop networks under this limitation. *Channel-sharing* may be achieved through *time-division multiplexing (TDM)*. The method is demonstrated by application to the generalized shuffle-exchange-based multihop architecture, called GEMNET, which was introduced in the previous chapter. Interesting characteristics of the channel-sharing approach are also highlighted.

Multicasting – the ability to transmit information from a single source node to multiple destination nodes – is becoming an important requirement in high-performance networks. Typical applications where multicasting is critical include the following: (1) wire services used by news agencies such as the Associated Press to distribute news reports from their bureaus to newspapers and radio stations throughout the world, (2) multiperson conferences, (3) video lectures where a single speaker can address a large audience, (4) "multipoint-LAN interconnection" which allows large companies to treat their many geographically distributed LANs as a single large network, and (5) the more recently developed collaborative systems. Multicasting is also required to make updates in replicated and distributed databases. Multiprocessor systems also require multicasting for cache coherency and message passing.

Multicasting, if improperly implemented, can be *bandwidth-abusive.* Channel-sharing is one approach toward efficient management of multicast traffic. This chapter develops a general modeling procedure for the analysis of multicast (point-to-multipoint) traffic in shared-channel, multihop WDM networks. The analysis is comprehensive in that it considers all components of delay that packets in the network experience – namely, synchronization, queueing, transmission, and propagation. The results show that, in the presence of multicast traffic, WDM networks with $w < N$ channels may actually perform better than networks with $w \geq N$ channels.

7.1 Introduction

The multihop technique discussed in Chapters 5 and 6 assumes that the number of channels in the system, w, is some integer multiple of N. However, this is not practical for systems with large N due to various device constraints such as fiber nonlinearities and power budget. (Recall the discussion on these topics from Chapter 2.) Furthermore, this chapter will show that it may not be desirable to operate the system with $w \geq N$ channels. So, for $w < N$, each of the w channels will be assumed to be shared by one or more nodes in a time-division multiplexed (TDM) fashion.[1]

We shall demonstrate later in this chapter that a multihop network, in which each node is equipped with only a single transmitter-receiver pair, and in which WDM channels are shared by nodal transmitters and receivers in TDM fashion, exhibit an interesting and anomalous delay[2] behavior when the number of wavelengths is varied (see Fig. 7.1). Within the practical region ($w \leq N$), for small values of w, the average packet delay initially decreases with increasing w; however, for high values of w, an increase in w increases the delay as well. For any given λ, there is an optimal value of w (denoted by w^*) for which the delay is minimized. Also, for any given λ, there are minimum and maximum values on the number of wavelengths (w_L and w_H,

[1] Random-access techniques, such as slotted-ALOHA, may also be employed for channel-sharing; however, these techniques require more processing and cannot support high throughput, and hence may be less preferable than TDM.

[2] The total delay experienced by a packet is the duration between the time of its arrival at a source node and the time it is delivered to the destination, after possibly being forwarded through intermediate nodes.

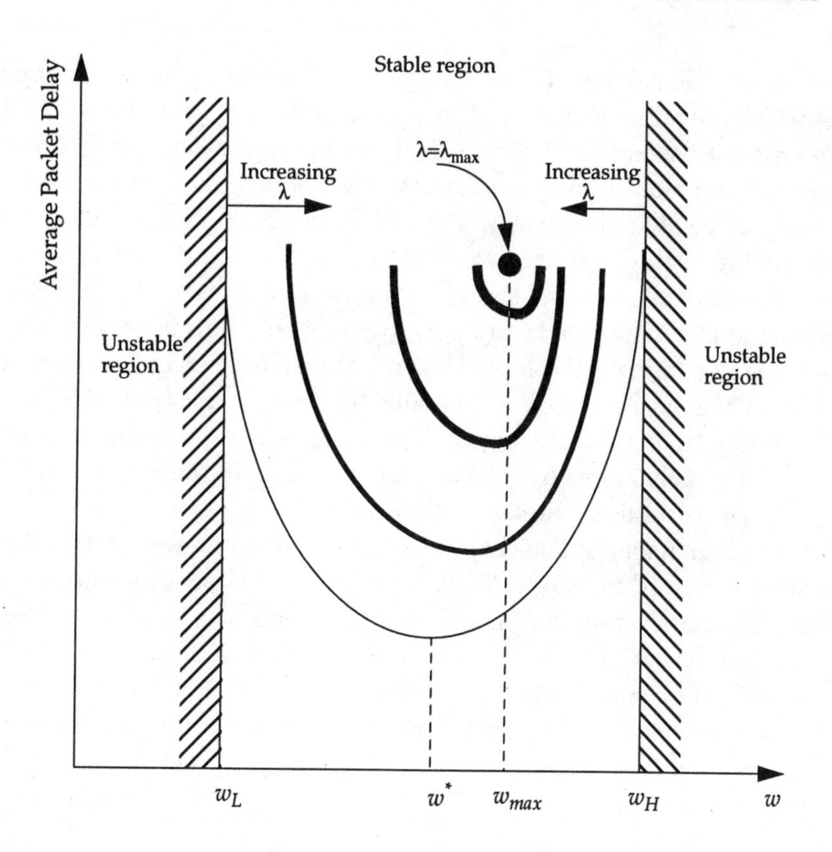

Figure 7.1 Delay behavior of a shared-channel, WDM, multihop network with changes in load λ and number of available wavelengths w.

respectively) that can be effectively used. Employing fewer than w_L or more than w_H channels leads to an unstable system. For fewer than w_L channels, the system does not have enough bandwidth to carry its offered traffic; however, with a larger number of channels (e.g., $w > w_H$), the network's logical connectivity gets sparser, so that each WDM channel may be carrying a lot of forwarded traffic and not leaving enough bandwidth for newly generated packets. Figure 7.1 also shows that, with increasing load, w_L increases and w_H decreases, i.e., they are collapsing toward each other, until, for a certain load (λ_{max}) there exists only one stable value of w (denoted by w_{max}). For the load λ_{max}, $w_L = w_H = w_{\text{max}} = w^*$. Characterization of this interesting behavior of shared-channel, multihop networks is one of the contributions of this chapter.

Current trends in networking applications indicate that there will be an increasing demand in future networks for multicasting – the capability of efficiently sending a stream of information from a single source to a number of destinations, m (for $1 \leq m \leq N$). (Of course, the case $m = 1$ is the trivial unicast communication case, and the case $m = N$ is the trivial broadcasting case.) Typical applications where multicasting is critical were outlined at the beginning of this chapter.

Interestingly, channel-sharing approaches are very well suited for multicasting applications in a local lightwave network. This is because our methods will entail sharing at both the transmitting and the receiving ends. The latter implies that many nodes will be able to simultaneously receive the same information from a single transmission, thereby increasing the *receiver throughput* (informally defined as the average rate of information received by nodes in the network) for multidestination traffic.

The remainder of this chapter is organized as follows. Section 7.2 provides a brief review of the literature on (1) multichannel, multihop optical networks, and (2) multicasting in WDM-based local optical networks. Section 7.3 describes a general multihop system, called Shared-Channel GEMNET, where the values of w and N are not very restrictive. Section 7.4 develops an analytical model to study the performance of multicast traffic in a generic, shared-channel, multihop system. Based on this analysis, the special case of unicast traffic is examined in Section 7.5 to demonstrate the interesting properties of the channel-sharing approach. Section 7.6 provides numerical results which prove that channel-sharing is beneficial in the context of multicast traffic.

7.2 Background

7.2.1 Channel-Sharing

Channel-sharing was introduced in Chapter 5, but some further elaboration and illustrations will be provided in this section.

The first extensive treatment of the multihop approach in WDM optical networks was provided in [Acam87]. This work considered the ShuffleNet logical topology, where each node in the network has P transmitter-receiver pairs. With such an arrangement, each node has a degree P, and networks of size $N = KP^K$ can be constructed with the number of channels $w = KP^{K+1}$, where K is the number of columns in the ShuffleNet. The idea of *channel-sharing* is achieved through the use of TDM; *all receivers in a common row of the ShuffleNet are assigned the same wavelength.* Figure 7.2 shows an eight-node network arranged as a logical ShuffleNet with $P = 2$. Part (a)

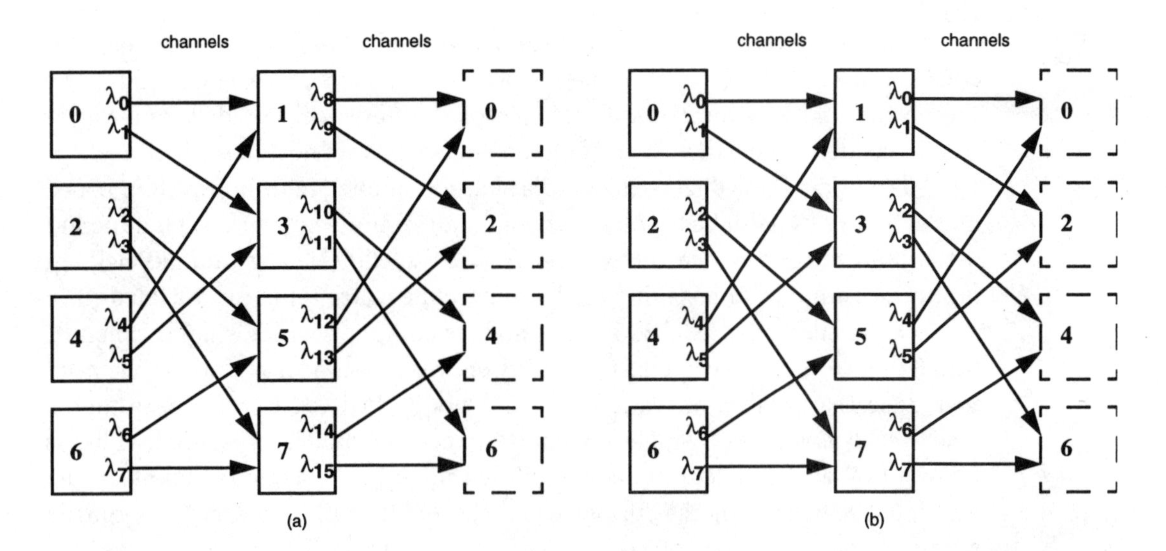

Figure 7.2 Logical assignment of wavelengths in an eight-node network arranged as a ShuffleNet with $K = 2$ and $P = 2$: (a) nonshared case where $w = 2N = 16$; (b) shared case where $w = 8 < 2N$.

of the figure uses $w = 16$, and each communicating transmitter-receiver pair has a unique wavelength. Part (b) considers the case where $w = 8$ (which is less than $2N$); notice that receivers in the same row are assigned the same wavelength. For example, both nodes 2 and 3 have receivers tuned to wavelength λ_1. Since transmitters at both nodes 0 and 1 are also tuned to the same wavelength, λ_1, these transmitters must take turns (e.g., in TDM fashion) to use this channel. Any packet of information that these two transmitters put onto the channel will be "heard" by both nodes 2 and 3. The limitation of this strategy is that each node requires two transmitter-receiver pairs even in the shared case. The approach that this chapter will present requires a single transmitter-receiver pair per node. The work in [Acam87] presents an analysis of the queueing delay which a packet experiences as it threads its way through multiple hops in the network. This analysis, which uses an $M/M/1$ queueing model at each node, while applicable for the non-shared case (when $w = 2N$), cannot be extended to the shared-channel case because of the discrete packet sizes used by the TDM sharing strategy.

Follow-up work in [ZhAc90] used a $Geom^{[x]}/D/1$ queueing system to represent the delay encountered at each node in the ShuffleNet. This model is more accurate when fixed-length packets are considered.

A shared-channel, shuffle-exchange-based multihop lightwave system was considered in [HlKa91] where the efficiency of the system was derived to be the inverse of the average hop distance of the network. No delay measures were considered and the effect of channel-sharing was not quantified.

The work in [Dowd92] considered a shared-channel, multihop system based on the binary hypercube topology where, in an N-node network, each node has n transmitters and n receivers where ($N = 2^n$). All of the logical links belonging to a common "dimension" of the hypercube share a unique wavelength.

In a recent study [KoGB95], each node is equipped with a single transmitter and one or more receivers. Our study shares the same motivation: the need to reduce the number of transceivers per node. In [KoGB95], a base model where each node has a single transmitter-receiver pair is presented. This is followed by an extended model where each node has a single transmitter and multiple receivers. In the former case, the number of wavelengths required is less than the number of nodes, N, in the network; for the latter case, the number of wavelengths required is equal to N. A "power" metric, defined as the ratio of the network's throughput to its delay, is employed in comparing the performance of systems with various degrees of sharing. However, the delay value that is employed in this performance metric considers only propagation delays but not the effect of frame synchronization and queueing; thus the power metric is valid only under light load.[3]

A graph-theoretic model of channel-sharing in multihop systems by introduced in [ToDV95] through the notion of *receiving graphs.* The analysis considered throughput performance and delay performance. While the delay value considered synchronization, this work did not quantify the effects of queueing.

7.2.2 Multicasting

Multicasting is also referred to as point-to-multipoint or simply multipoint communication. Multicasting has been studied extensively by both the computer networking and the telecommunication research communities. Various studies have addressed a wide variety of topics ranging from multicast

[3] In [KoGB93], it has also been discussed that, in a completely different type of system – single-hop – increasing the number of channels could reduce the throughput performance. Factors arguing for a reduced number of channels in single-hop systems are the following: (1) an increased number of channels implies a larger tuning range, which increases the tuning latency, and (2) since fewer stations would be sharing each channel, the probability that a component would be required to tune for every packet transmission increases with an increase with in the number of channels.

switching in the Asynchronous Transfer Mode (ATM) and network-level multicast path finding and routing algorithms (e.g., see [Waxm88]). Multicasting has been studied in the context of single-hop optical networks [BoMu95b, BoMu95c, RoAm94]. In both [BoMu95b] and [RoAm94], it was shown that, for systems with tunable receivers, the *receiver throughput* [BoMu95b] – defined to be the number of packets that are delivered in a single-transmission slot to various destinations in a passive-star network – can be greater than the transmitter throughput. It was also shown in [RoAm94] that receiver throughput can be much larger than transmitter throughput even in systems with tunable transmitters provided $w < N$ channels are used.

A multihop WDM network specifically for supporting multicast communication is reported in [TrMe95] where an algorithm for optimal node-placement (equivalently, wavelength assignment) in chain and ring networks was presented. An analysis of the average performance of multicasting in chains and rings was also presented in [TrMe95]; however, the results are not general because (1) they cannot be easily extended to networks with a larger degree, and (2) the cost measure considered only the hop distance. The results in [TrMe95] indicate that good node-placement algorithms must consider the multicast destination set size. Our current work only considers a random traffic model; therefore, node-placement algorithms will not provide any added benefits. While only hop distance was considered in [TrMe95] as a delay figure, our current work provides a more comprehensive assessment of delay.

Our multihop model, while employing a number of common features in the previous approaches discussed above, differs in the following ways:

- We assume that each node is equipped with only *one fixed transmitter and one fixed receiver.* This is not only cost-effective, but can also provide a good basis for comparison with single-hop systems, many of which are based on a single transmitter-receiver pair per node.

- The number of wavelengths w is less than N. The only requirement is that N should be an integral multiple of w. Again, this is not binding. Without this condition, our model will still function, although there may be some wastage in channel utilization. Also, should more than N channels be available, as in previous studies, our analysis will still hold.

- We illustrate our channel-sharing approach using the generalized, shuffle-exchange-based multihop architecture called GEMNET (see Chapter 6). With this architecture, it is possible to consider less restrictive values of N. That is, N need not be equal to KP^K, as required in pure ShuffleNet [Acam87], or 2^n, as required in the hypercube [Dowd92].

Our analysis has the following features:

- Since channel-sharing is effected through TDM, fixed length packets allow the use of $M/D/1$ queueing model in our evaluation.
- Nonzero propagation delays are considered.
- The effect of channel-sharing on the mean total delay experienced by a packet in the network can be quantified.
- Our delay expressions are general and are not specific to any architecture.
- Finally, our work considers the full effects of multicast traffic, i.e., the increased traffic due to forwarded packets and the average hop distance, both of which are functions of the destination set size.

7.3 Shared-Channel Multihop GEMNET

This section illustrates our sharing approach and its effect on the nodal degree in an example multihop network, viz., GEMNET.

Recall that, in a (K, M, P) GEMNET, N nodes – each of degree P – are arranged in a cylinder of K columns and M nodes per column so that nodes in adjacent columns are arranged according to a generalization of the shuffle-exchange connectivity pattern using directed links (see Fig. 6.1). Without channel-sharing, the number of channels required in a (K, M, P) GEMNET equals $N \times P = K \times M \times P$.

When each node can have only one fixed transmitter-receiver pair and $w < N$, *the degree of the network takes on a new meaning.* Channel-sharing through TDM implies a *higher logical nodal degree* (viz., ability to reach more nodes directly), albeit with a *lower capacity* on each of the *logical links.* The nodal degree is redefined as a logical quantity P, where $P = N/w$. For analytical convenience, N is assumed to be an integral multiple of w to ensure equal and fair sharing by all logical links.

Thus, we define a (K, M, P)–SC_GEMNET (Shared-Channel GEMNET) as a GEMNET having K columns and M rows, where M is an integral multiple, n, of P (i.e., $M = nP$) and $K = N/M = N/(nP)$.

As examples, Fig. 7.3 shows several 12-node ($N = 12$) SC_GEMNETs with different values of w. Fig. 7.3(a) shows the trivial 12-channel case. Since each node has only one fixed transmitter and one fixed receiver, the only SC_GEMNET that can be constructed is a ring (i.e., $K = 12, M = 1$). For the other trivial case, viz., the one-channel case, which is not shown in the figure, the resulting topology is a fully connected network, with all 12 nodes time-sharing the bandwidth available on the single channel. Parts (b) through (e) of Fig. 7.3 show the other four cases which allow for "perfect sharing," viz., $w = 6, 4, 3, 2$. In each of these figures, both a logical topology and the corresponding TDM frames on the various channels are shown. With reference to the portions of the figures representing the TDM frames, the numbers in the "slots" indicate the transmitting nodes and the numbers in parentheses indicate the nodes which have their receivers tuned to the channel which the frame represents. Note that, for each value of w, multiple SC_GEMNETs may exist. For example, in part (c) for the $w = 4$ case, it is also possible to construct a $K = 4, M = 2$ SC_GEMNET.

In deriving the average hop distance $\bar{h}$ of the (K, M, P)-GEMNET, its diameter $D = \lceil \log_p(M) \rceil + K - 1$ is used. Given D, an upper bound for the average hop distance $\bar{h}_{\max}$ was derived in Chapter 6 to be

$$\bar{h}_{\max} = \frac{MK\left(D + \frac{1}{2} - \frac{K}{2}\right) - K\left(\frac{P^{D-K+1}-1}{P-1}\right)}{M \times K - 1} \tag{7.1}$$

In determining a lower bound for the average hop distance, we need to define L to be the last column in which P^i *new* nodes could be covered starting from any arbitrary node. With this definition and also defining $Z = \lfloor (D - K)/K \rfloor$, we have $L = D - K$ if $P^{(D-K)\bmod K} \times [(P^K)^{Z+1} - 1]/[P^K - 1] \leq M$; otherwise, we have $L = D - K - 1$. The average value of the minimum hop distance can now be defined as follows:

$$\bar{h}_{min} = \sum_{i=0}^{L} iP^i + \sum_{i=L+1}^{L+K} i\left(M - P^{i \bmod K} \sum_{j=0}^{\lfloor \frac{1}{K} \rfloor - 1} P^{Kj}\right) \tag{7.2}$$

It has been shown in Chapter 6 that, in general, two-column GEMNETs (i.e., $K = 2$) have the best performance in terms of the average hop distance. Therefore, in constructing logical topologies for the multihop WDM network, it is, in general, best to consider $K = 2$ wherever possible and use the next higher value of K as dictated by the requirement $K = N/M = N/(nP) =$

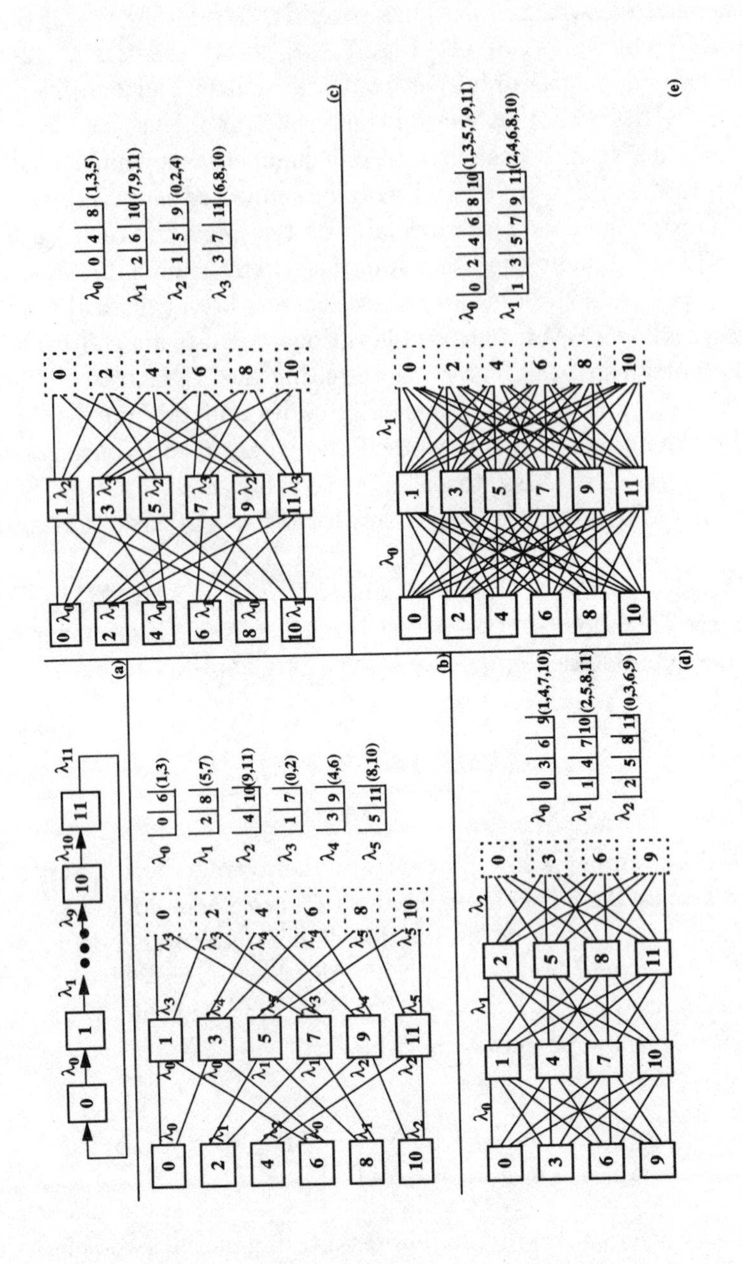

Figure 7.3 Twelve-node SC_GEMNETs, along with corresponding timing diagrams, for various values of w: (a) $w = 12$; (b) $w = 6$; (c) $w = 4$; (d) $w = 3$; and (e) $w = 2$. (Note: unless otherwise shown, all links are directed from left to right.)

w/n, where n is an integer which equals the number of channels required to provide connectivity between two successive columns of the SC_GEMNET.

It is interesting to consider what happens when the available number of channels, w, is not an exact divisor of N. Such a system can be treated in two ways: (1) $w' < w$ channels may be used such that N/w' nodes share a channel with all nodes getting an equal share of the bandwidth, and $w - w'$ channels are wasted; (2) alternatively, all of the available channels may be used by employing different TDM frame lengths on different channels, i.e., with unequal sharing. For example, consider a system with $N = 100$. If $w = 50$, each channel is shared by two nodes, and the TDM frame length in each channel is two slots – one slot allocated to each of the two nodes that transmit on that channel. If $w = 51$, there are two possibilities. First, only $w' = 50$ channels may be used, and the fifty-first channel may be wasted resulting in a loss of efficiency of $1/51 \approx 2\%$. Alternatively, 98 of the nodes may share 49 channels – 2 nodes per channel, and the last two nodes may use the remaining two channels exclusively. The unfairness problem becomes more pronounced if $w = 99$. The first solution of using only 50 channels is quite wasteful in this case resulting in a loss of efficiency of approximately 50%. On the other hand, 98 nodes may be allowed exclusive use of one channel, and the remaining two nodes may share the remaining channel. For the remainder of this chapter, we will try to consider only parameters which lead to perfect sharing of channels. For cases where this is not possible, we will approximate the number of channels in the system to the closest integer value which leads to perfect sharing.

The SC_GEMNET lends itself to multicasting applications, because sharing at the receiving ends reduces the total number of transmissions required for multicasting. The next section develops an analytical technique to study the delay performance of multicast traffic in general, shared-channel optical networks.

7.4 Performance Evaluation

Let us now derive an expression for the average delay encountered by a packet before it is delivered to all of its destinations. We assume that communication is time-slotted with a slot being equal to the transmission time of a packet. For example, if a 1-Gbps channel is used with a 53-byte packet (or cell) as defined in the ATM standard, then the slot duration is 0.424 μs. All other delays in the system are normalized to this slot duration. It is assumed that intermediate nodes are capable of examining the headers of the packets they receive to determine whether the local node is a destination. This examination will also indicate to the local node whether a packet needs to be forwarded. Note that a packet that is forwarded on a single channel may be received by multiple nodes; one or more of these nodes may "consume" the packet and/or forward the packet on its outgoing channel. The following notations are used in the derivation:

- N: Number of nodes in the system.

- w: Number of channels in the system. w is chosen such that N/w is an integer. This ensures that all nodes have the same amount of bandwidth allocated to them.

- R: Round-trip propagation delay between a node and the passive star. We assume that all nodes are equidistant from the passive star. If needed, delay lines can be installed so that R is the same for all nodes.

- F: The frame length is equal to N/w. Since TDM is used on each of the channels and N/w is the number of nodes which share a single channel, a frame comprises F slots in which different nodes get to transmit.

- m: Multicast destination set size. All packets in the network are multicast packets, and each one of them has a destination set size specified by a constant[4] m. Also, the arrival rate of fresh packets to each node is the same and the m destinations are randomly selected from the N nodes.

- $\bar{h}_m$: The average number of transmissions required to deliver information to the m destinations of a multicast packet. This term, derived later in this section, is required to estimate the total amount of traffic (both fresh and forwarded) that the network will be required to handle.

- $\bar{g}_m$: The average number of hops separating the source from each one of the destinations of a multicast packet. This term is needed to quantify the average packet delay in the network, and is described in greater detail later in this section.

[4]We could also allow the destination set size to be governed by one of the following distribution: geometric with parameter m and truncated at $N-1$, or uniform between 1 and $N-1$.

- λ: The average rate of arrival of fresh packets to a node per slot. This implies that the average arrival rate of packets to the entire network in one slot is $N\lambda$. The arrival process is Poisson, and is independently and identically distributed for all network nodes. Furthermore, all packets are equally likely to be destined to any one of the network nodes. Both fresh and forwarded packets arriving at a node are maintained in an infinite buffer before being processed.

- λ': The average rate of packet arrivals per node per frame, including both fresh packets and forwarded packets. This composite arrival rate, λ', may be derived as follows. In any one frame, i.e., F slots, the total number of fresh packets that arrive to the network is $\lambda NF = \lambda N(N/w)$. Each of these packets is transmitted an average of $\bar{h}_m$ times within the network. Therefore, under steady-state conditions, there will be $\lambda N(N/w)\bar{h}_m$ packets transmitted in the network in each frame. Note that a tagged node, say node A, will only observe $\lambda N(N/w)\bar{h}_m(1/w)$ packets since its receiver is only tuned to one of the w channels being used. Furthermore, since there are N/w nodes listening to the same channel, the packet will actually be received by only one of these nodes. That is, under the symmetric assumption that a packet being transmitted on any wavelength is equally likely to be picked up by any one of the nodes with its receiver tuned to that particular wavelength, the total arrival rate to any arbitrary node in the network is given by

$$\lambda' = \frac{\lambda N\left(\frac{N}{w}\right)\bar{h}_m\left(\frac{1}{w}\right)}{\left(\frac{N}{w}\right)} = \lambda N\bar{h}_m\left(\frac{1}{w}\right) \tag{7.3}$$

With the above notation, we can now define an expression for the average delay encountered by a packet before it is received by all of its destinations. Each copy of a packet encounters the following delay components before it is delivered to its destination: (1) queueing delay at the source node and possibly at one or more intermediate nodes; (2) transmission delay along each hop; and (3) propagation delay on each link that the packet traverses.

The queueing delay component is composed of two subcomponents: (1) half-frame-length delay required by a packet to synchronize with the appropriate outgoing slot in the frame belonging to the outgoing channel on which the node transmits (this is given by $0.5F$); and (2) the actual $M/D/1$ queueing delay that the packet experiences at a node after the synchronization. The latter is a function of λ' and is given by

$$Q = \frac{F\lambda'}{2(1-\lambda')} \tag{7.4}$$

The second delay component is given by $\bar{g}_m$ since there will be a one-unit transmission delay on each hop along the packet's route. The third component is a product of $\bar{g}_m$ and R. Therefore, the total delay that a packet experiences is given by

$$T = \bar{g}_m \left[\frac{F}{2} + \frac{F\lambda'}{2(1-\lambda')} + 1 + R\right] \tag{7.5}$$

where $\bar{g}_m$ is the average number of hops separating a source from its m destinations.

Recall from Eqn. (7.3) that λ' in Eqn. (7.5) is a function of $\bar{h}_m$. We now need to derive expressions for $\bar{h}_m$ and $\bar{g}_m$ in order to be able to obtain numerical instances of Eqn. (7.5). For obtaining values of $\bar{h}_m$, we use a modified version of the lower bound for the cost of multicasting in an N-node network with degree P [TrBM96]. This bound is similar to the Moore bound in that it assumes an ideal, (P, K) Moore graph, where P is the degree and K is the diameter, and it specifies the expected number of transmissions required to transmit information to m randomly chosen nodes in such a graph. The only restriction of this bound is that the number of nodes $N = N_K$, and N_K is given by

$$N_K = \frac{P^{K+1}-1}{P-1} \tag{7.6}$$

Equation (7.6), with k substituted for K, also specifies the number of nodes N_K, in a subtree rooted at the kth level of the complete, ideal Moore graph of degree P. Figure 7.4 depicts a $P = 3, k = 2$ tree with 13 nodes.

Figure 7.5 illustrates the differences between $\bar{h}_m$ and $\bar{g}_m$, and also provides an example of the effects of sharing on multicast traffic in a $(3, 2)$ complete, Moore graph. The source node is node 1, and the destination nodes are nodes 2, 3, 6, and 7. In the nonshared case, the system requires 12 channels, whereas the shared-channel case requires only 4 channels. We must note here that these two numbers cannot be carried over directly to any realizable network such as the SC_GEMNET, because this figure is only illustrating the receiver graph model, has only one source that can reach all destinations, and does not indicate the effect of sharing at the transmitting ends. The average

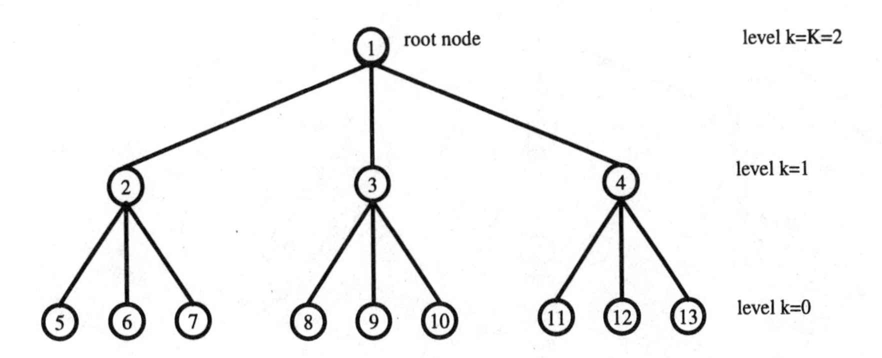

Figure 7.4 A (3,2) complete Moore graph.

hop distance to the destinations is the same in both cases and is equal to $(1+1+2+2)/4 = 1.5$ (since nodes 2 and 3 are one hop away from the source, and nodes 6 and 7 are two hops away from the source). The average number of transmissions in the nonshared case is four since four channels are used to reach the destinations. In the channel-sharing approach, only two transmissions – one each on two channels – are required.

For the nonshared case, the bound $\bar{h}_m$ is given by a recursive expression in terms of $h_{m,k}$ – the expected number of transmissions required to reach m destinations in a k-level tree. In order to express $h_{m,k}$ succinctly, we introduce the following notation:

$$f(x) = \begin{cases} 0, & \text{when } x = 0 \\ 1, & \text{otherwise} \end{cases} \tag{7.7}$$

Let the vector $\vec{L} = [l_1, l_2, \ldots, l_P]$ represent the number of destinations in each of the P subtrees below a node at level k. That is, l_i is the number of destinations in the ith subtree rooted at level $k-1$, and the notation

$$\sum_{1 \le x \le P, (\vec{L} \in X)} l_x = \mu \tag{7.8}$$

refers to an instance of $\vec{L}$ such that the P components of the vector $\vec{L}$ sum to μ, for $0 \le l_x \le \mu, 1 \le x \le P$, and $\mu \ge 0$. Using the indicator function defined in Eqn. (7.7) and the notation specified in Eqn. (7.8), we obtain

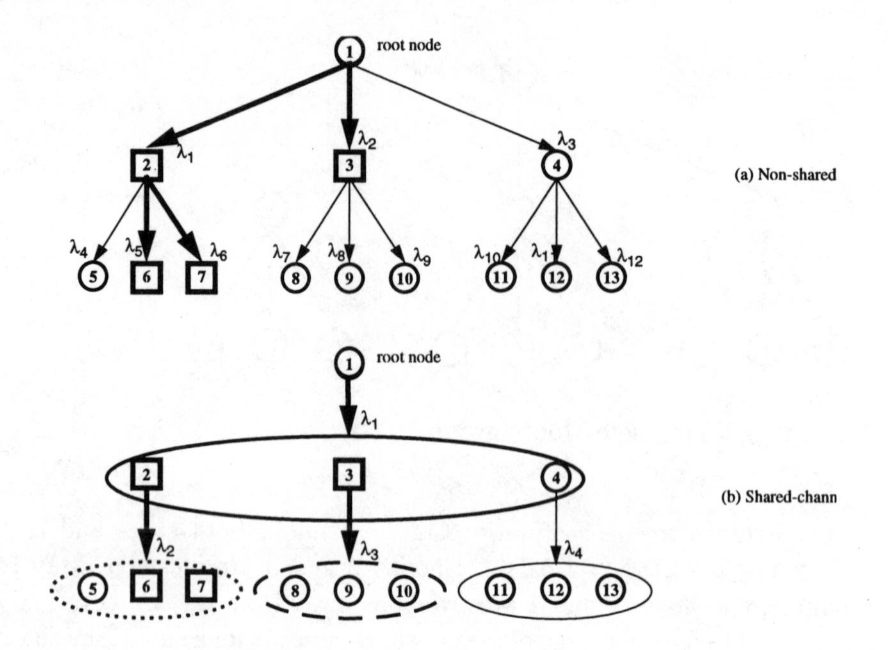

Figure 7.5 Effects of sharing on multicasting in a (3,2) complete, Moore graph. The source node is node 1 and the destination nodes are nodes 2, 3, 6, and 7: (a) nonshared case; (b) shared-channel case.

$$
\begin{aligned}
E(C_1) &= \frac{m}{N_k} \sum_{\sum_{1\le x\le P, (\vec{L}\in X)} l_x = m-1} \left[\binom{m-1}{l_1, l_2, \ldots, l_P} \right. \\
&\left. \times \prod_{i=1}^{P} \prod_{j=0}^{l_i-1} \frac{N_k - j}{N_{k-1} - 1 - \left(i_P + \sum_{k=1}^{P-1} l_k\right)} \left(\sum_{i=1}^{P} f(l_i) + h_{l_i,k-1} \right) \right] \qquad (7.9)
\end{aligned}
$$

which is the cost of transmitting information to nodes in a subtree rooted at the kth level if the root happens to be one of the destinations, and

$$
\begin{aligned}
E(C_2) &= \left(1 - \frac{m}{N_k}\right) \sum_{\sum_{1\le x\le P, (\vec{L}\in X)} l_x = m} \left[\binom{m}{l_1, l_2, \ldots, l_P} \right. \\
&\left. \times \prod_{i=1}^{P} \prod_{j=0}^{l_i-1} \frac{N_k - j}{N_{k-1} - 1 - \left(i_P + \sum_{k=1}^{P-1} l_k\right)} \left(\sum_{i=1}^{P} f(l_i) + h_{l_i,k-1} \right) \right] \qquad (7.10)
\end{aligned}
$$

which is the cost of transmitting information to m nodes in a subtree rooted at the kth level if the root is not one of the destinations. In Eqn. (7.9), we are essentially taking the weighted average of the costs of transmitting to $m-1$ destinations [m destinations in the case of Eqn. (7.10)] over all possible ways in which the destination set can be selected from the $N_k - 1$ nodes below the root node in the subtree rooted at the kth level.

We can now express $h_{m,k}$ in terms of Eqns. (7.7) through (7.10) as

$$h_{m,k} = E(C_1) + E(C_2) \tag{7.11}$$

for $1 \leq k \leq K$, where the base cases for the recursion are $h_{1,0} = 0$ and $h_{0,k} = 0$ for $1 \leq k \leq K$, and we recall that by definition

$$\bar{h}_m = h_{m,K} \tag{7.12}$$

When channel-sharing is employed, as depicted in Fig. 7.5(b), a transmission by a node rooted at the kth level will be heard by all the nodes immediately below this node. The fact that fewer transmissions are required due to channel-sharing is expressed by slightly modifying the cost terms in Eqns. (7.9) and (7.10), giving us

$$\begin{aligned} E(C_1) &= \left(\frac{m}{N_k}\right) \sum_{\substack{\sum_{1 \leq x \leq P, (\vec{L} \in X)} l_x = m-1}} \left[\binom{m-1}{l_1, l_2, \ldots, l_P} \right. \\ &\left. \times \prod_{i=1}^{P} \prod_{j=0}^{l_i - 1} \frac{N_k - j}{N_{k-1} - 1 - \left(i_P + \sum_{k=1}^{P-1} l_k\right)} \left(\sum_{i=1}^{P} f(m-1) + h_{l_i, k-1} \right) \right] \end{aligned} \tag{7.13}$$

and

$$\begin{aligned} E(C_2) &= \left(1 - \frac{m}{N_k}\right) \sum_{\substack{\sum_{1 \leq x \leq P, (\vec{L} \in X)} l_x = m}} \left[\binom{m}{l_1, l_2, \ldots, l_P} \right. \\ &\left. \times \prod_{i=1}^{P} \prod_{j=0}^{l_i - 1} \frac{N_k - j}{N_{k-1} - 1 - \left(i_P + \sum_{k=1}^{P-1} l_k\right)} \left(\sum_{i=1}^{P} f(m) + h_{l_i, k-1} \right) \right] \end{aligned} \tag{7.14}$$

respectively. The expected number of transmissions is, therefore, given by Eqns. (7.11) and (7.12); the terms $E(C_1)$and $E(C_2)$ in Eqn. (7.11) are specified by Eqns. (7.13) and (7.14), respectively.

For both the nonshared and the shared-channel cases, the average number of hops, $\hat{g}_m$, separating a source from its m destinations is given by a similar expression, in terms of $g_{m,k}$ as follows

$$g_{m,k} = E(C_3) + E(C_4) \tag{7.15}$$

where

$$E(C_3) = \left(\frac{1}{m}\right)\left(\frac{m}{N_k}\right) \sum_{\sum_{1\leq x\leq P,(\vec{L}\in X)} l_x = m-1} \left[\binom{m-1}{l_1, l_2, \ldots, l_P}\right.$$
$$\left. \times \prod_{i=1}^{P} \prod_{j=0}^{l_i-1} \frac{N_k - j}{N_{k-1} - 1 - \left(i_P + \sum_{k=1}^{P-1} l_k\right)} \left(\sum_{i=1}^{P} l_i + g_{l_i,h-1}\right)\right] \tag{7.16}$$

$$E(C_4) = \left(\frac{1}{m}\right)\left(1 - \frac{m}{N_k}\right) \sum_{\sum_{1\leq x\leq P,(\vec{L}\in X)} l_x = m} \left[\binom{m}{l_1, l_2, \ldots, l_P}\right.$$
$$\left. \times \prod_{i=1}^{P} \prod_{j=0}^{l_i-1} \frac{N_k - j}{N_{k-1} - 1 - \left(i_P + \sum_{k=1}^{P-1} l_k\right)} \left(\sum_{i=1}^{P} l_i + g_{l_i,k-1}\right)\right] \tag{7.17}$$

and

$$\bar{g}_m = g_{m,K} \tag{7.18}$$

The base cases for the recursion in Eqn. (7.15) are $g_{1,0} = 0$ and $g_{0,k} = 0$ for $1 \leq k \leq K$. Note that, since $\bar{g}_m$ simply measures the expected distance to the destinations, it is the same for both the shared-channel and the nonshared cases.

Equation (7.5) may be rewritten, in terms of λ, w, R, N, and m, by using the expressions for $\bar{h}_m$ and $\bar{g}_m$ given by Eqns. (7.11) and (7.15), respectively, as follows

$$T(N, \lambda, w, R, m) = \bar{g}_m \left[\frac{N}{2w} + \frac{N^2(\lambda \bar{h}_m)}{2w(w - \lambda N \bar{h}_m)} + 1 + R\right] \tag{7.19}$$

By inspection, Eqn. (7.19) clearly indicates that, for reasonable values of N, the total average packet delay will increase with increasing R and increasing λ. However, it is not very apparent what happens as the number of channels, w, is increased because two effects may be noticed. First, the hop distance goes up because the degree of each node is inversely proportional to w. Second, both the frame synchronization and queueing delays decrease because the capacity available to each node on its outgoing link increases as a result of reduced sharing.

Note that, in the point-to-point case, that is, when $m = 1$, the average number of transmissions and the average hop distance is the same, that is, $\bar{g}_1 = \bar{h}_1$. Under this condition, $\bar{h}_1$ is simply given by the Moore bound and may be written as

$$\bar{h}_1 = \frac{1}{N}\left[(K+1)\left(N - \frac{P^{K+1}-1}{P-1}\right) + \sum_{i=1}^{K} iP^i\right] \qquad (7.20)$$

where $K = \lfloor \log_P\{[N(P-1)/P] + 1\} - 1 \rfloor$. The work in [TrBM96] has numerically verified that Eqns. (7.11) and (7.15) do indeed collapse to Eqn. (7.20) for $m = 1$. If the more specific SC_GEMNET is employed, then the upper and lower bounds, specified by Eqns. (7.1) and (7.2), respectively, may be used in the place of $\bar{h}_1$.

We also note that, when the nodal degree $P = 1$, i.e., when N dedicated (nonshared) channels are used, leading to a ring topology, $\bar{h}_m$ can be obtained through an alternate method based on order statistics. Such an expression was derived for this restrictive case in [TrMe95], which gives

$$\bar{h}_m = 1 - \left(\frac{1}{m+1}\sum_{i=1}^{K}\frac{1}{i}\right) \qquad (7.21)$$

Again, the general equations for $\bar{h}_m$, Eqns. (7.11) through (7.14), collapse to the limiting case given by Eqn. (7.21).

We note again that the expressions derived in this section are general in that they are not restricted to any specific topology, and, therefore, provide bounds for any network topology. We also note that the expressions provide exact values only for cases where $N = N_k$.

7.5 Illustrative Examples: Unicast Traffic

The above analysis for multicast traffic can be specialized for the unicast-traffic case by substituting the average hop distance $\bar{h}$ in place of the corresponding quantity $\bar{g}_m$ in Eqn. 7.5 for multicast traffic (see [TrMu95]). This unicast-traffic analysis brings out some interesting properties of channel-sharing which will be highlighted in this section. Specifically, we will present our results in this section in three parts. First, via a 12-node example, we study the behavior of the various delay components with changes in w. Then, we consider the same 12-node network under various load conditions. Finally, we investigate larger networks to predict how channel sharing may perform in a more realistic scenario.

7.5.1 Twelve-Node Example: Various Delay Components

Figure 7.6 shows the various delay components encountered by a typical packet as the number of wavelengths is varied in a 12-node network where the round-trip propagation delay $R = 2$ and the load $\lambda = 0.05$ packets per slot per node. For these parameters, if ATM cells (53 bytes long) are used with each channel operating at 100 Mbps, $R = 2(4.24) = 8.48$ μs, which corresponds to a node-to-star distance of a little less than a kilometer, and the load offered by a node is 5 Mbps leading to a network-wide loading of 60 Mbps.

Note that, although continuous curves are shown in the figure, only the points where $N/w = \lfloor N/w \rfloor$ (or $w/N = \lfloor w/N \rfloor$) are meaningful because of our desire to employ a fair scheme where bandwidth is equally divided among all nodes. These valid points are highlighted on the curves. At other points, that is, when $N/w \neq \lfloor N/w \rfloor$, we have assumed that the bandwidth is equally shared by the nodes only for the sake of analysis.

We first consider the case $w < N$ and make the following observations.

- We notice that, when the number of wavelengths w is very small (one or two), the queueing delays are very large because all 12 nodes have to share the channels.

- For small values of w, the TDM frames are so long that the frame synchronization delay is also significant, and we can expect this frame synchronization delay to become very large for larger values of N. For instance, when $w = 1$, all 12 nodes are sharing a single channel (the TDM case), and the frame synchronization delay is, on average, six slots; the corresponding propagation delay is only two since all packets only travel across one hop.

- When the number of channels is equal to the number of nodes, each node has only one outgoing link leading to a logical ring network, and the full capacity of a link is available to each node; however, at this point, the mean hop-distance of the network, which in this case is equal to 5.5 leading to an average propagation delay of 11 slots, tends to increase the average delay.

- Somewhere between these two extremes, the total average delay has a minimum value. Specifically, for the example network in Fig. 7.6, the total average delay which is the sum of the synchronization, queueing, transmission, and propagation delays, is minimized when $w = 4$.

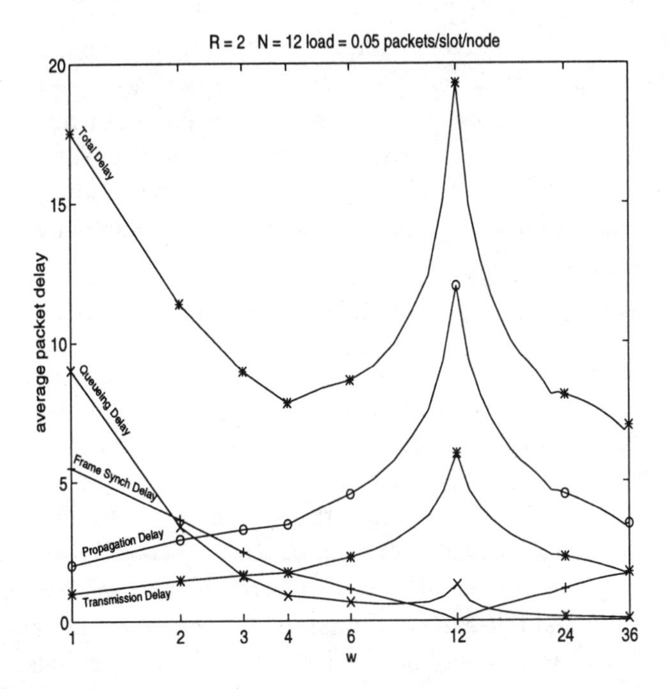

Figure 7.6 Various delay components vs. number of wavelengths in a 12-node network with $R = 2$ and $\lambda = 0.05$.

In Fig. 7.6, we have also plotted the delay for the cases $N < w < 3N$. In the figure, there are only two such points which are realizable in a fair manner: $w = 24$ and $w = 36$. We note, again, that these two cases are more expensive and require that each node have two and three transmitter-receiver pairs, respectively. When we also consider the $w > N$ case, the following can be observed.

- For the value of load we are considering, we notice that the delay for the 24-channel case ($w = N > 2N$) is higher than that for the four-channel ($w < N$) case. The reason is that the four-channel case has a degree of three, while the 24-channel case has a degree of two albeit with a higher capacity.

- Increasing w further to 36 gives us the case where the degree is three with a larger capacity on each link in the virtual topology than in the four-channel case. At this point, however, the delay is only slightly less than the four-channel case. This decrease is due to the smaller queueing and synchronization delays, which result from having dedicated channels.

Thus, it appears that, under certain load conditions, having $w < N$ channels with a single transmitter-receiver pair at each node is preferable to having $w > N$ channels with more transmitter-receiver pairs per node. We must caution the reader, though, that under heavier load conditions, the 36-channel system is likely to perform better. For the rest of this section, we will only study systems which have a single transmitter-receiver pair per node.

7.5.2 Twelve-Node Example: Delay Behavior with Variation in λ

An even more interesting result, as was promised in Fig. 7.1, can be observed in Fig. 7.7, wherein we depict the total delay for various loads in the same 12-node network. Here again, only those points where $N/w = \lfloor N/w \rfloor$ are meaningful, and these points are marked in the figure. For instance, at a load of $\lambda = 0.25$ packets per slot per node, the system is unstable at all values of where equal sharing is possible (i.e., $w = 1, 2, 3, 4, 6, 12$) with each node equipped with only one transmitter and one receiver. For $\lambda = 0.25$, it is possible to maintain a stable system if nine (or ten) channels are available. In this case, six of the channels may be shared equally by the 12 nodes such that they each have access to a capacity equivalent to half a channel. The other channels may also be distributed equally to the various nodes; however, each node would then require an additional transmitter-receiver pair to utilize the additional available capacity.

When w is very small, the system becomes unstable. We note that, as the load is increased, the shape of the curve becomes a more pronounced trough. For example, if only one wavelength is to be used, we have the trivial TDM case where all 12 nodes have to share the wavelength. In this case, the capacity available to each node is limited to $1/12 = 0.0833$: that is, loads

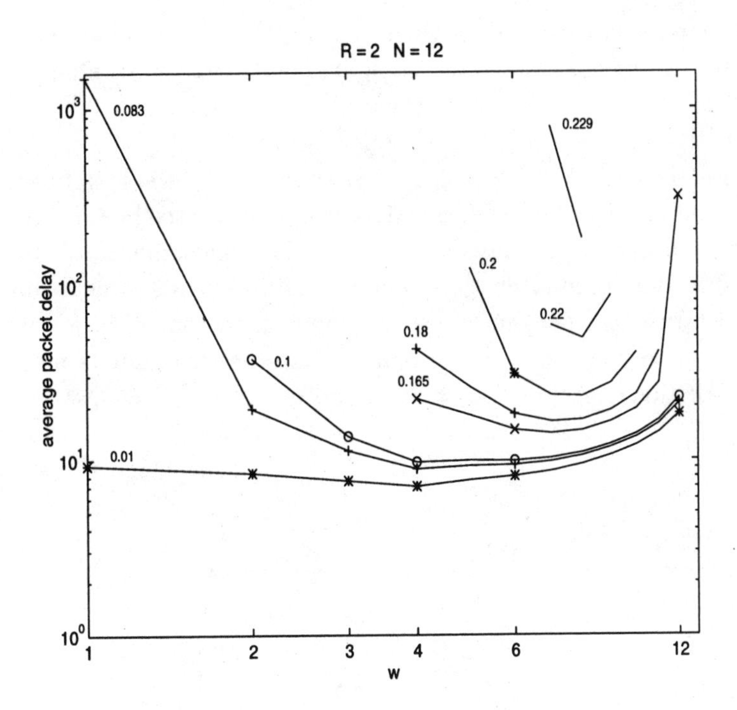

Figure 7.7 Average delay vs. number of wavelengths in a 12-node network with $R = 2$ and various load values. Marked points indicate values of admitting equal sharing of the available bandwidth among the 12 nodes when each node has a single transmitter-receiver pair.

higher than 0.0833 packets per slot per node leads to queueing instability.[5] On the other hand, when 12 channels are used, the average hop-distance is 5.5 and $\lambda' \approx 5.5\lambda$, implying that the system becomes unstable when subjected to a load $\lambda > 1/5.5 \approx 0.182$ packets per slot per node (corresponding to a network-wide throughput of 218.4 Mbps if the channel rate is 100 Mbps). The total delay is also affected by the propagation delay at this value of w. For example, for $R = 1$, the average propagation delay experienced by a packet is 5.5 slots and the number of slots expended for transmitting the packet over 5.5 hops is also 5.5 slots.

Figure 7.8 also shows the optimal number of wavelengths, w^*, that minimizes the total delay for various values of λ and three values of R (1, 10,

[5] We remind the reader that, although $\lambda = 0.0833$ packets per slot per node appears to be a small load, it corresponds to a network-wide throughput of 99.96 Mbps if the channel rate is 100 Mbps.

100). The curves corresponding to $R = 1$ and $R = 0$ are identical; therefore only the $R = 1$ curve is shown in the figure. Note that increasing R indicates a smaller w^*. For any given load, there are lower and upper bounds, w_L and w_H, respectively, on the range of w which admits stability. These bounds are depicted for the case $N = 12$ in Fig. 7.8. From the figure, it is observed that as the load increases, w_L needs to be larger to counter the instability resulting from increased queueing delays at each node. At the same time, w_H needs to decrease in order to counter the queueing instability resulting from a larger number of forwarded packets. The maximum load that such a network can support, $\lambda_{\max}$, corresponds to the point where the two bounds meet, that is, at $w_{\max}$. For this example $\lambda_{\max} = 0.23$ and $w_{\max} = N/2 = 6$.

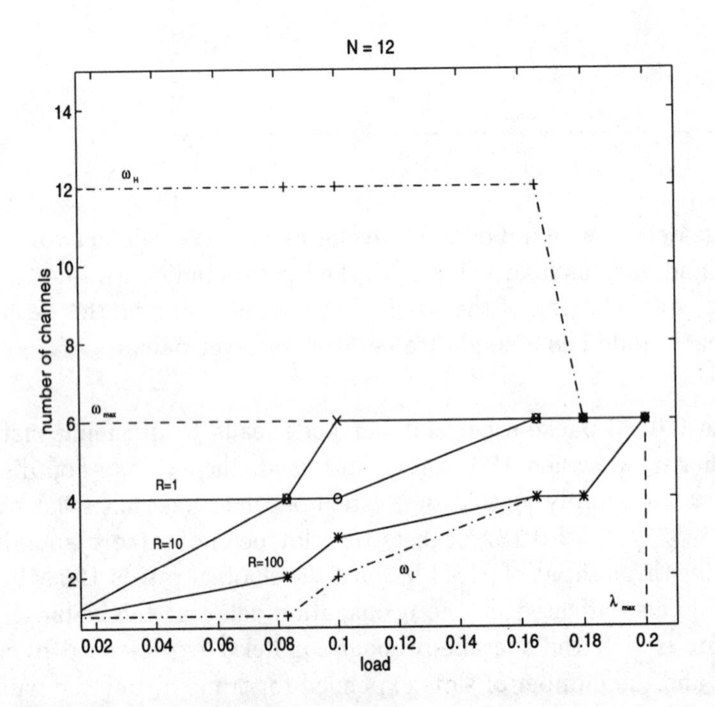

Figure 7.8 Upper (w_H) and lower (w_L) bounds on the number of channels admitting stability and optimal number of channels (w^*) vs. the load in a 12-node network. Note that the curves for $R = 1$ and $R = 0$ are identical; therefore only one of them is shown.

7.5.3 Delay Behavior in Larger Networks

This subsection discusses the results for $N = 120$. These results are shown in Figs. 7.9 through 7.11, and in Table 7.1. Figure 7.9 shows the average packet delay as a function of w for various values of load. It is clear that increasing λ or R increases the delay. When $\lambda = 0.05$ packets per slot per node, having $w \leq 10$ results in an unstable system since there just isn't enough capacity to satisfy the load. For all cases, there are one or more optimum values of w that result in the lowest total delay. As predicted in Fig. 7.1, almost all the curves in Fig. 7.10 show that the delay initially decreases as the number of wavelengths increases; after the critical value, w^*, the delay starts increasing again. Note that, for large values of R and small λ, the curves provide a range of possible values for w^*. Particularly for the case $\lambda = 0.0001$ and $R = 100$, we see that there is a range of values of w for which the delay, T, does not change appreciably. We would obviously choose w^* from the lower end of this range.

In Fig. 7.11, we depict the upper (w_H) and lower (w_L) bounds on the range of w, which can support various loads; we also depict the variation of w^* with λ and R. We would be more interested, in general, in the lower bound since we would like to use as few channels as possible; furthermore, the upper bound may not be achievable due to device technology limitations. For a 120-node network, a load of 0.08 packets per slot per node (or a network-wide throughput of 960 Mbps at channel rates of 100 Mbps) can be achieved using 30 channels. Using more than 60 channels leads to queueing instability due to forwarded traffic. We also note that w^* is more sensitive to R when λ is small. At higher values of λ, all of the curves in the figure converge to $w = N/2$, and R does not significantly affect w^*.

Table 7.1 provides values of w^* for various values of load. The table shows that, at least for low load conditions, the delay performance can be improved if w is reduced with increasing R. For large R ($R = 100$), every hop implies a delay of 100 slots, so minimizing the number of hops is critical; this can be achieved by sharing fewer channels.

Finally, in Table 7.2, we provide values of w^*, w_L, w_H, and $w_{\max}$ for various values of N and λ. The trends seem to indicate that w^* increases logarithmically with N for small values of load; for larger values of load, we see that $w^* = w_{\max} = N/2$ seems to provide the best choice. The latter is consistent with the results reported in [KoGB93], where it was shown that using $w = N/2$ maximizes the throughput. Specifically, for $N = 1200$, and $\lambda = 0.01$, we see that we can obtain a stable system for w as low as 12. For a

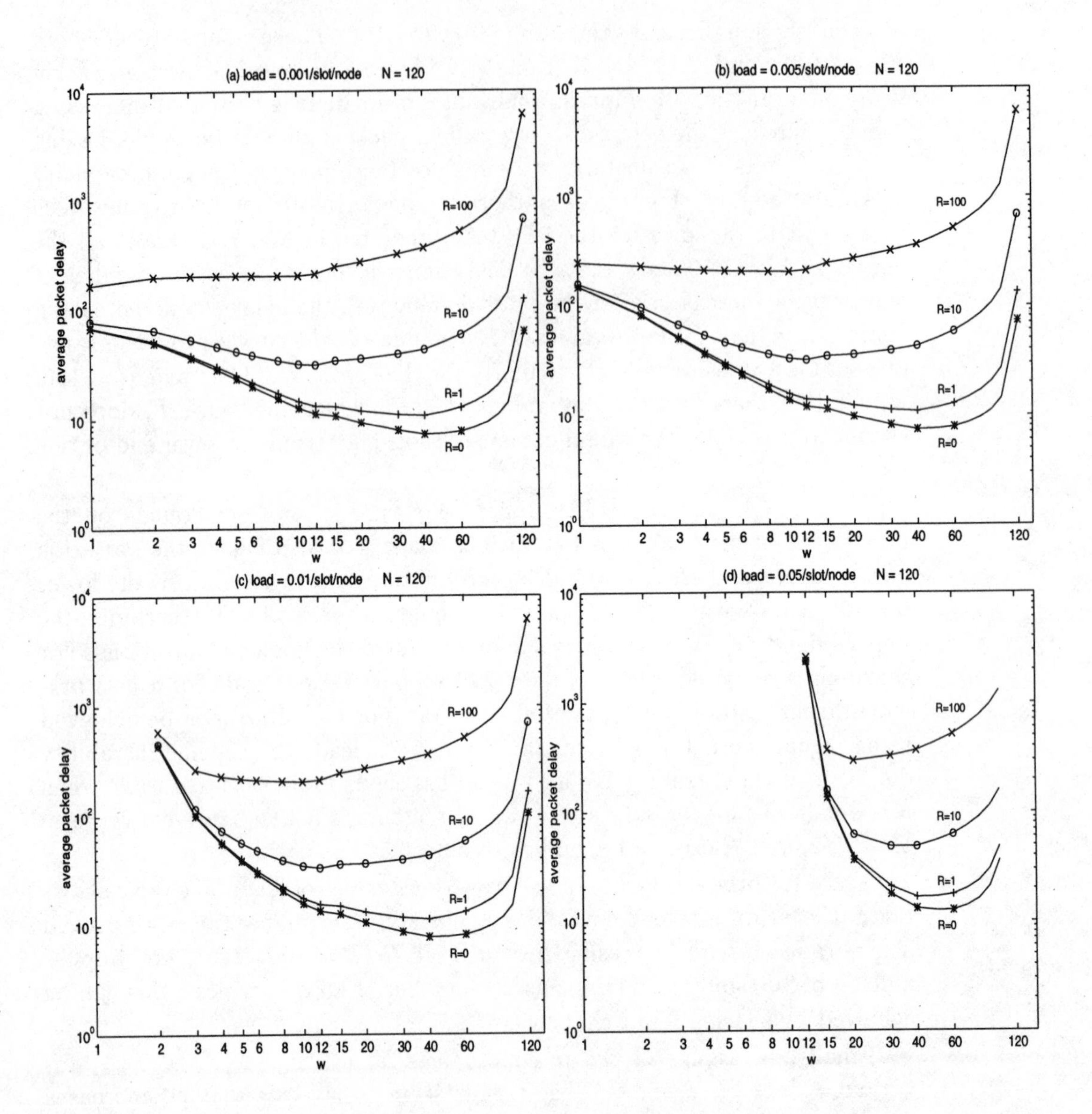

Figure 7.9 Average delay, T, vs. number of wavelengths, w, for various values of propagation delay in a 120-node network: (a) $\lambda = 0.001$; (b) $\lambda = 0.005$; (c) $\lambda = 0.01$; (d) $\lambda = 0.05$.

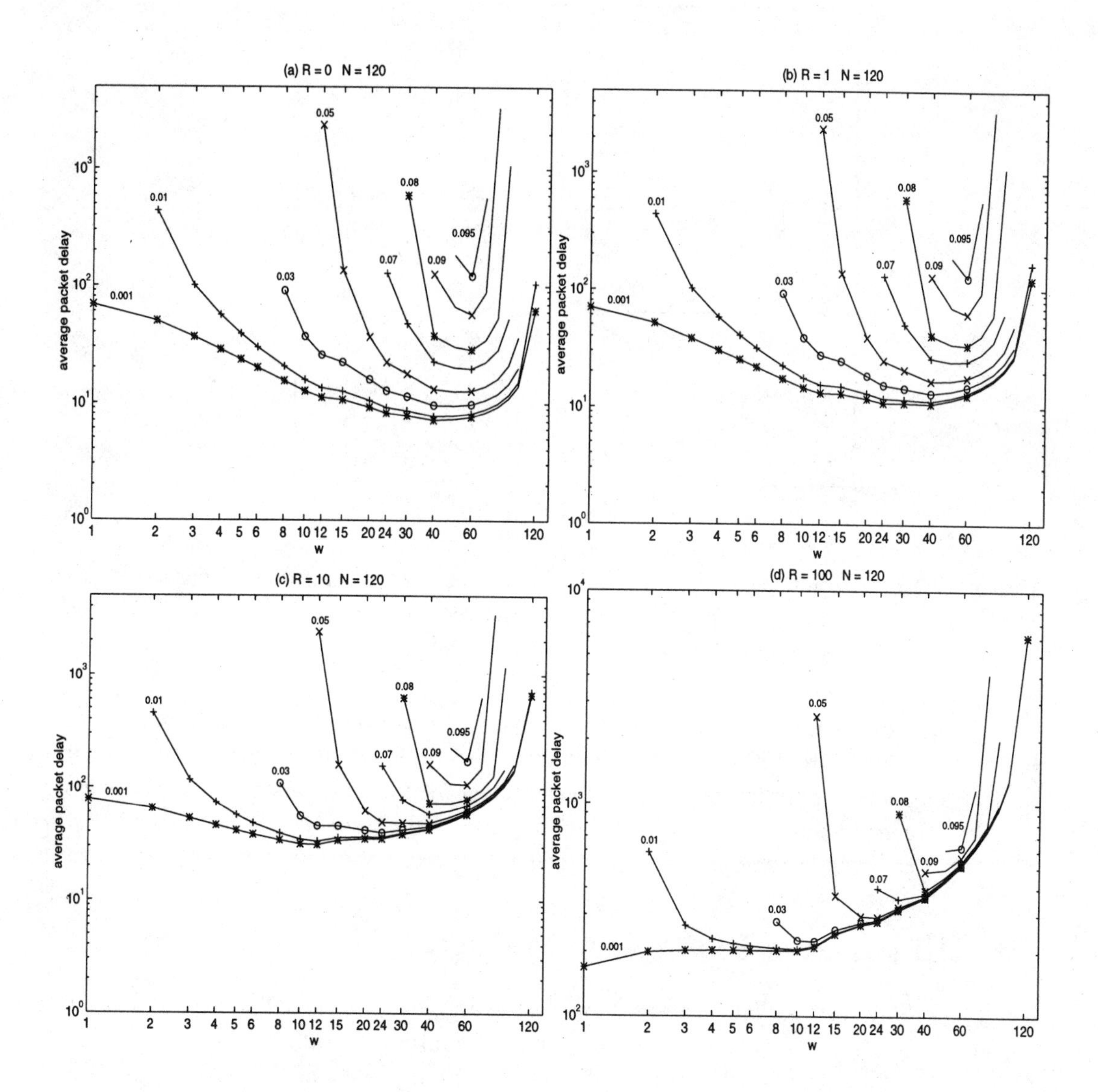

Figure 7.10 Average delay, T, vs. number of wavelengths, w, for various values of load in a 120-node network: (a) $R = 0$; (b) $R = 1$; (c) $R = 10$; (d) $R = 100$.

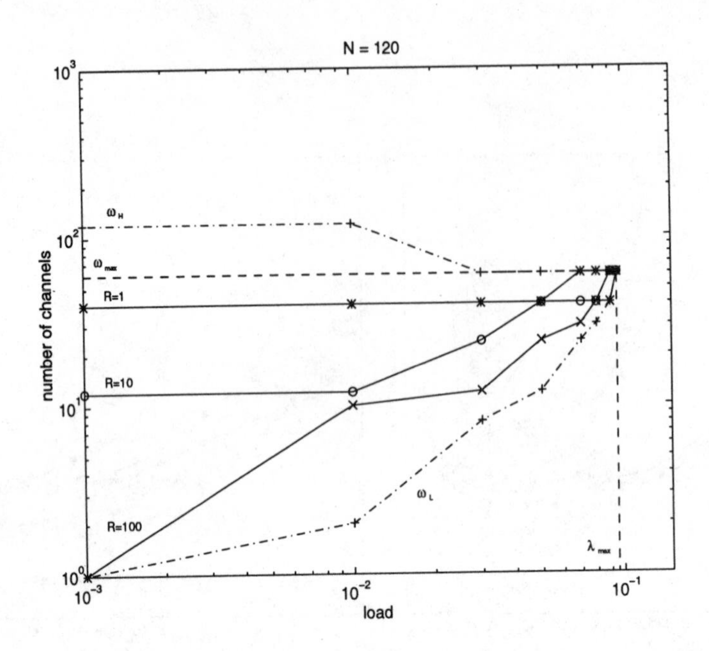

Figure 7.11 Upper (w_H) and lower (w_L) bounds on the number of channels admitting stability and optimal number of channels (w^*) vs. load in a 120-node network.

more realistic, local area network (LAN) example, we see that increasing N from 60 to 120 nodes does not increase either w^* or w_L for loads below 0.03 packets per slot per node. These above results indicate the WDM networks are indeed scalable in a graceful manner: *large networks do not always require an equally large number of wavelengths.*

7.6 Illustrative Examples: Multicast Traffic

This section will present numerical examples to demonstrate the effectiveness of channel-sharing in the presence of multicast traffic. We consider seven different networks with approximately 40 nodes each:

- System A: A 39-node network employing a single transmitter-receiver pair at each node and 39 dedicated channels; each node has a degree of one, and the entire bandwidth of a channel is available to each node. The only logical topology that can be realized practically with these specifications is a 39-node ring network.

	$\lambda = 0.001$	$\lambda = 0.005$	$\lambda = 0.01$	$\lambda = 0.05$
$R = 0$	40	40	40	40
$R = 1$	24	24	24	40
$R = 10$	12	12	12	24
$R = 100$	10	10	10	20

Table 7.1 Optimal number of wavelengths w^* for $N = 120$ and various values of λ and R.

- System B: A 39-node network with three transmitter-receiver pairs at each node and 117 dedicated channels; each node has a degree of three, and the bandwidth available to each node is 3 times the bandwidth of a channel. A (13, 3, 3)-GEMNET can be constructed to match these specifications.

- System C: A 39-node network employing a single transmitter-receiver pair at each node and 13 channels – three nodes sharing each channel such that each node has a degree of three. A (13, 3, 3)- SC_GEMNET can be constructed to match these specifications.

- System D: A 42-node ring network, employing a single transmitter-receiver pair at each node and 42 channels.

- System E: A (7, 6, 6)-GEMNET, employing six transmitter-receiver pairs at each node and $42 \times 6 = 252$ channels, leading to a nodal degree of six.

- System F: A (7, 6, 6)-SC_GEMNET, employing a single transmitter-receiver pair at each node and seven channels, leading to a nodal degree of six.

- System G: A (1, 42, 42)-SC_GEMNET, employing a single transmitter-receiver pair at each node and one channel. This is the classical TDM system.

In analyzing $\bar{h}$ and $\bar{g}$ for Systems A through C, we approximate the number of nodes in the network to 40. This is because the extended Moore bound is applicable only to complete Moore graphs. Exact values, considering only a 39-node incomplete Moore graph, can be derived with some effort. However, the delay figures which would be obtained from such an exact evaluation would only be smaller (marginally) than the results that we have obtained,

λ	$N = 12$			$N = 24$			$N = 60$			$N = 120$			$N = 600$			$N = 1200$			$N = 2400$		
	w_L	w^*	w_H	w_L	w^*	w_H	w_L	w^*	w_H	w_L	w^*	w_H	w_L	w^*	w_H	w_L	w^*	w_H	w_L	w^*	w_H
0.001	1	1	12	1	1	24	1	6	60	1	12	120	1	75	600	2	120	1200	5	200	1200
0.010	1	1	12	1	4	24	2	6	60	2	12	120	12	75	300	24	120	600	48	300	1200
0.030	1	1	12	1	6	24	3	6	60	8	24	60	50	120	300	120	300	600	300	600	1200
0.053	1	4	12	2	6	24	6	15	30	15	40	60	120	200	300	300	400	600	1200	1200	1200
0.055	1	4	12	2	6	24	10	15	30	20	40	60	150	200	300	400	400	600			
0.060	1	4	12	3	6	24	10	15	30	20	40	60	200	300	300	600	600	600			
0.068	1	4	12	3	6	24	10	15	30	24	40	60	300	300	300						
0.070	1	4	12	3	6	24	10	20	30	24	40	60									
0.080	1	4	2	4	6	24	12	20	30	30	40	60									
0.090	2	4	12	4	6	12	15	30	30	40	60	60									
0.095	2	4	12	4	8	12	15	30	30	60	60	60									
0.100	2	4	12	6	8	12	20	30	30												
0.110	2	4	12	6	8	12	30	30	30												
0.120	2	4	12	6	12	12	30	30	30												
0.140	3	4	12	8	12	12															
0.150	3	4	12	12	12	12															
0.190	4	6	6																		
0.220	6	6	6																		
w_{max}		6			12			30			60			300			600			1200	

Table 7.2 w^*, w_L, w_H, and $w_{\max}$ for $R = 10$ and various values of N and λ.

and will not make a difference to the conclusions that we draw. For Systems D through F, again for analytical convenience, we approximate the number of nodes in the network to 43 for the evaluation of $\bar{h}_m$ and $\bar{g}_m$, since with $K = 2$, and $P = 6$, we get $N_k = 43$. This approximation also will not make a difference to the conclusions that we draw.

We assume that the traffic arrival rate $\lambda = 0.01$ packets per slot per node, and the propagation delay $R = 10$. The latter corresponds to a node-to-star distance of about 1 km, a reasonable assumption for a LAN if we assume that the network operates at a 1-Gbps rate with 1000 bit packets. Under these conditions, we evaluated the delay for the various systems for different values of destination set size[6] m. These results are tabulated in Table 7.3.

Multicast set size m	System A	System B	System C	System D	System E	System F	System G
1	232.5	30.1	33.0	250.2	21.4	26.7	47.2
5	235.0	30.1	33.7	253.3	21.4	28.0	47.2
15	236.0	30.2	34.6	254.5	21.4	29.7	47.2
25	236.3	30.2	35.1	254.8	21.4	30.0	47.2
30	236.3	30.2	35.2	254.9	21.4	30.0	47.2
35	236.6	30.3	35.2	255.0	21.4	30.0	47.2

Table 7.3 Delay, T, in slots, for the seven example systems.

Let us consider Systems A, B, and C first. We note that System A with 39 dedicated channels has the worst performance of Systems A through C, because the only network that can be logically implemented for this case is a ring network, which can have significant delays due to increased hop distance. Our shared-channel scheme with only 13 channels and a degree of three performs as well as the 117-channel system also with degree three. We remind the reader that the comparison is not fair and is stacked against the shared-channel case since it uses fewer channels and/or transceivers than the dedicated-channel cases (Systems A and B); however, it does perform better in terms of delay.

[6] Calculation of λ' assumes that all nodes originated packets with m destinations. We also obtained results for cases where m was distributed (1) in truncated geometric fashion, and (2) uniformly between 1 and $N-1$. The trends with different distributions were similar to the case where m was fixed. Therefore, the other results are not tabulated.

The results for Systems D through F are also shown in Table 7.3. We used the same value of λ, which means that the network-wide load is greater for the 42-node network. As can be expected, the delay for System D, a 42-node ring network, is larger than the delay for System A, a 39-node ring network. The surprising result is that, *even though System F is larger than System C, it has lower average delay*; this is due to the increased sharing that is achieved with a degree of six. System F has delay values almost twice that of System E, the dedicated channel system with the same degree; however, it is at least 6 times cheaper in terms of channels and transmitter-receiver costs.

Finally, increasing the degree further by introducing greater sharing leads to the classical TDM system in System G, which has a constant delay value of 47.2 for all values of m. This large delay is chiefly due to the large synchronization and queueing delays, since the sum of the transmission and propagation delays amounts to only 11 (10 for propagation and 1 for transmission) slots for this case. Thus, there is a trade-off between the hop distance and the M/D/1 queueing delay as w is varied. With a very small number of channels ($w < N$), the hop distance is very small, but the queueing delay is large; conversely, when $w = N$, the total delay can be large chiefly due to increased hop distance. If the load $\lambda = 0.01$ per node per slot, $w = 7$ is a good choice for a network size $N = 42$.

Figure 7.12 depicts the delay vs. load for all seven systems for four different values of m ($m = 1, 5, 20, 36$). Here, again, it is clear that the delay is highest for the ring systems (Systems A and D), and the maximum throughput is achieved by expensive, dedicated channel systems (Systems B and E). Note that the x-axis refers to the nodal throughput. If these throughput values are converted to network throughputs and normalized by the number of channels used in each system, we will obtain the per-channel throughput. The performance of the dedicated channel systems (Systems B and E) in terms of the per-channel throughput will not be as impressive. Not quite obvious in this set of figures is the rate at which the maximum throughput drops as m is increased for the various systems: the drops are highest for the dedicated channel systems (B and E), strengthening the case for channel-sharing. We clarify this final argument further in Fig. 7.13, where we depict the ratio of the maximum throughput for each system at a given m to its throughput at $m = 1$. The drops in the throughputs for the shared-channel systems (Systems C and F) are less than the drops in the throughputs for the corresponding dedicated-channel systems with the same degree (Systems B and E). The figure also shows that there is no drop in throughput for the TDM case, because the hop distance is 1 for all values of m. In terms of this throughput drop with increasing m, the ring systems (Systems A and D) perform better than the shared-channel systems (Systems C and F); however, this increase has to be considered along with the almost two orders of magnitude difference in the delay between the two sets of systems as seen in Fig. 7.12.

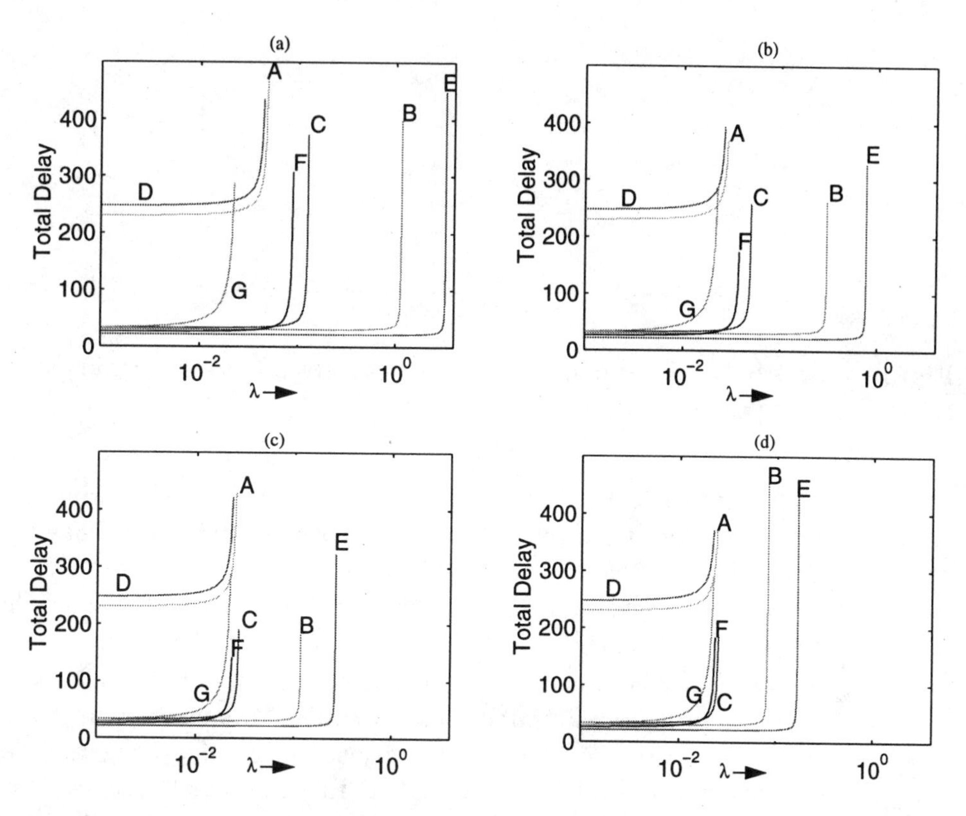

Figure 7.12 Delay vs. load for Systems A through F: (a) $m = 1$; (b) $m = 5$; (c) $m = 20$; (d) $m = 36$.

7.7 Summary

The multihop method of operation seems to be the logical choice for passive-star-based WDM LANs if wavelength-agile transceivers are not available or prohibitively costly. Motivated by the limitation on the number of channels that can be supported in such a WDM LAN and the increasing demand for

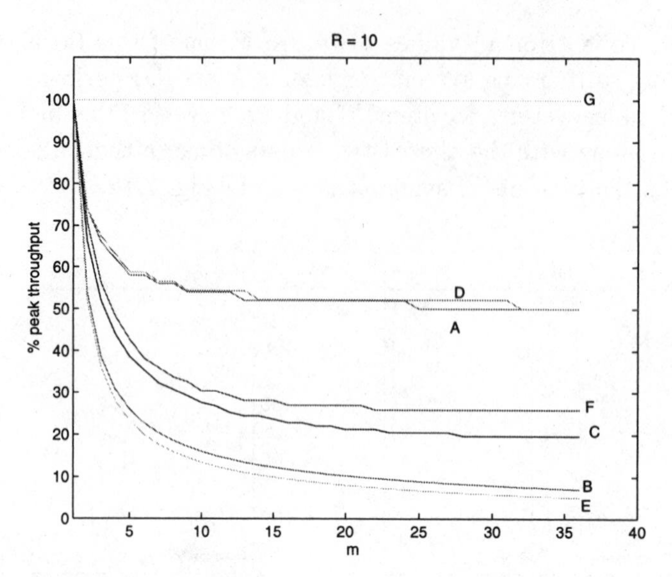

Figure 7.13 Ratio of maximum throughput at a given m to the maximum throughput at $m = 1$.

multicasting capability, we considered the use of channel-sharing in such a system. We presented a general method of constructing shared-channel, multihop WDM LANs. The method was illustrated using the GEMNET architecture as the target logical topology. Intuitively, it appears that channel-sharing may be particularly effective in the presence of multicast traffic.

We provided a general analytical method for modeling such a system in the presence of multicast traffic. Our analysis allows us to compute the average packet delay in the network for various values of propagation delay, load, and number of available wavelengths. Having a small number of channels is not only a technology requirement, but, based on our results, may actually be desirable from the system's performance point of view. Our illustrative examples indicate that channel-sharing is an effective technique for supporting unicast as well as multicast traffic.

Our analytical method has other merits: its application is not restricted to optical systems. The expression we have derived to analyze multicast performance is a generalization of the Moore bound. This general expression subsumes the Moore bound, and it should prove to be a useful bound in the comparison of topologies and algorithms for multicasting in various contexts.

This work may be extendible toward a comparative study involving shared-channel, multihop systems and shared-channel, single-hop systems.

Exercises

7.1. The statement is made that "channel-sharing through TDM implies a higher logical nodal degree (viz., ability to read more nodes directly), albeit with a lower capacity on each of the logical links." Explain this statement.

7.2. Show that the de Bruijn graph has minimum diameter among all graphs with the same number of nodes and maximum out degree.
Hint: Consider the Moore bound.

7.3. If we increase the available channels in a network, then the bandwidth will increase. Therefore, packet delay will decrease. Is this statement correct? Explain.

7.4. Explain why for a given load, the number of channels is bounded from below by w_L and from above by w_H.

7.5. Explain why $w_{\max} = N/2$.

7.6. Consider a network of six nodes. Construct a SC_GEMNET logical topology for each of the following cases:
(a) Number of channels = 6
(b) Number of channels = 3
(c) Number of channels = 2
Assume a $FT - FR$ system in each case.

7.7. Given a network with eight nodes and six available channels, give a possible SC_GEMNET configuration.

7.8. Construct a (2,4,2) SC_GEMNET with $w = 3$, and show a possible TDM frame structure.

7.9. Draw the 18-node SC_GEMNET with three channels.

7.10. Find all possible channel-sharing configurations for a network with 10 nodes. Consider the situation in which w is an exact divisor of 10. Compute the stable load range for each configuration, as well as $\lambda_{\max}$ and $w_{\max}$. Find w^* with load $\lambda = 0.15$. Assume $R = 1$ and only unicast traffic is present.

7.11. Look at Fig. 7.1. Explain the occurrence of the unstable regions.

7.12. Figure 7.6 shows all delay components, except for queueing, peaking at $w = 12$ channels. Explain this phenomenon. Also explain why the case of $w = 24$ is so similar in performance to the case of $w = 6$.

7.13. Consider an 8-node, (2,2) SC-ShuffleNet, with $w = 4$, $\lambda = 10^4$ packets per slot per node, packet transmission time = 1 μs, and the distance from each node to the star coupler is 10 km. There is no multicast traffic. Find the average packet delay.

7.14. Consider a (3,4,2) SC_GEMNET without multicast traffic. Packets arrive at each node according to a Poisson process with rate $\lambda = 2 \cdot 10^5$ pkt/s. The distance between each node and the star coupler is 10 km. There are six wavelengths used in the system. Slot length is equal to 1 μs. Find the average packet delay.

7.15. Consider a (2,2) SC-ShuffleNet with $w = 4$. Packets arrive at each node according to a Poisson process with rate $\lambda = 10^5$ pkt/s. Each packet is multicast to exactly two destinations ($m = 2$). The distance between each node and the star coupler is 5 km. Slot length is equal to 1 μs. Find the average packet delay.

7.16. Consider a (2,6,3) SC-GEMNet with $w = 4$. Packets arrive at each node according to a Poisson process with rate $\lambda = 1 \cdot 10^5$ pkt/s. Each packet is multicast to exactly two destinations ($m = 2$). The distance between each node and the star coupler is 5 km. Slot length is equal to 1 μs. Find the average packet delay.

7.17. Consider a (3,8,2) SC-GEMNet with $w = 12$. Packets arrive at each node according to a Poisson process with rate $\lambda = 4 \cdot 10^4$ pkt/s. Each packet is broadcast to all nodes ($m = 23$). The distance between each node and the star coupler is 5 km. Slot length is equal to 1 μs. Find the average packet delay.

7.18. What is the difference between $\overline{h}_m$ and $\overline{g}_m$?

7.19. Calculate $\overline{g}_2$ for the network in Fig. 7.14.

7.20. Consider the graph in Fig. 7.15. Suppose node 1 is the source and nodes 8, 9, 15, 16, and 17 are the destinations. Find $\overline{h}_m$ and $\overline{g}_m$. Suppose we have a choice between the nonshared and the shared-channel structures. How many channels and transmissions are required in each case?

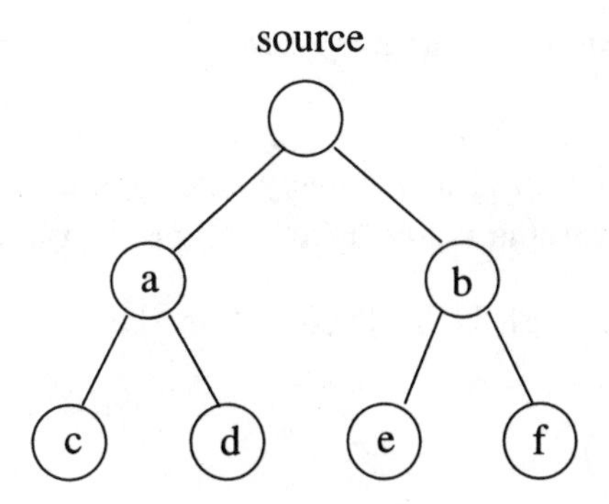

Figure 7.14 A (2,2) complete Moore graph.

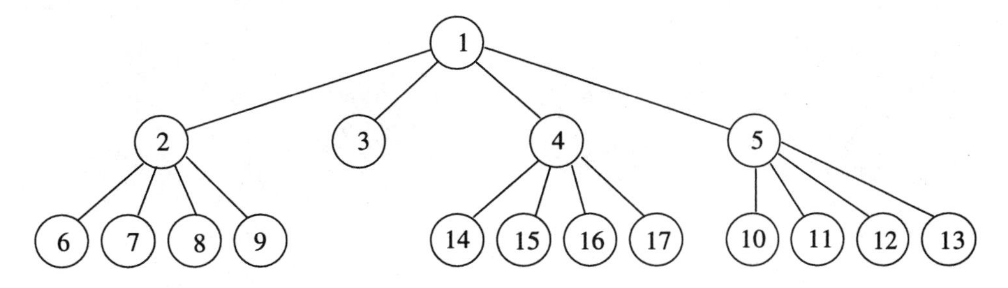

Figure 7.15 A (4,2) incomplete Moore graph.

7.21. Consider a 12-node SC_GEMNET with $w = 3$. The source node is node 0, and the multicast group consists of nodes 2, 4, 6, and 10. Find $\overline{h}_m$ and $\overline{g}_m$. Compare with the nonshared case.

7.22. Consider a (2,8,4) SC_GEMNET. Node 0 wants to send a packet to nodes 2, 3, 10, and 15. Assuming that one hop counts for a delay of one unit, what is the total delay in the case of (i) unicasting, (ii) optimal multicasting? What is the number of transmissions in each case?

7.23. Consider two networks – a ring network of four nodes having four channels, and a (2,2,2) SC_GEMNET with two channels. Let the capacity of each channel be C. Assuming that the load offered at each node is λ and is uniformly distributed, which network is able to sustain a higher value of λ? Assume that there is no self-traffic. For the SC_GEMNET, assume shortest-path routing.

7.24. Is it always possible to achieve a reduction in the average number of transmissions using channel-sharing in the case of multicasting? In particular, can the extent of reduction as suggested by the Moore graph always be realized?

7.25. In Table 7.3, Systems E and F have lower delay than Systems B and C, respectively, even though Systems E and F are larger. Explain.

7.26. Which of the Systems D through G in Table 7.3 is the best? Explain your choice.

7.27. Derive Eqn. (7.5).

Part III

Wavelength-Routed (Wide-Area) Optical Networks

Chapter 8

Elements of Virtual Topology Design

This chapter explores design principles for next-generation optical wide-area networks, employing wavelength-division multiplexing (WDM) and targeted to nationwide and global coverage. This optical network exploits wavelength multiplexers and optical switches in routing nodes, so that an arbitrary virtual topology may be embedded on a given physical fiber network. The virtual topology, which is operated as a packet-switched network and which consists of a set of all-optical "lightpaths," is set up to exploit the relative strengths of both optics and electronics – viz., packets of information are carried by the virtual topology "as far as possible" in the optical domain using *optical circuit switching*, but packet forwarding from lightpath to lightpath is performed via *electronic packet switching*, whenever required.

We formulate the virtual topology design problem as an optimization problem with one of two possible objective functions: (1) for a given traffic matrix, minimize the network-wide average packet delay (corresponding to a solution for *present traffic demands*), or (2) maximize the scale factor by which the traffic matrix can be scaled up (to provide the maximum capacity upgrade for *future traffic demands*). Since simpler versions of this problem have been shown to be NP-hard, we resort to heuristic approaches. Specifically, we employ an iterative approach which combines "simulated annealing" (to search for a good virtual topology) and "flow deviation" (to optimally route the traffic – and possibly bifurcate its components – on the virtual topology). We illustrate our approaches by employing experimental traffic statistics collected from NSFNET.

In this chapter, we do not consider the number of available wavelengths to be a constraint. This constraint as well as additional relaxations are studied in Chapter 9 so that the entire virtual topology design problem can be *linearized*, and it can, hence, be *solved optimally*. Also, for now, we ignore the routing of lightpaths and wavelength assignment for these lightpaths, a topic which is examined in Chapter 10.

8.1 Introduction

This chapter examines an "optical" wide-area WDM network which utilizes wavelength multiplexers and optical switches in a routing node so that an arbitrary virtual topology can be embedded on a given physical fiber network. The virtual topology, which is packet switched and which consists of a set of "all-optical lightpaths," is set up to exploit the relative strengths of both optics and electronics – viz., packets of information are carried by the virtual topology "as far as possible" over the same wavelength in the optical domain (i.e., there is no wavelength conversion in a lightpath), but packet forwarding from lightpath to lightpath is performed via electronic packet switching, whenever required. Optical circuit switching at a node is achieved by using a wavelength-routing switch (WRS), which is capable of optically bypassing a lightpath from an input fiber to an output fiber, without any electronic processing. Because there is no wavelength conversion in the WRS, the wavelength of the lightpath stays the same in the output fiber as it was in the input fiber.

This architecture is a combination of the well-known "single-hop" and "multihop" approaches, and it attempts to exploit the characteristics of both. A "lightpath" in this architecture provides "single-hop" communication. However, by employing a limited number of wavelengths, it may not be possible to set up "lightpaths" between all user pairs; as a result, "multihopping" between "lightpaths" may be necessary. In addition, when the prevailing traffic pattern changes, a different set of "lightpaths" forming a different "multihop" virtual topology may be more desirable. A networking challenge is to perform the necessary reconfiguration with minimal disruption to the network's operations [LaHA94, RoAm94b]. In this architecture, using wavelength multiplexers provides the advantage of higher aggregate system capacity due to spatial reuse of wavelengths and supports a large number of users, given a limited number of wavelengths. Specifically, we investigate the overall design, analysis, and optimization of a nationwide WDM network consistent with device capabilities, e.g., aimed at upgrading the NSFNET.

We formulate the problem as an optimization problem which can optimally select a virtual topology subject to transceiver (transmitter and receiver) and wavelength constraints, with one of two possible objective functions: (1) for a given traffic matrix, minimize the network-wide average packet delay (corresponding to a solution for *present traffic demands*), or (2) maximize the scale factor by which the traffic matrix can be scaled up (to provide the maximum capacity upgrade for *future traffic demands*). Since the objective functions are nonlinear and since simpler versions of the problem have been shown to be NP-hard, we resort to heuristic approaches. Specifically, we employ an iterative approach which combines "simulated annealing" (to search for a good virtual topology) and "flow deviation" (to optimally route the traffic – and possibly bifurcate its components – on the virtual topology). We illustrate our approaches by employing experimental traffic statistics collected from NSFNET.

Section 8.2 explains the system architecture including motivation, general problem statement, and an illustrative example. The problem is formulated as a combinatorial optimization in Section 8.3. Since the problem is NP-hard, heuristic solutions are developed in Section 8.4. Results of applying experimental data (collected from NSFNET) to our algorithms are discussed in Section 8.5.

8.2 System Architecture

8.2.1 Motivation

Consider the NSFNET T1 backbone in Fig. 8.1.[1] Information is transferred over this backbone as packets (of possibly variable sizes) at a rate of 1.544

[1]The NSFNET has been upgraded from time to time by adding new nodes and/or links, and the backbone we show in Fig. 8.1 existed in the early 1990s. It was replaced with a faster T3 backbone in the 1992-93 time frame, and with even faster backbones subsequently. However, we proceed with the T1 backbone since we have obtained a significant amount of traffic statistics from this configuration, and we will employ these statistics to illustrate our proposed "optical" solutions to the design of a wide-area WDM network.

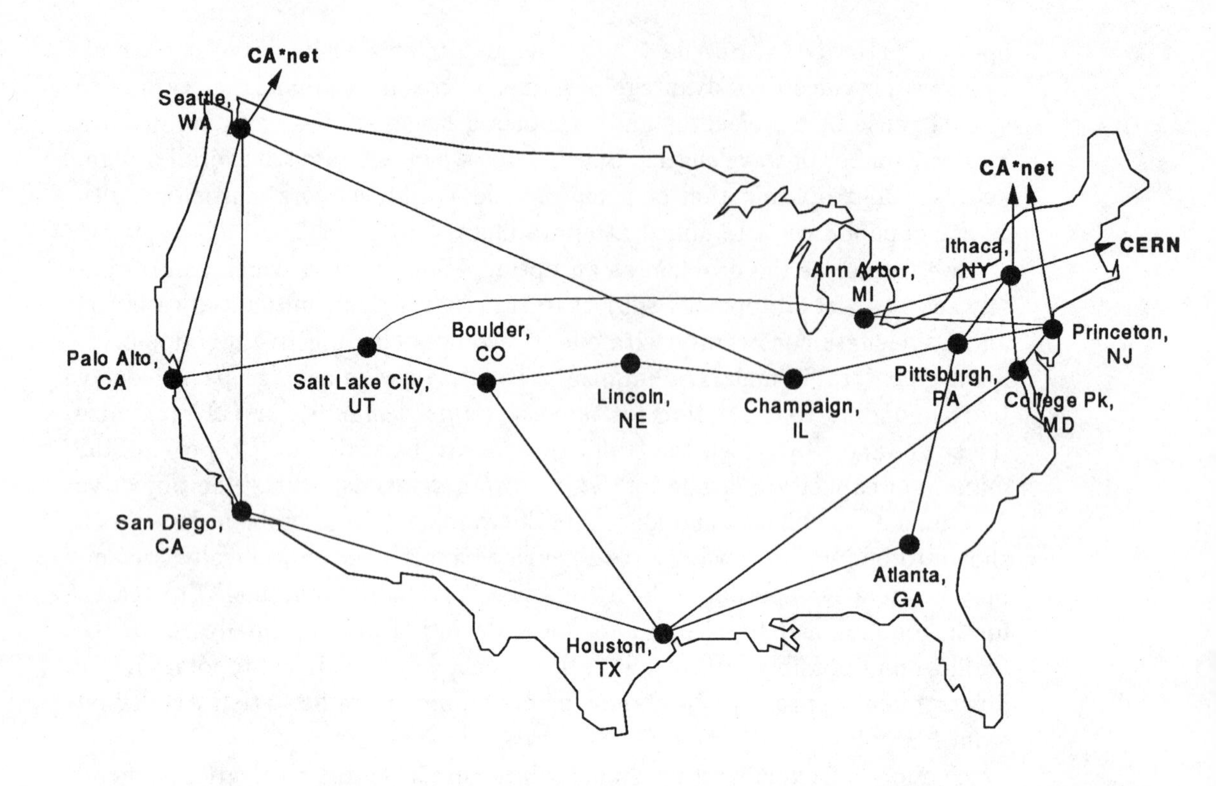

Figure 8.1 NSFNET T1 backbone, 1991. (©Merit Network, Inc.)

Mbps per link. The backbone consists of over a dozen nodes, each containing one or more computers operating as electronic packet switches. These switches are connected with one another via optical fibers to form an irregular mesh structure.[2] Store-and-forward packet switching is performed at the network nodes. That is, a packet traveling from a source node to a destination node may have to pass through zero or more intermediate nodes, and at each such intermediate node, the packet has to be completely received (stored in memory), its header has to be processed by the intermediate node to determine on which of the node's outgoing links this packet must be forwarded, and the packet may have to wait at this node longer if that corresponding outgoing link is busy due to the transmission of other packets. Although fiber is employed

[2]A structure or graph employed to interconnect a number of nodes, possibly computers or processing elements, is known as a mesh, and it is referred to as irregular if it does not possess any well-defined connectivity pattern.

to connect the nodes, the fiber's tremendous transmission bandwidth is not exploited since data transmission on each fiber link is performed only at T1 (1.544 Mbps) rate, and on a single wavelength.

While WDM networks are being investigated, the Asynchronous Transfer Mode (ATM) solution for high-speed networks is making rapid progress. Because of the investment made in the development of ATM, we expect that nationwide and global backbone networks such as NSFNET will also soon be supporting ATM. Although many networking issues will evolve with time, we expect that the underlying physical network will continue to be a fiber plant on which our WDM solutions will still be applicable.

Accordingly, our examination of optical networks is fueled by the following criteria and observations: (1) Any future technology must be incrementally deployable. That is, instead of scrapping an existing and operational wide-area fiber-based network such as NSFNET, we examine how such networks can be upgraded to support (and exploit the capabilities of) WDM. In this regard, we show later how one might replace (or upgrade) the existing packet switches in NSFNET with wavelength-routing switches (WRS), or alternately, how one might embellish an electronic switch with an optical component to transform it to a WRS. (2) We must also explore how the WDM solution can be used to upgrade an existing ATM solution. (3) In addition, we must explore how the WDM solution can support both ATM and non-ATM services, based on WDM's data and protocol transparency property.[3] However, the last point is beyond the scope of this chapter. In the following subsections, we provide a general problem statement, followed by an illustrative example using the NSFNET T1 topology.

8.2.2 General Problem Statement

The problem of embedding a desired virtual topology on a given physical topology (fiber network) is formally stated below. We are given the following inputs to the problem:

1. A physical topology $G_p = (V, E_p)$ consisting of a weighted undirected graph, where V is the set of network nodes, and E_p is the set of links

[3]Since multiple WDM channels on the same fiber can be operated independently, they can carry data at different rates and formats (including some analog and some digital channels, if desired); also, the protocols controlling the data transfers over different channels can be different, so that one can establish independent subnetworks operating on different sets of WDM channels over the same fiber plant. This is referred to as WDM's *transparency* property.

connecting the nodes. Undirected means that each link in the physical topology is bidirectional. Nodes correspond to network nodes (packet switches) and links correspond to the fibers between nodes; since links are undirected, each link may consist of two fibers or two channels multiplexed (using any suitable mechanism) on the same fiber. Links are assigned weights, which may correspond to physical distances between nodes. A network node i is assumed to be equipped with a $D_p(i) \times D_p(i)$ wavelength-routing switch (WRS), where $D_p(i)$, called the physical degree of node i, equals the number of physical fiber links emanating out of (as well as terminating at) node i.[4]

2. Number of wavelength channels carried by each fiber = M.

3. An $N \times N$ traffic matrix, where N is the number of network nodes, and the (i, j)-th element is the average rate of traffic flow from node i to node j. Note that the traffic flows may be asymmetric, i.e., flow from node i to node j may be different from the flow from node j to node i.

4. The number of wavelength-tunable lasers (transmitters) and wavelength-tunable filters (receivers) at each node.

Our goal is to determine the following:

1. A virtual topology $G_v = (V, E_v)$ as another graph where the out-degree of a node is the number of transmitters at that node and the indegree of a node is the number of receivers at that node. The nodes of the virtual topology correspond to the nodes in the physical topology. Each link between a pair of nodes in the virtual topology corresponds to a direct all-optical "lightpath" between the corresponding nodes in the physical topology. (Noting that each such link of the virtual topology may be routed over one of several possible paths on the physical topology, an important design issue is "optimal routing" of *all lightpaths* so that the constraint on having a limited number of wavelengths per fiber is satisfied. This problem is treated in Chapter 10.)

2. A wavelength assignment for lightpaths, such that if two lightpaths share a common physical link, they must necessarily employ different wavelengths.

---

[4]Note that $D_p(i)$ includes the fiber(s) corresponding to local connections, i.e., for attaching an electronic router to the WRS (shown later in this chapter). There are wavelength-related and cost-related issues which affect the decision on the number of fibers chosen to connect the local node to the local switch, and these issues are discussed in Section 8.2.3.

3. The sizes and configurations of the WRSs at the intermediate nodes. Once the virtual topology is determined and the wavelength assignments have been performed, the switch sizes and configurations follow directly.

Communication between any two nodes now takes place by following a path (a sequence of lightpaths) from the source node to the destination node on the virtual topology. Each intermediate node in the path must perform (1) an opto-electronic conversion, (2) electronic routing (possibly ATM switching, if needed), and (3) electrooptic forwarding onto the next lightpath.

8.2.3 An Illustrative Example

We employ an illustrative example to demonstrate how WDM can be used to upgrade an existing fiber-based network. Using a slightly modified version of the NSFNET in Fig. 8.1 as an example (see Fig. 8.2), we demonstrate how a hypercube (Fig. 8.3) can be embedded as a virtual topology over this physical topology (to be discussed shortly). We also assume an undirected virtual topology comprising bidirectional lightpaths in this example. In general, the virtual topology may be a directed graph, as our formulation in Section 8.3 will assume.

The NSFNET's T1 backbone (Fig. 8.1) has 14 nodes, with some links connecting these nodes to one another and some links connecting to the "outside world." For illustration purposes, we supplement this physical topology by adding two fictitious nodes, AB and XY, to capture the effect of NSFNET's connections to Canada's communication network, CA*net. Node XY is connected to Ithaca (NY) and Princeton (NJ) nodes of NSFNET, while node AB is connected to the Seattle (WA) and Salt Lake City (UT) nodes, where we have employed the last link as a fictitious one to render the physical topology richer and fault tolerant. (Note that, to provide fault tolerance, each node in the NSFNET is connected to at least two other nodes, and we did not want to violate this policy for node AB.) Thus, we get the modified physical topology, hereafter referred to as the physical topology, of Fig. 8.2.

The nodal switching architecture consists of an optical component and an electronic component. The optical component is a wavelength-routing switch (WRS), which can optically bypass some lightpaths, and which can locally terminate some other lightpaths by directing them to node's electronic component. The electronic component is an electronic packet router (which may be an ATM switch), which serves as a store-and-forward electronic overlay on top of the optical virtual topology. Figure 8.4 provides a schematic diagram

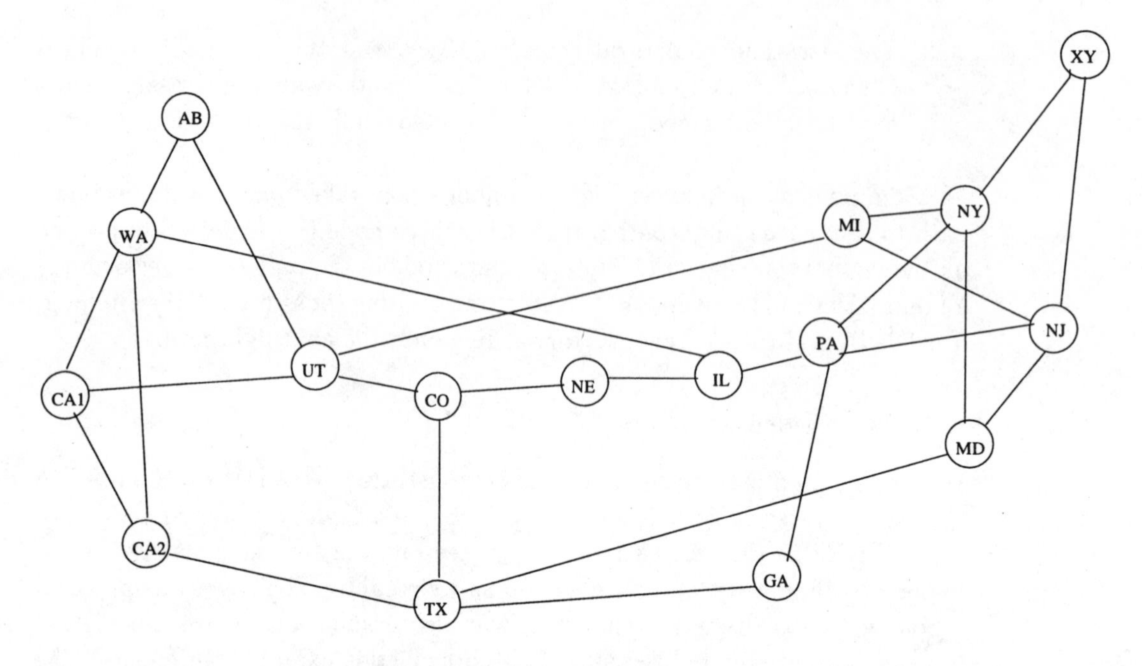

Figure 8.2 Modified physical topology.

of the architecture of the Utah node (UT) in our illustrative example.

The design of a WRS can take several forms. An attractive choice is to employ an array of optical space-division switches, one per wavelength, between the demux and mux stages (see Fig. 8.4). These switches can be reconfigured under electronic control, e.g., to adapt the network's virtual topology on demand. One approach would be to have the local lasers/filters normally operated on fixed wavelengths, but a facility to tune them to different wavelengths must be provided.

The virtual topology chosen for our illustration is a 16-node hypercube (e.g., Fig. 8.3), although algorithms for arbitrary virtual topology embedding will be studied later in this chapter. Two embeddings which result from the optimization criteria for hypercube embedding studied in [MRBM94] are shown in Figs. 8.5 and 8.6. One of these embeddings (Fig. 8.6) assumes that all of the local laser-filter pairs at a node operate on different wavelengths, while the other (Fig. 8.5) does not.

Note that each virtual link in the virtual topology of Fig. 8.3 is a "lightpath" (or a "clear channel") with electronic terminations at its two ends only. For example, the CA1-NE virtual link in Fig. 8.6 could be set up as an

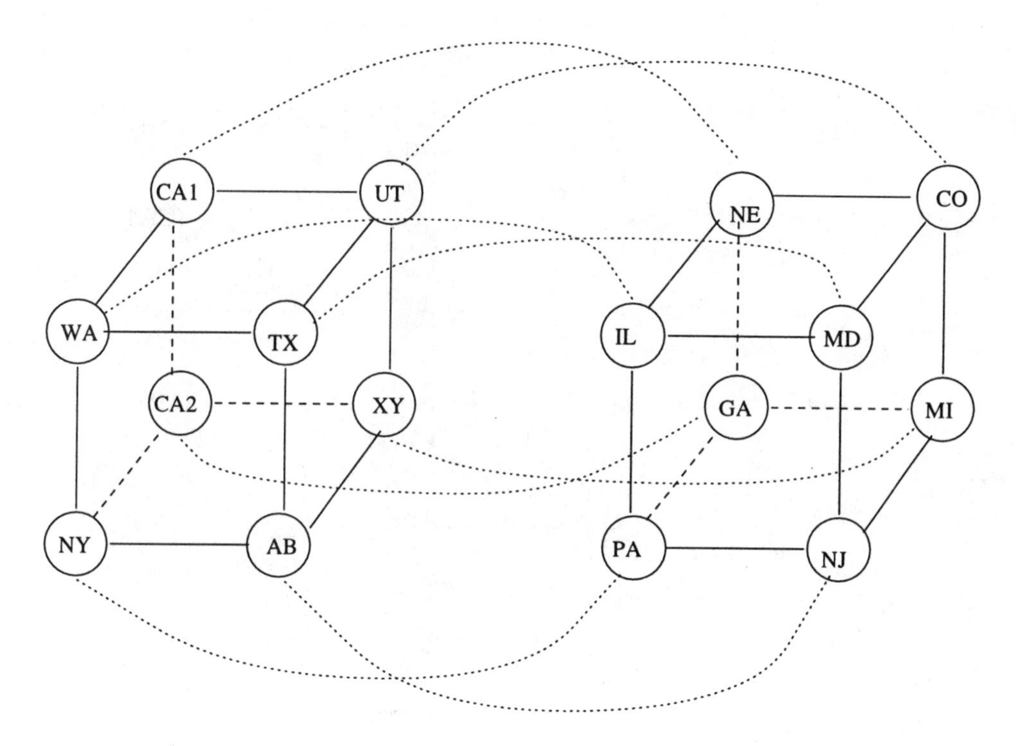

Figure 8.3 A 16-node hypercube virtual topology embedded on the NSFNET physical topology.

all-optical channel on one of several possible wavelengths on one of several possible physical paths, e.g., CA1-UT-CO-NE, or CA1-WA-IL-NE, or others (see Fig. 8.6). According to the solution in [MRBM94], the first path is chosen on wavelength 2 for this CA1-NE lightpath. This means that the WRSs at the UT and CO nodes must be properly configured to establish this CA1-NE lightpath. For example, the switch at UT must have wavelength 2 on its fiber to CA1 connected to wavelength 2 on its fiber to CO. Since we have assumed in this example that connections are bidirectional (note that the virtual topology in this illustration is also an undirected graph), the CA1-NE connection implies two directed lightpaths, one from CA1 to NE and the other from NE to CA1.

The solutions in Figs. 8.5 and 8.6 require a maximum of five and seven wavelengths per fiber, respectively, by employing shortest-path routing of lightpaths on the physical topology. If only one fiber is used to connect the local node to the local WRS, each of the lightpaths emanating from (and terminating at) that node would need to be on a different wavelength to avoid

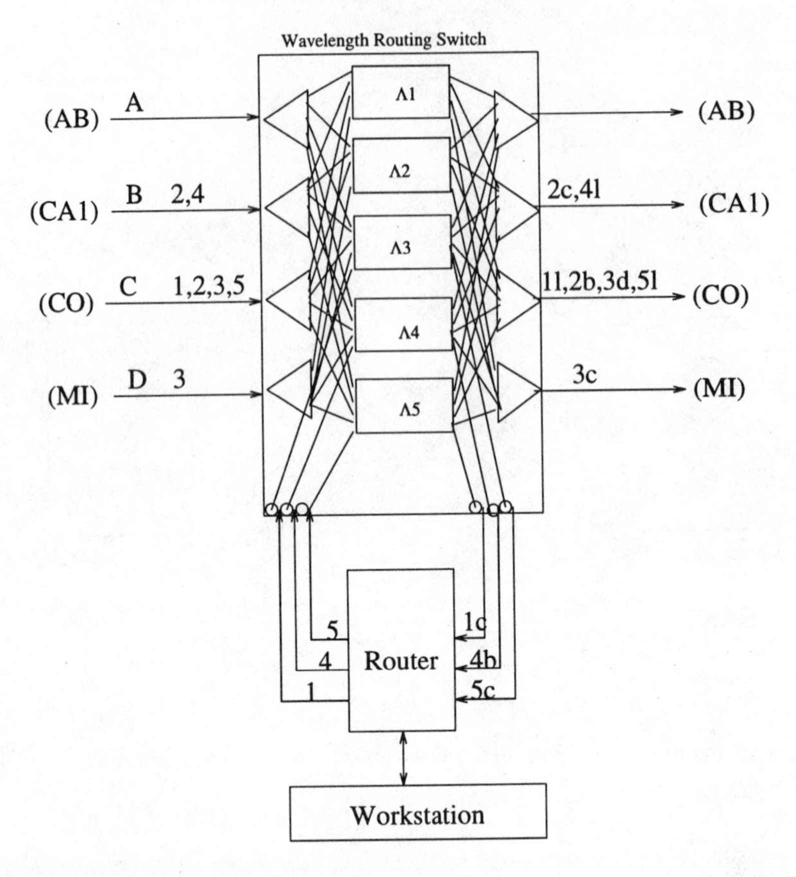

Figure 8.4 Details of the Utah (UT) node.

wavelength conflicts on the local fiber (as in the solution in Fig. 8.6 for the entire network, and in Fig. 8.4 for the UT node); accordingly, this solution needs more wavelengths to embed a virtual topology. If multiple fibers are used to connect the local node to the local WRS, multiple lightpaths may emerge from a node on the same wavelength, and hence fewer wavelengths would be needed, e.g., see the corresponding solution in Fig. 8.5 for the same hypercube virtual topology. However, in both solutions, not all wavelengths on all fibers may be utilized, e.g., in Fig. 8.6, only wavelengths 2 and 4 are used on the CA1-UT fiber, while wavelengths 1, 2, 3, and 5 are used on the UT-CO fiber.

For the solution in Fig. 8.6, the details of one of the nodes, viz., the one at UT, are shown in Fig. 8.4. Note that this switch has to support four incoming fibers plus four outgoing fibers, one each to nodes AB, CA1, CO,

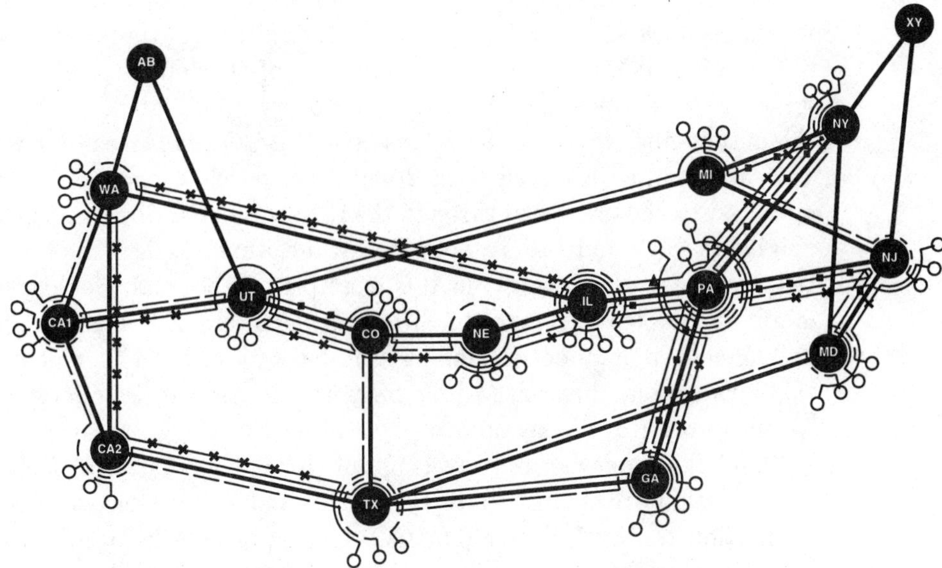

Figure 8.5 The physical topology with embedded wavelengths corresponding to an optimal solution (more than one transceiver at any node can tune to the same wavelength).

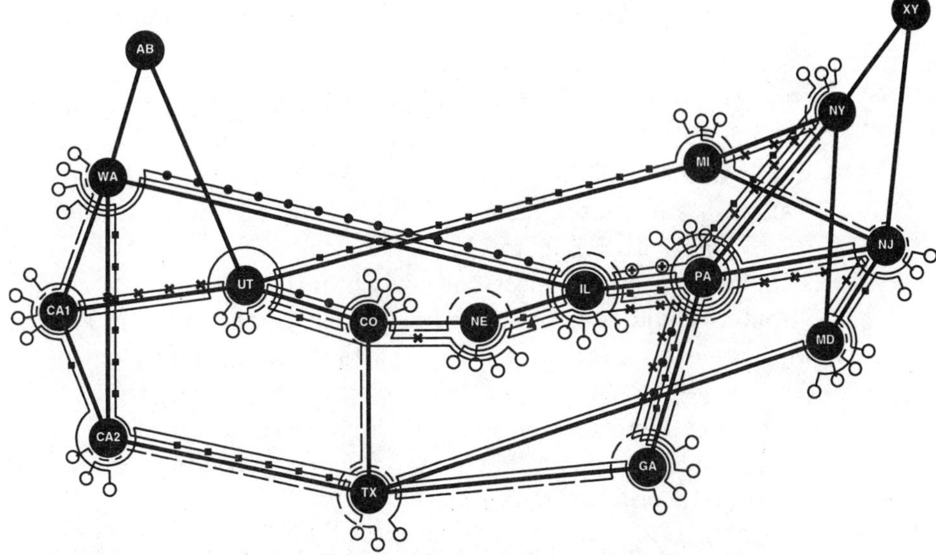

Figure 8.6 The physical topology with embedded wavelengths corresponding to an optimal solution (all transceiver pairs at any node must be tuned to different wavelengths).

and MI, as dictated by the physical topology. In general, each switch also interfaces with four lasers (inputs) and four filters (outputs), with each laser-filter pair dedicated to accommodate each of the four virtual links which a node has to support on the virtual topology. The labels "1*l* 2*b* 3*d* 5*l*" on the output fiber to CO indicate that the UT-CO fiber uses four wavelengths 1, 2, 3, and 5, with wavelengths 2 and 3 being "clear channels" (i.e., optical-circuit-switched channels) through the UT switch and directed to the physical neighbors CA1 and MI, respectively, while wavelengths 1 and 5 connect to two local lasers. However, in this example, the virtual topology embedded is an "incomplete" hypercube with nodes AB and XY considered nonexistent. Hence some nodes, including UT, have fewer than four neighbors. For this illustration, three laser-filter pairs at UT need to be operated – one on wavelength 1 (for connection to physical neighbor CO, to support the lightpath UT-TX); another on wavelength 4 (for connection to physical neighbor CA1, to support the lightpath UT-CA1); and a third on wavelength 5 (for connection to physical neighbor CO, to support the lightpath UT-CO); see Fig. 8.6. Labels on the switch's output ports indicate which wavelength on which input fiber or local laser is connected to which wavelength on which output fiber or local filter. (Note that, just like the physical topology, the virtual topology chosen in this illustration also has *bidirectional* links, although this is not a requirement, i.e., a virtual topology with unidirectional links may also be embedded.)

Figure 8.4 shows a few more interesting issues of this architecture. The box labeled "Router" is an electronic switch which takes information from terminated lightpaths (1*c* 4*b* 5*c*) as well as a local source (labeled "Workstation"), and routes them – via electronic packet switching – to the local lasers (lightpath originators) and the local destination. We reiterate that the "Router" can be any electronic switch, including an ATM switch. Also, the "Router" could have been the existing electronic switch in an electronic network (before upgrading to WDM), with its input and output ports connected directly to the incoming and outgoing fibers, respectively, at the switching node. The nonrouter portions of the node architecture in Fig. 8.4 are the optical embellishments that may be incorporated to upgrade the electronic switch to incorporate a WRS.

The WRS associated with the Utah switch would be different in the solution corresponding to Fig. 8.5. Since we would have multiple fibers connecting the electronic router to the WRS, the size of the WRS would need to be a 7×7 switch instead of the 4×4 switch used in the solution corresponding to

Fig. 8.6. However, since the solution corresponding to Fig. 8.5 requires fewer wavelengths, the number of space-division switches inside the largest WRS would reduce from seven to five. There are cost implications associated with the WRS design and these costs may influence the solution that is adopted.

8.3 Formulation of the Optimization Problem

We formulate the problem as an optimization problem, using principles from multicommodity flow for physical routing of lightpaths and traffic flow on the virtual topology, and using the following notation:

1. s and d used as subscript or superscript denote *source* and *destination* of a packet, respectively,

2. i and j denote *originating and terminating nodes, respectively, in a lightpath*, and

3. m and n denote *endpoints of a physical link that might occur in a lightpath.*

8.3.1 Given:

- Number of nodes in the network = N.

- Maximum number of wavelengths per fiber = M (a system-wide parameter).

- Physical topology P_{mn}, where $P_{mn} = P_{nm} = 1$ if and only if there exists a direct physical fiber link between nodes m and n, where m, $n = 1, 2, 3, \ldots, N$; $P_{mn} = P_{nm} = 0$ otherwise (i.e., fiber links are assumed to be bidirectional).

- Distance matrix, viz., fiber distance d_{mn} from node m to node n. For simplicity in expressing packet delays, d_{mn} is expressed as a propagation delay (in time units). Note that $d_{mn} = d_{nm}$ since fiber links are bidirectional, and $d_{mn} = 0$ if $P_{mn} = 0$.

- Number of transmitters at node $i = T_i$ ($T_i \geq 1$).
  Number of receivers at node $i = R_i$ ($R_i \geq 1$).

- Traffic matrix λ_{sd} which denotes the average rate of traffic flow from node s to node d, with $\lambda_{ss} = 0$ for $s, d = 1, 2, \ldots, N$. [Additional assumptions are that packet interarrival durations at node s and packet lengths are exponentially distributed, so standard M/M/1 queueing results can be applied to each network link (or "hop") by employing the independence assumption on interarrivals and packet lengths due to traffic multiplexing at intermediate hops. Also, by knowing the mean packet length (in bits per packet), the λ_{sd} can be expressed in units of packets per second.]
- Capacity of each channel = C (normally expressed in bits per second, but converted to units of packets per second by knowing the mean packet length).

8.3.2 Variables:

- Virtual topology: The variable $V_{ij} = 1$ if there exists a lightpath from node i to node j in the virtual topology; $V_{ij} = 0$ otherwise. Note that this formulation is general since lightpaths are not necessarily assumed to be bidirectional, i.e., $V_{ij} = 1 \not\Rightarrow V_{ji} = 1$.
- Traffic routing: The variable λ_{ij}^{sd} denotes the traffic flowing from node s to node d and employing V_{ij} as an intermediate virtual link. Note that traffic from node s to node d may be "bifurcated" with different components taking different sets of lightpaths.
- Physical topology route: The variable $p_{mn}^{ij} = 1$ if the fiber link P_{mn} is present in the lightpath for virtual link V_{ij}; $p_{mn}^{ij} = 0$ otherwise.
- Wavelength color: The variable $c_k^{ij} = 1$ if a lightpath from originating node i to terminating node j is assigned the color k, where $k = 1, 2, \ldots, M$; $c_k^{ij} = 0$ otherwise.

8.3.3 Constraints:

- On virtual topology connection matrix V_{ij}:

$$\sum_j V_{ij} \leq T_i \quad \forall i \tag{8.1}$$

$$\sum_i V_{ij} \leq R_j \quad \forall j \tag{8.2}$$

The equalities in Eqns. (8.1) and (8.2) hold if all transmitters at node i and all receivers at node j are in use.

- On physical route variables p_{mn}^{ij}:

$$p_{mn}^{ij} \leq P_{mn} \tag{8.3}$$

$$p_{mn}^{ij} \leq V_{ij} \tag{8.4}$$

$$\sum_m p_{mk}^{ij} = \sum_n p_{kn}^{ij} \quad if\ k \neq i, j \tag{8.5}$$

$$\sum_n p_{in}^{ij} = V_{ij} \tag{8.6}$$

$$\sum_m p_{mj}^{ij} = V_{ij} \tag{8.7}$$

- On virtual topology traffic variables λ_{ij}^{sd}:

$$\lambda_{ij}^{sd} \geq 0 \tag{8.8}$$

$$\sum_j \lambda_{sj}^{sd} = \lambda_{sd} \tag{8.9}$$

$$\sum_i \lambda_{id}^{sd} = \lambda_{sd} \tag{8.10}$$

$$\sum_i \lambda_{ik}^{sd} = \sum_j \lambda_{kj}^{sd} \quad if \quad k \neq s, d \tag{8.11}$$

$$\sum_{s,d} \lambda_{ij}^{sd} \leq V_{ij} * C \tag{8.12}$$

- On coloring of lightpaths c_k^{ij}:

$$\sum_k c_k^{ij} = V_{ij} \tag{8.13}$$

$$\sum_{ij} p_{mn}^{ij} . c_k^{ij} \leq 1 \quad \forall m, n, k \tag{8.14}$$

8.3.4 Objective: Optimality Criterion

(1) Delay minimization:

$$\text{Minimize}: \quad \sum_{ij}\left[\sum_{sd}\lambda_{ij}^{sd}\left(\sum_{mn}p_{mn}^{ij}d_{mn} + \frac{1}{C - \sum_{sd}\lambda_{ij}^{sd}}\right)\right] \tag{8.15}$$

(2) Maximizing offered load (equivalent to minimizing maximum flow in a link):

$$\min\left[\max\left(\sum_{sd}\lambda_{ij}^{sd}\right)\right] \equiv \max\frac{C}{\left[\max\left(\sum_{sd}\lambda_{ij}^{sd}\right)\right]} \quad \forall i,j \tag{8.16}$$

8.3.5 Explanation of Equations

The above equations are based on principles of conservation of flows and resources (transceivers, wavelengths, etc.) as well as on conflict-free routing, e.g., two lightpaths that share a fiber should not be assigned the same wavelength. Equations (8.1) and (8.2) ensure that the number of lightpaths emanating out of and terminating at a node are at most equal to that node's out-degree and in-degree, respectively. Equations (8.3) and (8.4) constrain the problem so that p_{mn}^{ij} can exist only if there is a physical fiber and a corresponding lightpath present. Equations (8.5) through (8.7) are the multicommodity equations that account for the routing of a lightpath from its origin to its termination. Equations (8.8) through (8.12) are responsible for the routing of packet traffic on the virtual topology, and they take into account the fact that the combined traffic flowing through a channel cannot exceed the channel capacity. Equation (8.13) requires that a lightpath be of one color only. Equation (8.14) ensures that the colors used in different lightpaths are mutually exclusive over a physical link.

Equations (8.15) and (8.16) represent two possible objective functions. In Eqn. (8.15), in the innermost brackets, the first component corresponds to the propagation delays on the links mn which form the lightpath ij, while the second component corresponds to delay due to queueing and packet transmission on lightpath ij (using a M/M/1 queueing model for each lightpath). If we assume shortest-path routing of the lightpaths over the physical topology, then the p_{mn}^{ij} values become deterministic. If, in addition, we neglect queueing delays, the optimization problem in Eqn. (8.15) reduces to minimizing $\sum_{sd}\sum_{ij}\sum_{mn}\lambda_{ij}^{sd}p_{mn}^{ij}d_{mn}$ which is a mixed-integer linear program in which the V_{ij} and the c_k^{ij} variables need to have integer solutions, while the λ_{ij}^{sd} variables do not.

The objective function in Eqn. (8.16) is also nonlinear and it minimizes the maximum amount of traffic that flows through any lightpath. This corresponds to obtaining a virtual topology which can maximize the offered load to the network if the traffic matrix is allowed to be scaled up. We choose this optimization for our algorithms in Section 8.4 because our purpose is to demonstrate how to upgrade the capacity of existing fiber-based networks by employing WDM.

8.4 Algorithms

8.4.1 Subproblems

The optimization problem in Section 8.3 is NP-hard, since several subproblems of this problem are NP-hard. The problem of optimal virtual topology design can be partitioned into the following four subproblems, which are not necessarily independent:

1. determine a good virtual topology, viz., which nodal transmitter should be directly connected to which nodal receiver,

2. route the lightpaths over the physical topology,

3. assign wavelengths optimally to the various lightpaths (this problem has been shown to be NP-hard in [ChGK93]), and

4. route packet traffic on the virtual topology (as in any packet-switched network).

Subproblem 1 addresses how to properly utilize the limited number of available transmitters and receivers. Subproblems 2 and 3 deal with proper usage of the limited number of available wavelengths, and will be addressed in Chapter 10. Subproblem 4 minimizes the effect of store-and-forward (queueing plus transmission) delays at intermediate electronic hops. The remainder of this chapter will address Subproblems 1 and 4.

8.4.2 Previous Work

The problem of designing optimal virtual topologies has been studied before [BaFG90, LaAc91]. Our formulation is more general in the sense that we accommodate many of the physical connectivity constraints which were not considered earlier. In general, the optimal virtual topology problem has been conjectured to be NP-hard, which means that the problem cannot be solved optimally for large problem sizes, unless one resorts to some form of exhaustive search. One instance of this problem has been formulated as a mixed integer linear program which gets difficult to solve with increasing problem size [LaAc91]. Accordingly, heuristic approaches have been employed to solve these problems [BaFG90, LaAc91].

Related work on these problems can be found in [ChGK93, MRBM94, RaSi95, ZhAc94]. Embedding of a packet-switched virtual topology on a physical fiber plant in a switched network was first introduced in [ChGK93], and this network architecture was referred to as a *lightnet.* Some algorithms to embed a hypercube virtual topology were provided in [ChGK93, MRBM94]. The work in [RaSi95] proposes a virtual topology design for packet-switched networks. The average hop distance is minimized, which automatically increases the network traffic supported. The work in [RaSi95] uses the physical topology as a subset of the virtual topology, employing algorithms for maximizing the throughput subject to bounded delay characteristics.

8.4.3 Our Solution Approach

To obtain a thorough understanding of the problem, we concentrate on Subproblems 1 and 4 above, i.e., *for the purposes of this chapter, we do not consider the number of available wavelengths to be a constraint.* In the expanded problem, both the number of wavelengths and their exact assignments are critical, and they are accommodated in Chapters 9 and 10. Specifically, we employ an iterative approach consisting of "simulated annealing" to search for a good virtual topology (Subproblem 1), in conjunction with the "flow deviation" algorithm for optimal (possibly "bifurcated") routing of packet traffic on the virtual topology (Subproblem 4). Also, although the virtual topology can be an undirected graph, we consider lightpaths to be bidirectional in our solution here since most (Internet) network protocols rely on bidirectional paths and links. In addition, we consider Optimization Criterion (2) of Eqn. (8.16) (maximizing offered load) for our illustrative solution below, mainly because we are interested in upgrading an existing fiber-based network to a WDM solution.

We start with a random configuration (virtual topology) and try to find a good virtual topology through simulated annealing by using node-exchange (similar to branch-exchange [LaAc91]) techniques. Then, we *scale up* the traffic matrix to ascertain the *maximum throughput* that can be accommodated by the virtual topology, using flow deviation for packet routing over the virtual topology. For a given traffic matrix, the flow-deviation algorithm minimizes the network-wide packet delay by properly distributing the flows on the virtual links (to reduce the effect of large queueing delays).

We have used measured data over the NSFNET backbone as our sample traffic matrix. We scale up each entry in the traffic matrix by a constant scaleup factor and verify if the offered load from the scaled-up traffic matrix can be accommodated by the virtual topology. Our goal is to design the virtual topology that can accommodate the maximum traffic scaleup. This provides an estimate of the maximum throughput we can expect from the current fiber network if it were to support WDM, and if future traffic characteristics were to model present-day traffic characteristics except for the traffic intensities to grow by a constant scale factor. While it is difficult to predict future traffic characteristics, we believe that our approach provides a reasonable framework for analysis and design.

8.4.4 Simulated Annealing

Simulated annealing (along with genetic algorithms) has been found to provide good solutions for complex optimization problems [AaKo89]. In the simulated annealing process, the algorithm starts with an initial random configuration for the virtual topology. Node-exchange operations are used to arrive at neighboring configurations. In a node-exchange operation, adjacent nodes in the virtual topology are examined for swapping, e.g., if node i is connected to nodes j, a, and b, while node j is connected to nodes p, q, and i in the virtual topology, after the node-exchange operation between nodes i and j, node i will be connected to nodes p, q, and j, while node j will be connected to nodes a, b, and i. Neighboring configurations which give better results (lower average packet delay) than the current solution are accepted automatically. Solutions which are worse than the current one are accepted with a certain probability which is determined by a system control parameter. The probability with which these failed configurations are chosen, however, decreases as the algorithm progresses in time so as to simulate the "cooling" process associated with annealing. The probability of acceptance is based on a negative exponential factor and is inversely proportional to the difference between the current solution and the best solution obtained so far.

The initial stages of the annealing process examine random configurations in the search space so as to obtain different initial starting configurations without getting stuck at a local minimum as in a greedy approach. However, as time progresses, the probability of accepting bad solutions goes down, and the algorithm settles down into a minimum after several iterations. The state become "frozen" when there is no improvement in the objective function of the solution even after a large number of iterations. For further information on simulated annealing, see [AaKo89].

8.4.5 Flow Deviation Algorithm

By properly adjusting link flows, the flow deviation algorithm [FrGK73] provides an optimal algorithm for minimizing the network-wide average packet delay. However, traffic from a given source to a destination may be bifurcated, i.e., different fractions of it may be routed along different paths in order to minimize the packet delay. If the flows are not balanced, then excessively loading of a particular channel may lead to large delays on that channel and thus have a negative influence on the network-wide average packet delay. The algorithm is based on the notion of *shortest-path flows* which first calculates the linear rate of increase in the delay with an infinitesimal increase in the flow on any particular channel. These "lengths" or "cost rates" are used to pose a shortest-path flow problem (which can be solved using one of several well-known algorithms such as Dijkstra's algorithm, Bellman-Ford algorithm, etc.) and the resulting paths represent the "cheapest" paths on which some of the flow may be deviated. An iterative algorithm determines *how much* of the original flow needs to be deviated. The algorithm continues until a certain performance tolerance level is reached.

8.5 Experimental Results

The simulated annealing algorithm as well as the flow deviation algorithm were both used to derive results for the virtual topology design problem, viz., to study Subproblems 1 and 4 outlined in Section 8.4.1. The traffic matrix employed for this mapping is an actual measurement of the traffic on the T1 NSFNET backbone for a 15-minute period (11:45 pm to midnight on January 12, 1992, EST). The raw traffic matrix showing traffic flow in bytes per 15-minute intervals between network nodes is shown in Table 8.1.[5] Nodal

[5]NSFNET backbone data was collected by the `nnstat` program and made available to us by the National Science Foundation (NSF) through its Merit partnership. The traffic matrix shown in Table 8.1 is not exactly the same as the one used in our previous work [MRBM94]. Discrepancies can be attributed to different ways of filtering the raw data. The

Traffic Matrix (multiply by 1000 to get bytes per 15-minute interval)														
	WA	*CA1*	*CA2*	*UT*	*CO*	*TX*	*NE*	*IL*	*PA*	*GA*	*MI*	*NY*	*NJ*	*MD*
WA	531	2682	1171	272	1966	88	538	2490	342	185	3118	967	442	1914
CA1	7191	391	6101	3013	5864	2618	3988	15497	1145	2141	7993	10314	5524	7759
CA2	1092	4757	4	4661	851	3637	866	8567	1003	462	5164	621	1392	2158
UT	702	621	1364	0	191	61	70	288	200	326	1311	1216	239	697
CO	12277	15999	1902	344	36	404	1078	6223	2402	1792	7211	11856	1318	2176
TX	184	1654	343	552	340	0	261	269	88	387	606	482	154	696
NE	3701	6201	10231	448	2204	790	0	11418	1983	2196	15403	9333	2367	16388
IL	1495	23455	21035	852	2822	267	9708	32	4395	3301	9006	7116	2020	8890
PA	8493	1994	3735	601	2499	681	2507	6102	0	3962	11069	14761	4567	6314
GA	186	4193	1026	374	2234	948	499	5708	685	14	3632	2617	1270	1437
MI	1117	3761	5830	507	945	1299	1879	3789	2048	2512	4550	5967	3228	3719
NY	3123	13184	1987	1462	4300	715	1732	5732	3960	2943	21164	7425	2800	6597
NJ	3937	5534	1860	754	842	85	449	2440	11768	3569	6918	7921	707	5220
MD	8191	22701	5429	2296	8928	3182	3270	9185	3061	166	12970	13760	6275	12163

Table 8.1 Traffic matrix (in bytes per 15-minute interval).

distances used are the actual geographical distances and they are not shown here to conserve space. Initially, it was assumed that each node could set up at most four bidirectional lightpath channels, but later more experiments were conducted to study the effect of having higher nodal degree. The number of wavelengths per fiber was assumed to be large enough so that all possible virtual topologies could be embedded.[6]

Some of our results are tabulated in Table 8.2. For each experiment, the maximum scaleup achieved is tabulated along with the corresponding individual delay components, the maximum and minimum link loading as well as the average hop distance.[7] Since the aggregate capacity for the carried traffic is fixed by the number of links in the network, reducing the average hop distance can lead to higher values of load that the network can carry. The queueing delay was calculated using a standard M/M/1 queueing system with a mean packet length calculated from the measured traffic (133.54 bytes per packet) and a link speed of 45 Mbps. *We assume infinite buffers at all nodes.* The "cooling" parameter for the simulated annealing is updated after every 100 acceptances using a geometric parameter of value 0.9. A state is considered "frozen" when there is no improvement over 100 consecutive trials.

8.5.1 Physical Topology as Virtual Topology (No WDM)

Our goal was to obtain a fair estimate of what optical hardware can provide in terms of extra capabilities. In this experiment, we start off with just the existing hardware (as in any electronic, packet-switched network), comprising fiber and point-to-point connections using a single bidirectional lightpath channel per fiber link, i.e., *no WDM is used.* Using flow deviation, the maximum scaleup that could be achieved was found to be 49 (only integer values of the scaleup were considered). The link with the maximum traffic (WA-IL)

---

matrix we are currently using is more accurate. Also, the traffic matrix in Table 8.1 shows that several nodes have "self-traffic." This is due to the fact that nodes in the NSFNET are actually gateways connecting to regional networks, so some of the intra-regional traffic at each node also showed up in our measurements.

[6] If we limit the number of wavelengths supported on each fiber, it might not be possible to set up all possible lightpaths in a virtual topology. There are established bounds on the minimum number of wavelengths required to set up arbitrary virtual topologies for a given number of nodes [Pank92].

[7] The average hop distance for packets is an important figure of merit for the topology chosen. It not only has a direct bearing on the queueing delays that a packet suffers, but more importantly, it determines the maximum offered load for the network. The product of the average hop distance and the offered load in the network equals the aggregate traffic load on all the links.

Parameter	*Physical Topology (No WDM)*	*Multiple Pt-to-Pt Links (No WRS)*	*Arbitrary Virtual Topology (Full WDM)*
Maximum scale up	49	57	106
Avg. Pkt. Delay	11.739 ms	12.5542 ms	17.8146 ms
Avg. Prop. Delay	10.8897 ms	10.9188 ms	14.4961 ms
Queueing Delay	0.8499 ms	1.6353 ms	3.31849 ms
Avg. Hop Distance	2.12186	2.2498	1.71713
Max. Link Loading	98%	99%	99%
Min. Link Loading	32%	23%	71%

Table 8.2 Summary of experimental results.

was loaded at 98%, while the link with the minimum traffic (NY-MD) was at 32%. (These numbers are truncated to show their integer part only.) These values serve as a basis for comparison as to what can be gained in terms of throughput by adding extra WDM optical hardware, viz., tunable transceivers and wavelength routing switches at nodes.

8.5.2 Multiple Point-to-Point Links (No WRS)

The goal of the next experiment was to determine how much throughput we could obtain from the network without adding any photonic switching capability at a node, but by adding extra transceivers (up to four) at each node, i.e., *WDM is used on some links, but no WRS capability is employed at any node.* The initial network had 21 links in the physical topology (see Fig. 8.2). Using extra transceivers at the nodes, we set up extra links on the paths NE-CO, NE-IL, WA-CA2, CA1-UT, MI-NJ, and NY-MD. These lightpaths are chosen manually. Different combinations were considered and the choice of channels which provided the maximum scaleup was chosen. Given 14 nodes, each with a nodal degree of four, we should have been able to have 28 channels. However, the GA node is connected only to TX and PA, both of which are physically connected to four nodes already; hence, they do not have any free transceivers to create an extra channel to GA. Thus, we could add only six new channels. In this case, the maximum scaleup was found to be 57. We found that the two NY-MD channels had a minimum load of only 23%, while the UT-MI channel had a maximum load of 99%.

8.5.3 Arbitrary Virtual Topology (Full WDM)

In the final experiment, we assumed *full WDM with all nodes equipped with WRSs*, i.e., it may now be possible to set up lightpaths between any two nodes. We did not constrain the number of wavelengths supported in each fiber, so that all graphs of degree four were candidates for possible virtual topologies; also, all lightpaths were assumed to be routed over the shortest path on the physical topology. Starting off with a random initial topology, we used simulated annealing to get the best virtual topology. The experiment ran on an unloaded Sparc 10 for approximately three days. The best virtual topology, which is shown in Table 8.3, provided a maximum scaleup of 106. Clearly, the increased scaleup demonstrates the benefits of the WDM-based virtual-topology approach. Now, the minimum loading was on link UT-TX at 71%, while all the other links were above 98% loading.

Source	*Neighbors*
WA	CA1, CA2, MI, UT
CA1	WA, CO, IL, TX
CA2	WA, PA, NE, GA
UT	WA, TX, IL, MD
CO	CA1, MD, NE, GA
TX	CA1, UT, GA, NJ
NE	CA2, CO, IL, MI
IL	CA1, UT, NE, PA
PA	CA2, IL, NY, NJ
GA	CA2, CO, TX, NY
MI	WA, NE, NY, NJ
NY	PA, GA, MI, MD
NJ	TX, PA, MI, MD
MD	CO, NY, NJ, UT

Table 8.3 Virtual topology for nodal degree $P = 4$ and best scaleup (106).

8.5.4 Comparisons

The various delay characteristics [viz., overall average packet delay (FD), average propagation delay encountered by each packet (PD), average queueing delay experienced by each packet (QD), and the mean hop distance (HD)], as functions of the scaleup (throughput) for the above three experiments are shown in Figs. 8.7 through 8.9. The scaleup provides an estimate of the throughput in the network. We note from these figures that the propagation delay is the dominant component of the packet delay. Also, at light loads, the average propagation delay faced by packets in NSFNET is a little over 9 ms (for the given traffic matrix), and this serves as a lower bound on the average packet delay. As a basis for comparison, the coast-to-coast, one-way propagation time in the U.S. is nearly 23 ms. Thus, on an average, each packet travels about 40% of the coast-to-coast distance.

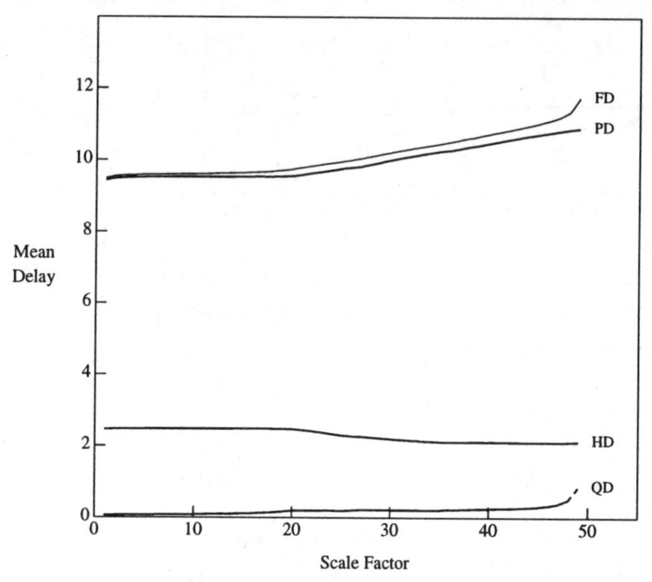

Figure 8.7 Delay vs. throughput (scaleup) characteristics with no WDM, i.e., physical topology as virtual topology.

Note that the average queueing delay increases slightly with increasing traffic until the scaleup nearly reaches its maximum value, after which there is a sharp increase in queueing delay. The propagation delay also increases slightly with increasing scaleup as more traffic is deviated away from shortest path routes by the flow deviation algorithm. One interesting feature is that the average hop distance decreases as the traffic load is increased; this is again because of the flow deviation algorithm which will deviate traffic onto longer links, which might increase the propagation delay encountered by a packet (compared to shortest-path routing), but helps to decrease the average hop distance. Note that our target network in Fig. 8.1 has only 14 nodes and is not very dense; it has a hop distance of a little over 2.0 under the no-WDM case,

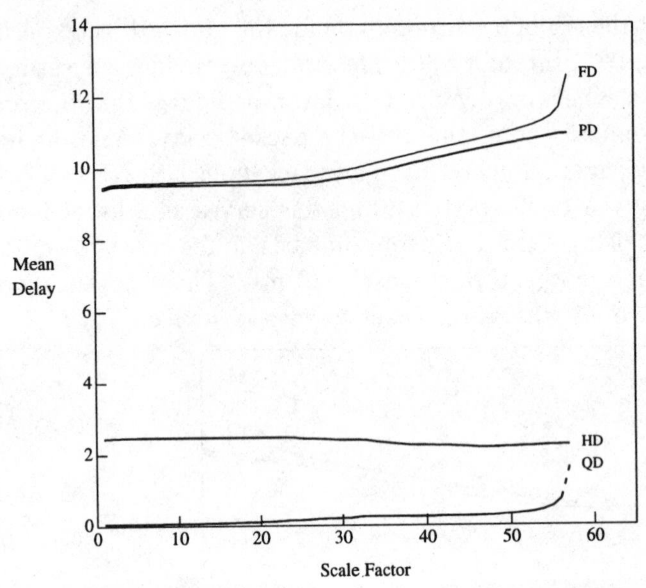

Figure 8.8 Delay vs. throughput (scaleup) characteristics with WDM used on some links, but no WRSs, i.e., multiple point-to-point links are allowed on the physical topology.

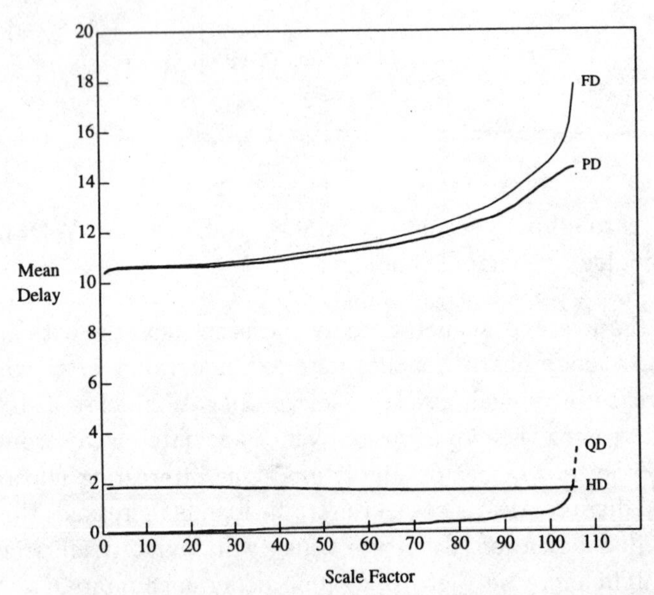

Figure 8.9 Delay vs. throughput (scaleup) characteristics with full WDM on some links and a WRS at each node, i.e., arbitrary virtual topologies are allowed.

so incorporating WDM can only slightly reduce the hop distance which has a lower bound of 1.0 under the fully connected virtual-topology case. Larger networks – which have a much larger hop distance to start with under no WDM – are expected to provide more-dramatic improvements with WDM.

Figure 8.10 plots the different aggregate average packet delays for the three different schemes. The throughput advantage of having no WDM vs. using WDM on a few links but no WRS vs. employing full WDM and WRSs is again clear from this figure. But we notice that the delay in the first two cases is lower than that when using a virtual topology. This is because, in the full WDM case, the shortest path along the physical topology cannot always be chosen because of the virtual topology embedding, so that some packets may have to travel longer distances, in general. However, the scaleup in the virtual topology is much more than that for the other two schemes; so the addition of switches at intermediate nodes to perform wavelength routing provides a significant improvement in throughput for the network.

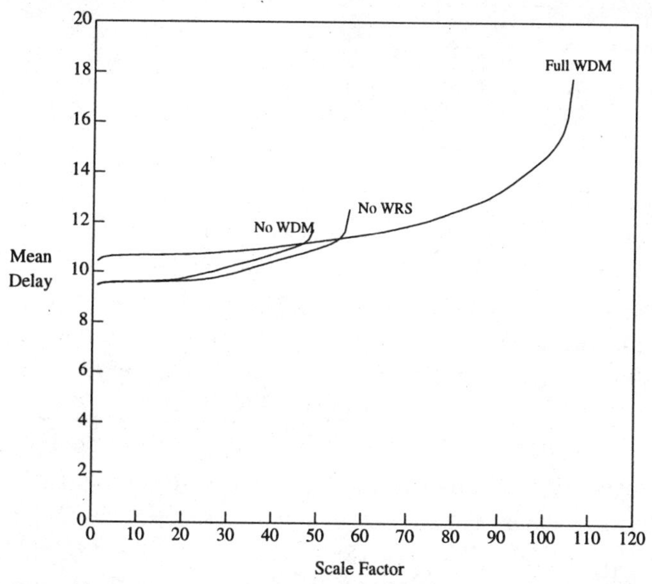

Figure 8.10 Delay vs. throughput (scaleup) characteristics for different virtual topologies.

8.5.5 Effect of Nodal Degree and Wavelength Requirements

So far, the nodal degree (P) was four. Now, we consider full WDM (with a WRS at each node), and increase the nodal degree to five and six. We find that the maximum scaleup increased nearly proportionally with increasing nodal degree. Actually, with the scaleup of 106 for $P = 4$ as a baseline, proportional increase in scaleup for $P = 5$ and 6, would yield 132.5 and 159, respectively. However, in our experiments, the observed maximum scaleups for $P = 5$ and 6 were higher, viz., 135 and 163, respectively (refer to Table 8.4). This is due to the fact that, as the nodal degree is increased, the average hop distance of the virtual topology is reduced, which provides the extra improvement in the scaleup. Minimizing hop distance can be an important optimization problem, and is studied in Chapter 9.

Transceivers/node	*Scaleup*
4	106
5	135
6	163

Table 8.4 Traffic scaleups for different nodal degrees.

Although no constraints on wavelengths per fiber were imposed in this study, we also examined the wavelength requirements to set up a virtual topology using shortest-path routing of lightpaths on the physical topology. Assuming no limit on the supply of wavelengths, but with the wavelength constraints as outlined before (Section 8.3), the maximum number of wavelengths required for embedding the best virtual topology (which provided the maximum scaleup) with degree $P = 4$, 5, and 6 were found to be 6, 8, and 8 wavelengths, respectively. The corresponding distributions of the number of wavelengths used in each of the 21 fiber links of the NSFNET (see Fig. 8.1) are plotted in Fig. 8.11. We find that, with increasing nodal degree, i.e., with an increasing number of lightpaths to be supported, the average number of wavelengths a fiber needs to support increases. However, due to the combination of reasons such as desired virtual topology, shortest-path routing of lightpaths, and wavelength constraints, it may so happen that there is no link on the physical topology that employs all of the required wavelengths. This happened for our $P = 6$ experiment, i.e., although eight wavelengths were required to embed the virtual topology, no physical link carried all eight wavelengths.

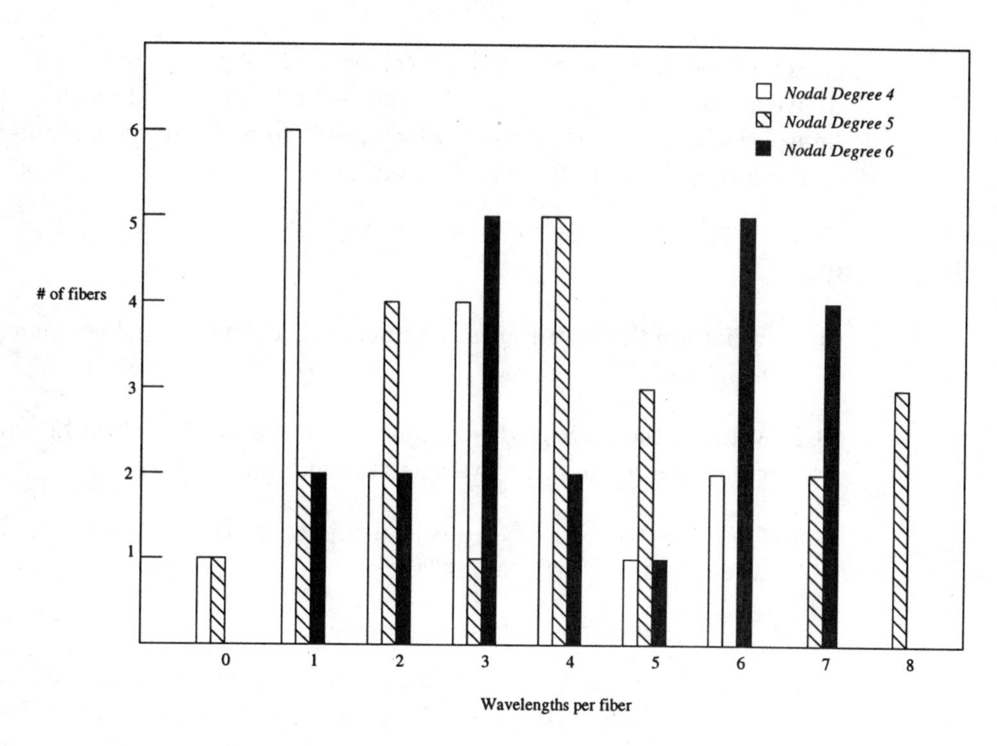

Figure 8.11 Distributions of the number of wavelengths used in each of the 21 fiber links of the NSFNET for the virtual topology approach with nodal degree P = 4, 5, and 6.

8.6 Summary

This chapter explored design principles for next-generation optical wide-area networks, employing wavelength-division multiplexing (WDM) and targeted to nationwide coverage. We showed that such a WDM-based network architecture can provide a high aggregate system capacity due to spatial reuse of wavelengths. Our objective was to investigate the overall design, analysis, upgradability, and optimization of a nationwide WDM network consistent with device capabilities.

The virtual topology optimization problem discussed in this chapter serves as an illustration, and it is a first step toward a robust and versatile WDM WAN solution. A significant amount of room exists for developing improved approaches and algorithms, and a number of research issues must be addressed in this regard. An interesting avenue of research is to study how routing and wavelength assignment of lightpaths can be combined with the choice of virtual topology and its corresponding packet routing in order to arrive at an optimum solution. Dynamic establishment and reconfiguration of lightpaths is an important issue which needs to be thoroughly studied. Some of these topics are studied in Chapters 9 and 10.

Exercises

8.1. What are the advantages of embedding a virtual topology on a physical topology?

8.2. What is the wavelength-continuity constraint? Which of the constraint equations relate to the wavelength-continuity constraint?

8.3. Consider the NSFNET physical topology shown in Fig. 8.1. Remove nodes WA, CO, NE, and GA.

(a) Draw the new physical topology.

(b) Set up lightpaths on the new topology so that the resulting virtual topology is the Petersen graph. In your virtual topology, what is the maximum number of wavelengths used on any link in the network?

(c) Show the details of the UT (Utah) switch (similar to the one in Fig. 8.4.

(d) Assume a uniform traffic matrix, i.e., equal amount traffic between any two nodes. Assume that packets are routed via the shortest path. Calculate the average packet hop distance when packets are routed over the physical topology and when packets are routed over the virtual topology.

8.4. Give a set of lightpaths which embeds a 4×4 Manhattan Street Network with bidirectional links onto the NSFNET physical topology shown in Fig. 8.2. How many wavelengths are required for this embedding?

8.5. For the physical NSFNET topology shown in Fig. 8.2, find a logical ring embedding which uses only one wavelength.

8.6. The embedding shown in Fig. 8.6 assumes that all of the local laser-filter pairs at a node operate on different wavelengths. What are the advantages of this design? What are the disadvantages?

8.7. In Fig. 8.10, at low loads, the average delay in a network employing full WDM is more than the average delay in a network with no WDM. Why?

Chapter 9

Virtual Topology: LP, Cost, Reconfiguration

For the same wavelength-routed network setting as in Chapter 8, this chapter presents an *exact linear programming* formulation for the complete virtual topology design, including choice of the constituent lightpaths, routes for these lightpaths, and intensity of packet flows through these lightpaths. By making a shift in the objective function to minimal hop distance and by relaxing the wavelength-continuity constraints (i.e., assuming wavelength converters at all nodes), we demonstrate that the entire optical network design problem can be *linearized* and hence solved *optimally.*

The objective function is to minimize the average packet hop distance which is inversely proportional to the total network throughput under balanced flows through the lightpaths. The linear problem formulation can be used to design a balanced network, such that the utilization of both transceivers and wavelengths is maximized, thus reducing cost of the network equipment. We analyze the trade-offs in budgeting of resources (transceivers and switch sizes) in the optical network, and demonstrate how an improperly designed network may have low utilization of one of these resources. We also use the linear problem formulation to provide a reconfiguration methodology in order to adapt the virtual topology to changing traffic conditions.

9.1 Introduction

This chapter presents the design of a lightpath-based optical network as a *linear* optimization problem, and uses the problem formulation to derive an *exact* optimal network design. The formulation presented here is a modified version of the problem formulation given in Chapter 8, which – though complete – contained nonlinear equations, and was difficult to solve exactly. We simplify the objective function to minimize the average packet hop distance[1] (which is inversely proportional to the network throughput under balanced network flows and which is a linear objective function).

If the channel (lightpath) capacity is C, the number of lightpaths is L, and the average packet hop distance is H, then the network throughput is bounded by:

$$T \leq \frac{CL}{H} \tag{9.1}$$

Therefore, minimizing H and maximizing the network throughput are equivalent in an asymptotic sense when the equality can be satisfied.[2] A linear program (LP) formulation is developed to minimize H, the average packet hop distance, for a virtual-topology-based, wavelength-routed network. The LP provides a complete specification to the virtual topology design, routing of the constituent lightpaths, and intensity of packet flows though the lightpaths.

Section 9.2 presents the mathematical problem formulation as an LP. This section also discusses some of our underlying assumptions that can drastically reduce the size of the problem, and consequently minimize the running time of the solution. Section 9.3 presents two simple heuristics with fast running times whose performance compares favorably with the performance of the optimized solution (obtained from the LP solution). Heuristics become important when the size of the problem becomes larger than what an LP solver can handle, or when the optimization needs to be achieved in real time; the proposed heuristics will be demonstrated to perform well with respect to the optimal bound (viz., LP output). Section 9.4 examines trade-offs that affect the quality of the solution, and discusses resource-budgeting strategies for transmission/reception as well as switching equipment. This approach allows the design of a balanced network, which can provide "optimal" net-

---

[1] The average packet hop distance is defined as the number of lightpaths that a packet has to traverse on average, and is a function of the virtual topology.

[2] In a non-WDM network, the number of channels is given by M, the number of fiber links in the network. Therefore, the non-WDM network throughput is given by $T_p \leq CM/H_p$, where H_p denotes the average packet hop distance in the non-WDM network.

work throughput along with high utilization of transceivers and wavelengths. Section 9.5 proposes an algorithm which can be used for virtual topology reconfiguration. The proposed algorithm computes a new virtual topology from an existing virtual topology, such that the new virtual topology is optimal with respect to the changing traffic patterns; among all such optimal virtual topologies, the algorithm selects the topology which is "closest" in structure to the current virtual topology. Section 9.6 provides numerical simulation results obtained from solving the above formulation for the NSFNET topology (see Fig. 9.1). From our LP output, two sample embeddings of a virtual topology over the NSFNET for two and five wavelengths are shown in Fig. 9.2; they indicate that, even though full wavelength conversion capabilities are assumed at all nodes, in reality, only a few wavelength converters are needed in the network anyway.

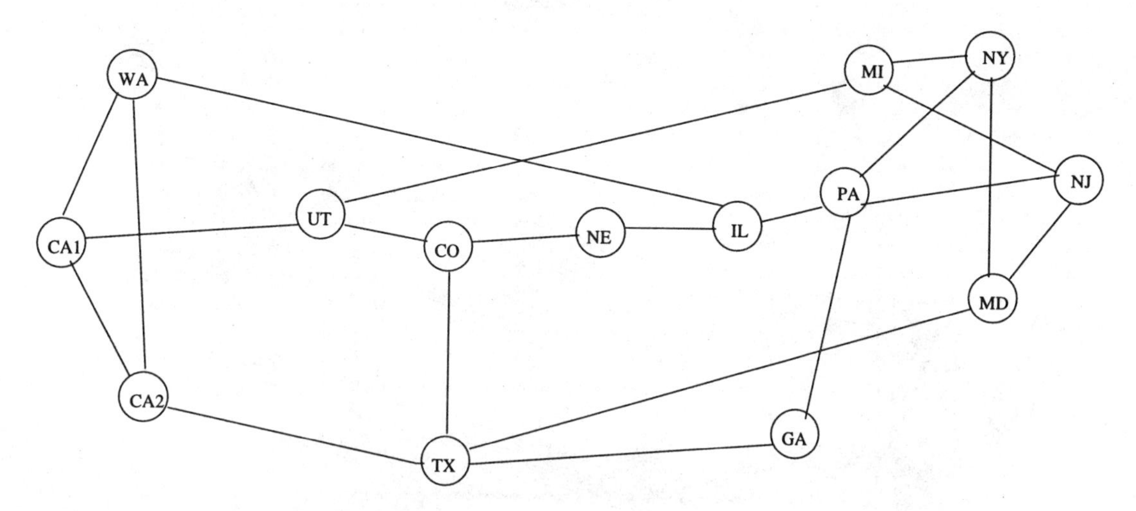

Figure 9.1 NSFNET T1 backbone.

For clarity of exposition, this chapter deals in detail with one LP approach to virtual topology design. A summary of other approaches to virtual topology design in a WDM WAN may be found in Chapter 12, Section 12.3.

9.2 Problem Specification

Much of the notation and the constraints used in this section are borrowed from Chapter 8. They are repeated here for completeness of the problem specification. New material, relative to that in Chapter 8, include a new *linear*

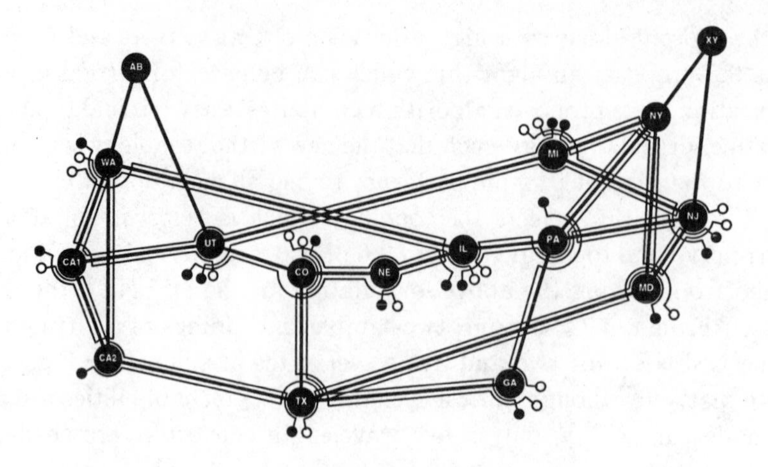

(a) two-wavelength solution

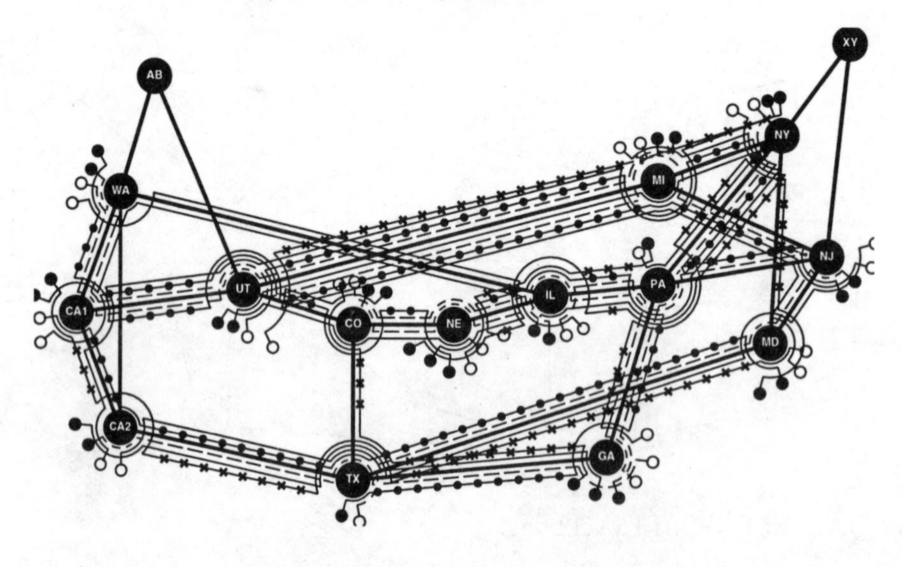

(b) five-wavelength solution

Figure 9.2 Optimal solution for a two-wavelength and a five-wavelength network. Each physical link consists of two unidirectional fibers carrying transmissions in opposite directions (hence, each wavelength may appear twice on any link in the diagrams; their signal propagation directions are opposite to each other in such cases). Wavelength 0 is used to embed the physical topology over the virtual topology, so the Wavelength-0 lightpaths are not shown explicitly in these diagrams to preserve clarity. Note: ∘ = transmitter; • = receiver.

objective function (Eqn. (9.2)), a constraint to bound the lightpath length (Eqn. (9.17)), a constraint to bound the maximum loading per channel (Eqn. (9.26)), a constraint to incorporate the physical topology as part of the virtual topology (Eqn. (9.16)), all of the simplifying assumptions in Section 9.2.2 (to make the problem linear and tractable), and several other generalizations (multiple fibers between nodes, multiple lightpaths between node pairs, etc.).

9.2.1 Linear Formulation

We formulate the problem as an optimization problem, using principles from multicommodity flow for physical routing of lightpaths and traffic flow on the virtual topology, and using the following notation:

- s and d used as subscript or superscript denote *source* and *destination* of a packet, respectively,
- i and j denote *originating* and *terminating* nodes, respectively, in a lightpath, and
- m and n denote *endpoints of a physical link* that might occur in a lightpath.

Given:

- Number of nodes in the network = N.
- Maximum number of wavelengths per fiber = W (a system-wide parameter).
- Physical topology P_{mn} denotes the number of fibers interconnecting node m and node n. $P_{mn} = 0$ for nodes which are not physically adjacent to each other. $P_{mn} = P_{nm}$ indicates that there are an equal number of fibers joining two nodes in different directions. Note that there may be more than one fiber link connecting adjacent nodes in the network. $\sum_{m,n} P_{mn} = M$ denotes the total number of fiber links in the network.
- Fiber length matrix, viz., fiber distance d_{mn} from node m to node n. For simplicity in expressing packet delays, d_{mn} is expressed as a propagation delay (in time units). Note that $d_{mn} = d_{nm}$, and $d_{mn} = \infty$ if $P_{mn} = 0$.

- Shortest-path delay matrix D where D_{sd} denotes the delay (sum of propagation delays only) over the shortest path between nodes s and d.

- Lightpath length bound $\alpha, 1 \leq \alpha < \infty$, bounds the delay over a lightpath (and hence the length of the lightpath) between two nodes i and j, with respect to the shortest path delay D_{ij} between them, i.e., the maximum permissible propagation delay over the lightpath between the two nodes i and j is αD_{ij}.

- Number of transmitters at node $i = T_i$ ($T_i \geq 1$). Number of receivers at node $i = R_i$ ($R_i \geq 1$). In general, we would assume that $T_i = R_i, \forall i$, although this is not a strict requirement.

- Traffic matrix Λ_{sd} which denotes the average rate of traffic flow (in packets per second) from node s to node d, with $\Lambda_{ss} = 0$ for $s, d = 1, 2, \ldots, N$.

- Capacity of each channel = C (normally expressed in bits per second, but converted to units of packets per second by knowing the mean packet length).

- Maximum loading per channel = $\beta, 0 < \beta < 1$. β restricts the queueing delay on a lightpath from getting unbounded by avoiding excessive link congestion. *We do not incorporate queueing delays explicitly in the problem formulation, under the assumption that they are negligible for suitably chosen values of* β. Also, previous results in Chapter 8 indicate that queueing delays are negligibly small compared to propagation delays for a large network setting as in Fig. 9.1, except under *extremely heavy loading.*

Variables:

- Virtual topology: The variable V_{ij} denotes the number of lightpaths from node i to node j in the virtual topology. Note that the current formulation is general, since lightpaths are not necessarily assumed to be bidirectional, i.e., $V_{ij} = 0$ does not imply that $V_{ji} = 0$. As an example, Fig. 9.2 contains a lightpath from CA1 to TX, but not in the reverse direction. Therefore, $V_{CA1,TX} = 1$. Moreover, there may be multiple lightpaths between the same source-destination pair, i.e., $V_{ij} > 1$, for the case when traffic between nodes i and j is greater than a single lightpath's capacity (C).

- Traffic routing : The variable λ_{ij}^{sd} denotes the traffic flowing from node s to node d and employing V_{ij} as an intermediate virtual link. Note that traffic from node s to node d may be "bifurcated," with different components flowing through different sets of lightpaths.

- Physical topology route : The variable p_{mn}^{ij} denotes the number of lightpaths between nodes i and j being routed though fiber link mn. For example, since the lightpath from CA1 to TX passes through CA2, the variables $p_{CA1,CA2}^{CA1,TX} = 1$ and $p_{CA2,TX}^{CA1,TX} = 1$.

Objective: Optimality Criterion

$$\textbf{Minimize}: \quad \frac{1}{\sum_{s,d} \Lambda_{sd}} \sum_{i,j} \sum_{s,d} \lambda_{ij}^{sd} \tag{9.2}$$

The objective function minimizes the average packet hop distance in the network. This is a linear objective function because $\sum_{i,j} \sum_{s,d} \lambda_{ij}^{sd}$ is a linear sum of variables, while $\sum_{s,d} \Lambda_{sd}$ is a constant for a given traffic matrix. The two objective functions used in Chapter 8 were (1) minimization of the average packet delay over the network, and (2) minimization of the maximum traffic flow in a lightpath; both were nonlinear.

Constraints:

- On virtual topology connection matrix V_{ij}:

$$\sum_j V_{ij} \leq T_i \quad \forall i \tag{9.3}$$

$$\sum_i V_{ij} \leq R_j \quad \forall j \tag{9.4}$$

$$V_{ij} \quad \text{int} \tag{9.5}$$

The above equations ensure that the number of lightpaths emerging from a node is constrained by the number of transmitters at that node, while the number of lightpaths terminating at a node are constrained by the number of receivers at that node. V_{ij} variables can only hold integer values. If V_{ij} has a value greater than 1, it means that there is more than one lightpath between the particular source-destination pair. These lightpaths may follow the same route or different routes through the network.

- On physical route variables p_{mn}^{ij}:

$$\sum_m p_{mk}^{ij} = \sum_n p_{kn}^{ij} \quad if\ k \neq i, j \tag{9.6}$$

$$\sum_n p_{in}^{ij} = V_{ij} \tag{9.7}$$

$$\sum_m p_{mj}^{ij} = V_{ij} \tag{9.8}$$

$$\sum_{i,j} p_{mn}^{ij} \leq WP_{mn} \tag{9.9}$$

$$p_{mn}^{ij} \quad \text{int} \tag{9.10}$$

Equations (9.6) through (9.8) are multicommodity-flow-based equations governing the routing of lightpaths from source to destination. Equations (9.9) ensure that the number of lightpaths flowing through a fiber link does not exceed W. Note, however, that the equations do not ensure the wavelength-continuity constraint (under which the lightpath is assigned the same wavelength on all the fiber-links through which it passes). In the absence of wavelength-continuity constraints in the equations, the solution obtained from our current formulation may require the network to be equipped with wavelength converters.

- On virtual topology traffic variables λ_{ij}^{sd}:

$$\sum_j \lambda_{sj}^{sd} = \Lambda_{sd} \tag{9.11}$$

$$\sum_i \lambda_{id}^{sd} = \Lambda_{sd} \tag{9.12}$$

$$\sum_i \lambda_{ik}^{sd} = \sum_j \lambda_{kj}^{sd} \quad if \quad k \neq s, d \tag{9.13}$$

$$\lambda_{ij}^{sd} \leq \Lambda_{sd} V_{ij} \tag{9.14}$$

$$\sum_{s,d} \lambda_{ij}^{sd} \leq \beta V_{ij} C \tag{9.15}$$

Equations (9.11) through (9.13) are multicommodity-flow equations governing the flow of traffic through the virtual topology. Note that the routing of traffic from a given source to a given destination may be "bifurcated." Equations (9.14) ensure that traffic can only flow through an existing lightpath, while Eqns. (9.15) specify the capacity constraint in the formulation.

- Optional constraints:

  1. Physical topology as a *subset* of the virtual topology:

$$P_{mn} = 1 \quad \Rightarrow \quad V_{mn} = 1, p_{mn}^{mn} = 1 \tag{9.16}$$

  2. Bound lightpath length:

$$\sum_{m,n} p_{mn}^{ij} d_{mn} \quad \leq \quad \alpha D_{ij},\ for\ K\ alternate\ paths \tag{9.17}$$

Equations (9.16) and (9.17) are optional, and may be incorporated to ensure bounded packet delays in the network. These equations reduce the solution space of the problem, and could theoretically affect the optimality of the solution. However, in general, their effect on the solution quality is found to be minimal for most networks of interest.

Equations (9.16) embed the physical topology as a subset of the virtual topology, i.e., every link in the physical topology is also a lightpath in the virtual topology, in addition to which there are lightpaths which span multiple fiber links, e.g., Fig. 9.2(a) demonstrates a virtual topology embedding for a two-wavelength solution. This approach for choosing lightpaths can satisfy packets with the tightest delay constraints [RaSi96]. The lightpaths corresponding to the physical topology may also be used to route network control messages efficiently, and this approach can simplify network management. For these equations to be valid, $T_n \geq \delta_n^o$ and $R_n \geq \delta_n^i$, where δ_n^o and δ_n^i denote the physical number of fibers emerging from and terminating at node n, respectively.

Equations (9.17) may be used to restrict the enumerated p_{mn}^{ij} variables to be only among those present in K alternate shortest paths from i to j, where $K \geq 1$. These equations prevent long and convoluted lightpaths, i.e., lightpaths with an unnecessarily long route instead of a much shorter route, from occurring. The value of K may be selected by the network designer. We choose $K = 2$ in our experiments reported in Section 9.6, and found this choice to work well (relative to higher values of K).

9.2.2 Simplifying Assumptions

This section outlines some simplifying assumptions to make the problem more tractable.

- *Wavelength-continuity constraints for a lightpath are intentionally ignored* in the current formulation, which only ensures that the total number of lightpaths routed through a fiber is less than or equal to W. Adding wavelength-continuity constraints to the above set of equations significantly increases the complexity of the problem,[3] e.g., if the variable $c_k^{ij} = 1$ signifies that a lightpath from node i to node j is assigned the wavelength k (where $k = 1, 2, \ldots, W$), the relevant equations are as follows:

$$\sum_k c_k^{ij} = V_{ij} \tag{9.18}$$

$$\sum_{ij} p_{mn}^{ij} . c_k^{ij} \leq 1 \quad \forall m, n, k \tag{9.19}$$

  Equations (9.19) are nonlinear because they involve the product of two variables. Therefore, we intentionally ignore the wavelength assignment of lightpaths in the current problem formulation,[4], assuming that the wavelength-assignment problem will be solved separately, based on the lightpath routes obtained through the current formulation, or that wavelength converters are available at the routing nodes.

- *Queueing delays are also intentionally ignored*, partly to simplify (linearize) the objective function, and also because it has been observed that the propagation delay dominates the overall network delay in nationwide optical networks as in Fig. 9.1. The exact optimization function for delay minimization is as follows (see Chapter 8):

$$\textbf{Minimize :} \quad \sum_{ij} \left[\sum_{sd} \lambda_{ij}^{sd} \left(\sum_{mn} p_{mn}^{ij} d_{mn} + \frac{1}{C - \sum_{sd} \lambda_{ij}^{sd}} \right) \right] \tag{9.20}$$

  This is a nonlinear equation because it involves the product of two variables, λ_{ij}^{sd} and p_{mn}^{ij}, and also because the term $1/(C - \sum_{sd} \lambda_{ij}^{sd})$ is nonlinear in $\{\lambda_{ij}^{sd}\}$.

[3] The wavelength assignment problem has been demonstrated to be NP-complete [ChGK93].

[4] The efficacy of using wavelength converters in the routing nodes is currently an active area of research. It has been shown that, in many situations, networks with sparse wavelength conversion have performance nearly equivalent to that of networks with full wavelength conversion (see Chapter 11).

- *The number of variables and equations in the formulation are reduced.* The number of variables and equations in the original problem formulation grows as $O(N^4)$, and can very easily overwhelm today's state-of-the-art computing facilities. To make the problem more tractable, we reduce the number of constraints by pruning the search space. Pruning is based on tracking a limited number of alternate shortest paths, denoted by K, between source-destination pairs, such that the selected routes are within a constant factor ($\alpha \geq 1$) of the shortest-path distance between the given source-destination pair. We assume that traffic flow will only use the lightpaths which interconnect nodes present in these alternate paths, i.e., all values of λ_{ij}^{sd} are not enumerated. Likewise, lightpaths may only be routed through one of a few permissible routes, i.e., all possible values of p_{mn}^{ij} are not enumerated. Since these assumptions are incorporated during the generation of the problem formulation, it helps reduce the total number of equations and variables. The amount of pruning (hence, the value of K) required is a function of the size of the problem that can be solved in "reasonable time" by the chosen LP solver. In our experiments, we used the package *lpsolve* [Berk94], running on an unloaded DEC-Alpha, but were still restricted to using two alternate shortest paths for the network in Fig. 9.1 in order to get reasonable running times.

  To understand the pruning process, let us consider the NSFNET topology in Fig. 9.1. If we consider two alternate shortest paths between any source-destination pair, then the two alternate paths from node (CA) to node (IL) may be CA1-UT-CO-NE-IL and CA1-WA-IL. Then, the enumerated variables for lightpath routing are as follows:

$$p_{CA1,UT}^{CA1,IL}, p_{UT,CO}^{CA1,IL}, p_{CO,NE}^{CA1,IL}, p_{NE,IL}^{CA1,IL}, p_{CA1,WA}^{CA,IL}, p_{WA,NE}^{CA1,IL}$$

  Likewise, the enumerated variables for packet routing are as follows:

$$\lambda_{CA1,UT}^{CA1,IL}, \lambda_{CA1,CO}^{CA1,IL}, \lambda_{CA1,NE}^{CA1,IL}, \lambda_{CA1,IL}^{CA1,IL}, \lambda_{UT,CO}^{CA1,IL}, \lambda_{UT,NE}^{CA1,IL}, \lambda_{UT,IL}^{CA1,IL},$$

$$\lambda_{CO,NE}^{CA1,IL}, \lambda_{CO,IL}^{CA1,IL}, \lambda_{NE,IL}^{CA1,IL}, \lambda_{CA1,WA}^{CA1,IL}, \lambda_{WA,IL}^{CA1,IL}.$$

- The current formulation allows bifurcated routing of packet traffic. To specify nonbifurcated routing of traffic, we use new variables γ_{ij}^{sd} which are only allowed to take binary values, and the equations are suitably modified. Under nonbifurcated routing, Eqns. (9.11) through (9.15) become:

$$\gamma_{ij}^{sd} \in 0,1 \tag{9.21}$$

$$\sum_j \gamma_{sj}^{sd} = 1 \tag{9.22}$$

$$\sum_i \gamma_{id}^{sd} = 1 \tag{9.23}$$

$$\sum_i \gamma_{ik}^{sd} = \sum_j \gamma_{kj}^{sd} \quad if \quad k \neq s,d \tag{9.24}$$

$$\gamma_{ij}^{sd} \leq V_{ij} \tag{9.25}$$

$$\sum_{s,d} \gamma_{ij}^{sd} \Lambda_{sd} \leq \beta V_{ij} C \tag{9.26}$$

The objective function becomes:

$$\textbf{Minimize :} \quad \frac{1}{\sum_{s,d} \Lambda_{sd}} \left(\sum_{i,j} \sum_{s,d} \gamma_{ij}^{sd} \Lambda_{sd} \right) \tag{9.27}$$

We only used bifurcated routing in our experiments, since it was found that nonbifurcated routing of packet traffic significantly increased the running time of the optimization solution. The increase in running time is primarily due to the computation of the product terms in Eqns. (9.26) and in the modified objective function. Note, however, that these equations are also strictly linear.

9.3 Heuristic Approaches

This section presents two heuristic approaches that allow us to solve large problem instances of the virtual-topology design problem, in order to minimize the average packet hop distance. Heuristics become important when the problem formulation becomes large due to an increase in the physical size of the network, and becomes difficult to solve by traditional LP methods due to computational constraints. Results of these heuristics compare favorably with the optimal result obtained by solving the exact problem formulation, for small to medium-sized networks that can be solved exactly by the LP method.

To ensure that the results from the heuristics are comparable with those from the optimization formulation, we do not impose wavelength-continuity constraints on the lightpath routing, although the heuristics can easily accommodate this feature without any sacrifice in their running time.

- **Maximizing Single-Hop Traffic.** This simple heuristic attempts to establish lightpaths between source-destination pairs with the highest Λ_{sd} values, subject to constraints on the number of transceivers at the two end nodes, and the availability of a wavelength in some path connecting the two end nodes. Pseudo-code for this heuristic follows.

```
procedure MaxSingleHop(void)
  while (not done)
    Find Λs'd' = Max (Λsd)
     if ((free transmitter available at s') AND
       (free receiver available at d') AND
       (free wavelength available in any alternate path from s' to d'))
       begin
           Establish lightpath between s' and d'
           Λs'd' = Λs'd' − C
       end
     endif
  endwhile
```

- **Maximizing Multihop Traffic.** In a packet-switched network, the traffic carried by a link may include forwarded traffic as well as traffic originating from that node. Intuitively, it seems that any lightpath establishment heuristic which accounts for the forwarded traffic that the lightpath will carry should provide better performance than a heuristic which only tries to maximize the single-hop traffic. This intuition led to the derivation of the current heuristic. The heuristic starts with the physical topology as the initial virtual topology, and attempts to add more lightpaths one by one. The performance of this heuristic is found to be slightly better than that of the previous heuristic (see Section 9.6).

  Let H_{sd} denote the number of electronic hops needed to send a packet from source s to destination d. The heuristic attempts to establish lightpaths in decreasing order of $\Lambda_{sd}(H_{sd}-1)$, subject to constraints on the number of transceivers at the two end nodes, and the availability of a wavelength in some path connecting the two end nodes.[5] After each lightpath is established, H_{sd} values are recalculated, as traffic flows might have changed due to the new lightpath, in order to minimize the average packet hop distance. This algorithm allows only a single lightpath to be established between any source-destination pair. Pseudo-code for this heuristic is provided below.

```
procedure MaxMultiHop(void)
  Initial Virtual Topology = Physical Topology
  while (not done)
    Compute H_sd ∀s, d
    Find Λ_s'd'(H_s'd' − 1) = Max (Λ_sd(H_sd − 1))
      if ((free transmitter available at s') AND
        (free receiver available at d') AND
        (free wavelength available in any alternate path from s' to d'))
        begin
            Establish lightpath between s' and d'
        end
      endif
  endwhile
```

9.4 Network Design: Resource Budgeting and Cost Model

This section discusses some of the network design principles that can be derived from the linear problem formulation. We present a cost model for the network design, in terms of the costs for the transmission as well as the switching equipment, which may be used in conjunction with the optimization formulation, to derive a minimum cost solution.

9.4.1 Resource Budgeting

It is intuitive that, in a network with a very large number of transceivers per node, but with very few wavelengths per fiber and few fibers between node pairs, a large number of transceivers may be unused because some lightpaths may not be establishable due to wavelength constraints. Similarly, a network with few transceivers but a large number of available wavelengths may have

[5] This approach is similar to the longest-lightpath-first strategy proposed in [ChGK93]; however, in that case, the lightpaths to be established were known a priori. In the current problem specification, we need to select the lightpaths in addition to routing them.

a large number of wavelengths unutilized because the network is transceiver-constrained.

This mismatch in transceiver vs. wavelength utilization has a direct impact on the cost of the network. The number of wavelengths supported in the network determines the cost of the switching equipment.[6] Likewise, the number of transceivers per node determines the cost of the terminating equipment. We would like to balance these network resources, in order to maximize the utilizations of both the transceivers and the wavelengths in the network. Resource budgeting becomes an important issue when we attempt to optimize network design with constrained total cost.

A very simple analysis leads to some insights into the resource-budgeting problem. Given a physical topology, and a routing algorithm for lightpaths, we can determine the average length of a lightpath (in terms of the number of fiber links traversed by a lightpath, averaged over all source-destination pairs in the network); let the average length of a lightpath be denoted by H_P. If there are M fiber links in the network, each supporting W wavelengths, then the maximum number of lightpaths that can be supported is MW/H_P, assuming uniform utilization of wavelengths on all fiber links.[7] Therefore, the number of transceivers per node should be approximately:

$$T_i = R_i \approx \frac{MW}{NH_P} \tag{9.28}$$

in order to get a balanced network. Our optimization-based and heuristic-based network simulations on the NSFNET reinforce this conjecture (see Section 9.6).

9.4.2 Network Cost Model

Resource budgeting in the network has a direct impact on the cost of setting up the network. In this chapter, we only consider equipment costs.[8] Our

[6] A WRS (see Fig. 9.3) with δ input ports and δ output ports (including the ports used to connect to the local node used for electronic termination of lightpaths), supporting W wavelengths, requires W $\delta \times \delta$ wavelength-insensitive optical switches. Increasing the number of fibers interconnecting two nodes increases δ, and increasing W increases the number of crosspoint switching elements required; thus, in both cases, the cost of the switching equipment would increase.

[7] Note, however, that in many cases, the routing on the network is dependent on the wavelength congestion in the network, and any static routing policy may not yield very accurate results.

[8] A lot of fiber has already been buried in the ground, and hence, for our purposes here, the cost of laying fiber is assumed to be zero.

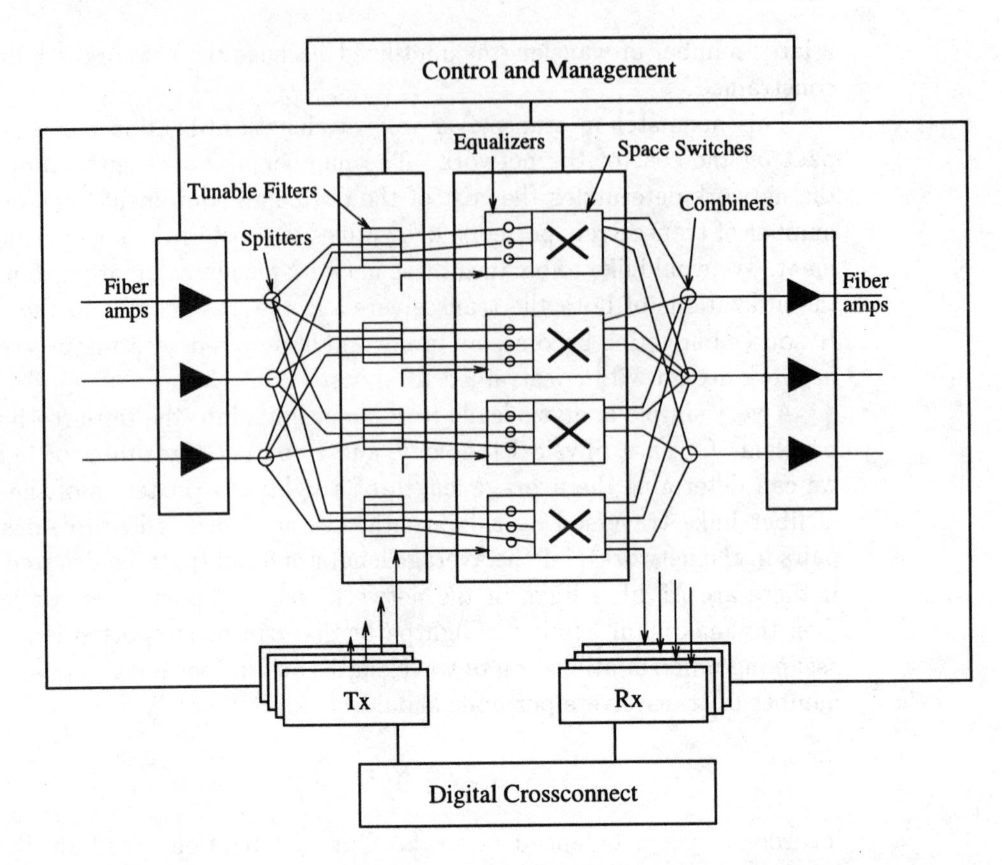

Figure 9.3 Transport node in the RACE WDM optical network architecture.

model of the WRS is based on the prototype used in the RACE project (see Fig. 9.3) [Hill93]. This cost model is valid for $W \geq 2$ wavelengths.

- Let C_t be the cost of a transceiver. Then, the aggregate network-wide equipment cost for transceivers is $C_t(\sum_i T_i + \sum_i R_i)$.

- Let C_m be the cost of a multiplexer or a demultiplexer. Then, the aggregate cost of multiplexers/demultiplexers in the network is given by $C_m(2M + \sum_{i=1}^N \lceil T_i/W \rceil + \sum_{j=1}^N \lceil R_j/W \rceil)$, where $(\sum_{i=1}^N \lceil T_i/W \rceil + \sum_{j=1}^N \lceil R_j/W \rceil)$ denotes the cost of providing (de)multiplexers for the local access ports needed to launch or terminate lightpaths.

- Let C_x be the cost of a 2×2 optical crosspoint switching element. Then, the cost of a switch with δ input and δ output ports built from

2×2 optical switching elements arranged in a multistage interconnection network (MIN) is $C(\delta) = C_x \delta \log \delta/2$. There is a MIN switch per wavelength in a WRS; hence, the cost for node m with degree $\delta_m = \sum_{n=1}^{N} P_{mn} + \lceil T_m/W \rceil$ is $WC(\delta_m)$. The total network cost is therefore $W \sum_{m=1}^{N} C(\delta_m)$.

- The cost of a wavelength converter may also be incorporated in the current model; we, however, choose to ignore the converter cost in this chapter, due to lack of a clear consensus regarding the exact architecture (and, hence, cost) of optical switches based on wavelength converters. Also, very few wavelength converters are needed anyway [SuAS96], and as our illustrative example in Section 9.6 will show.

To summarize, the total cost of a wavelength-routed optical network, excluding fiber-layout cost, may be expressed as

$$\mathbf{C} = C_t \left(\sum_i T_i + \sum_i R_i \right) + C_m \left(2M + \sum_{i=1}^{N} \left\lceil \frac{T_i}{W} \right\rceil + \sum_{j=1}^{N} \left\lceil \frac{R_j}{W} \right\rceil \right) + WC_x \left(\sum_{m=1}^{N} \delta_m \log \delta_m \right) /2 \quad (9.29)$$

for $W \geq 2$.

Using the above equation, and assuming that $C_t = \$5,000, C_m = \$100$, and $C_x = \$1,000$, Table 9.1 demonstrates the equipment cost of building a wavelength-routed optical network embedded on the NSFNET of Fig. 9.1.[9] The network cost does not increase monotonically with the number of wavelengths because of the $\lceil T/W \rceil$ and the $\lceil R/W \rceil$ component in the above equation. Also, since we use 2×2 crosspoint switching elements to build the MIN, it is not possible to build a MIN switch of odd degree, i.e., δ_m *odd*. Therefore, a node with $\delta_m = 5$ would require a three-stage MIN with three crosspoint elements per column. The costs given in Table 9.1 are independent of the utilizations of wavelengths and transceivers in the network, and

[9] We remark that the costs shown in Table 9.1 are meant to provide only a comparative understanding of how the system cost scales with the number of transceivers and wavelengths, based on rough estimates of current market values for C_t, C_m, and C_x. The actual cost of upgrading the network in Fig. 9.1 to accommodate WDM will probably be orders of magnitude higher than those shown in Table 9.1, when costs for the electronics, other supporting optics (such as preamplifiers and signal equalizers), installation, and maintenance are also taken into account.

in several improperly designed cases, a significant amount of the resources may be underutilized. Thus, it is easy to observe from Table 9.1 that, given a cost constraint, it is possible to iteratively try different combinations of wavelengths and transceivers, in order to optimize the network performance for a given network cost.

Network Equipment Cost (in $1000)							
	Wavelengths						
Transceivers	2	3	4	5	6	7	8
4	806.0	922.0	835.2	901.2	967.2	1033.2	1099.2
5	980.8	1062.0	1178.0	1041.2	1107.2	1173.2	1239.2
6	1120.8	1202.0	1318.0	1434.0	1247.2	1313.2	1379.2
7	1323.6	1392.8	1458.0	1574.0	1690.0	1453.2	1519.2
8	1463.6	1532.8	1598.0	1714.0	1830.0	1946.0	1659.2
9	1650.4	1672.8	1804.8	1854.0	1970.0	2086.0	2202.0
10	1790.4	1905.6	1944.8	1994.0	2110.0	2226.0	2342.0
11	2093.2	2045.6	2084.8	2216.8	2250.0	2366.0	2482.0
12	2233.2	2185.6	2224.8	2356.8	2390.0	2506.0	2622.0
13	2424.0	2394.4	2487.6	2496.8	2628.8	2646.0	2762.0
14	2564.0	2534.4	2627.6	2636.8	2768.8	2786.0	2902.0
15	2786.8	2674.4	2767.6	2776.8	2908.8	3040.8	3042.0
16	2926.8	3057.2	2907.6	3069.6	3048.8	3180.8	3182.0

Table 9.1 Cost of upgrading the NSFNET using WDM.

9.5 Virtual Topology Reconfiguration

A major advantage of an optical network is that it may be able to reconfigure its virtual topology to adapt to changing traffic patterns. Some reconfiguration studies on optical networks have been reported before [BiGu92, LaHA94, RoAm94b]; however, these studies assumed that the new virtual topology *was known a priori*, and were concerned with the cost and sequence of branch-exchange operations to transform from the original virtual topology to the new virtual topology. We propose a methodology to *obtain the new virtual topology*, based on optimizing a given objective function, as well as minimizing the changes required to obtain the new virtual topology from the current virtual topology.

The linear problem formulation in Section 9.2 can help us derive new virtual topologies from existing virtual topologies. In the ideal situation, given a small change in the traffic matrix, we would prefer for the new virtual topology to be largely similar to the previous virtual topology, in terms of the constituent lightpaths and the routes for these lightpaths, i.e., we would prefer to minimize the changes in the number of WRS configurations needed to adapt from the existing virtual topology to the new virtual topology. More formally, it would be preferable if a large number of the V_{ij} and the P_{mn}^{ij} variables retain the same values in the two solutions, without compromising the quality of the solution (in terms of minimizing the average packet hop distance).

Let us consider the snapshot of two traffic matrices, Λ_{sd}^1 and Λ_{sd}^2, taken at two not-too-distant time instants. We assume that there is a certain amount of correlation between these two traffic matrices. Given a certain traffic matrix, there may be many different virtual topologies, each of which has the same optimal value with regard to the objective function, i.e., Eqn. (9.2). Usually, an LP solver will terminate after it has found the first such optimal solution. Our reconfiguration algorithm finds the virtual topology corresponding to Λ_{sd}^2 which matches "closest" with the virtual topology corresponding to Λ_{sd}^1 (based on our above definition of "closeness").

9.5.1 Reconfiguration Algorithm

We perform the following sequence of actions:

- Generate linear formulations $\mathcal{F}(1)$ and $\mathcal{F}(2)$ corresponding to traffic matrices Λ_{sd}^1 and Λ_{sd}^2, respectively, based on the formulation in Section 9.2.

- Derive solutions $\mathcal{S}(1)$ and $\mathcal{S}(2)$, corresponding to $\mathcal{F}(1)$ and $\mathcal{F}(2)$, respectively. Denote the variables' values in $\mathcal{S}(1)$ as $V_{ij}(1), P_{mn}^{ij}(1), \lambda_{ij}^{sd}(1)$, and those in $\mathcal{S}(2)$ as $V_{ij}(2), P_{mn}^{ij}(2), \lambda_{ij}^{sd}(2)$, respectively. Let the value of the objective function for $\mathcal{S}(1)$ and $\mathcal{S}(2)$ be OPT_1 and OPT_2, respectively.

- Modify $\mathcal{F}(2)$ to $\mathcal{F}'(2)$ by adding the new constraint:

$$\frac{1}{\sum_{s,d} \Lambda_{sd}} \sum_{i,j} \sum_{s,d} \lambda_{ij}^{sd} = OPT_2 \tag{9.30}$$

This ensures that all the virtual topologies generated by $\mathcal{F}'(2)$ would be optimal with regard to the objective function.

- The new objective function for $\mathcal{F}'(2)$ is:

$$\textbf{Minimize}: \quad \sum_{ij}\sum_{mn} \mid p_{mn}^{ij}(2) - p_{mn}^{ij}(1) \mid \tag{9.31}$$

We could also have used the following objective function:

$$\textbf{Minimize}: \quad \sum_{ij} \mid V_{ij}(2) - V_{ij}(1) \mid \tag{9.32}$$

Note that the mod operation, $\mid x \mid$, is a nonlinear function. If we assume that p_{mn}^{ij} and V_{ij} can only take on binary values, then Eqns. (9.31) and (9.32) become linear, i.e., if $V_{ij}(1) = 1$, then $\mid V_{ij}(2) - V_{ij}(1) \mid \equiv (1 - V_{ij}(2))$, else if $V_{ij}(1) = 0$ then $\mid V_{ij}(2) - V_{ij}(1) \mid \equiv V_{ij}(2)$. Hence, $\mathcal{F}'(2)$ may be solved directly using an LP solver.

Also, note that $\mid p_{mn}^{ij}(2) - p_{mn}^{ij}(1) \mid$ also implies that $\mid V_{ij}(2) - V_{ij}(1) \mid$. Therefore, Eqn. (9.31) is a stronger condition than Eqn. (9.32). Hence, we chose Eqn. (9.31) for our simulation studies on reconfiguration at the end of Section 9.6.

9.6 Illustrative Examples

This section presents numerical examples of the network design problem, using the NSFNET T1 backbone (Fig. 9.1) as our physical topology. The NSFNET consists of 14 nodes connected in a mesh network. Each of its links are bidirectional, i.e., for each link, there is a pair of unidirectional fibers which carry transmissions in opposite directions and which join physically adjacent nodes, i.e., $P_{mn} = P_{nm} = 1$. Each node consists of an optical WRS along with multiple transceivers for optical origination and termination of lightpaths. The number of transmitters is assumed to be equal to the number of receivers, and is the same for all nodes.

The traffic matrix is randomly generated, such that a certain fraction F of the traffic is uniformly distributed over the range $[0, C/a]$ and the remaining traffic is uniformly distributed over $[0, C\Upsilon/a]$, where C is the lightpath channel capacity, a is an arbitrary integer which may be 1 or greater, and Υ denotes the average ratio of traffic intensities between node-pairs with high traffic values and node-pairs with low traffic values. This model allows us to generate traffic patterns with varying characteristics.

Figures 9.4, 9.5, and 9.6 plot system characteristics averaged over 25 different virtual topologies, each corresponding to an independent traffic matrix, obtained with the parameters $C = 1250, a = 20, \Upsilon = 10, F = 0.7, \beta = 0.8, K = 2$, and $\alpha = 2$. T_i and R_i were assumed to be equal for all nodes, and were allowed values between 4 and 8. W was allowed to take values between 1 (no WDM) and 7.

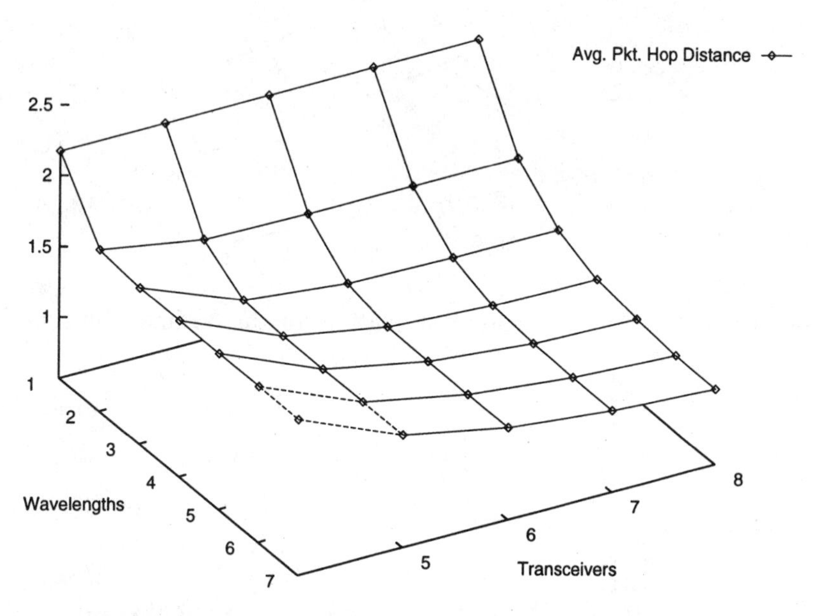

Figure 9.4 Average packet hop distance for the optimal solution.

Figure 9.4 plots the average packet hop distance for optimal virtual topologies given a different number of transceivers per node, and a different number of supported wavelengths in the system. The average hop distance in the network is a function of the number of lightpaths set up in the network, which directly depends on the number of transceivers and wavelengths supported. The case corresponding to one wavelength in the system corresponds to today's point-to-point network (no WDM). As expected, the average hop distance decreases with a balanced increase in the number of transceivers and wavelengths in the network. Increasing transceivers without adding extra wavelengths marginally improves the quality of the solution. For more than six transceivers, and more than four wavelengths, the performance improvement

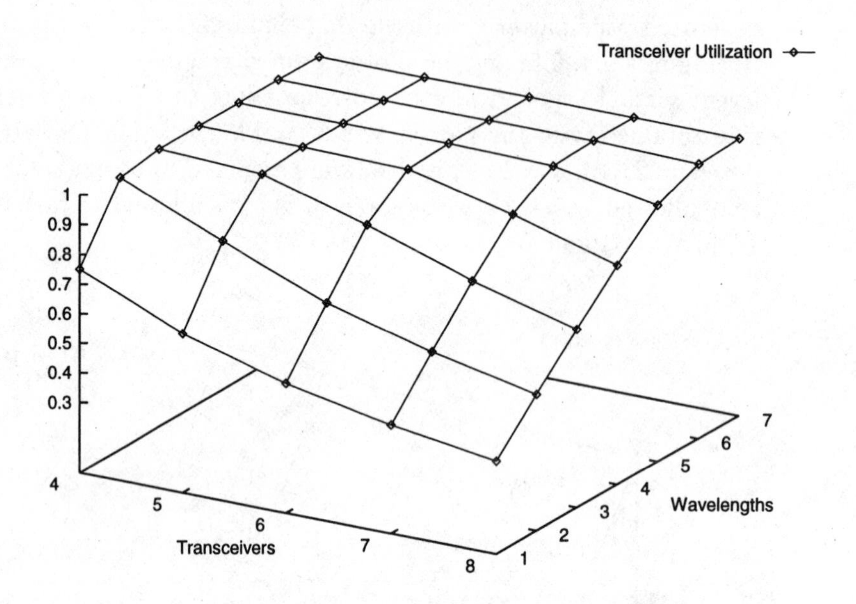

Figure 9.5 Average transceiver utilization for the optimal solution.

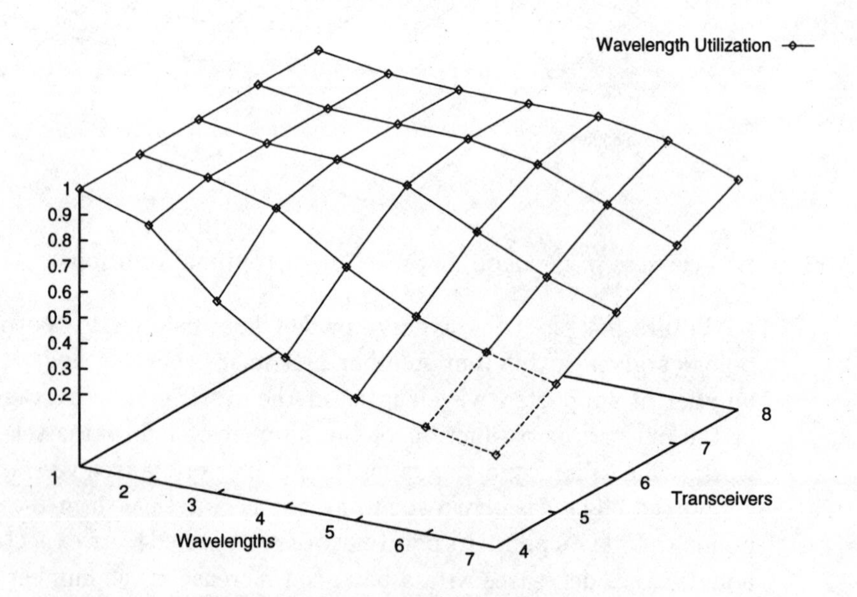

Figure 9.6 Average wavelength utilization for the optimal solution.

is marginal for the network in Fig. 9.1.

Figure 9.5 plots the transceiver utilization for different values of the number of wavelengths in the system, and number of transceivers at a node. Figure 9.6 plots the wavelength utilization for the same set of experiments. As one would expect, Fig. 9.5 quantitatively demonstrates that the transceiver utilization decreases as the number of wavelengths is reduced and/or the number of transceivers is increased. Similarly, Fig. 9.6 demonstrates that the wavelength utilization decreases when the number of wavelengths is increased and/or the number of transceivers is reduced. These results confirm our original hypothesis that it is necessary to obtain the correct balance between transceivers and wavelengths in the system in order to properly utilize both of these expensive resources. Given a cost constraint, resource-budgeting trade-off becomes an important issue, in order not to underutilize transceivers and wavelengths in the system.

Based on the results in Figs. 9.4, 9.5, and 9.6, a good operating point for the chosen system parameters appears to be *four wavelengths and six transceivers* since one would like to keep the system cost down as well as achieve low average hop distance and high utilizations (close to unity!) of transceivers and wavelengths.

Figure 9.2 demonstrates virtual topology solutions for the NSFNET, given a specific traffic matrix, and for $T_i = R_i = 5$, and for $W = 2$ and 5. Since there are 14 nodes in the network, and five transceivers per node, a maximum of $14 \times 5 = 70$ lightpaths may be established in the network. In the two-wavelength solution, only 59 lightpaths could be established, out of which 42 lightpaths constituted the physical topology embedding over the virtual topology. In the five-wavelength solution, all of the 70 lightpaths could be established, so that the transceiver utilization is 100% as opposed to a transceiver utilization of less than 85% for the two-wavelength case. It is also evident that the wavelength utilization in the two-wavelength case is much higher than that in the five-wavelength case (96% vs. 57%). Therefore, increasing the number of wavelengths in the system might increase the transceiver utilization, but tends to decrease the wavelength utilization in the system.

The wavelength assignment of lightpaths in the two solutions was done arbitrarily, and was not based on any optimal algorithm. In the current wavelength assignment, *only two wavelength converters are needed at the NY node*, in order to establish all the lighpaths. This vindicates our original assumption that sparse wavelength conversion may be sufficient to ensure good virtual topologies. Both of these solutions also assume that there may be multiple lightpaths on the same wavelength, emerging from a particular node. These solutions may require larger switch sizes (specifically, there may need to be multiple fibers from the local node to the local WRS) (see Chapter 8), and can increase the cost of the WRS beyond that presented in the cost model in Section 9.4.2.

Table 9.2 tabulates the average hop distance for the two heuristic approaches as compared to the optimal solution obtained in Fig. 9.4. The same sample of 25 traffic matrices were used to evaluate the performance of the heuristics. Figure 9.7 also plots these performance results for a four-wavelength system. As expected, the average hop distance decreases with an increase in the number of transceivers in the system, with the heuristics performing a little poorly relative to the LP's optimal solution which can be treated as a lower bound. Also, the heuristic which maximizes the multihop traffic is found to perform slightly better than the heuristic which maximizes single-hop traffic for smaller number of transceivers

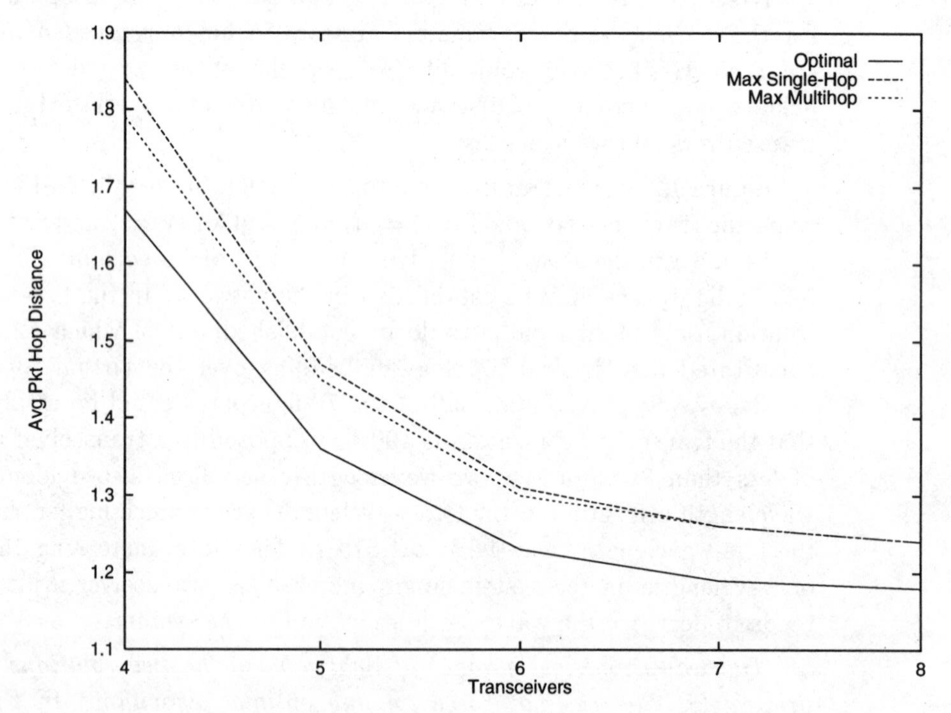

Figure 9.7 Comparison of heuristic algorithms for a four-wavelength network.

We demonstrate the reconfiguration capabilities of the problem formulation in Section 9.5.1. We generate two sequences of 25 traffic matrices

(a) Linear Problem Solution							
	Wavelengths						
Transceivers	1	2	3	4	5	6	7
4	2.17	1.71	1.67	1.67	1.67	1.67	1.67
5	2.17	1.58	1.39	1.36	1.36	1.36	1.36
6	2.17	1.56	1.31	1.23	1.22	1.22	1.22
7	2.17	1.56	1.29	1.19	1.15	1.14	1.14
8	2.17	1.56	1.29	1.18	1.13	1.10	1.10
(b) Maximizing Single-Hop Traffic Heuristic							
	Wavelengths						
Transceivers	1	2	3	4	5	6	7
4	2.41	1.91	1.84	1.84	1.84	1.84	1.84
5	2.41	1.77	1.52	1.47	1.46	1.46	1.46
6	2.41	1.75	1.41	1.31	1.37	1.27	1.27
7	2.41	1.74	1.39	1.26	1.20	1.20	1.17
8	2.41	1.74	1.39	1.24	1.17	1.14	1.12
(c) Maximizing Multihop Traffic Heuristic							
	Wavelengths						
Transceivers	1	2	3	4	5	6	7
4	2.41	1.87	1.80	1.79	1.79	1.79	1.79
5	2.41	1.72	1.50	1.45	1.44	1.44	1.44
6	2.41	1.70	1.41	1.30	1.27	1.26	1.27
7	2.41	1.69	1.39	1.26	1.20	1.18	1.17
8	2.41	1.69	1.38	1.24	1.16	1.13	1.12

Table 9.2 Average packet hop distance for different virtual topology establishment algorithms.

each, with the same set of statistical parameters as used before. In the first sequence, exactly 20% of the entries in successive traffic matrices in the sequence are forced to differ. In the second sequence, 80% of the entries differ. The traffic sequence is created by generating an initial traffic matrix, and then swapping a fraction (either 20% or 80% depending on the chosen sequence) of nondiagonal entries in the traffic matrix. The algorithm in Section 9.5.1 was applied to this traffic sequence, in order to generate virtual topologies in a network with eight transceivers per node, and eight wavelengths per fiber. Figure 9.8 plots the fraction of lightpath additions and deletions as observed over the sequence of 25 traffic matrices, for 20% and 80% changes in the traffic matrix. The fraction of common lightpaths between two successive traffic matrices remains fairly uniform throughout the entire sequence. As expected, the number of deleted lightpaths and added lightpaths increases when the difference between consecutive traffic matrices gets larger.

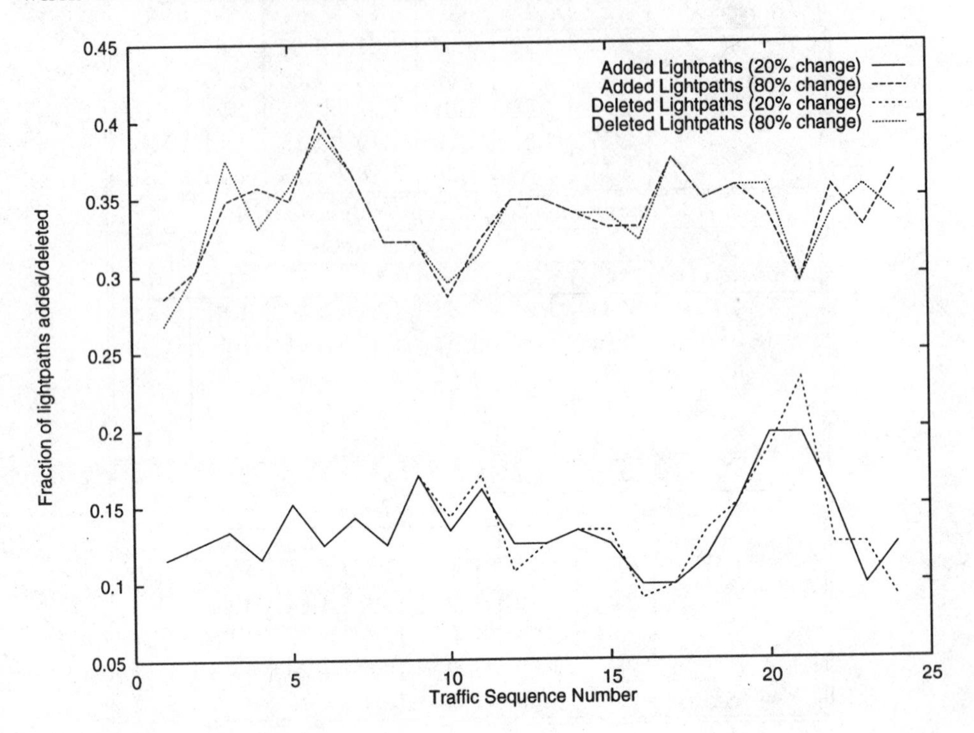

Figure 9.8 Reconfiguration statistics.

9.7 Summary

This chapter presented a linear programming formulation to derive an exact, minimal-hop-distance solution to the virtual topology design problem in a wavelength-routed optical network, in the absence of wavelength-continuity constraints. The problem formulation is general, and can be used to derive a complete virtual topology solution, including choice of the constituent lightpaths, routes for these lightpaths, and intensity of packet flows through these lightpaths. We showed that adding wavelength-continuity constraints and queueing delays makes the problem nonlinear. We used simplifying assumptions to make the problem tractable. We also proposed two simple heuristics and demonstrated that these heuristics perform well with respect to the optimal solution.

We studied resource-budgeting trade-offs in the allocation of transceivers per node, and wavelengths per fiber. A simple analysis in Section 9.4.1 [Eqn. (9.28)] provided an approximate bound regarding the number of transceivers that can be supported in a network with W wavelengths. We demonstrated how we can equip the network with an optimal balance of transceivers and wavelengths, in order to derive minimal-hop-distance solutions, along with high utilization of both transceivers and wavelengths.

We proposed an exact reconfiguration procedure which, for a changed traffic matrix, searches through all possible *optimal* virtual topologies, in order to obtain a solution which shares the maximum number of lightpaths with the previous virtual topology. The solution to the reconfiguration algorithm generates a virtual topology which minimizes the amount of reconfiguration that needs to be performed, in order to adapt the virtual topology to the new traffic matrix. We believe that reconfiguration in wide-area optical networks is an important issue that needs to be studied further. We would prefer to have algorithms, which can perform network reconfiguration locally, without a global knowledge of the network state. The sequence of actions that need to be performed to effect a reconfiguration is an important problem in network management, and is an open problem for further research.

Exercises

9.1. In this chapter, average packet hop distance is used as the objective function. Why?

9.2. Heuristics `MaxSingleHop` and `MaxMultiHop` assign lightpaths based on traffic flows. `MaxSingleHop` adjusts the traffic matrix after assigning a lightpath, while `MaxMultiHop` does not. Why?

9.3. For this exercise, consider the physical topology shown in Fig. 9.9 and the traffic matrix shown below. Assume two channels per link, where each channel has a capacity of five units. Use the `MaxSingleHop` heuristic to set up lightpaths. Assume that each node has sufficient number of transceivers. The traffic matrix is:

$$T = \begin{bmatrix} 0 & 4 & 5 & 0 & 0 \\ 0 & 0 & 0 & 0 & 0 \\ 1 & 0 & 0 & 2 & 3 \\ 0 & 0 & 0 & 0 & 0 \\ 0 & 0 & 0 & 0 & 0 \end{bmatrix}$$

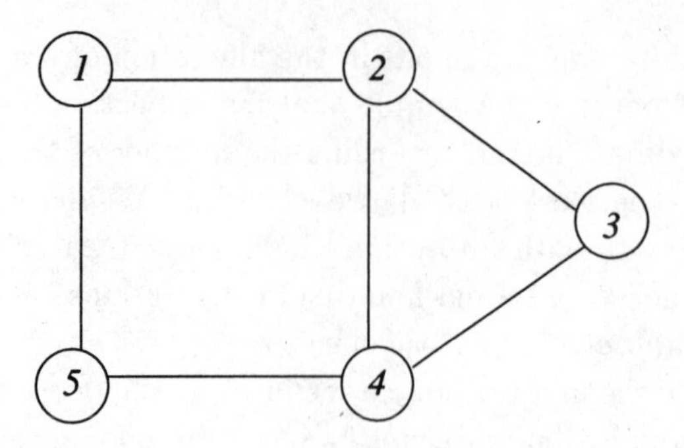

Figure 9.9 Physical network topology.

9.4. Draw the virtual topology of the network shown in Problem 9.3. Calculate V_{ij}, λ_{ij}^{sd}, and p_{mn}^{ij} for this network.

9.5. Derive Eqn. (9.1). Show that the inequality holds for the network shown in Problem 9.3. When will the equality hold?

9.6. Consider the network shown Fig. 9.10 with two transmitters and two receivers per node, two wavelengths, and the following traffic matrix:

$$\Lambda = \begin{bmatrix} 0 & 2 & 2 & 1 & 4 & 3 \\ 1 & 0 & 4 & 4 & 3 & 2 \\ 3 & 2 & 0 & 6 & 2 & 1 \\ 1 & 1 & 2 & 0 & 1 & 7 \\ 3 & 5 & 2 & 1 & 0 & 1 \\ 2 & 4 & 5 & 3 & 2 & 0 \end{bmatrix}$$

Determine a set of lightpaths using the `MaxSingleHop` heuristic.

9.7. For the network in Problem 9.6, run the `MaxMultiHop` heuristic to determine a set of lightpaths.

9.8. Assume that the network equipment cost budget is $1,000,000. Using Table 9.1, find the network configurations (i.e., number of transceivers and wavelengths) that can be supported. Which network configuration maximizes the total network throughput?

9.9. Assume the Petersen graph as the physical topology. Assume that the network can support 10 wavelengths. Now assume that we embed a complete graph on 10 nodes as the virtual topology. (Use shortest-path

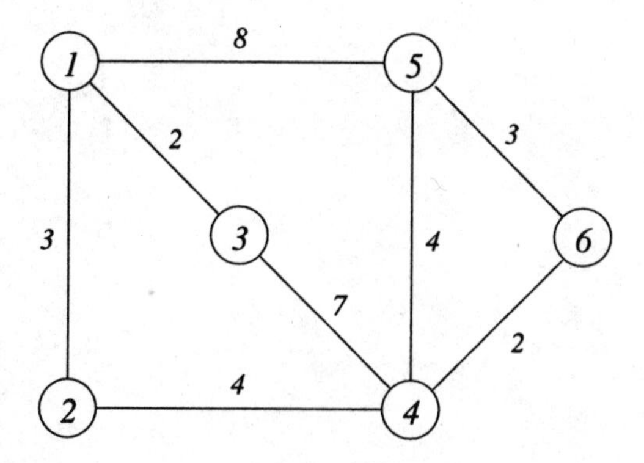

Figure 9.10 Physical network topology.

routing to embed the virtual topology). Using the cost model described in Section 9.4, compute the cost to build the network. Which type of component has the highest networkwide cost?

Chapter 10

Routing and Wavelength Assignment

This chapter considers large optical networks in which nodes employ wavelength-routing switches which enable the establishment of wavelength-division multiplexed (WDM) channels, called lightpaths, between node pairs. It develops a practical approach to solve routing and wavelength assignment (RWA) of lightpaths in such networks. A large RWA problem is partitioned into several smaller subproblems, each of which may be solved independently and efficiently using well-known approximation techniques. A multicommodity flow formulation combined with randomized rounding is employed to calculate the routes for lightpaths. Wavelength assignments for lightpaths are performed based on graph-coloring techniques. Representative numerical examples indicate the accuracy of our algorithms.

10.1 Introduction

It is expected that a wavelength-routed optical network will be deployed mainly as a backbone network for large regions, e.g., for nationwide or global coverage. End-users – to whom the architecture and operation of the backbone will be transparent except for significantly improved response times –

will attach to the network through a wavelength-sensitive switching/routing node, as shown, for example, in Fig. 10.1. An end-user in this context need not necessarily be a terminal equipment, but the aggregate activity from a collection of terminals – including those that may possibly be feeding in from other regional and/or local subnetworks – so that the end-user's aggregate activity on any of its transmitters is close to the peak electronic transmission rate.

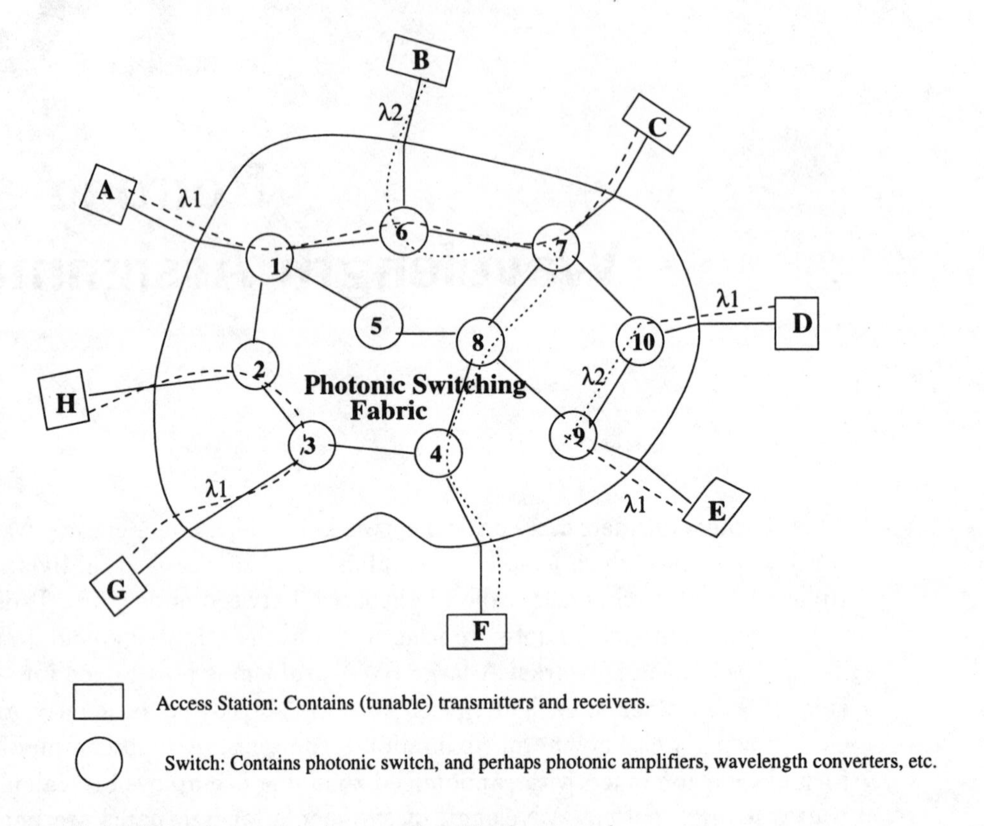

Figure 10.1 Lightpath routing in an all-optical network.

End-users communicate with one another via *all-optical (WDM) channels*, which are referred to as *lightpaths* or *connections*. Although we have used the term *lightpath* so far, we will use the terms *lightpath* and *connection* interchangeably in this chapter, since the term *connection* may make more sense in some of the problem settings now, e.g., in order to establish a "connection" between a source-destination (sd) pair, we need to set up a "lightpath" between them. A lightpath may span multiple fiber links, e.g., to provide a "circuit-switched" interconnection between two nodes which may have a heavy traffic flow between them and which may be located "far" from each other in the physical fiber network topology. Each intermediate node in the lightpath essentially provides an all-optical bypass facility to support the lightpath.

In an N-node network, if each node is equipped with $N - 1$ transceivers [transmitters (lasers) and receivers (filters)] and if there are enough wavelengths on all fiber links, then every node pair could be connected by an all-optical lightpath, and there is no networking problem to solve. However, it should be noted that the network size (N) should be scalable, transceivers are expensive so that each node may be equipped with only a few of them, and technological constraints dictate that the number of WDM channels that can be supported in a fiber be limited to W (whose value is a few tens today, but is expected to improve with time and technological breakthroughs). Thus, only a limited number of lightpaths may be set up on the network.

The complete set of lightpaths was considered to form a *virtual topology* in Chapters 8 and 9, and packet traffic had to be routed over it. In the present chapter, the offered traffic is itself "circuit-oriented," i.e., the offered traffic consists of a set of connections such that each connection requires the full bandwidth of a lightpath in order for it to be routed between its corresponding source-destination pair.

Once a set of lightpaths has been chosen or determined, we need to route each such lightpath in the network and assign a wavelength to it. This is referred to as the *routing and wavelength assignment (RWA)* problem.

Formally, the RWA problem can be stated as follows. Given a set of lightpaths that need to be established on the network, and given a constraint on the number of wavelengths, determine the routes over which these lightpaths should be set up and also determine the wavelengths that should be assigned to these lightpaths so that the maximum number of lightpaths may be established. While shortest-path routes may be most preferable, note that this choice may have to be sometimes sacrificed, in order to allow more lightpaths to be set up. Thus, RWA algorithms generally allow several alternate routes for each lightpath that needs to be established. Lightpaths that cannot be set up due to constraints on routes and wavelengths are said to be blocked, so the corresponding network optimization problem is to minimize this blocking probability.

In this regard, note that, normally, a lightpath operates on the same wavelength across all fiber links that it traverses, in which case the lightpath is said to satisfy the *wavelength-continuity constraint.* Thus, two lightpaths that share a common fiber link should not be assigned the same wavelength. However, if a switching/routing node is also equipped with a wavelength converter facility, then the *wavelength-continuity constraints* disappear, and a lightpath may switch between different wavelengths on its route from its origin to its termination.

For such a network, the work in [RaSi94] develops an integer linear programming (ILP) formulation of the RWA problem as a multicommodity flow problem, from which it derives any generic RWA algorithm's performance bounds – an upper bound on the carried traffic (number of lightpaths established), or, equivalently, a lower bound on the lightpath blocking probability. The approach is extendible to accommodate wavelength conversion as well [RaSi94]. It can be proved that the wavelength assignment problem can be shown to be equivalent to the graph-coloring problem [ChGK93] and is hence NP-complete. A number of heuristics exist on how to obtain good solutions to the RWA problem [ChGK92, ChGK93, RaSi94, SBJS93, ZhAc94].

In this chapter, we employ well-known approximation algorithms to solve the RWA problem. Our approach consists of practical algorithms which enable us to solve the RWA problem for large network sizes. Our final objective is to minimize the number of wavelengths that will be needed to carry a certain number of connections in the network, given a certain physical topology.

The RWA problem is decomposed into four different subproblems and each subproblem is solved independently with the results of one stage fed in as the input to the next stage. We formulate a linear program (LP) (using the idea of multicommodity flow in a network), which is based on the physical topology as well as the set of connections to be routed, and use a general-purpose LP solver to derive solutions to this problem. Since the general form of the LP can easily overwhelm the capabilities of today's state-of-the-art computing facilities, even for moderate sized networks (of, say, 10 nodes with 4 connections per node), we develop simple specialized techniques to drastically prune the size of the LP (in terms of both the number of variables and the number of equations it needs to handle) so that solutions to large instances of the problem (e.g., a few hundred nodes) can be obtained. Specifically, this pruning approach is based on tracking only a limited number of alternate breadth-first paths between source-destination pairs to reduce the size of the LP formulation. We then use a technique called *randomized rounding* (explained in Section 10.2.3) to convert the fractional flows provided by the LP solution to integer flows through the physical fiber links. Once the flows through the fiber links have been determined, *sequential coloring algorithms* are used to assign wavelengths to the lightpaths by taking into account the wavelength-continuity constraints. This method of subdividing the overall problem into smaller subproblems, each of which may be solved efficiently, allows practical solutions to the RWA problem for networks with a large number of nodes.

The solution to our linear program formulation gives a lower bound on the number of wavelengths we would need in the network to route the given set of connections.[1] How close the final result is to this value determines the efficacy of the algorithms proposed. Our proposed solutions turn out to be very close to the lower bound for large networks.

This chapter initially examines the problem for a known set of connections that need to be routed – referred to as the static lightpath establishment (SLE) problem [ChGK93]. Then, simple heuristics are employed to extend our approach and obtain good results for the case of dynamic lightpath establishment (DLE).

The remainder of the chapter is organized as follows. Section 10.2 formulates the RWA problem and describes our tools and techniques to solve this problem for large networks. Section 10.3 describes the results for an experimental network with 100 nodes; the results are given for both the SLE as well as the DLE problems.

Just like Chapter 9, for clarity of exposition, this chapter deals in detail with one approach to solve large instances of the RWA problem. A summary of other approaches to accommodate circuit-switched traffic in a WDM WAN may be found in Chapter 12, Section 12.2.

10.2 Problem Formulation

Significant work has been reported on combinatorial formulations, in terms of using mixed-integer linear programs, for solving the RWA problem [RaSi94, ZhAc94]. However these formulations, when applied to solving large problems, become computationally expensive in spite of using sophisticated techniques such as branch-and-bound methods. The algorithms in the next few sections elaborate on how it is possible to use approximation techniques to arrive at optimal results for the RWA problem in large networks.

[1] Because of our pruning the search space, there might be a small discrepancy between the absolute lower bound and our result; however, we can make the difference arbitrarily small by increasing the number of alternate paths.

10.2.1 Solution Approach

Our objective is to minimize the number of wavelengths needed to set up a certain set of lightpaths for a given physical topology. Our approach employs approximation results based on a combinatorial formulation to arrive at optimally close results to the lower bound for the number of wavelengths required in the network. The RWA problem, without the wavelength-continuity constraint, can be formulated as a straightforward multicommodity flow problem with integer flows in each link. This corresponds to an integer linear program (ILP) with the objective function being to minimize the flow in each link, which, in turn, corresponds to minimizing the number of lightpaths passing through a particular link.

Let λ_{sd} denote the traffic (in terms of a lightpath) from any source s to any destination d. We consider at most one lightpath from any source to any destination; hence $\lambda_{sd} = 1$ if there is a lightpath from s to d; otherwise $\lambda_{sd} = 0$. Let F_{ij}^{sd} denote the traffic (in terms of number of lightpaths) that is flowing from source s to destination d on link ij. The linear programming formulation is written as follows:

$$\textbf{Minimize}: \quad F_{max} \tag{10.1}$$

such that

$$F_{max} \geq \sum_{s,d} F_{ij}^{sd} \quad \forall \; ij \tag{10.2}$$

$$\sum_{i} F_{ij}^{sd} - \sum_{k} F_{jk}^{sd} = \begin{cases} \lambda_{sd} & \text{if } s = j \\ -\lambda_{sd} & \text{if } d = j \\ 0 & \text{otherwise} \end{cases} \tag{10.3}$$

$$\lambda_{sd} = 0, 1 \tag{10.4}$$

$$F_{ij}^{sd} = 0, 1 \tag{10.5}$$

This problem is NP-complete [EvIS76] but it can be approximated successfully using *randomized rounding*, which is outlined below in Section 10.2.3.

10.2.2 Problem Size Reduction

If we consider the general multicommodity formulation, the number of equations and the number of variables in the formulation grow rapidly with the size of the network. For example, let us assume that there are 10 nodes, 30 links (ij pairs), and an average of 4 connections per node, i.e., 40 connections (sd pairs) need to be set up on the network.

In the simplest and most general formulation, the number of λ_{sd} variables is $10 \times 9 = 90$, since there are 90 sd pairs. The number of F_{ij}^{sd} variables will be 90 sd pairs $\times$ 30 ij pairs = 2,700. The number of equations will be 3,721.[2] Thus, even for a small problem, we observe that the number of variables and the number of equations are very large, and these numbers grow proportionally with the square of the number of nodes.

A smarter solution can be obtained by only considering the λ_{sd} variables that are 1, thus reducing the number of λ_{sd} variables from 100 to 40. This can eliminate all of Eqns. (10.4). Also, this approach will reduce the number of F_{ij}^{sd} variables to be $40 \times 30 = 1{,}200$. This approach is more specific to the particular instance of lightpaths that need to be set up, since it has to take into account the lightpaths that need to be established.

Further reduction of the number of variables can be achieved if we assume that a particular lightpath will not pass through all of the ij links. If we can determine the links which have a good probability of being in the path through which a lightpath may pass, we can only consider those links as the F_{ij}^{sd} variables for that particular sd pair. Thus, if on an average, a lightpath sd passes through seven links, there will be approximately $40 \times 7 = 280$ F_{ij}^{sd} variables. We employ an *extended* breadth-first search to obtain a set of alternate, shortest paths between a given source-destination pair. The links constituting these alternate paths are then used as part of the LP formulation. Also, as part of the randomized rounding algorithm (to be discussed in Section 10.2.3 below), we can relax the integrality constraints, hence getting rid of all of Eqns. (10.5). Using this approach, we will be left with a total of 351 equations. This will comprise of Eqn. (10.1), 30 instances of Eqns. (10.2), and 320 instances of Eqns. (10.3). Since there are on an average of seven links considered per connection, we need to enumerate Eqns. (10.3) for eight nodes (six intermediate nodes and the two end nodes) per connection, for each of 40 connections.

Hence, using knowledge which is specific to a particular set of lightpaths, we can drastically reduce the size of the LP problem formulation and make it tractable for large networks.

---

[2]There are one instance of Eqn. (10.1), 30 instances of Eqns. (10.2), 900 instances of Eqns. (10.3), 90 instances of Eqns. (10.4), and 2,700 instances of Eqns. (10.5).

10.2.3 Randomized Rounding

Randomized rounding has been studied extensively in [RaTh87]. The relation of an ILP to its fractional relaxation has been the subject of considerable interest. Such efforts fall into two categories: (1) showing existence results for feasible solutions to an ILP in terms of the solution to its fractional relaxation and (2) using the information derived from the solution of the relaxed problem in order to construct a provably-good solution to the original ILP.

The *randomized rounding* technique is applicable to a class of 0-1 ILPs and yields results in both of the categories listed above. The technique is probabilistic, i.e., with high probability, the algorithm will provide an integer solution in which the objective function takes on a value close to the optimum of the rational relaxation. This is a sufficient condition to show the near-optimality of our 0-1 solution since the optimal value of the objective function in the relaxed version is better than the optimal value of the objective function in the original 0-1 integer program. This technique has been effectively used in multicommodity flow problems.

In a general, undirected, multicommodity flow problem, we are given an undirected graph $G(V, E)$. Let there be k commodities that need to be routed. In an instance of the problem, various vertices are the sites of *sources* and *sinks* for a particular commodity. One unit of flow is to be conveyed from each source s_t to each destination d_t through the edges in E. Each edge $e \in E$ has a capacity $c(e)$ which is an upper limit on the total amount of flow in E. The flow of each commodity in each edge must be 0 or 1. The objective function is to minimize the common capacity in each link, while still realizing unit flows for all commodities. The general integral problem is known to be NP-complete [EvIS76], although the nonintegral version can be solved using linear programming methods [Karp72] in polynomial time.

In our formulation of the problem, each commodity corresponds to a lightpath from a source node to a destination node, the capacity is the number of wavelengths supported in each fiber, and the objective function is to minimize the number of wavelengths needed to accommodate all of the requests, so as to maximize the spare capacity in the network.

The algorithm consists of the following three major phases:

1. solving a nonintegral multicommodity flow problem,
2. path stripping, and
3. randomized path selection.

1. *Nonintegral Multicommodity Flow.* We relax the requirement of the 0-1 flows to allow fractional flows in the interval [0,1]. The relaxed capacity-minimization problem can be solved by a suitable linear programming method. If the flow for each commodity i on edge $e \epsilon E$ is denoted by $f_i(e)$, a capacity constraint of the form

$$\sum_{i=1}^{k} f_i(e) \leq C$$

is then satisfied for each edge in the network, where C is the optimal solution to the nonintegral, edge-capacity optimization problem. The value of C is also a lower bound on the best possible integral solution.

2. *Path Stripping.* The main idea in this phase is to convert the edge flows for each commodity i into a set τ_i of possible paths which could be used to realize the flow of that commodity. Initially, τ_i is empty. For each commodity i, the following steps need to be performed:

a. Discover a loop-free, depth-first, directed path $e_1, e_2, \ldots, e_p$ from the source to the destination.

b. Let $f_m = \min \; f_i(e_j)$ where $1 \leq j \leq p$. For $1 \leq j \leq p$, replace $f_i(e_j)$ by $f_i(e_j) - f_m$. Add the path $e_1, e_2, \ldots, e_p$ to τ_i along with its weight f_m.

c. Remove any edges with zero flow from the set of edges that carry any flow for commodity i. If there is nonzero flow leaving s_i, repeat Step b. Otherwise, continue for next commodity i.

Upon termination, the sum of the weights of all the paths in τ_i is 1. Path stripping gives us a set of paths τ_i that may carry the flow for commodity i in the optimal case.

3. *Randomization.* For each i, cast a $\mid \tau_i \mid$ die with face probabilities equal to the weights of the paths in τ_i. Assign to commodity i, the path whose face comes up.

It has been shown in [RaTh87] that, provided $C \geq 2 \ln \mid E \mid$, the integer capacity of the solution produced by the above procedure does not exceed

$$C + \sqrt{3C \ln \frac{\mid E \mid}{\epsilon}}$$

where $0 \leq \epsilon \leq 1$ with probability at least $1 - \epsilon$.

Our formulation of the problem allows the F_{ij}^{sd} variables to take on fractional values. These values are then used to find the fractional flow through each of a set of alternate paths. A coin-tossing experiment is used to select the path over which to route the lightpath λ_{sd} based on the probability of the individual paths. This technique can be used to solve large problems for which solving the original integer linear program would have been computationally very expensive.

10.2.4 Graph Coloring

Once a path has been chosen for each connection, the number of lightpaths going through any physical fiber link defines the *congestion* on that particular link. Now, we need to assign wavelengths to each lightpath such that any two lightpaths that pass through the same physical link are assigned different wavelengths.

If the intermediate switches do not have the capability to perform wavelength conversion, a lightpath is constrained to operate on the same wavelength throughout its path. This *wavelength-continuity constraint* may reduce the effective utilization of the wavelengths in the network, because a lightpath that needs to be set up may not find a free wavelength *of the same color* in *all* of the physical fiber links it passes through, even though these link may have free wavelengths. This problem may be alleviated by the use of wavelength converters in a switching node. (Recall that a wavelength converter may all-optically switch a signal arriving on a wavelength on an input fiber to a different free wavelength on an output fiber corresponding to the next physical link.)

Assigning wavelength colors to different lightpaths, so as to minimize the number of wavelengths (colors) used under the wavelength-continuity constraint, reduces to the graph coloring problem, as stated below.

1. Construct a graph $G(V, E)$, so that each lightpath in the system is represented by a node in graph G. There is an undirected edge between two nodes in graph G if the corresponding lightpaths pass through a common physical fiber link.

2. Color the nodes of the graph G such that no two adjacent nodes have the same color.

This problem has been shown to be NP-complete, and the minimum number of colors needed to color a graph G (called the chromatic number $\chi(G)$ of the graph) is difficult to determine. However, there are efficient *sequential graph coloring* algorithms which are optimal in the number of colors used.

In a *sequential graph coloring* approach, vertices are sequentially added to the portion of the graph already colored, and new colorings are determined to include each newly adjoined vertex. At each step, the total number of colors necessary is kept at a minimum. It is easy to observe that some particular sequential vertex coloring will yield a $\chi(G)$ coloring. To see this, let A_i be the vertices colored i by a $\chi(G)$ coloring of G. Then, for any ordering of the vertices $V(G)$, which has all members of A_i before any member of A_j for $1 \leq i \leq j \leq \chi(G)$, the corresponding sequential coloring will be a $\chi(G)$ coloring.

It is easy to note that, if $\Delta(G)$ denotes the maximum degree in a graph, then $\chi(G) \leq \Delta(G) + 1$. However, intuitively, if a graph has only a few nodes of very large degree, then coloring these nodes early will avoid the need for using a very large set of colors. This gives rise to the following theorem:

Theorem: Let G be a graph with $V(G) = v_1, v_2, \ldots, v_n$ where $\deg(v_i) \geq \deg(v_{i+1})$ for $i = 1, 2, \ldots, n-1$. Then $\chi(G) \leq \max_{1 \leq i \leq n} \min \{i, 1 + \deg(v_i)\}$. Determination of a sequential coloring procedure corresponding to such an ordering will be termed the *largest-first* algorithm. The proof is straightforward and can be found in [MaMI72].

A closer inspection of the sequential coloring procedure shows that, for a given ordering $v_1, v_2, \ldots, v_n$ of the vertices of a graph G, the corresponding sequential coloring algorithm could never require more than k colors where

$$k = \max_{1 \leq i \leq n} \{1 + \deg_{<v_1, v_2, \ldots, v_n>}(v_i)\}$$

and $\deg_{<v_1, v_2, \ldots, v_n>}(v_i)$ refers to the degree of node v_i in the vertex-induced subgraph denoted by $< v_1, v_2, \ldots, v_n >$. The determination of a vertex ordering that minimizes k was derived in [Matu72] and can be found in the following procedure:

1. For $n = | V(G) |$, let v_n be chosen to have minimum degree in G.

2. For $i = n-1, n-2, \ldots, 2, 1$, let v_i be chosen to have minimum degree in
$< V(G) - v_n, v_{n-1}, \ldots, v_{i+1} >$.

For any vertex ordering $v_1, v_2, \ldots, v_n$ determined in this manner, we must have

$$\deg_{<v_1,v_2,\ldots,v_i>}(v_i) = \min_{1 \leq j \leq i} \deg_{<v_1,v_2,\ldots,v_i>}(v_j)$$

for $1 \leq i \leq n$, so that such an ordering will be termed a *smallest-last* (SL) vertex ordering. The fact that any smallest-last vertex ordering minimizes k over the $n!$ possible orderings is shown in [Matu72].

We employ the *smallest-last* coloring algorithm to color our lightpaths. There may exist better algorithms for graph coloring; however, we chose this algorithm for the sake of simplicity and because our contribution is a methodology on how we can solve large RWA problems efficiently rather than to propose the best possible method to solve each subproblem.

10.3 Illustrative Examples

For illustration purposes, consider a randomly generated physical topology consisting of 100 nodes, with each node having a physical nodal degree uniformly distributed between two and five. All links are unidirectional, and it turns out that there are 357 directed links in our simulation of this network.

Our model of offered traffic is the following. A set of lightpaths needs to be established between randomly chosen source-destination (*sd*) pairs, so that an *sd* pair can have zero or one lightpath, and all *sd* pairs are treated equally. Associated with each (source) node, we identify a quantity d which is the average number of lightpath connections the node will source. Thus, in an N-node network, the probability that a node will have a lightpath with each of the remaining $(N-1)$ nodes equals $d/(N-1)$. Note that d can also be considered to be the average "logical degree" (hereafter, referred to simply as "degree") of a node.

We assume – as do other RWA algorithms – that there are enough transceivers at the access nodes to accommodate all of the lightpath requests that need to be established. That is, no lightpath request will be blocked due to lack of transceivers at the access nodes.

10.3.1 Static Lightpath Establishment (SLE)

First, let us consider the static lightpath establishment (SLE) problem. Given the physical topology and the set of connections to be routed, generate the linear program to obtain a lower bound on the number of wavelengths required for the RWA problem. To reduce the size of the linear program formulation, we consider a set of K alternate, shortest paths between a given source-destination pair. Only the links which constitute these alternate paths are used as the F_{ij}^{sd} variables. An extended breadth-first search algorithm is used to derive the set of K alternate shortest paths between a source-destination pair.

The linear program formulation is solved by a LP solver called *lpsolve* [Berk94] and the resultant output (containing the flow values for the F_{ij}^{sd} variables) is used as input for the randomized rounding algorithm. The value of the objective function denotes the lower bound on congestion that can be achieved by any RWA algorithm. Using the values of the individual variables F_{ij}^{sd}, we use the path stripping technique and the randomization technique outlined in Section 10.2.3 to assign physical routes for the different lightpaths. Once this procedure is completed, we obtain the congestion on the different links in the network.

Once the lightpaths have been assigned physical routes, we need to assign a wavelength to each individual lightpath. This is performed by generating the conflict graph for a lightpath; each lightpath corresponds to a node in a conflict graph G, and lightpaths that pass through a common physical link are adjacent nodes in the graph G. Coloring the nodes of graph G, such that adjacent nodes get different colors, corresponds to assigning wavelengths correctly to the lightpaths. We use the smallest-last vertex coloring as discussed earlier in Section 10.2.4.

A set of numerical results for the 100-node random network is shown in Table 10.1. The LP experiments were run on an unloaded DEC-Alpha workstation and the time taken for each experiment provides relative guidance as to the time complexity of this approach.

In Table 10.1, note that, when the number of alternate paths is one, the lower bound exactly matches the congestion. This is expected because this case corresponds to shortest-path routing. For larger values of K, observe that the number of wavelengths needed to color all of the lightpaths is a little higher but very close to the maximum congestion in the network. The maximum network congestion gives the number of wavelengths we would need to have in the network, if the intermediate switching nodes were equipped with wavelength converters, so that there was no wavelength-continuity constraint for a lightpath.

The time taken to solve the linear program increases rapidly as the number of connections increase (corresponding to larger problem formulations). The

Avg. Deg. d	*Alt. Paths K*	*Connec-tions*	*LP vars.*	*LP eqns.*	*LP time (sec)*	*LP Lower Bound*	*Congestion per link*	*Wavelengths needed*
1	1	100	319	776	0.5	4.00	4	4
2	1	205	684	1246	1.4	8.00	8	8
3	1	300	1024	1681	2.7	10.00	10	10
4	1	400	1362	2119	4.4	11.00	11	11
10	1	984	3334	4675	22.0	22.00	22	22
20	1	1958	6692	9007	90.6	38.00	38	41
1	2	100	647	1004	2.7	3.00	3	4
2	2	205	1353	1710	28.0	4.33	5	6
3	2	300	2001	2358	100.7	6.00	7	7
4	2	400	2678	3035	271.9	7.00	8	8
10	2	984	6618	6975	2585.1	12.00	14	17
20	2	1958	13219	13576	9113.7	22.00	24	29
1	3	100	962	1219	18.8	2.50	3	4
2	3	205	2003	2155	215.2	3.75	5	6
3	3	300	2931	2988	545.2	5.00	7	8
4	3	400	3923	3880	1205.6	6.50	8	10
10	3	984	9655	9028	-	-	-	-
20	3	1958	19270	17669	-	-	-	-
1	4	100	1257	1414	52.4	2.50	4	4
2	4	205	2607	2555	420.0	3.67	5	6
3	4	300	3811	3569	1225.2	4.67	7	8
4	4	400	5102	4661	2253.7	5.50	8	9
10	4	984	12521	10915	-	-	-	-

Table 10.1 Sample numerical results for static lightpath establishment (SLE) on a 100-node random network.

table entries which are *nil* correspond to the case when the LP solver failed to give a solution due to lack of memory.

Note that, as the number of connections in the network increases (as in the case where there are 20 connections per node and $K = 2$ alternate paths for each connection), the difference between the LP lower bound and the maximum network congestion continues to stay small; however, the difference between the network congestion and the number of wavelengths needed to color the lightpaths has increased. This indicates that we might have to use better coloring algorithms which use backtracking techniques to achieve better performance. Figure 10.2 shows the effect of the nodal degree d (which is proportional to the number of connections in the network) on the wavelength congestion, if we consider $K = 2$ alternate paths. The correspondence is linear in this case.

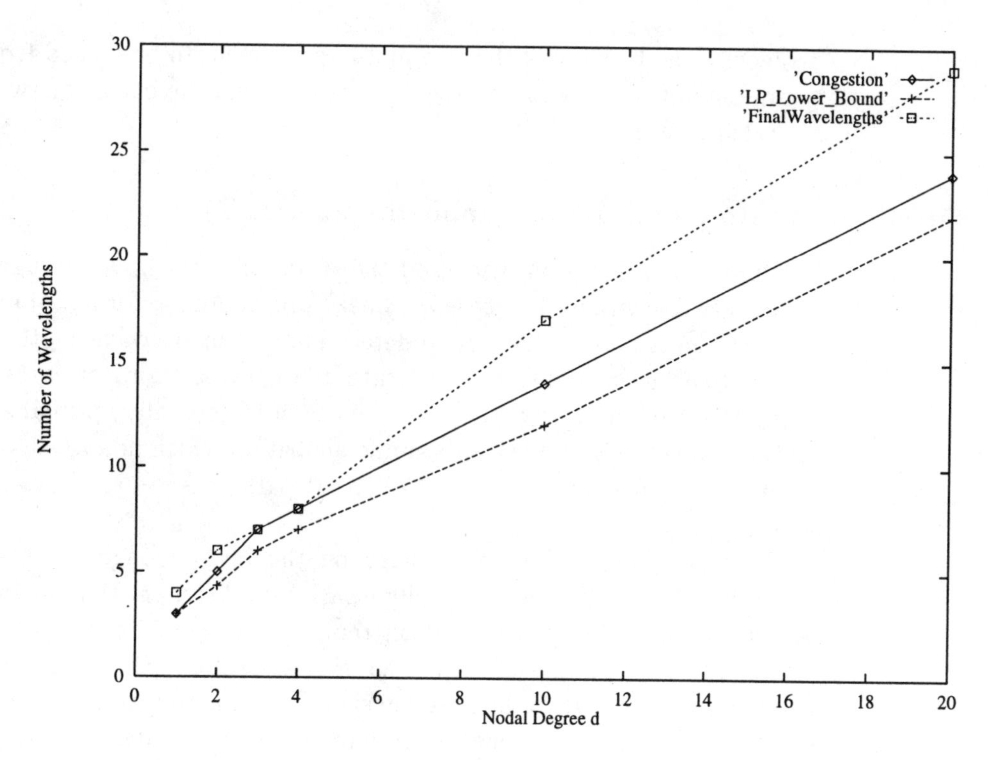

Figure 10.2 Effect of nodal degree d (for $K = 2$ alternate paths) on wavelength routing.

As expected, there is a sharp decrease in the number of wavelengths needed when we change from pure shortest-path routing to considering $K = 2$ alternate paths. This is because we go from a situation in which there can be no congestion balancing when $K = 1$, to a situation where we can balance the flow of lightpaths by using alternate paths ($K > 1$). As we go from two alternate paths to three or more alternate paths, we would still expect some improvement, although the rate of improvement goes down. The LP lower bound is a monotonically decreasing function as the number of paths is increased. However, when K is large, with a lot of choices, randomized rounding may choose a nonoptimal path with nonzero probability. This may explain why we notice that the congestion and maximum wavelengths needed when we use four alternate paths is slightly higher than the case when we use two or three alternate paths. In large networks with higher number of connections, we would expect such occurrences to be rare due to averaging effects. Because of the large time complexity required for solving larger LP problems, and because of the minimal improvement to be gained from having a large number of alternate paths, we recommend the choice of two or three alternate paths.

10.3.2 Dynamic Lightpath Establishment (DLE)

The previous subsection discussed the static problem in which lightpath requests are known in advance. Our linear programming formulation and subsequent algorithms enabled us to determine how many wavelengths would be needed to accommodate all of the lightpath requests. However, if the lightpath requests are dynamic, we cannot make use of the globally optimal algorithms and we would need to have a dynamic algorithm which adaptively routes the incoming traffic connections over different paths based on congestion on the different links.

We use a simple heuristic based on the Least Congested Path (LCP) algorithm [ChYu94] for dynamic lightpath establishment (DLE). In the LCP algorithm, a lightpath is routed on the least congested path from among a set of alternate paths between a source-destination pair. The wavelength allocated on this path is the first wavelength that is free among all of the links in this path, where wavelengths are numbered arbitrarily.

The results (with regard to congestion and number of wavelengths) in this study will depend on the order in which the connections arrive. For the DLE case, the network is assumed to be wide-sense nonblocking, i.e., existing connections are *not* rerouted so as to accommodate a new connection that is being blocked due to lack of free wavelengths. The set of connections are kept the same as in the static case. This approach provides a fair comparison between the static and the dynamic cases. A point to note is that the congestion results using LCP routing are very close to the optimal results obtained in the static optimization case, after running the randomized rounding algorithm. This is understandable since the LCP algorithm tries to adaptively minimize the congestion as each connection arrives.

It is expected that, as the congestion increases in the network, the number of wavelengths needed in the dynamic case will be more than the wavelengths needed in the static case. In the static case, the coloring algorithm assigned wavelengths to the lightpaths in a specific order so as to minimize the number of wavelengths needed. This method of assigning a wavelength to a lightpath is not possible in the dynamic case where lightpaths arrive in a random order. Thus, under the wavelength-continuity constraint, optimal allocation of wavelengths may not be possible in the dynamic case.

To alleviate the above problem, there are known techniques by which it may be possible to *reassign* existing connections to unused wavelengths in order to optimally allocate resources for new connections. This problem has been studied in the graph-theoretical case in [MaMI72]. The algorithm uses a fundamental operation called *color interchange* which corresponds to two lightpaths interchanging the wavelengths they use so as to create spare capacity in the network. This approach corresponds to global synchronization across four nodes (the endpoints of these two lightpaths) as well as the intermediate switches, and may be difficult to achieve in wide-area networks.

The results in the dynamic case are given in Table 10.2. The set of lightpaths is the same as those in the static case; however the order in which the lightpaths arrive is changed randomly. Each experiment is analyzed over 10 random arrival patterns for the lightpaths.

Average Degree d	*Alternate Paths K*	*LP Lower Bound*	*Congestion*		*Wavelengths*	
			Min	*Max*	*Min*	*Max*
1	1	4.00	4	4	4	5
2	1	8.00	8	8	8	8
3	1	10.00	10	10	10	11
4	1	11.00	11	11	11	12
1	2	3.00	3	4	4	5
2	2	4.33	5	6	6	7
3	2	6.00	7	8	8	10
4	2	7.00	8	10	10	11
1	3	2.50	3	4	4	4
2	3	3.75	5	6	6	7
3	3	5.00	6	8	8	9
4	3	6.50	8	9	10	11
1	4	2.50	3	4	4	4
2	4	3.67	5	6	5	7
3	4	4.67	6	7	8	9
4	4	5.50	7	8	10	11

Table 10.2 Sample numerical results for dynamic lightpath establishment (DLE) using the LCP routing scheme on the same 100-node random network as in the static case. Same set of lightpaths as in SLE is considered, but lightpaths are made to arrive randomly. Results in this table are averaged over 10 random arrival patterns of lightpaths.

Observe in Table 10.2 that the results in the DLE problem are very close to the lower bound for number of wavelengths achievable, for the given set of connections. Another important fact is that the variance in the results is very low (over the 10 sample runs that were conducted for each set of parameters). This leads us to believe that the LCP routing algorithm is quite a suitable choice for dynamic routing of lightpaths, because of its simplicity and adaptive properties.

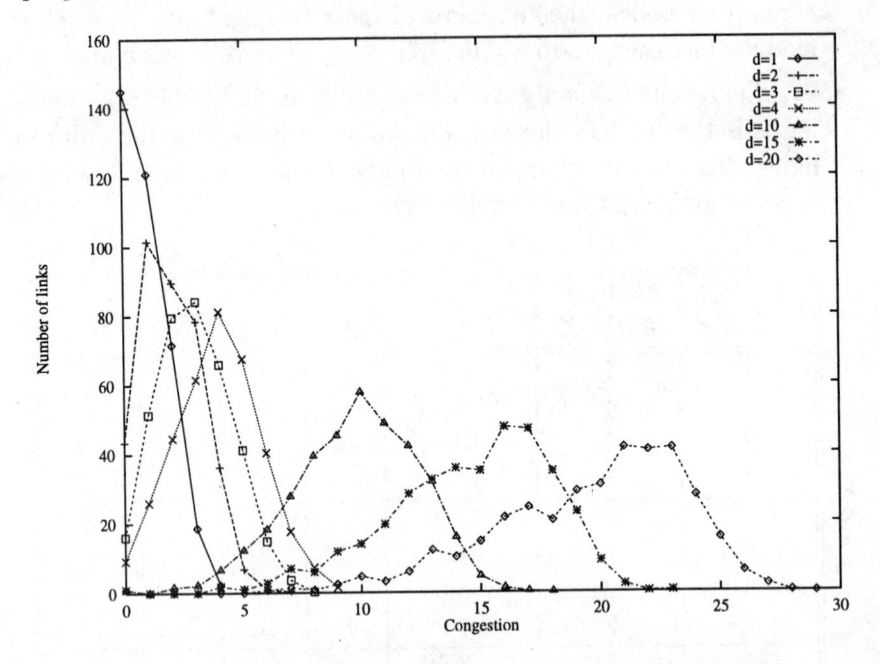

Figure 10.3 Effect of number of connections on link congestion.

Figure 10.3 shows the distribution of the congestion on the network's links, viz., the number of links that have a particular congestion value, with increase in the nodal degree d. As expected, the congestion in the links increases with increasing nodal degree, i.e., the distribution shifts to the right with increasing number of connections.

10.4 Summary

This chapter studied large instances of the routing and wavelength assignment (RWA) problem as applied to a wavelength-routed, all-optical network. The objective was to minimize the number of wavelengths needed, given a certain set of lightpath requests that were to be satisfied on a given physical topology. The problem was partitioned into (1) routing of a lightpath over the physical fiber links, and (2) assigning wavelengths to each lightpath so that no two lightpaths passing through the same physical fiber link are assigned the same wavelength. Algorithms were proposed which can solve the RWA problem for large networks. The problem was studied for the static case (where all lightpath requests are available in advance) as well as the dynamic case (where lightpath requests arrive and need to be established one by one). A linear programming formulation was used along with good approximation techniques to solve the static lightpath establishment (SLE) problem. Our algorithms provided results which were very close to the lower bound for the number of wavelengths that will be needed to establish a given set of lightpaths. Simple heuristics were used for the dynamic case; the results obtained were also close to the aforementioned lower bound. These algorithms can be used for the design and analysis of large all-optical networks.

Exercises

10.1. In the linear programming formulation in Section 10.2.1, we minimize $F_{\max}$. What is $F_{\max}$? Why should we minimize $F_{\max}$?

10.2. Let G be a graph with $V(G) = v_1, v_2, \ldots, v_n$, where $\deg(v_i) \geq \deg(v_{i+1})$ for $i = 1, 2, \ldots, n-1$. Prove that $\chi(G) \leq \max_{1 \leq i \leq n} \min(i, 1 + \deg(v_i))$.

10.3. Given a graph $G = (V, E)$. Define

$$k = \max_{1 \leq i \leq n} 1 + \deg_{<v_1, v_2, \ldots, v_n>}(v_i) \tag{10.6}$$

where $v_1, v_2, \ldots, v_n \in V$ and $< v_1, v_2, \ldots, v_n >$ is a vertex ordering. Define *smallest-last (SL)* vertex ordering $< v_1, v_2, \ldots, v_n >$ to be the vertex ordering such that $\deg(v_{i+1}) \leq \deg v_i$, for $1 \leq i \leq n$. Show that, over all the $n!$ possible vertex orderings, the SL vertex ordering minimizes the value of k.

10.4. We know that the static lightpath establishment (SLE) problem is NP-complete. Show a simple transformation that transforms a SLE problem into a graph coloring problem.

10.5. Consider the network shown in Fig. 10.1. Let the connection requests be as follows:

B-H, A-E, B-D, D-F, B-F, C-E, C-H, A-G, A-C
Set up lightpaths to satisfy the above connection requests using at most three wavelengths per link. Assume no wavelength conversion.

10.6. Consider the network in Fig. 10.1, and the following lightpaths:

(a) C-7-8-9-E
(b) A-1-5-8-9-E
(c) H-2-1-5-8-7-C
(d) B-6-7-8-9-E
(e) A-1-6-7-10-D
(f) G-3-2-1-6-B
(g) H-2-3-4-F

Color the lightpaths using the minimum number of wavelengths.

10.7. Consider the NSFNET physical topology shown in Fig. 9.1. Remove the nodes WA, CO, NE, and GA. Suppose the connection requests are as shown in Fig. 10.4. What is the minimum number of wavelengths needed to satisfy all the connection requests?

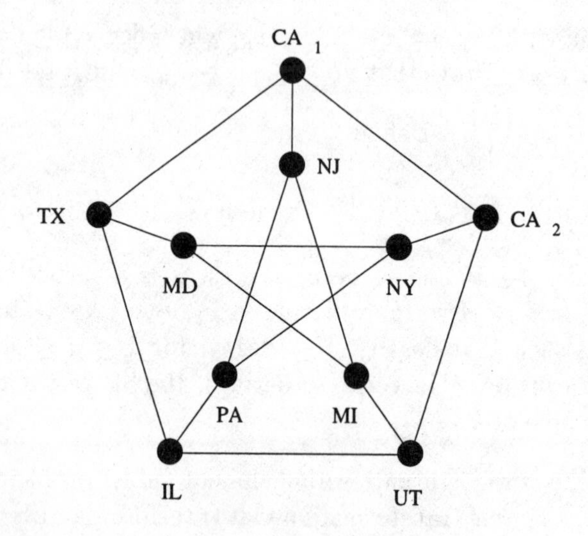

Figure 10.4 Connection requests.

Chapter 11

Wavelength Conversion

Wavelength conversion has been proposed for use in wavelength-division multiplexed networks to improve efficiency. The enabling technologies for building wavelength converters and the corresponding switch designs were reviewed in Chapter 2, Section 2.6. After briefly re-reviewing what wavelength conversion is, its objectives, and the corresponding switch designs, this chapter will examine the network performance issues which need to be addressed before wavelength conversion can be incorporated effectively. A review of various analytical models that have been employed to assess the performance benefits in a wavelength-convertible network is provided.

11.1 Introduction

Consider the network in Fig. 11.1. It shows another view of wavelength-routed network containing two *WDM crossconnects* (S1 and S2) and five *access stations* (A through E). A WDM crossconnect is essentially the same as a wavelength-routing switch (WRS) that we have discussed so far. However, so far, each WRS (crossconnect) had only one electronic attachment (access station), and the combination of the WRS plus electronic attachment had been referred to as a node. Figure 11.1 is a generalization showing that more than one "access station" may be attached to a WDM crossconnect. Thus, this model can perform a clear demarkation between the "all-optical portion" and the electronic portion of the network, as can be seen in Fig. 11.1.

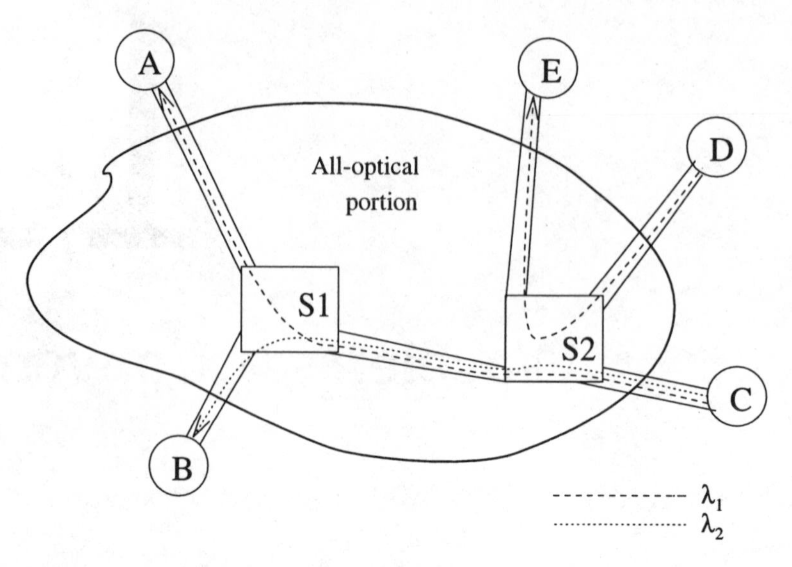

Figure 11.1 An all-optical wavelength-routed network.

In Fig. 11.1, three lightpaths have been set up (C to A on wavelength λ_1, C to B on λ_2, and D to E on λ_1). To establish a lightpath, we sometimes may prefer that the *same* wavelength be allocated on all of the links in the path. Recall that this requirement is known as the *wavelength-continuity constraint.* This constraint distinguishes the wavelength-routed network from a "circuit-switched" network which has no such constraints since the latter blocks a connection only when there is no capacity along any of the links in the path assigned to the connection.

Now, consider the example in Fig. 11.2(a) with $W = 2$. Two lightpaths have been established in the network: (1) between node 1 and node 2 on wavelength λ_1 and (2) between node 2 and node 3 on wavelength λ_2. Now, suppose a lightpath between node 1 and node 3 needs to be set up. Establishing such a lightpath is impossible even though there is a free wavelength on each of the links along the path from node 1 to node 3. This is because the available wavelengths on the two links are *different.* Thus, a wavelength-routed network *with wavelength-continuity constraint* may suffer from higher blocking as compared to a circuit-switched network.

It is easy to eliminate the wavelength-continuity constraint, if we can *convert* the data arriving on one wavelength along a fiber link into another

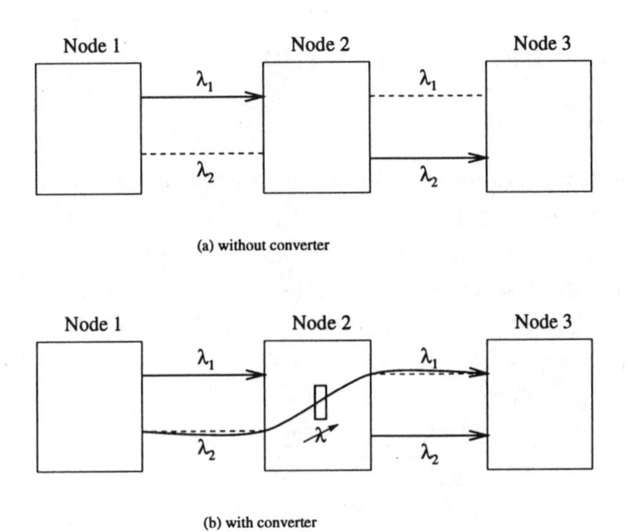

Figure 11.2 Wavelength-continuity constraint in a wavelength-routed network.

wavelength at an intermediate node and forward it along the next fiber link. Such a technique is actually feasible and is referred to as *wavelength conversion.* In Fig. 11.2(b), a wavelength converter at node 2 is employed to convert data from wavelength λ_2 to λ_1. The new lightpath between node 1 and node 3 can now be established by using wavelength λ_2 on the link from node 1 to node 2 and then by using wavelength λ_1 to reach node 3 from node 2. Notice that a single lightpath in such a *wavelength-convertible* network can use a different wavelength along each of the fiber links in its path. Thus, wavelength conversion may improve the efficiency in the network by resolving the wavelength conflicts of the lightpaths.

This study will examine the effects of wavelength converters in wavelength-routed networks. A note on terminology: Wavelength converters have been referred to in the literature as wavelength shifters, wavelength translators, wavelength changers, and even frequency converters. Throughout this study, we refer to these devices as wavelength converters.

Section 11.2 reviews the basic principles of wavelength conversion and the corresponding switch designs. Section 11.3 highlights the network design, control, and management issues for effectively using wavelength conversion. The approaches taken to tackle some of these issues are highlighted and new problems in this area are introduced. Section 11.4 discuss various approaches that have been proposed to quantify the benefits of wavelength conversion.

11.2 Basics of Wavelength Conversion

11.2.1 Wavelength Converters

A wavelength converter's function is to convert data on an input wavelength onto a possibly different output wavelength among the N wavelengths in the system (see Fig. 11.3). In this figure and throughout this section, λ_s denotes the input signal wavelength; λ_c, the converted wavelength; λ_p, the pump wavelength; f_s, the input frequency; f_c, the converted frequency; f_p, the pump frequency; and CW, the continuous wave generated as the signal.

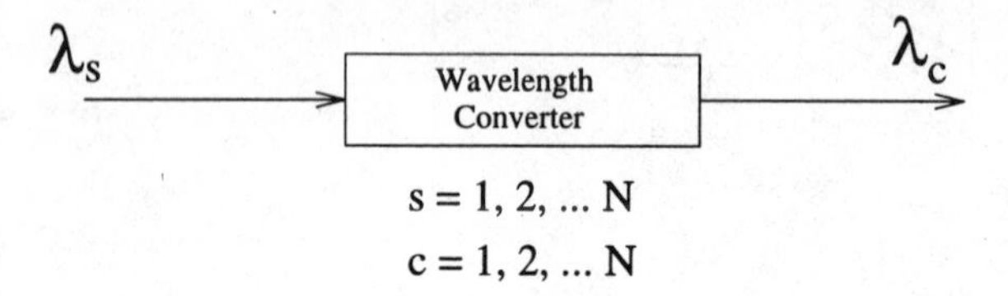

Figure 11.3 Functionality of a wavelength converter.

An ideal wavelength converter should possess the following characteristics [DMJD96]:

- transparency to bit rates and signal formats,
- fast setup time of output wavelength,
- conversion to both shorter and longer wavelengths,
- moderate input power levels,
- possibility for same input and output wavelengths (i.e. no conversion),
- insensitivity to input signal polarization,
- low-chirp output signal with high extinction ratio[1] and large signal-to-noise ratio, and
- simple implementation.

A classification of wavelength conversion schemes was provided in Chapter 2, Section 2.6.

[1]The *extinction ratio* is defined as the ratio of the optical power transmitted for a bit "0" to the power transmitted for a bit "1."

11.2.2 Switches

As wavelength converters become readily available, a vital question comes to mind: Where do we place them in the network? An obvious location is in the switches (i.e., crossconnects) in the network. A possible architecture of such a wavelength-convertible switching node is the dedicated wavelength convertible switch (see Fig. 11.4, from [LeLi93]). In this architecture, each wavelength along each output link in a switch has a *dedicated* wavelength converter, i.e., an $M \times M$ switch in an N-wavelength system requires MN converters. The incoming optical signal from a link at the switch is first wavelength demultiplexed into separate wavelengths. Each wavelength is switched to the desired output port by the nonblocking optical switch. The output signal may have its wavelength changed by its wavelength converter. Finally, various wavelengths combine to form an aggregate signal coupled to the outbound link.

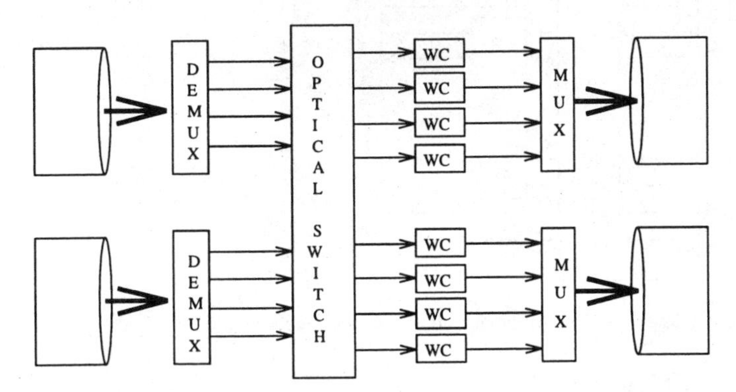

Figure 11.4 A switch which has dedicated converters at each output port for each wavelength.

However, the dedicated wavelength-convertible switch is not very cost efficient since all the wavelength converters may not be required all the time. An effective method to cut costs is to share the converters. Two architectures have been proposed for switches sharing converters [LeLi93]. In the share-per-node structure (see Fig. 11.5(a)), all the converters at the switching node are collected in the converter bank. (A converter bank is a collection of a few wavelength converters, each of which is assumed to have identical characteristics and can convert any input wavelength to any output wavelength.) This bank can be accessed by any of the incoming lightpaths by appropriately configuring the larger optical switch in Fig. 11.5(a). In this architecture, only the wavelengths which require conversion are directed to the converter

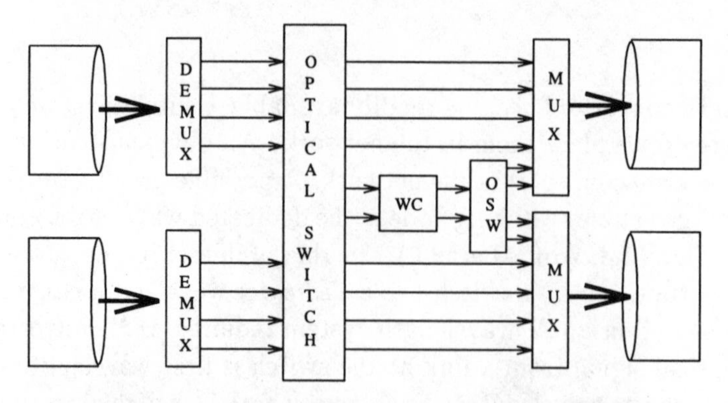

(a) Share-per-node wavelength-convertible switch architecture

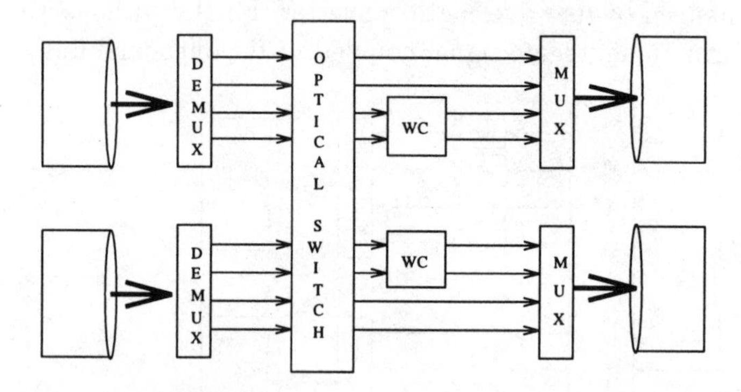

(b) Share-per-link wavelength-convertible switch architecture

Figure 11.5 Switches which allow sharing of converters.

bank. The converted wavelengths are then switched to the appropriate outbound link by the second optical switch. In the share-per-link structure (see Fig. 11.5(b)), each outgoing link is provided with a dedicated converter bank which can be accessed only by those lightpaths traveling on that particular outbound link. The optical switch can be configured appropriately to direct wavelengths toward a particular link, either with conversion or without conversion.

When opto-electronic wavelength conversion is used, the functionality of the wavelength converter can be performed at the access stations instead of at the switches. The share-with-local switch architecture the simplified network access station architecture fall under this category (see Chapter 2).

11.3 Network Design, Control, and Management Issues

11.3.1 Network Design

Network designs must evolve to incorporate wavelength conversion effectively. Network designers must choose not only among the various conversion techniques described in Section 2.6, but also among the several switch architectures described in Section 11.2.2. An important challenge in the design is to overcome the limitations in using wavelength conversion technology. These limitations fall into the following three categories:

1. *Limited availability of wavelength converters at the nodes.* As long as wavelength converters remain expensive [Yoo96], it may not be economically viable to equip all the nodes in a WDM network with them. Some effects of sparse conversion (i.e., having only a few converting switches in the network) have been examined [SuAS96]. An interesting question which has not been answered is *where* (optimally?) to place these few converters in the network.

2. *Sharing of converters.* Even among the switches capable of wavelength conversion, it may not be cost effective to equip all the output ports of a switch with this capability. Designs of switch architectures have been proposed (see Section 11.2.2) which allow sharing of converters among the various signals at a switch. It has been shown in [LeLi93] that the performance of such a network saturates when the number of converters at a switch increases beyond a certain threshold. An interesting problem is to quantify the dependence of this threshold on the routing algorithm used and the blocking probability desired.

3. *Limited-range wavelength conversion.* Four-wave-mixing-based all-optical wavelength converters provide only a limited-range conversion capability. If the range is limited to k, then an input wavelength λ_i can only be converted to wavelengths $\lambda_{\max(i-k,1)}$ through $\lambda_{\min(i+k,N)}$, where N is the number of wavelengths in the system (indexed 1 through N). Analysis shows that networks employing such devices, however, compare favorably with those utilizing converters with full-range capability, under certain conditions [YLES96].

Other wavelength converter techniques too have some limitations. As seen in Section 2.6, the wavelength converter using SOAs in XGM mode suffers greater degradation when the input signal is up-converted to a signal of equal or longer wavelength than when it is down-converted to a shorter wavelength. Moreover, since the signal quality worsens after multiple such conversions, the effect of a cascade of these converters can be devastating. The implications of such a device on the design of the network need to studied further.

Apart from efficient wavelength-convertible switch architectures and their optimal placement, several other design techniques offer promise. Networks equipped with multiple fibers on each link have been considered for potential gains [JeAy96] in wavelength-convertible networks and suggested as a possible alternative to conversion. This work will be reviewed in greater detail in Section 11.4. Another important problem is the design of a fault-tolerant wavelength-convertible network. Such a network could reserve capacity on the links to handle disruptions due to link failure caused by a cut in the fiber. Quantitative comparisons need to be developed for the suitability of a wavelength-convertible network in such scenarios.

11.3.2 Network Control

Control algorithms are required in a network to manage its resources effectively. An important task of the control mechanism is to provide routes to the lightpath requests while maximizing a desired system parameter, e.g., throughput. Such routing schemes can be classified into *static* and *dynamic* categories depending on whether the lightpath requests are known a priori or not. These two categories are described below.

1. *Dynamic Routing.* In a wavelength-routed optical network, lightpath requests arrive at random between source-destination pairs and each lightpath has a random holding time after which it is torn down. These lightpaths need to be set up dynamically between source-destination pairs by determining a route through the network connecting the source to the destination and assigning a free wavelength along this path. Two lightpaths which have at least a link in common cannot use the same wavelength. Moreover, the same wavelength has to be assigned to a path on all of its links. This is the wavelength-continuity constraint described in Section 11.1. This routing and wavelength assignment (RWA) problem was studied in Chapter 10.

   However, if all switches in the network have full wavelength conversion (see Fig. 11.4), the network becomes equivalent to a circuit-switched telephone network [RaSi95]. Routing algorithms have been proposed for use in wavelength-convertible networks. In [LeLi93], the routing algorithm approximates the cost function of routing as the sum of individual costs due to using channels and wavelength converters. For this purpose, an auxiliary graph is created [BaSB91] and the shortest-path algorithm is applied on the graph to determine the route. In [ChFZ96], an algorithm with provably optimal running time has been provided for such a technique. Algorithms have also been studied which use a *fixed path* or *deterministic* routing [RaSi95]. In such a scheme, there is a fixed path between every source-destination pair in the network. Several RWA heuristics have been designed based on which wavelength to assign to a lightpath along the fixed path [BaSB91, MoAz96a, MoAz96b] and which, if any, lightpaths to block selectively. However, design of efficient routing algorithms which incorporate the limitations in Section 11.3.1 still remains an open problem.

2. *Static Routing.* In contrast to the dynamic routing problem described above, the static RWA problem assumes that all the lightpaths that are to be set up in the network are known initially. The objective is to maximize the total throughput in the network, i.e., the total number of lightpaths which can be established simultaneously in the network. An upper bound on the carried traffic per available wavelength has been obtained (for a network with and without wavelength conversion) by relaxing the corresponding integer linear program (ILP) [RaSi95]. Several heuristic-based approaches have been proposed for solving the static RWA problem in a network without wavelength conversion [ChBa96]. Again, efficient algorithms which incorporate the limitations in Section 11.3.1 for a wavelength-convertible network are still unavailable.

11.3.3 Network Management

Issues arise in network management regarding the use of wavelength conversion to promote interoperability across subnetworks managed by independent operators. Wavelength conversion supports the distribution of network control and management functionalities into smaller subnetworks by allowing flexible wavelength assignments within each subnetwork [Yoo96]. As shown in Fig. 11.6, network operators 1, 2, and 3 manage their own subnetworks and may use wavelength conversion for communication *across* subnetworks.

A related network interconnection problem has also been examined in [SoAz97]. This work shows that wavelength conversion for communication

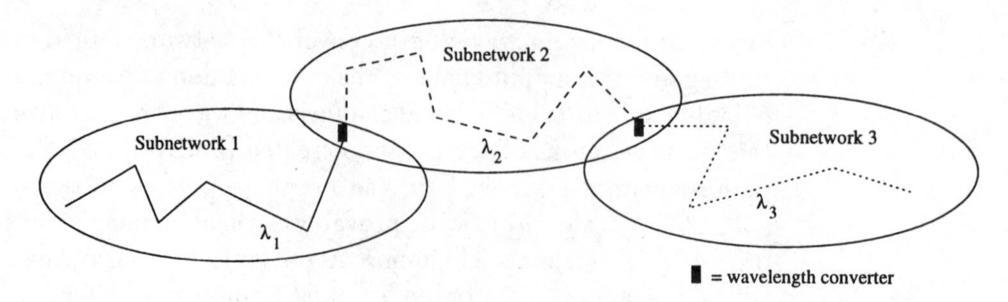

Figure 11.6 Wavelength conversion for distributed network management.

across subnetworks offers very little benefit when the subnets are broadcast stars.

11.4 Benefit Analysis

As mentioned in Section 11.1, wavelength conversion eliminates the wavelength-continuity constraint in wavelength-routed WDM networks. In fact, a wavelength-routed network with full conversion at all nodes behaves essentially like a circuit-switched telephone network. The wavelength assignment algorithm in such a network becomes trivial because all the wavelengths can be treated equivalently, and wavelengths used on successive links along a path can be independent of one another. In addition to reducing blocking, converters can improve fairness by allowing many long distance calls which would have been blocked otherwise due to the wavelength-continuity constraint.

Efforts have been made to quantify the benefits of wavelength conversion. Such attempts involve either probabilistic approaches or they employ deterministic algorithms on specific network topologies. These studies have shown that the benefits of wavelength conversion are greater in a mesh network than in a ring or a fully connected network [KoAc96a]. In the following subsections, we first present a probabilistic approach to quantify the benefits (e.g., decrease in connection blocking probability) of using wavelength conversion, followed by a brief review of several of benefits-analysis studies.

11.4.1 A Probabilistic Approach to Wavelength-Conversion Benefits Analysis

This development is due to the work in [BaHu96], and is based on standard series independent link assumptions, i.e., a connection (lightpath) request sees a network in which a wavelength's usage on a fiber link is statistically independent of other fiber links and other wavelengths. However, this model generally tends to overestimate the blocking probability.

Let there be W wavelengths per fiber link, and let ρ be the probability that a wavelength is used on any fiber link. (Since ρW is the expected number of busy wavelengths on any fiber link, ρ is also the "fiber utilization" of any fiber.) We will consider an H-link path for a connection from node A to node B that needs to be set up.

First, let us consider a network *with* wavelength converters. The probability P_b' that the connection request from A to B will be blocked equals the probability that, along this H-link path, *there exists a fiber link with all of its W wavelengths in use*, so that

$$P_b' = 1 - \left(1 - \rho^W\right)^H \tag{11.1}$$

Defining q to be the achievable utilization for a given blocking probability in a wavelength-convertible network, we have

$$q = \left[1 - (1 - P_b')^{1/H}\right]^{1/W} \approx \left(\frac{P_b'}{H}\right)^{1/W} \tag{11.2}$$

where the approximation holds for small values of P_b'/H.

Next, let us consider a network *without* wavelength converters. The probability P_b that the connection request from A to B will be blocked equals the probability that, along this H-link path, *each wavelength is used on at least one of the H links*, so that

$$P_b = \left[1 - (1 - \rho)^H\right]^W \tag{11.3}$$

Defining p to be the achievable utilization for a given blocking probability in a network without wavelength conversion, we have

$$p = 1 - \left(1 - P_b^{1/W}\right)^{1/H} \approx -\frac{1}{H} \ln\left(1 - P_b^{1/W}\right) \tag{11.4}$$

where the approximation holds for large values of H, and for $P_b^{1/W}$ not too close to unity. Observe that the achievable utilization is inversely proportional to the "length of the lightpath connection" (H), as expected.

Define $G = q/p$ to be a measure of the benefit of wavelength conversion, which is the increase in (fiber or wavelength) utilization for the same blocking probability. From Eqns. (11.2) and (11.4), after setting $P_b = P'_b$, we get

$$G \approx H^{1-(1/W)} \frac{P_b^{1/W}}{-\ln\left(1 - P_b^{1/W}\right)} \tag{11.5}$$

where the approximation holds for small P_b, large H, and moderate W so that $P_b^{1/W}$ is not too close to unity.

Observe that, if $H = 1$ or $W = 1$, then $G = 1$, i.e., there is no difference between networks with and without wavelength converters in these cases.

It is also reported in [BaHu96] that the gain increases as the blocking probability decreases, but this effect is small for small values of P_b. Also, as W increases, G also increases until it peaks around $W \approx 10$ (for $q \approx 0.5$), and the maximum gain is close to $H/2$. After peaking, G decreases, but very slowly. Generally, it is found that, for a moderate to a large number of wavelengths, the benefits of wavelength conversion increases with the "length" of the connection, and decreases (slightly) with an increase in the number of wavelengths.

While this was a simple analysis, more detailed and rigorous treatment of the benefits of wavelength conversion can be found in the references cited in the following subsection.

11.4.2 A Review of Benefit-Analysis Studies

1. *Bounds on RWA algorithms with and without wavelength converters* [RaSi95]. Upper bounds on the carried traffic (i.e.,, lower bounds on the blocking probability) in a wavelength-routed WDM network are derived in [RaSi95]. The generalized routing and wavelength assignment (RWA) problem (both the static and the dynamic cases) are formulated as integer linear programs (ILP) with the objective of maximizing the number of lightpaths that are successfully routed. The formulation is similar to a multicommodity flow problem with integer flows through the links. The upper bound is obtained by relaxing the integrality constraints in the formulation. A similar bound was also obtained for networks with full wavelength conversion at all nodes. The bound is shown to be achievable asymptotically by a fixed RWA algorithm using a large number of wavelengths. A heuristic shortest-path RWA algorithm for dynamic routing is provided which employs a set of shortest paths

and assigns the first available free wavelength to the lightpath requests. The wavelength reuse factor – which is defined as the maximum offered traffic per wavelength for which the blocking probability can be made arbitrarily small by using sufficiently large number of wavelengths – is found to increase by using wavelength converters in large networks.

2. *Probabilistic model with independent link load assumption* [KoAc96a]. An approximate analytical model is developed for a (deterministic) fixed-path wavelength-routed network with an arbitrary topology, both with and without wavelength conversion. This model is then used along with simulations to study the performance of three exemplary networks: the nonblocking centralized switch, the two-dimensional torus network, and the ring network. The traffic loads on the different links in the network are assumed to be independent. The wavelength occupancy probabilities on the links are also assumed to be independent. A wavelength assignment strategy is employed in which a lightpath is assigned a wavelength at random from among the available wavelengths in the path. The blocking probability of the lightpaths is used to study the performance of the network. The benefits of wavelength conversion are found to be modest in networks such as the nonblocking centralized switch and the ring; however, wavelength conversion is found to significantly improve performance of large two-dimensional-torus networks. The analytical model employed in this study cannot be applied to a ring network because the very high load correlation along the links of a path in a ring network invalidates the independent link load assumption.

3. *Probabilistic model without independent link load assumption* [BaHu96]. A model which is more analytically tractable than the ones in [Birm96, KoAc96a] is provided in this study; however, it uses more simplistic traffic assumptions. The link loads are not assumed to be independent; however, the assumption is retained that a wavelength is used on successive links independent of other wavelengths. The concept of *interference length* (L), i.e., the expected number of links shared by two sessions which share at least one link, is introduced. Analytical expressions for the link utilization and the blocking probability are obtained by considering an average path which spans H (*average hop distance*) links in networks with and without wavelength conversion. The gain (G) due to wavelength conversion is defined as the ratio of the link utilization with wavelength conversion to that without wavelength con-

version for the same blocking probability. The gain is found to be directly proportional to the effective path length (H/L). A larger switch size (Δ) tends to increase the blocking probability in networks without wavelength conversion. The model used in [BaHu96] is applicable to ring networks unlike the work in [KoAc96a], and correctly predicts the low gain in utilizing wavelength conversion in ring networks.

4. *Probabilistic model for a class of networks* [Birm96]. Paper [Birm96] provides an approximate method for calculating the blocking probability in a wavelength-routed network. The model considers Poisson input traffic and uses a Markov chain model with state-dependent arrival rates. Two different routing schemes are considered: *fixed routing*, where the path from a source to a destination is unique and is known beforehand; and *least loaded routing* (LLR), an alternate-path scheme where the route from source to destination is taken along the path which has the largest number of idle wavelengths. Analysis and simulations are carried out using fixed routing for networks of arbitrary topology with paths of length at most three hops, and using LLR for fully connected networks with paths of one or two hops. The blocking probability is found to be larger without wavelength conversion. However, this method is computationally intensive and is tractable only for networks with a few nodes.

5. *Multifiber networks* [JeAy96]. The benefits of wavelength conversion in a network with *multiple* fiber links are studied in [JeAy96], by extending the analysis presented in [BaHu96] to multifiber networks. Multifiber links are found to reduce the gain obtained due to wavelength conversion, and the number of fibers is found to be more important than the number of wavelengths for a network. A heuristic is also provided to solve the capacity assignment problem in a wavelength-routed network without wavelength conversion where the multiplicity of the fibers is sought to be minimized. Multiple lightpaths between a source-destination pair is allowed in the network, with each lightpath utilizing a separate wavelength channel. It is concluded that a mesh network enjoys a higher utilization gain with wavelength conversion for the same traffic demand than a ring or a fully connected network.

6. *Sparse wavelength conversion* [SuAS96]. Paper [SuAS96] quantifies the effects of sparse wavelength conversion (see Section 11.3.1), where only a few of the nodes in the network are capable of full wavelength con-

version and the remaining nodes do not support any conversion at all. (But it must be noted that the performance model in [SuAS96] applies equally well to systems with full conversion and systems without conversion.) The analytical model in [SuAS96] improves on the model proposed in [KoAc96a] by relaxing the wavelength independence assumption while retaining the link-load independence assumption, thus incorporating, to a certain extent, the correlation between the wavelengths used in successive links of a multilink path. This paper shows that, in most cases, only a small fraction of the nodes has to be equipped with wavelength conversion capability for good performance. Also, in general, converters are more effective when the number of wavelengths is substantial and when the load in the network is low. This is especially true in networks with a high degree of connectivity, like the mesh-torus, but with fairly large average hop distances. It is also concluded that, in a wide-area network (WAN) with an irregular topology and a large number of links, the connectivity and the number of wavelengths are much more important than the availability of wavelength conversion.

7. *Limited-range wavelength conversion* [YLES96]. The effects of limited-range wavelength conversion (see Section 11.3.1) on the performance gains achievable in a network are considered in [YLES96]. The model used in this work captures the functionality of certain all-optical wavelength converters (e.g., those based on four-wave mixing) whose conversion efficiency drops with increasing range. The analytical model employs both link-load independence and wavelength independence assumptions. The routing algorithm used in the simulations is a fixed one. The wavelength assignment algorithm attempts to minimize the number of converters used and breaks ties by choosing the input wavelength with the lowest index at the converters. Simulations are conducted on a unidirectional ring and a mesh-torus network. The results obtained indicate that significant improvement in the blocking performance of the network is obtained when limited-range wavelength converters with as little as one-quarter of the full range are used. Moreover, converters with just half of the full-conversion range deliver almost all of the performance improvement offered by an ideal full-range converter.

11.5 Benefits of Sparse Conversion

11.5.1 Goals

The goals of this study are summarized below. In most cases, it may be undesirable (e.g., due to budget constraints) to deploy "full" wavelength-conversion capabilities at all nodes, so some degree of sparse wavelength conversion may be desirable.

- Given that "full" wavelength conversion will be installed at a few nodes, what are the "best" nodes at which wavelength converters are placed?
- Different wavelength-converting-switch designs can utilize fewer wavelength converters effectively, so which switch design should be implemented in the network?
- In order to avoid under- or overutilizing the wavelength converters at each node, how many wavelength converters should be placed at each node?
- Make a general determination as to whether wavelength converters offer significant benefits to optical networks, i.e., does the reduction in blocking probabilities justify the increased costs due to deploying wavelength converters?
- Analyze how different traffic loads affect the need and/or desirability for wavelength converters.

While this section provides a summary of our sparse-conversion findings, much more detailed results on analytical models as well as detailed analysis and simulation examples can be found in [InMu96, Ines97], including results on a "interconnected rings" network topology that is very typical of the telecom network environment.

Three Degrees of Sparseness

- *Sparse nodal conversion:* This design attempts to reduce costs by only giving a few nodes in the network "full" conversion capabilities.
- *Sparse switch-output conversion:* This design attempts to reduce costs by limiting the number of wavelength converters at each node.

- *Sparse- (or limited-) range conversion:* In this environment, costs (both monetary and power costs) are reduced by limiting the conversion capabilities (or range, distance) of the wavelength converters. For some wavelength converter design technologies, e.g., wave-mixing-based approaches, it is more efficient, in terms of power maintenance, to convert to wavelengths that are closer (in terms of nm) than to those that are farther away [YLES96].

11.5.2 Simulator

We have developed a network simulator to study the benefits of wavelength conversion (including sparse wavelength conversion) in a wavelength-routed optical network. The simulation also assumes that there are enough transceivers at each node such that they never become a limitation.

The simulator is designed to be flexible enough to test all possible aspects of wavelength converters, such as traffic model, (sparse) switch design, arbitrary network topology with arbitrary set of nodes with conversion capabilities, and arbitrary routing and wavelength assignment (RWA) algorithms [InMu96, Ines97].

Our present study will assume Poisson arrivals, exponential holding times, and uniform (symmetric, balanced) traffic; $W = 5$ wavelengths per fiber link; and fixed-path routing (shortest path with respect to hops) for each connection, with one chosen randomly when multiple shortest paths exist.

11.5.3 Single Optical Rings

A study on unidirectional ring networks with "dynamic" traffic was conducted to assess the benefits wavelength converters. A varying number of nodes were given conversion capabilities and these nodes were spread out across the ring as evenly as possible. Each node that was given any converters was given full wavelength conversion capabilities (i.e., any wavelength entering a node can exit on any free wavelength on any output fiber). However, even with this "full-conversion" capability, our results tend to support earlier predictions based on "static" analysis [RaSa97] that wavelength converters have limited usefulness in a single optical ring (see Fig. 11.7).

When all connections to be established were known in advance, the work in [RaSa97] showed that *only one node* in the ring having "full" conversion capabilities was enough to satisfy all possible sets of requests that could have been established if there was full conversion capabilities at all nodes. Our

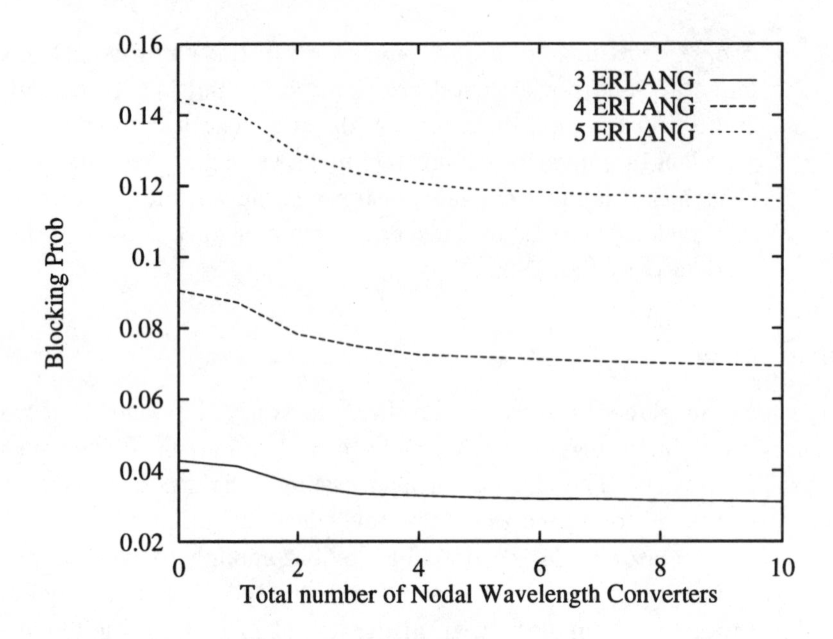

Figure 11.7 Blocking probabilities for different loads in a 10-node optical ring with sparse nodal conversion.

dynamic results (*where connections come and go*) in Fig. 11.7 show that one converter is not enough; however, for this 10-node case, more than two or three nodes having full conversion seems to yield only marginal benefits.

11.5.4 NSFNET

Our next network, the backbone of the NSFNET (see Fig. 11.8), allows wavelength converters to perform better than in a ring because it allows more "mixing" of traffic. (Additional results for a network of interconnected rings may be found in [InMu96, Ines97].

Sparse Nodal Conversion: Which Nodes Need Full Conversion?

Let us determine how much benefit can be achieved by giving a certain number of nodes in the NSFNET full conversion capabilities. We do this by determining the blocking probabilities, for certain traffic parameters, when one node is allowed full conversion, when two nodes are allowed full conversion, all the way up to allowing all nodes full conversion. The wavelength assignment strategy used is the First-Fit (FF) algorithm, which orders the wavelengths

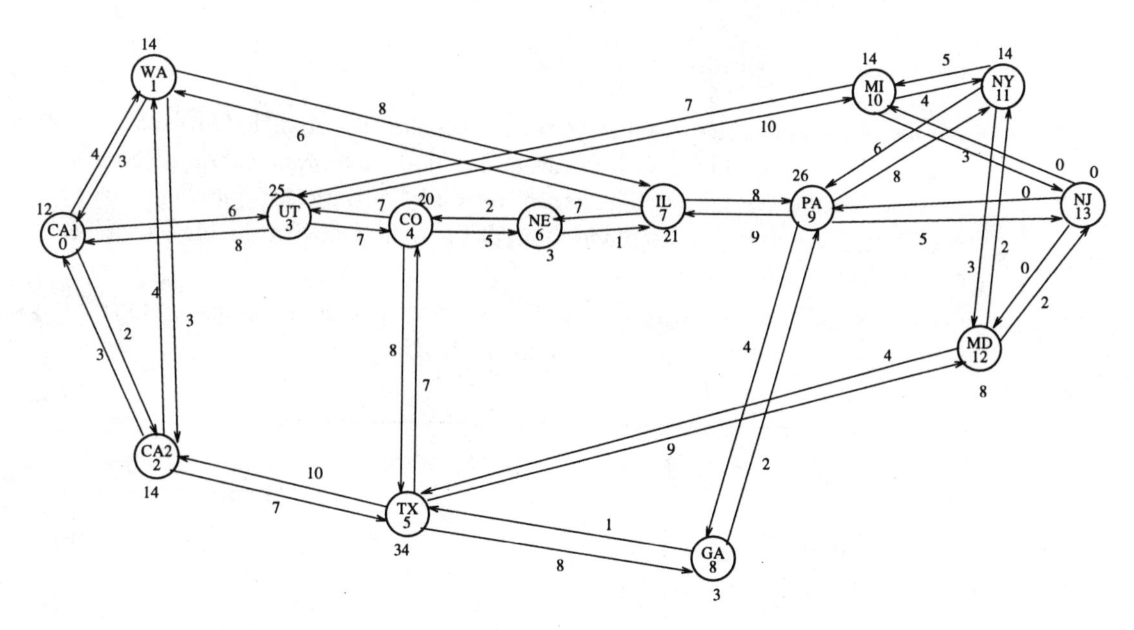

Figure 11.8 NSFNET with the number of convertible routes shown. A number on a link indicates how many source-destination paths passed through the previous node and possibly could have been converted. A number next to a node indicates how many source-destination paths pass through the node and can possibly be wavelength converted.

in some manner, and whenever there is a choice of wavelengths, it chooses the first available wavelength according to its ordering of the wavelengths. This wavelength assignment strategy has been shown to be fairly efficient (relative to a "Best-Fit (BF) Strategy" [InMu96, MoAz96a]) at utilizing the wavelengths and is fairly simple to implement.

Figure 11.9 shows the results of exhaustively searching all combinations of a given number of wavelength converter placements to determine the best blocking probabilities for a given number of wavelength converters placed in the NSFNET. Observe that this curve drops fairly rapidly and then approaches some asymptotic value. Thus, if a few carefully selected nodes are given full conversion, we can achieve almost the same low blocking probability when all nodes have full wavelength conversion.

Heuristic Wavelength Converter Placement

An examination of the exhaustive search results revealed that, to place wavelength converters, this search chose nodes that have *high average congestion of their output links.*[2] [3] Therefore, a good heuristic for placing C sets of wavelength converters is to place them at the C nodes with the highest average output link congestion.

Figure 11.9 also shows that the heuristic placement compares extremely favorably with the optimal placement in the NSFNET.

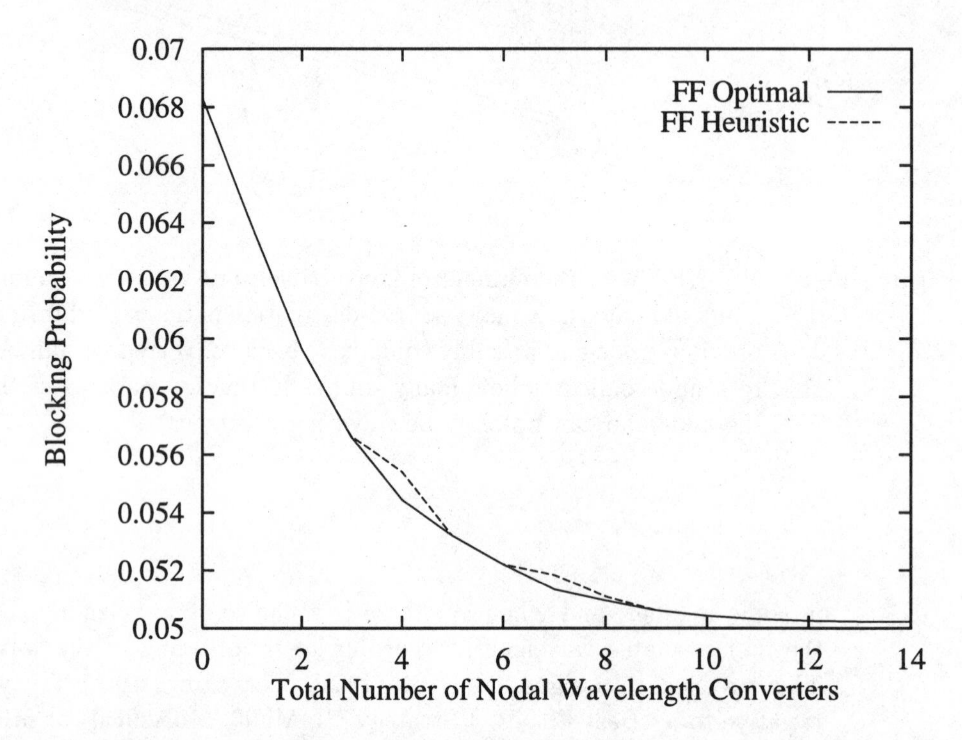

Figure 11.9 Blocking probabilities in the NSFNET for optimal and heuristic placement of wavelength converters (30 ERLANG load).

[2]Two other, but less dominant, factors that seemed to influence converter placements were: the distributions of the congestion on the output links of a node, and the average length of paths that pass through a node.

[3]These statements are made with respect to traffic that does not start at that node. Any traffic that is sourced or sinked by a node will never need to be converted by that node.

Traffic-Load Influences on the Benefit of Wavelength Converters

At light load, there is not much need for wavelength conversion since the few connections can find a route to their respective destinations. So, it was a common belief that the benefit of wavelength converters increases with increasing traffic. However, Figs. 11.10 and 11.11 demonstrate that this belief is only partially true.

Figure 11.10 reveals that the difference between the blocking probabilities with and without conversion becomes fairly constant after the traffic load reaches a certain value. This is because, at heavy load, the network is "capacity-limited," i.e., using wavelength converters can only squeeze through a limited number of additional connections beyond what no conversion allows. This also means that, as a percentage of total blocking, the benefit is decreasing. Thus, as can be seen from Fig. 11.11, there is an operating load at which the percentage gain from using wavelength converters is maximized.

Limited Wavelength Converters at each Node

It may be possible to operate a switch with only a few wavelength converters (as in Fig. 11.5) and still achieve most of the benefits of a switch that has full conversion (see Fig. 11.4). Let us now examine how many wavelength converters are actually needed/utilized at a switch.

Let us focus on one of the nodes in NSFNET (node 2), and study how its wavelength converters were utilized (for the same offered traffic as in Fig. 11.9, with all nodes having full conversion). The distribution of node 2's converter utilization is shown in Fig. 11.12. Observe that node 2 spends almost 95% of the time utilizing three or fewer wavelength converters. Since there are three output fibers and five wavelengths in the network, if node 2 had "full" conversion, it would have required 15 wavelength converters. Implementing a three-converter version of the switch in Fig. 11.5 (with three output fibers) would be very reasonable at node 2. Similarly, other nodes also utilize very few converters [InMu96, Ines97]. For additional related results, please see [Ines97].

Sparse- (or Limited-) Range Wavelength Conversion

This section briefly deals with issues regarding how far (in terms of distance between wavelengths) a wavelength converter needs to convert in order to allow the network to operate efficiently. Previous theoretical studies [YLES96] indicate that most of the reduction in blocking probabilities,

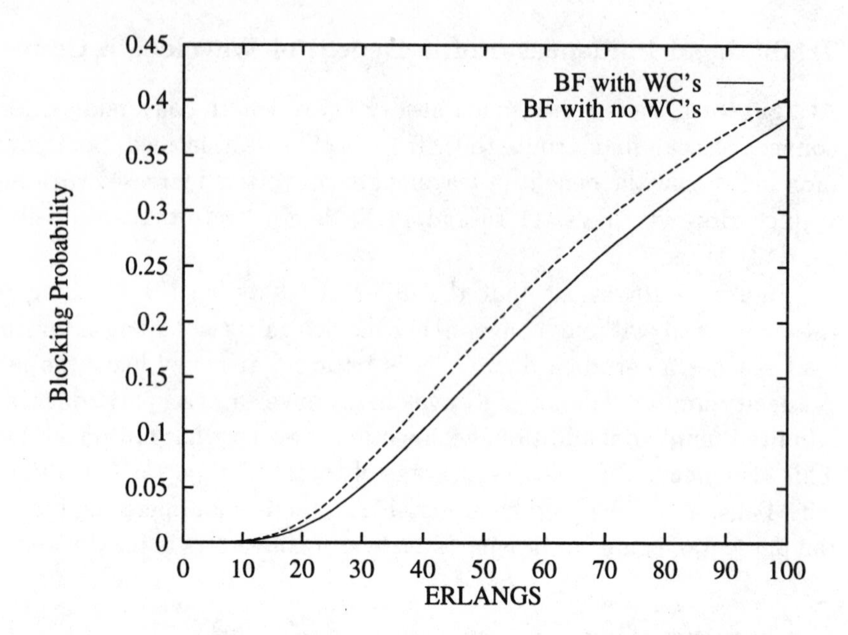

Figure 11.10 Comparison of blocking probabilities in the NSFNET when using full conversion and no conversion in the network with the Best-Fit algorithm.

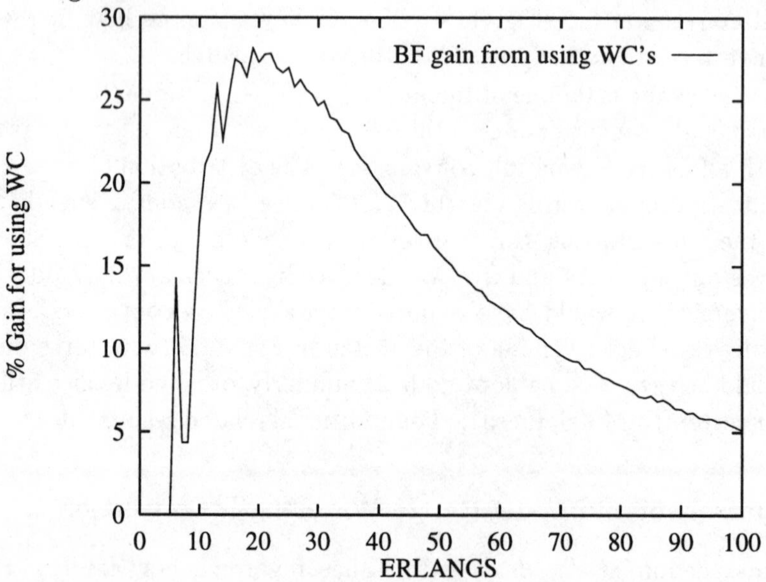

Figure 11.11 Percent gain in the NSFNET from using full-conversion at every node as opposed to no conversion in the network.

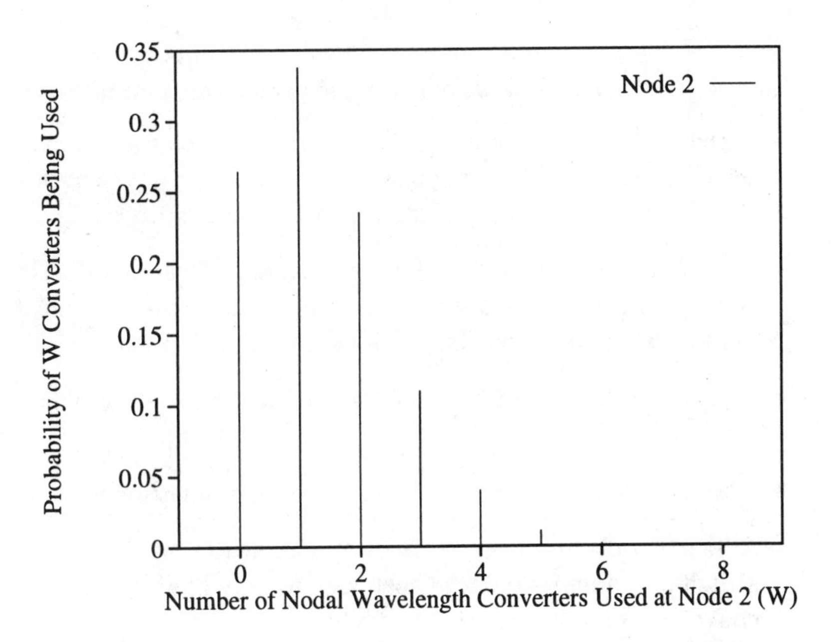

Figure 11.12 Distribution of the number of wavelength converters utilized at node 2 in the NSFNET (30 ERLANG load).

from using full-conversion-capable converters, can be achieved by using only limited-conversion-capable converters.

Detailed analytical and simulation models in [Ines97], as applied to mesh networks such as the NSFNET (Fig. 11.8) and a network of interconnected rings, corroborate the above observations. Specifically, for a seven-wavelength mesh network, it is found that a conversion distance of only one (i.e., approximately 25% conversion facility when taking into account the wavelength "boundaries" as well) yields approximately 75% of the benefits of full conversion! And a conversion distance which is half of the full range yields approximately 90% of the benefits of full conversion. Please see [Ines97] for details.

11.6 Summary

This chapter examined the various facets of the wavelength conversion: from its *incorporation* in a wavelength-routed network design to its *effect* on efficient routing and management algorithms to a *measurement* of its potential benefits under various network conditions.

Some of the important results that were highlighted by our simulation-based case study of sparse wavelength conversion are summarized below:

- A network needs a mixing of traffic for wavelength converters to be beneficial (i.e., single rings benefit little from wavelength converters, while graphs with higher connectivity benefit more).
- A network with sparse wavelength conversion, whether it is sparse nodal conversion or sparse output conversion, can achieve almost the same benefit as network that has "full" conversion capabilities.
- Simple heuristics can be employed to efficiently place wavelength conversion capabilities.
- Traffic load can influence the benefit of wavelength conversion.
- A shared-output-wavelength-conversion switch appears to be the best switch based on its reduced cost (due to only having sparse-wavelength-conversion) and reasonable flexibility.

While our understanding of wavelength conversion technology has improved during the past few years, several issues still remain unresolved especially in the context of efficient design mechanisms and routing protocols. Efforts in this area are needed to further the performance of such networks using intelligent routing and design methods. Another key issue which requires study is the design of a fault-tolerant wavelength-convertible network.

Exercises

11.1. What are the different methods in which we can increase the capacity of a WDM optical network? Also, mention the changes in the network that will be required to implement the method.

11.2. Suppose we have an optical network N_1 with one fiber between adjacent nodes in the physical topology and *four* wavelengths per fiber. Network N_1 does not allow wavelength conversion. Now consider another network N_2 with *four* fibers between adjacent nodes in the physical topology and one wavelength per fiber. Let N_3 be a network similar to N_1, but with full wavelength conversion. Assume that connection requests are set up dynamically. Let p_1, p_2, and p_3 be the average blocking probabilities of networks N_1, N_2, and N_3, respectively. What can we say about how p_1, p_2, and p_3 compare with one another?

11.3. For a network with *two* wavelengths, show an example of dynamic connection setup requests which can be satisfied with wavelength conversion and cannot be satisfied otherwise.

11.4. Given an optical network with the facility of recoloring existing lightpaths, show that such a network may block a connection request which could have been satisfied if wavelength conversion was allowed.

11.5. Explain why employing multiple fibers between nodes is better (i.e., results in lower blocking probabilities) than increasing the number of wavelengths?

11.6. Let N_1 and N_2 be two networks with the same physical topology and number of wavelengths per fiber. Wavelength conversion is not allowed in network N_1 while it is allowed in network N_2. Assume that we use the *least-congested path* routing scheme to satisfy dynamic connection requests. Let the blocking probabilities for a sequence of connection requests S be p_1 and p_2, for the networks N_1 and N_2, respectively. Is $p_1 > p_2$ for all S? If yes, prove it. If not, show an example of a network topology and sequence of connections S, such that $p_2 > p_1$.

11.7. Figure 11.11 shows a plot of the percent gain from using full conversion vs. the network load. Explain the local maximum in the plot.

11.8. Consider a node with two input fibers and two output fibers. There are four wavelengths that can be used in the system. For each of the following sets of connections, determine which node architecture – share-per-node, share-per-link, and dedicated converters – can support the connections.

(a) λ_1 from input 1 to output 1
λ_2 from input 1 to output 1
λ_3 from input 1 to output 2
λ_1 from input 2 to output 1
λ_2 from input 2 to output 1
λ_3 from input 2 to output 2
λ_4 from input 2 to output 2

(b) λ_1 from input 1 to output 2
λ_2 from input 1 to output 2
λ_3 from input 1 to output 2
λ_1 from input 2 to output 1

λ_2 from input 2 to output 2
λ_3 from input 2 to output 1
λ_4 from input 2 to output 1

11.9. Consider a path consisting of six links. Each link supports up to four wavelengths, and average link utilization is 0.5. Calculate the blocking probability with and without wavelength conversion. What is the gain for a blocking probability of 0.8?

11.10. Given a path consisting of five links, suppose full wavelength conversion is allowed between the second and third links, but not at any other location along the path. There are five wavelengths in the system, and the average link utilization is 0.6. What is the blocking probability?

11.11. Calculate the pass-through traffic for each node in the network shown in Fig. 11.13). Assume uniform loading.

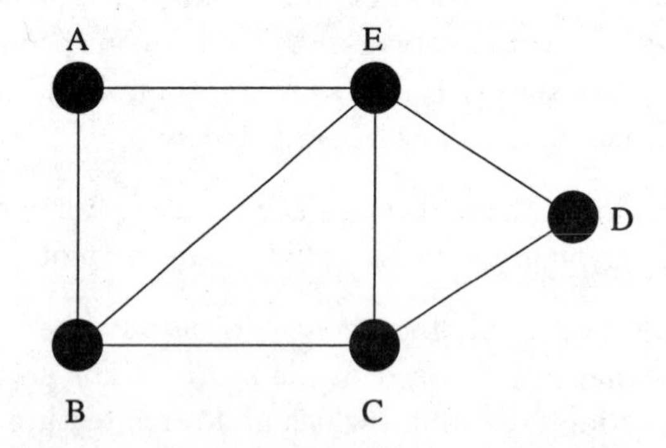

Figure 11.13 Network with uniform loading.

Chapter

12

Additional Topics on Wavelength Routing

While Chapters 8 through 11 studied various topics on wavelength-routed networks, they did not necessarily capture all of the important results on wavelength routing. The objective of the present chapter is to fill up these "gaps" and present other appropriate pieces of work that have been reported in the literature. Specifically, this chapter will review other work dealing with circuit-switched approaches, packet-switched approaches, virtual topology reconfiguration algorithms, WDM network control and management, amplification-related issues (only briefly, and to be expanded upon in Chapter 15), and testbed proposals. For the sake of completeness, this chapter will start with a reiteration of the basic principles in a wavelength-routed network.

12.1 Introduction

The general physical architecture of a WDM WAN is shown in Fig. 10.1. Depending on the characteristics of the photonic switches used in the switching fabric, and on the employment of wavelength converters in the network, the network architecture may be different. This chapter will examine network architectures based on the following network configurations: (1) networks employing only the nonreconfigurable wavelength routing switch (e.g., the waveguide grating router (WGR), termed *nonreconfigurable networks*, (2) networks employing only the reconfigurable wavelength routing switch (WRS), termed *reconfigurable networks*, (3) networks employing a combination of reconfigurable wavelength routing switches and wavelength converters, termed *wavelength-convertible networks*, and (4) networks employing the linear divider and combiner (LDC) as a switch, termed *linear lightwave networks* (LLNs). (For all of these switching devices, please refer to Chapter 2).

The parameters of interest in the analysis, design, and implementation of such networks include the following.

- *Hardware requirements*: In addition to fiber links, the hardware required to implement, maintain, and manage optical networks includes: photonic switches, transmitters and receivers at network nodes, photonic amplifiers, wavelength converters, etc.

- *Number of wavelengths*: Due to technological limitations, such as the amplification range of photonic amplifiers, interference and crosstalk between adjacent wavelength channels, fiber nonlinear effects such as Stimulated Raman Scattering, Stimulated Brillouin Scattering, short tuning range of tunable devices, etc., the number of wavelengths that can be used in the design of the network (i.e., to provide the desired level of connectivity between nodes) is limited. In particular, *scalability* considerations require that the design of the network should be such that the number of wavelengths needed to provide the desired connectivity should be independent of the number of nodes in the network.

- *Performance measures*: Depending upon the type of traffic (circuit- or packet-oriented), typical performance measures used to evaluate the design of networks include blocking probability, congestion, delay, and aggregate network throughput.

  For circuit traffic, given certain connection arrival and connection holding-time characteristics, the probability that a connection request cannot be satisfied is the *blocking probability* of the network.

  Congestion of the network is the utilization of the *heaviest loaded link* in the network.

  Dilation in a circuit-switched network is the length of the longest circuit (or connection).

The network-wide *average delay* is the average delay experienced by packets in the network.

The network *throughput* is the number of packets transported between network nodes per unit time in a packet-switched network, while in a circuit-switched network, it is the number (or fraction) of connection requests successfully satisfied by the network in unit time.

In general, while blocking probability is an appropriate metric in circuit-switched networks, throughput and average packet delay are mostly applicable to packet-switched networks.

- *Traffic models*: For circuit-switched traffic, connection request (call) arrivals are typically modeled as a Poisson process, and connection holding times are assumed to be exponentially distributed. For packet-switched networks, the traffic characteristics are modeled by a traffic matrix, $\{\lambda^{ij}\}$, where λ^{ij} is the total number of packets generated from node i to node j in unit time. Other, possibly more-accurate traffic models such as self-similar traffic, Markov Modulated Poisson Process, etc., are gaining increasing attention as inputs for the network design.

12.2 Circuit-Switched Approaches

Wavelength routing in a switched WDM network was initially studied in [ChGK92, SBJS93]. These as well as other approaches are reviewed in this section. In a circuit-switched network, node u needs to perform the following sequence of tasks in order to send data to node v: (1) establish a lightpath from u to v by configuring the WRSs at the intermediate nodes in the lightpath, (2) transfer data using the lightpath, and (3) tear down the lightpath, since the lightpath is no longer needed, so that the WRSs along the lightpath route can be reconfigured.

12.2.1 LDC-Based Approach

In [SBJS93], the switch was considered to be a linear divider and combiner (LDC), which is a passive device capable of switching groups of wavelength (called *wavebands*) and which is based on linear splitting and combining of power. Routing issues for networks based on the LDC, known as linear lightwave networks (LLNs), were studied. Static and dynamic routing techniques were considered and their performance was compared. The static methods employed preassigned wavelength routing to achieve a high degree of reuse of the optical spectrum in different parts of the network. In the dynamic routing algorithms, reuse of wavelengths was achieved as a natural result of the dynamic behavior of the algorithms.

An important feature of the LDC-based approach is that light (power) from an input may be divided and directed toward more than one output, if so desired. Thus, the LDC can support *multicasting at the optical layer.* However, optical WDM network architectures that can exploit this feature have not received much attention so far.

12.2.2 Lightpath-Based Approach

The idea of using lightpaths as the basis of communication was first introduced in [ChGK92]. Lightpaths were routed using shortest paths between each pair of end nodes, and simple heuristics were used for the assignment of wavelengths to lightpaths in the *static case (when the lightpath requests were known a priori)* as well as in the *dynamic case (when the lightpath requests were generated online).* Two cases in which the underlying network did and did not employ wavelength conversion were also compared, and the difference in the number of wavelengths needed to establish lightpaths in the two cases was shown to be small. This work also showed the correspondence between the wavelength assignment problem and the graph-coloring problem. Subsequently, in [ChGK93], the idea was extended to employ electronic packet-switching over a lightpath-based virtual topology. A packet is transported optically over a lightpath, and electronic packet-forwarding may be required to switch the packet across multiple lightpaths (multihopping), from a source to a destination node.

Embedding of regular virtual topologies onto the physical topology was studied, and bounds on the required number of wavelengths were analyzed.

12.2.3 Multihop Virtual Circuits

A heuristic algorithm for effectively assigning a limited number of wavelengths among the access nodes of a multihop wavelength-routed WAN was studied in [ZhAc94]. The traffic considered was based on virtual circuits, similar to traffic in ATM networks. A virtual circuit, which could span multiple lightpaths, was used for end-to-end connectivity. Admission-control schemes were studied and the performance of the network was compared to that of an ideal-capacity centralized switch. A mixed-integer linear program (MILP) was proposed which maximized the single-hop traffic subject to constraints on the number of transceivers and the number of wavelengths. The problem formulation was computationally expensive to solve exactly and it needed to be approximately decomposed into iterative subproblems. However, even this decomposition was not sufficiently tractable for a large number of nodes. A polynomial-time heuristic, which would iteratively assign to each wavelength as many connections as possible without violating the physical constraints, was finally used. The optical connections between access station pairs with the largest traffic demands were given preference in the above heuristic technique.

Call admission schemes were studied based on alternate routing schemes (by finding K sets of alternate shortest paths ordered by the number of hops) between source-destination pairs. One scheme used the shortest path that contained free wavelengths from among the set of alternate routes. For paths with equal length, priority was given to the path with the minimum load. Results show that, for reasonable values of K, when the node capacity (which was assumed to be an integral multiple of the optical channel rate) was less than the number of wavelengths (which indicated the channel rate per fiber link) used in the network, the network performed almost the same as the centralized switch. The connection graphs needed to be reconfigured only after the occurrence of large changes in traffic patterns.

12.2.4 Routing of Session Traffic

Routing of session-based traffic in a reconfigurable WDM optical network has been studied in [NoTo95]. The network architecture is implicitly assumed to be broadcast. The paper evaluates the performance of a WDM reconfigurable network in a realistic environment and shows that the performance is comparable to that of a centralized ATM switch. The routing and reconfiguration of multimedia streams over the physical topology can be written as a linear integer programming problem. However, this formulation is difficult to solve for large networks. Hence, a variation of Dijkstra's shortest path algorithm is proposed as a routing heuristic and the solution is shown to be close to the upper bound, which obviates the need to pursue the more complete exact solution, which could be obtained from the linear integer program.

12.2.5 Bounds for the RWA Problem

Upper bounds on the carried traffic for connections in a wavelength-routed WDM network were derived in [RaSi94]. (This work also reviewed in Chapter

10.) The problem was formulated as an integer linear program (ILP) with the objective of maximizing the number of connections that are successfully routed. The formulation was similar to a multicommodity flow problem with integer flows through the links. Relaxing the integrality constraints led to an upper bound on the number of successful connections that could be routed. This work also employed the idea of using a set of paths through which connections could be routed and studied the problem for cases with and without wavelength converters. Better results were obtained by constructing an auxiliary graph, so that each node in the auxiliary graph corresponded to a path and two nodes were connected if the corresponding paths passed through a common physical fiber link. The objective function was to color the maximum number of nodes in the graph given a certain number of colors (these colors corresponded to the number of wavelengths in the original network). Heuristic algorithms were proposed and results were compared with the achievable bounds. The bound was shown to be achievable asymptotically by a fixed RWA algorithm for a large number of wavelengths.

12.2.6 Least-Congested-Path (LCP) Routing

An adaptive-routing rule has been studied for a WDM lightwave network [ChYu94]. Each switching node in the network can have wavelength converters that are used to resolve wavelength conflicts in multihop paths. As an example, a *fully connected*, seven-node network with 30 channels on each link was considered. Since there exists a link between every pair of nodes, a direct path consists of a direct link between the node pair. Alternate routes correspond to two-hop paths between node pairs. The routing rule consists of choosing a path from a set of alternate paths, such that this path contains the largest number of free channels on its constituent links, i.e., the path chosen will have maximum spare capacity after the new connection has been established. Since alternate routing introduces instability [Krup82], this work used a channel reservation parameter.[1] It was demonstrated that, for their fully connected network, the use of wavelength converters caused insignificant reduction in blocking probability at light loading, whereas at heavy loading, the wavelength conflict possessed a built-in flow control mechanism. Also reported was that, without any wavelength converter but with the use of optimal channel reservation parameters, the end-to-end blocking probability was very

[1] In channel reservation, a certain number of wavelengths per fiber are used only for direct connections, while the remainder of the wavelengths may be used for alternately routed lightpaths.

close to the case where abundant wavelength converters and optimal channel reservation parameters were used.

12.2.7 Wavelength-Conversion-Based Routing

Routing of optical channels in a wavelength-converter-based network has been studied in [LeLi93]. In a wavelength-routed network with wavelength converting devices at the intermediate nodes, there is no wavelength-continuity constraint across multiple fibers along the route of a lightpath. This can lead to better utilization of wavelengths.

The work in [LeLi93] uses a centralized dynamic routing scheme in which a central controller is responsible for collecting information from the network and for finding the best routes. The objective of the routing algorithm is to minimize the long-term call blocking probability. The optical signal may pass through the switching node in one of two types of paths: (1) wavelength-continuous path, and (2) wavelength-conversion path. If there is sharing of wavelength converters in the intermediate switching nodes, the circuit should, in general, be routed on a wavelength-continuous path to reduce wavelength conversion, unless resolving the wavelength conflict by wavelength conversion is required. The algorithm involves a graph transformation technique to capture the costs of resources used along the path, and shortest-path routing on this transformed graph. Results have also been shown for multicast routing. Simulation results are obtained for the ARPA2 network, and results show the performance gain by using wavelength converters in terms of blocking probability. Performance saturation, when the number of converters is greater than a threshold, implies that a limited number of converters is sufficient to provide good performance (as was also shown in Chapter 11).

12.2.8 Latin-Router-Based Routing

The routing and wavelength assignment problem has also been studied with the intention of maximizing the wavelength utilization at the switching devices [ChBa95]. This has been shown to lead to better overall network performance. The concept of degrees of freedom and Latin Squares was used in the algorithm to configure the wavelength routers. Three heuristic algorithms were developed to complete an isolated Latin Router in which some of the entries may have been preassigned. Based on these heuristics, algorithms were developed for establishing lightpaths in a network of interconnected routers.

12.2.9 Theoretical Results

This subsection summarizes some relevant theoretical results in the area of wavelength-routed optical networks. First, we define some terms that are relevant to this discussion.

- A *Nonblocking network (NBN)* is a network that allows connections between nodes to be established and terminated dynamically. We distinguish between three types of nonblocking networks.
  1. In a *rearrangeably NBN*, all existing connections can be rerouted when a new connection comes in.
  2. In a *wide-sense NBN*, once a route is established, it cannot be rerouted for the entire duration of the connection.
  3. In a *strict-sense NBN*, a connection request is always routed on the same path, independent of existing connections.
- A *permutation network* is a network that can route connections between node pairs, when the connections are a permutation of the network nodes.
- An *oblivious* routing scheme is one in which a connection (source, destination) is always routed using a fixed wavelength (independent of the rest of the connections).
- In a *partially oblivious* routing scheme , a connection (source, destination) is routed using any one of a fixed set of wavelengths.

We summarize some of the important results below [Barr93, Pank92].

- *Number of wavelengths:* In [BaHu94], a lower bound is derived on the number of wavelengths required to support a set of traffic states in any wavelength-routed network. It is shown that permutation routing in a nonreconfigurable network requires $O(\sqrt{N})$ wavelengths, where N is the number of nodes in the network.

  It is also shown in [Barr93] that $\lceil N/2 \rceil + 2$ wavelengths are sufficient for oblivious permutation routing in nonreconfigurable networks.
- *Existence and construction of nonreconfigurable permutation networks:* It is shown in [ABCR94, Barr93] that there *exist* permutation networks

that require $O(\sqrt{N \log N})$ wavelengths. The work in [ABCR94] also shows how to construct such a network, and the construction requires $\sqrt{N}p(N)$ wavelengths, where $p(N) = 2^{(\log N)^{0.8+o(1)}}$.

- *Nonblocking reconfigurable networks:* It is shown in [ABCR94] that one can construct a rearrangeably NBN with N reconfigurable wavelength routers and $O(\log N)$ wavelengths, and one can construct a wide-sense NBN with $O(\log N)$ wavelengths.

- *Bounding wavelengths via congestion and dilation:* The *congestion* in a fiber link is the number of lightpaths that are passing through this fiber link. The *dilation* of a lightpath in the network is the maximum number of physical fiber links traversed by the lightpath. The following results on wavelength requirements, based on the congestion and dilation properties of routing algorithms, appeared in [ABCR94].

  Let the congestion and dilation of a given routing scheme be c and d, respectively. An easy observation is that cd wavelengths are sufficient to achieve the routing scheme. The result is made stronger as follows. For any set of routing requests with congestion c, $2c\sqrt{N}$ wavelengths are sufficient to satisfy the requests, and there exist networks and message routing requests with congestion c and dilation d that require $\Omega(c \times \min\{d, \sqrt{N}\})$ under any routing scheme.

- *Permutation routing in reconfigurable networks:* In any network with N nodes and maximum in-degree Δ, the permutation problem with oblivious routing requires at least $\sqrt{N}/\sqrt{2}\Delta$ wavelengths [Pank92].

  Permutation routing requires $W > 1/(2\Delta)\lfloor(\log \frac{N}{2})/(\log \Delta)\rfloor$.

  Rearrangeable NB permutation routing requires $O(\log^2 N)$ wavelengths and wide-sense NB permutation routing can be performed with $O(\log^3 N)$ wavelengths in popular interconnection networks such as ShuffleNet, de Bruijn graph, and hypercube [Pank92].

12.3 Packet-Switched Approaches

This section will review packet-switched approaches to optical network design, besides those that we examined in Chapters 8 and 9. These designs can be classified into two distinct approaches: (1) optical transmission over lightpaths with electronic packet switching at the end of the lightpaths [BoFB94,

ChGK93, RaSi96], and (2) optical packet switching at the intermediate nodes [ChFu94, Haas93, LeLi95]. The former approach is the preferable approach today because of the difficulty in building optical packet switches due to lack of optical memory.

12.3.1 Logical Topologies for Electronic Packet-Switched Networks

A set of lightpaths can be used as a *virtual topology* over which a packet-switched network may be embedded (as was demonstrated in Chapters 8 and 9). The problem of establishing an optimal virtual topology, for a given physical topology, based on the traffic intensities between nodes, in a WDM-based switched network has also been studied [RaSi96]. The objective function was to maximize the network throughput subject to some bounded average packet delay. The logical topology was constrained by the number of wavelengths supported in each fiber and by the nodal transceiver degree. This work derives a lower bound on the flow of traffic in each link based on the total traffic, the average hop distance in the virtual topology, and the number of links in the topology. It also formulates a mixed integer linear program (MILP) to minimize the maximum congestion in a link subject to constraints. The delay bound is formulated as a uniform bound on the average delay between every source-destination pair, i.e., the average delay between any source-destination pair is restricted to at most a constant times the worst case propagation delay between any source-destination pair in the network. However, this formulation is computationally very expensive for networks with a large number of nodes. Hence, a heuristic technique is used that attempts to maximize single-hop traffic. This work also proposes to have the physical topology embedded in the virtual topology so as to be able to route traffic with the tightest delay bounds through the links in the physical channel (MILP). It shows that, not imposing delay constraints on the topology results in topologies that are "unnatural," in that nodes that are physically located in close proximity may have to use long convoluted paths to communicate, and imposing the delay constraints does not significantly reduce the congestion in the examples considered. It also gives examples in which randomly chosen topologies perform better than heuristics that maximize the single-hop traffic. This is expected because the heuristic does not take into account the store-and-forward traffic.

12.3.2 Deflection Routing Networks

Deflection routing has been studied in optical networks in [BoFB94] because deflection routing can be implemented with modest packet-buffering requirements. However, a difficulty with deflection routing is due to its global time slotting, which might pose significant challenges. The work in [BoFB94] discusses approaches to the implementation of both slotted and unslotted deflection routing in optical networks. Improperly designed unslotted deflection-routing networks can suffer from severe congestion. To stabilize the network and to eliminate congestion, it is essential to provide an access mechanism to control link loading and to purge stale packets. It is also shown that further improvements to performance could be achieved by adding recirculating delay loops that could provide temporary buffering of packets. A comparison of k-buffered slotted and unslotted deflection-routed networks suggested that the unslotted network might provide higher throughput than the corresponding slotted network, when the inefficiency introduced by aligning packets to time slots is taken into account.

12.3.3 Optical Packet-Switch Design

A WDM-based optical packet switch which can shift optical packets to any free wavelength in their destined outbound links has been proposed [LeLi95]. Wavelength conversion can reduce the packet-dropping probability which exists due to wavelength contentions. However, not all packets need conversion and the work in [LeLi95] propose an optimization algorithm so as to minimize the number of packets which need to have wavelength conversion. It derives the maximum number of conversions needed in a packet switch for a given switch size and a given number of wavelengths, and develops a switch control algorithm to minimize the number of wavelength conversions, while maintaining the same minimum number of dropped packets. Simulation results show that the packet dropping probability decreases significantly with wavelength conversion and that very good performance can be achieved with relatively few conversions.

12.4 Reconfiguration in WDM Networks

In environments where traffic demands change over time, it is desirable to have the network connectivity dynamically respond to these changes. This section reviews proposals for reconfiguration in optical networks. Most of the existing work on reconfiguration in WDM-based systems has been (implicitly) proposed for a LAN/MAN environment using a broadcast medium such as a passive-star coupler [GaLZ92, LaAc94].

12.4.1 Passive-Star-Based (LAN) Algorithms

Branch Exchange in Reconfiguration

The reconfiguration phase in a broadcast environment in which the network traffic changes slowly and predictably over time has been studied [LaAc94]. The reconfiguration phase is responsible for taking the network from a particular connection diagram (virtual topology) to a new precomputed connection diagram.

One extreme form of reconfiguration is to simultaneously retune all transceivers involved in the logical links that need to be changed. This approach may lead to substantial network outage and a significant portion of the network may be unusable during the network-transition phase. The other approach is to use *branch-exchange* operations that minimize the portion of the network unavailable during the transition.

In a branch-exchange reconfiguration phase, two optical channels A and B interconnecting nodes P-Q and S-T, respectively, would be changed to create new optical channels A' and B', interconnecting nodes P-T and S-Q, respectively. Another important issue is to shorten the reconfiguration phase as much as possible. The work in [LaAc94] proposes the construction of an auxiliary graph that comprises the new links that need to be established and the old links that need to be torn down. Finding the minimum sequence of branch-exchange operations maps onto the problem of finding a decomposition of the auxiliary graph into the largest number of vertex-disjoint cycles.

Three heuristic, polynomial-time algorithms are proposed that seek a maximum decomposition of the auxiliary graph into the vertex-disjoint cycles. The same problem of cycle decomposition in the auxiliary graph has also been studied by others [BiGu92]. For networks consisting of up to 40 stations, theoretical and simulation results have shown that, when a randomly selected diagram is to be changed to another randomly chosen diagram, the average number of branch-exchange operations required grows linearly with the size of the network.

Physical Interconnection of Passive-Star Networks

The problem of optimally interconnecting passive-star couplers in a hierarchical architecture has been studied in [GaLZ92]. The goal of this work is to discuss the reconfiguration issues in a network with multiple interconnected stars. The *system reconfigurability* represents the number of nodes which can potentially be accessed by each node in a physical topology. It determines the degree of freedom for transmitter/receiver reassignments when the system virtual topology is reconfigured. For a given multistar-based physical topology, a system is *fully reconfigurable* when each node can potentially access all of the other nodes. A system is *partially reconfigurable* when each node can potentially access only part of the nodes. This work discusses a fiber assignment algorithm that is able to achieve the maximum reconfigurability possible in the network.

The work in [SoAz97] also addresses some problems related to the physical interconnection of passive stars.

Reconfiguration Criteria in Multihop WDM Networks

One of the most important concerns in the design of reconfigurable optical networks is to determine *when* and *how* the virtual topology should react to changing traffic patterns [RoAm94b]. This work models the penalty – in terms of packet loss – due to the reconfiguration phase on network performance. The reconfiguration penalty is taken into account in the design of the reconfiguration policies. The reconfiguration policy to be used, and consequently, the frequency of reconfiguration, is determined by the extent of packet loss. The problem is formulated as a Markov decision process (MDP). Two costs are considered: the fraction of packets lost as the network reconfigures from one interconnection pattern to another, and the distance that connections are routed over. Associated with each state transition in the model is a decision to reconfigure the network, thus defining the reconfiguration policy. Finding the optimal reconfiguration policy in a reasonably large network leads to state- and decision-space explosion and is not tractable for practical situations. Hence, heuristics are developed in [RoAm94b] to make decisions similar to the decisions obtained from the optimal policy.

12.4.2 WAN Algorithms

Circuit Rerouting Algorithm for WDM WAN

Rerouting of optical channels in a circuit-switched WDM optical network has been studied in [LeLi94a]. Due to the wavelength-continuity constraint in an optical network without wavelength converters, a new connection may be blocked even if bandwidth is available between the origin and the destination. This is because the *same* wavelength may not be free on all of the fiber links along the path of a lightpath. To alleviate this situation, it may be possible to *reroute existing connections, or to change the wavelength they operate on.* This approach might free up wavelength resources, thus enabling the new connection to be set up. However, rerouting will disrupt the transmissions on the rerouted connections, so minimization of the incurred disruptions, i.e., the number of rerouted circuits or the disruption period of existing connections, should be incorporated in the design of the rerouting methodology. The goal is to derive new circuit migration schemes which would minimize the disruption period and to derive rerouting algorithms so as to select the fewest circuits to be rerouted.

The basic operation of rerouting a circuit consists of the ideas of Wavelength-Retuning (WR) and Move-To-Vacant (MTV). *Wavelength-Retuning* involves retuning the wavelength of a circuit while maintaining its path. WR has certain advantages because (1) it facilitates control since the old and the new routes share the same switching nodes, and (2) it simplifies the calculation of the optimal rerouting solution because only wavelength changes of an existing circuit are considered.

Move-To-Vacant reroutes a circuit to a vacant route with no other circuits. The advantages of MTV are that (1) it does not interrupt other circuits during rerouting because there are no other circuits on the new route of the rerouted circuit, and (2) it preserves message transmissions over the old route during the setup of the new route, thereby reducing disruption of the rerouted circuit (see Fig. 12.1). It is possible to perform parallel retuning operations. Parallel MTV-WR moves each of the rerouted circuits to a vacant wavelength on the same path in parallel. The work in [LeLi94a] also discusses a technique to minimize the weighted number of existing circuits to be rerouted. Finding the optimal solution with the parallel MTV-WR rerouting scheme is shown to be solvable in polynomial time.

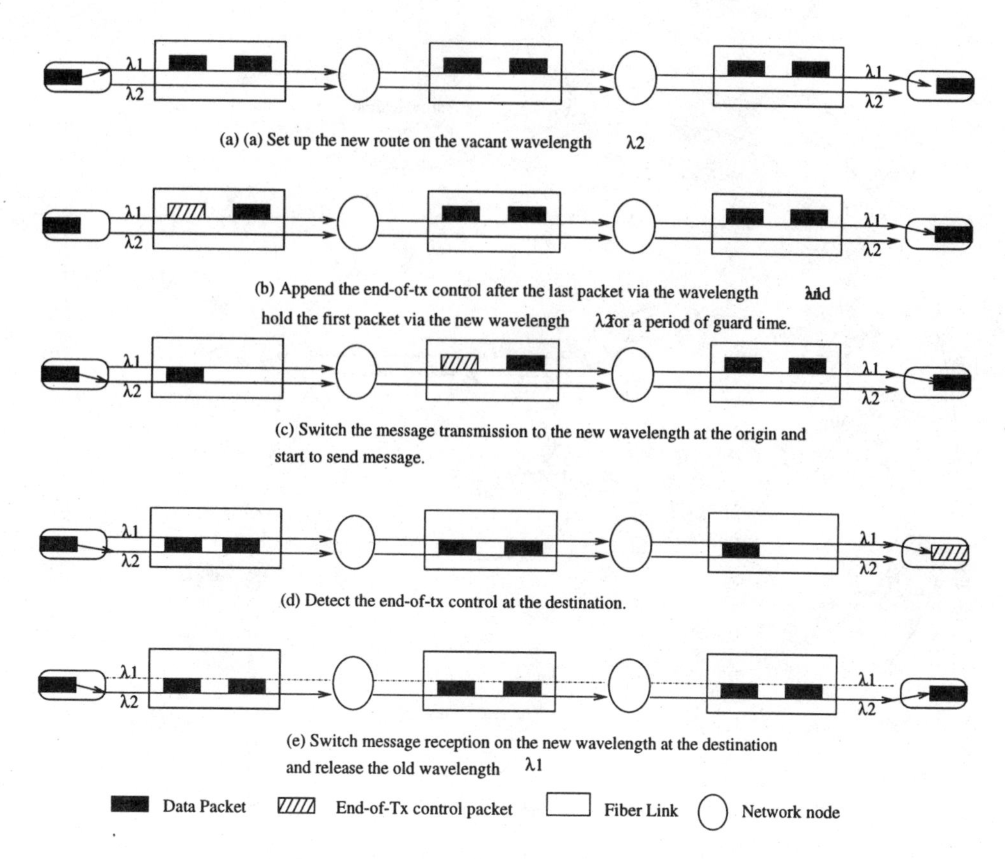

λ1
λ2
(a) (a) Set up the new route on the vacant wavelength λ2
(b) Append the end-of-tx control after the last packet via the wavelength λ1 and
hold the first packet via the new wavelength λ2 for a period of guard time.
(c) Switch the message transmission to the new wavelength at the origin and
start to send message.
(d) Detect the end-of-tx control at the destination.
(e) Switch message reception on the new wavelength at the destination
and release the old wavelength λ1
Data Packet
End-of-Tx control packet
Fiber Link
Network node

Figure 12.1 Implementation of the circuit-migration sequence.

12.5 WDM Network Control and Management

Distributed protocols for setting up, taking down, and maintaining the state of connections in a wavelength-routed optical network are studied in [RaSe96]. This work also examines protocols for recovering from link failures and wavelength failures on a link.

The architecture of the wavelength-routed optical network, shown in Fig. 12.2, consists of wavelength-routing switches (WRSs) interconnected by fiber links.

The WRS at each node is controlled by its own unique *controller*. The controllers communicate with one another using a controller communication network, which may operate either out-of-band or in-band.

A connection between two nodes in the network consists of a path in the network, with a wavelength assigned on each link along the path. A connection

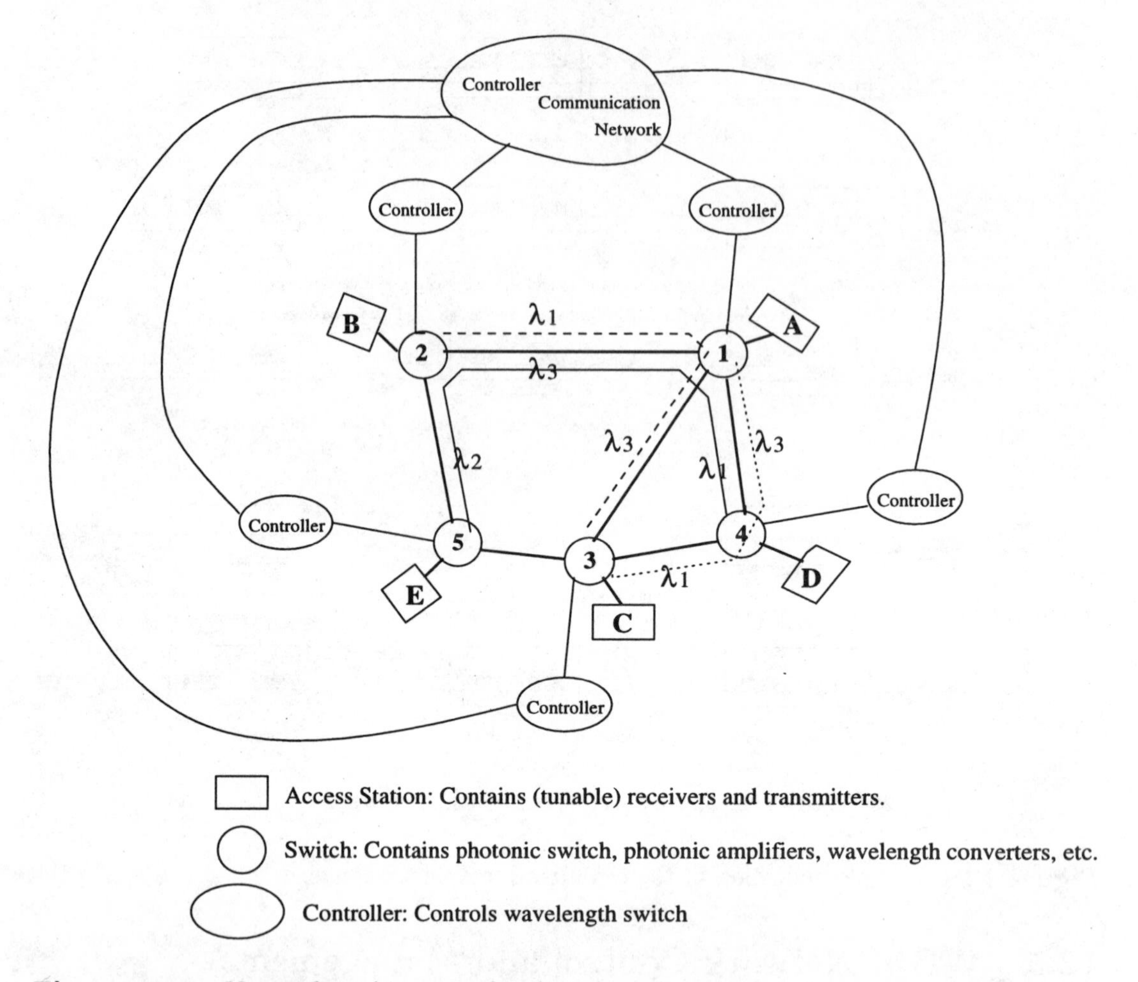

Figure 12.2 Network architecture for distributed control and management.

between two nodes may be set up and taken down dynamically. Connections may provide point-to-point trunks among nodes in the network, or they may be employed to realize a virtual topology. The work in [RaSe96] examines the distributed control of a wavelength-routed optical network. Distributed control protocols for connection management in nonoptical networks have been studied before, e.g., see [CiGS93].

12.5.1 State Information

The controller for a WRS maintains the state of the connections flowing through the switch in a *connection switch table* (CST). The state information is updated by a *connection update procedure.* In addition to the CST, the controllers maintain the topology of the network, and the availability of wavelengths on each link of the network. The network topology and wavelength information is maintained by a *topology update protocol.*

Connection Switch Table

Each connection is identified by a *connection-id* triplet: (origin of connection, destination of connection, sequence number). The connection-id is unique, and is assigned by the originator of the connection. At a switch, a connection may be in a *reserved* state, or in an *up* state. For example, the CST for node A in Fig. 12.2 is shown in Table 12.1.

connection-id	*input-port*	*output-port*	*state*
E,D,12	B,λ_3	D,λ_1	*up*
B,C,10	B,λ_1	C,λ_3	*up*
E,C,8	B,λ_2	C,λ_2	*reserved*
C,A,8	D,λ_3	I/O	*up*
-	D,λ_2	-	-

Table 12.1 The CST at node A.

CST Update Protocol

The CST update protocol informs the controllers of the connections passing through the WRS. Periodically, or when the CST is updated, the controller sends to each neighbor a list of connections that traverse the link shared between them. For example, controller A with the CST in Table 12.1 sends to controller C, the message ((λ_3,(B,C,10),*up*), (λ_2,(E,C,8),*reserved*)). To neighbor D, it sends the message ((λ_1,(E,D,12),*up*), (λ_3,(C,A,8),*up*)). The CST update protocol is used for updating connection status when a connection is taken down, and in situations where links fail, or when wavelengths over a link fail.

Topology Update

To set up a connection, a node needs to calculate the route for the connection, and a wavelength assignment for each link along the route. For calculating routes and wavelength assignment, each node maintains a topology database containing its knowledge of the state of the network.

Each controller periodically broadcasts a topology update message to all other controllers in the network. The topology update message contains the list of healthy links adjacent to the node, and their wavelength usage. For example, controller A with CST in Table 12.1 broadcasts the topology update message $(B,\lambda_1,\lambda_2,\lambda_3)$, (C,λ_2,λ_3), (D,λ_1,λ_3).

12.5.2 Connection Setup

The following procedure is executed to establish a connection between two nodes.

- *Route and wavelength determination*: The originator of the connection finds a route to the destination, and an assignment of wavelengths to links along the route.

- *Reservation*: The originator of the connection requests all controllers along the route of the connection to reserve the selected wavelengths on the links along the route.

- *Connection setup/release*: If reservation was successful at all controllers along the route, the connection is set up; otherwise, the reservation is released at all nodes along the route.

12.5.3 Connection Takedown and Update

To take down a connection, the originator of the connection sends a takedown message to each controller along the route of the connection. The CST update protocol ensures that the connection is taken down in situations of failures of nodes or links.

12.5.4 Fault Recovery

Wavelength Failure

One wavelength is reserved as a spare on each link of the network. When a controller detects the failure of a wavelength on a link, it updates its CST, finds some substitute path between the end-points of the link, and establishes a substitute connection between the end-points.

Link Failure

A spanning tree of spare links is maintained, and upon a link failure, connections are rerouted using links on the spanning tree.

12.6 Amplification-Related Issues

In optical networks, especially WANs, optical amplification is required to compensate for various losses such as fiber attenuation as well as coupling loss in the wavelength routers. Unfortunately, the gain spectra of rare-earth-doped amplifiers as well as semiconductor-laser amplifiers is nonflat over the fiber transmission windows at 1300 nm and 1500 nm, resulting in nonuniform amplification of the signals. Optical signals originating at various nodes at locations separated by large distances have a wide dynamic range among themselves when they arrive at a receiver. Thus, signal power equalization among different wavelength channels is required.

A compensation scheme using proper placement of amplifiers has been studied [LTGC94]. The proposed scheme achieves gain equalization by (1) proper placement of optical amplifiers, (2) assigning the laser emission power, and (3) assigning and adjusting the optical amplifier gain. Amplifier placement is based on calculating the signal power at each point in the network, and placing amplifiers whenever the signal level falls below a certain threshold. The optical gain for the amplifier is calculated to ensure that the total signal power does not lead to amplifier saturation. This problem will be examined in detail in Chapter 15, and an improved solution relative to the one in [LTGC94] will be provided.

Because the erbium-doped fiber amplifier's (EDFA's) bandwidth is limited, it has been proposed that the wavelengths outside the EDFA bandwidth be used. The work in [LeLi94b] discusses algorithms for both the static as well as the dynamic case. The idea is to send those messages that do not need amplification at wavelengths outside the EDFA bandwidth. This idea creates new constraints in designing routing protocols. Should a message be transmitted inside or outside the EDFA bandwidth? Which wavelength should be used? Which links should be used? This work discusses routing algorithms to solve these problems.

12.7 Systems Design Considerations

A large number of practical limitations encountered in the design of optical networks have been outlined in [Maho93].

The necessity of maintaining an adequate signal-to-noise ratio (SNR) limits the fiber distances and the number of nodes that can be traversed by an optical signal. Amplification is necessary for long-haul transmission. However, amplification introduces the following three significant problems, which must be carefully considered in the system design: (1) it introduces noise which degrades the SNR, (2) it causes amplified spontaneous emission (ASE) noise generated by the amplifiers which may cause problems in saturating subsequent amplifiers, and (3) the amplifier passband is not flat; thus, in a long cascade, the end-to-end bandwidth of the multiplexed path may be significantly reduced from that of a single amplifier. Even in special-purpose amplifiers which have been constructed to obtain a reasonably flat gain spectrum, the gain may vary by as much as 2 dB across the nominal 30 nm window. Some transport network applications may involve path lengths in excess of 1000 km, which would require a large number of amplifiers (e.g., ≥ 20). Concatenation of unequalized passbands will therefore lead to a much reduced end-to-end bandwidth of size possibly less than 30 nm.

Crosstalk is another important issue that needs to be dealt with in the design of optical networks. Crosstalk components arise not only from signals on adjacent wavelengths on the same fiber, but also from similar wavelengths on the other fibers when they pass through a common switch. A particularly severe form of crosstalk is caused by interference from a signal on the same wavelength on another fiber; this can build up very rapidly over a small number of nodes and severely impair performance. In the case of the RACE project [Hill93], analysis has shown that, to minimize the effect of crosstalk, the response of the filter must be greater than 20 dB down at the adjacent wavelength. To reduce crosstalk in optical switches, a wavelength dilated switch (WDS) has been proposed which can significantly reduce the crosstalk in 2×2 switches [ShCS93].

In the absence of compensation techniques, the distance that can be traversed by an optical signal in a fiber is limited by linear dispersion. Calculations show that, for networks with distances greater than 1,000 km, operating at 2.5 Gbps, it would be necessary to use dispersion-shifted fiber. Currently, there is much research into techniques for compensating dispersion, which will probably be necessary for the transmission of very high data rates over long distances. Fiber nonlinearity also effectively constrains the maximum number of wavelengths possible over a particular distance [Chra90]. Some of the important criteria for design of the topology in an optical network are the following.

1. *Minimize the wavelength separation. However, four-wave mixing must be considered when operating with dispersion-shifted fiber.*
2. *Minimize the number of wavelengths on long distance spans to minimize Stimulated Brillouin Scattering.*
3. *Operate with the lowest possible launch power per channel to minimize nonlinearities in the fiber.*
4. *Use high-saturation-power optical amplifiers to overcome amplified spontaneous emission effects.*
5. *Minimize optical component insertion losses to maximize range.*

12.8 Testbed Proposals

This section reviews some of the testbed proposals for optical WANs that are being implemented in Europe and the U.S. The current research on optical networking mainly focuses on two important issues: (1) development of better devices with lower crosstalk, finer tunability, more uniform gain spectrum, etc., and (2) development of algorithms for the optimal use of network resources. The networking issues are reviewed in the following subsections.

12.8.1 AT&T/MIT-LL/DEC All-Optical Network (AON) Architecture

The American Telephone and Telegraph Company (AT&T), Digital Equipment Corporation (DEC), and Massachusetts Institute of Technology Lincoln Laboratory (MIT-LL) formed a precompetitive consortium for ARPA to address the challenges of utilizing the evolving terahertz capability of optical fiber technology to develop a national information infrastructure capable of providing a flexible transport layer [Alex93]. The main motivations for this architecture are the following:

- The architecture should scale gracefully to accommodate thousands of nodes, and provide for a nationwide communication infrastructure.

- The architecture should be "future-proof," i.e., modular and flexible to incorporate future developments in technology.

The architecture, shown in Fig. 12.3, is based on WDM, and provides scalability through wavelength reuse and time-division-multiplexing (TDM) techniques. The architecture employs a three-level hierarchy. At the lowest level are Level-0 networks, each consisting of a collection of LANs. Each Level-0 network shares wavelengths internally, but there is extensive reuse of wavelengths among different Level-0 networks. A Level-1 network, which is a metropolitan area network, interconnects a set of Level-0 networks, and provides wavelength reuse among Level-0 networks via wavelength routers. The highest level is the Level-2 network, which is a nationwide backbone network that interconnects Level-1 networks, using wavelengths routers, wavelength converters, etc.

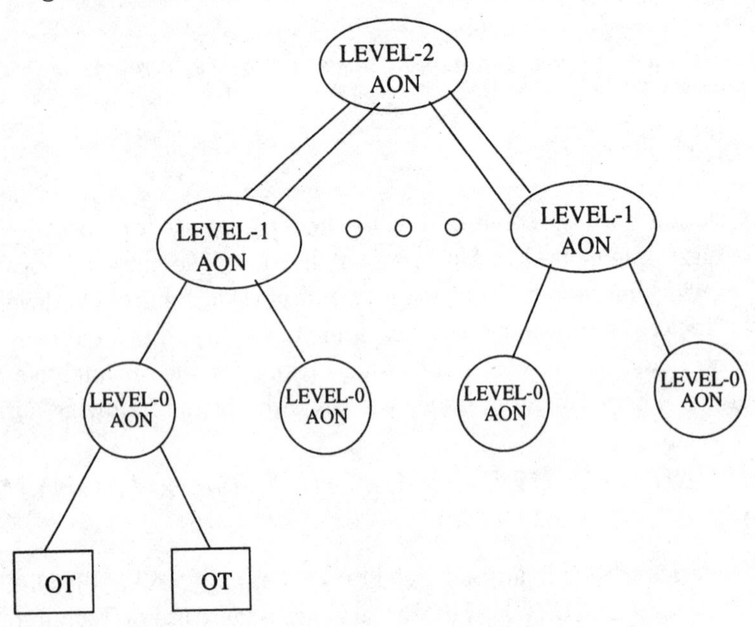

Figure 12.3 The AT&T testbed architecture.

The services provided by the architecture are classified as follows:

- Type A service provides a dedicated optical path for point-to-point, or point-to-multipoint communication. This would provide for Gbps circuit-switched digital or analog sessions.

- Type B service uses TDM over an optical path to provide circuit-switched sessions, with bandwidths in the range from a few Mbps to the full channel rate of a few Gbps.

- Type C service is packet-switched and would be used internally for control, scheduling, network management, and for user datagram services.

12.8.2 Bellcore's Optical Network Technology Consortium (ONTC)

Bellcore, Columbia University, and several other research laboratories and other network service providers are addressing key issues in the development of very high-capacity WANs [Brac93]. The proposed network is based on dense multiwavelength technology, and is scalable in terms of the number of networked users, the geographical range of coverage, and the aggregate network capacity. Of paramount importance to the issues of scalability are the notions of wavelength reuse and wavelength translation.

The proposed research encompasses device technologies, network architectures, and network control. The device technologies being investigated include the following:

- Use of acoustooptic tunable filters that form the WDM crossconnects that are at the core of the rearrangeable optical network.

- Integrated WDM laser (transmitter) array that are used to simultaneously generate a set of up to eight wavelengths at each access node.

- Integrated Opto-Electronic-Integrated Chip (OEIC) to simultaneously receive all wavelengths at each node.

Network-architecture-related issues include the following:

- The need for scalability to very large networks and the need for modular growth without sacrificing performance and survivability.

- Provisioning of fast, self-routed, packet-switching capability in such an environment.

- Determination of a practical approach to assign a channel to an access node, along with routing of optical connections between different node-pairs.

- Traffic management and fault management including admission control, access control, flow control, buffer and bandwidth management, and alternate routing.

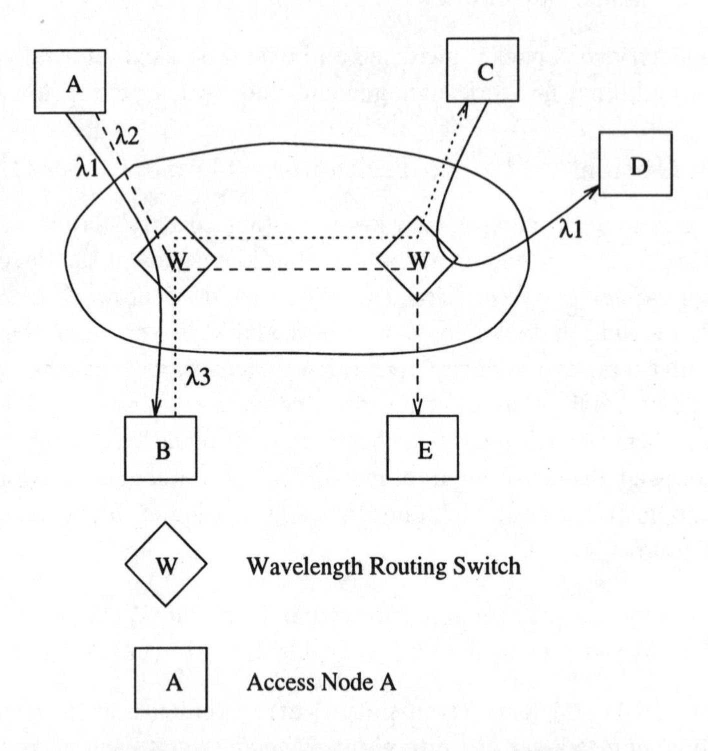

Figure 12.4 The Bellcore all-optical network architecture.

The network architecture, shown in Fig. 12.4, consists of an all-optical inner portion that contains wavelength-routing elements, and is similar to the architecture of reconfigurable networks outlined earlier in this chapter. The design of the network architecture is based on the following major components:

- Wavelength-routing switches to route optical signals, and to provide a high degree of wavelength reuse.

- Wavelength conversion at selected nodes in the network.

- Electronic packet routing overlaid on the all-optical network.

12.8.3 RACE MWTN Project

The Research and development in Advanced Communications technologies in Europe (RACE) is involved in employing integrated broadband communication across Europe, and it involves a number of network operators, manufacturers, and universities [Hill93].

The motivation for this research stems from the need to support both narrow-band (e.g., 64 kbps) and broad-band services in an efficient manner. Electronic switching systems are seen to have a capacity limit in the region of a few tens of Gbps. While it will be possible to increase this limit through the further development of high-speed electronic technology, there will be an increasing price to pay. By careful layering of a future transport network, it may be possible to minimize cost by introducing a new optical network layer with optical switching, controlled in an appropriate way through a network management system. Such an optical layer could be made transparent to data rate and format, and would be particularly suitable for bulk transport of high-bandwidth signals and services. This approach has been taken in the RACE-II "Multiwavelength Transport Network" (MWTN) project.

The network architecture employs the optical transport node, composed of the optical crossconnect, and digital switching (see Fig. 12.5) [Hill93]. The network architecture consists of a number of such interconnected optical transport nodes. The principal feature of this network structure is that the optical transport layer, which is composed of the optical crossconnect and digital switching, allows large blocks of capacity to be routed around the network without the need for opto-electronic conversion and processing.

The MWTN project will also explore the interaction between the optical network layer and the Synchronous Digital Hierarchy (SDH) layer. Within the program, selected components are being developed or supplied, subsystems designed and assembled, and network integration issues addressed.

12.8.4 Multiwavelength Optical Networking (MONET) Project

The Multiwavelength Optical NETworking (MONET) program was established by the U.S. Advanced Research Projects Agency (ARPA) to define, demonstrate, and help drive industry consensus on how to best achieve multiwavelength optical networking in the U.S. on a national scale that serves commercial and specialized government applications. Partners in the MONET consortium are: AT&T, Bellcore, Lucent Technologies, Bell Atlantic, BellSouth, Pacific Telesis, and SBC, with participation from the National Security

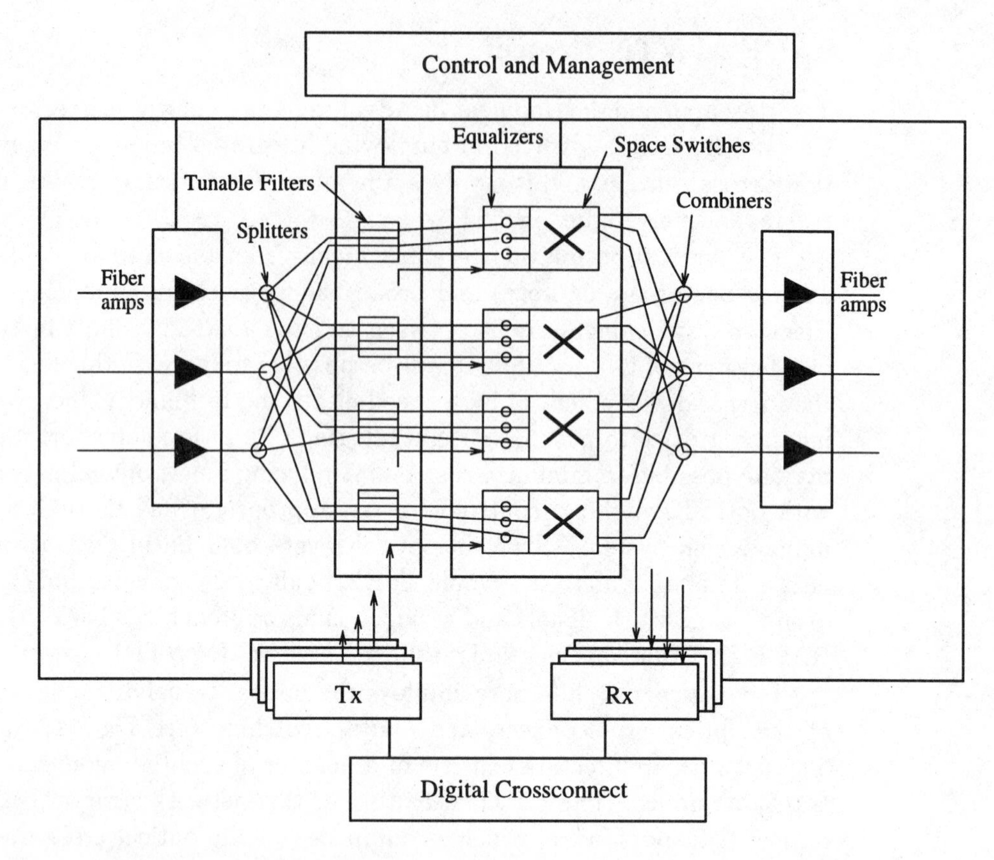

Figure 12.5 Transport node in the RACE architecture.

Agency and the Naval Research Laboratory.

Seven different types of network elements are identified in the MONET program, and they are outlined below:

1. Wavelength Terminal Multiplexer (WTM): A WTM performs multiplexing and demultiplexing of multiwavelength signals to and from a single fiber. It can also multiplex single-wavelength signals from client networks and create a multiwavelength signal to be transported over a MONET network.

2. Wavelength Amplifier (WAMP): A WAMP – i.e., an optical amplifier – should provide optical gain over all of the wavelengths within a fiber.

3. Wavelength Add/Drop Multiplexer (WADM): A WADM enables a node

to selectively remove signals on a wavelength from a fiber as well as insert an optical signal on the same wavelength on the fiber.

4. Fiber Crossconnect (FXC): A FXC performs fiber-to-fiber crossconnection and is transparent to signals within a fiber which may carry single or multiple wavelengths.

5. Wavelength Selective Crossconnect (WSXC): This is the same as a wavelength-routing switch (WRS) discussed throughout this document.

6. Wavelength Interchanging Crossconnect (WIXC): This is the same as a wavelength-convertible switch.

7. Wavelength Router (WR): A WR is equivalent to the WSXC with a fixed routing table. One implementation of a WR is through a waveguide grating router (WGR) discussed in Chapter 2.

The heavy involvement of telecommunication companies in the MONET program indicates that WDM technologies should have wide commercial applications in the national and global telecommunications infrastructure.

12.9 Summary

This chapter summarized other research contributions on wavelength routing that were not captured in Chapters 8 through 11. The emphasis in most of these investigations has been on the development of network architectures based on different optical components and devices. Issues relating to lightpath routing and management, switch design, wavelength bounds, etc., have been addressed.

The challenges in wavelength-routed optical network design can be summarized as follows.

- Design of a network architecture which can exploit the properties of WDM, subject to device and technological constraints.

- Seamless integration of the optical network layer with existing (or future) technologies, so as to be able to upgrade the network.

- Issues relating to routing, establishment, wavelength allocation, and management of lightpaths, so as to maximize wavelength reuse and provide fault-tolerance in the network.

Exercises

12.1. Suppose we wanted to reconfigure a (2,2) ShuffleNet into a (4,2,2) GEMNET. What is the minimum number of branch exchanges required to perform this task?

12.2. For the network in Fig. 12.2, show the Connection Switch Table for Node B.

12.3. Draw an LDC. Explain its operation. Show how you would use this device to construct a 4×4 crossconnect with multicasting capability. For the following connection pattern, give the values for $\alpha_i, 1 \leq i \leq n$, where n is the number of LDCs.

(a) $1 \rightarrow (3, 4)$ (a multicast connection)

(b) $3 \rightarrow 2$

(c) $4 \rightarrow 1$

12.4. Discuss the advantages and disadvantages of latin-router based wide-area networks.

12.5. Compare the Wavelength-Retuning (WR) and Move-To-Vacant (MTV) algorithms for rerouting circuits.

12.6. In wavelength-failure recovery, a controller detects the failure, finds a substitute path, and updates its CST. What else must the controller do?

12.7. What is the channel reservation paramenter in the LCP routing scheme? Why is this parameter important?

Part IV

Potpourri

Chapter

13

Multiwavelength Ring Networks

A fiber-optic ring network, such as a SONET ring network or a fiber distributed data interface (FDDI) network, can be operated over multiple wavelengths on its existing fiber plant consisting of point-to-point fiber links. Using wavelength-division multiplexing (WDM), the network nodes can be partitioned to operate over multiple subnetworks, with each subnetwork operating independently on a different wavelength, and intersubnetwork traffic forwarding performed by a bridge. For this multiwavelength version of a SONET ring or FDDI, which we refer to as wavelength distributed data interface (WDDI), we examine the necessary upgrades to the architecture of a node in the SONET ring or in FDDI, including its possibility to serve as a *bridge*. The main motivation behind this study is that, as network traffic scales beyond (the single-wavelength) SONET ring's or FDDI's information-carrying capacity, its multiwavelength version may be able to gracefully accommodate such traffic growth. A number of design choices exist in constructing a good multiwavelength ring network. Specifically, we investigate algorithms with which, based on prevailing traffic conditions, nodes can be partitioned into subnetworks in an optimized fashion. Our algorithms partition the nodes into subrings, such that the total traffic flow in the network and/or the network-wide average packet delay is minimized.

13.1 Introduction

Consider a fiber-optic ring network, such as a SONET ring or FDDI [FDDI90]. The network nodes are connected via point-to-point fiber links to form a ring. In FDDI, each link is operated at a single wavelength of 1300 nm, at a data rate of 100 Mbps. FDDI's dual counter-rotating rings provide reliability, reconfigurability, and real-time fault detection. Similar characteristics exist in the (single-wavelength) SONET ring as well.[1]

Emerging applications such as multimedia communications, medical imaging, and a proliferation of high-performance color-graphics workstations indicate that these services will soon overwhelm FDDI's information-carrying capacity. Therefore, an extremely desirable characteristic in FDDI (and in any other network) would be the ability to increase the available bandwidth in such a network on demand.

Developments in lightwave technology, particularly in the area of wavelength-division multiplexing (WDM), serve as a strong basis for providing growth capability in an existing FDDI network (as well as in SONET rings and most other fiber-based networks). The objective is to preserve as much of the FDDI and SONET ring standard (hardware and software) as possible, while offering enhanced capability through low-cost WDM devices.[2]

Specifically, this is done by segmenting the huge bandwidth capability of a single strand of fiber in its low-loss window (nearly 50 THz) into a number of nonoverlapping wavelength channels (say W), each operating at whatever rate one desires, e.g., at FDDI rate in order to preserve as much compatibility with the FDDI standard as possible. Although, theoretically, W can be very high, practical considerations such as channel spacing for reduced crosstalk, availability of suitable optical sources, cost, etc. limit W to a few tens. Now, by employing WDM, two neighboring ring nodes, which are attached via a physical point-to-point fiber link, can essentially consider themselves to be connected via W logical channels. System-wide, therefore,

---

[1] While most of our discussion in this chapter will refer to FDDI and its multiwavelength version, WDDI, we remark that these discussions apply equally to the SONET ring and its multiwavelength version as well.

[2] Our objective is to develop a network which is incrementally upgradable through WDM. By incremental, we mean that some subset of the nodes can be upgraded, while allowing other nodes to remain operational in their original form. Standard FDDI nodes utilize broadband (about 50-nm linewidth) LEDs operating in the 1300-nm band. Due to the wide spectral width of the LEDs, it is unlikely that additional wavelengths can be allocated in the 1300-nm band. Thus, additional wavelengths could be added in either the 850-nm band or the 1550-nm band or both.

each of the two counter-rotating rings of FDDI can be represented via W logical subrings. We refer to this multiple-wavelength version of FDDI as wavelength distributed data interface (WDDI). In the sequel, we consider only one of the physical rings since the other ring direction in FDDI is activated only during reconfiguration.

Out of the W channels now passing through a node's interface, a node can be made to logically exist on a subset m ($1 \leq m \leq W$) of these channels (subrings). For example, if we have a seven-node ring network with $W = 2$, then we can have nodes 1, 3, 5, and 7 logically connected to channel 1, while nodes 2, 4, 6, and 7 are connected to channel 2. This implies that we have two subrings with nodes 1, 3, 5, and 7 on subring 1 (and operated on wavelength λ_1), and nodes 2, 4, 6, and 7 on subring 2 (and operated on wavelength λ_2), as shown in Fig. 13.1. Bridging between subrings, as shown in Fig. 13.1, is accommodated by having node 7, the "bridge" node, attach itself logically to both subrings; hardware cost for the bridge, of course, will be higher. In general, if some of the subrings need to be operated independently, they need not be bridged to the other subrings.

The advantage of using such a WDDI network should be clear. Due to traffic growth, if the total load offered by the seven nodes in Fig. 13.1 exceeded 100 Mbps, then the FDDI network would not be able to accommodate the load. However, partitioning the nodes into multiple subrings would allow multiple parallel transmissions to occur (one transmission per subring, each at 100 Mbps), so that the aggregate network capacity can well exceed 100 Mbps.

How to optimally partition the FDDI (or SONET ring) nodes into multiple subrings, given a traffic matrix, is the problem that is addressed in this chapter. By optimality, one might mean *keeping the cross-traffic, viz., the traffic through the bridge, at a minimum*, so that not much queueing and/or processing occurs at the bridge. Another reason for keeping the cross-traffic at a minimum is to utilize the available bandwidth more effectively, since packets that pass through the bridge traverse two rings, and therefore, use up twice as much resources as the intraring packets. Alternately, one might wish to equalize the flows on the subrings as much as possible in order to perform *load balancing* because the larger the flow is on a subring, the higher will be the packet delay for that subring. As another alternative, optimization might mean *minimizing the network-wide mean packet delay* by operating the ring in multiple partitions. In all of the above cases, we would like to demonstrate that operating the ring in multiple subrings, rather than in one large ring, would reduce the link flows and/or delays *significantly*, thereby allowing future growth capability.

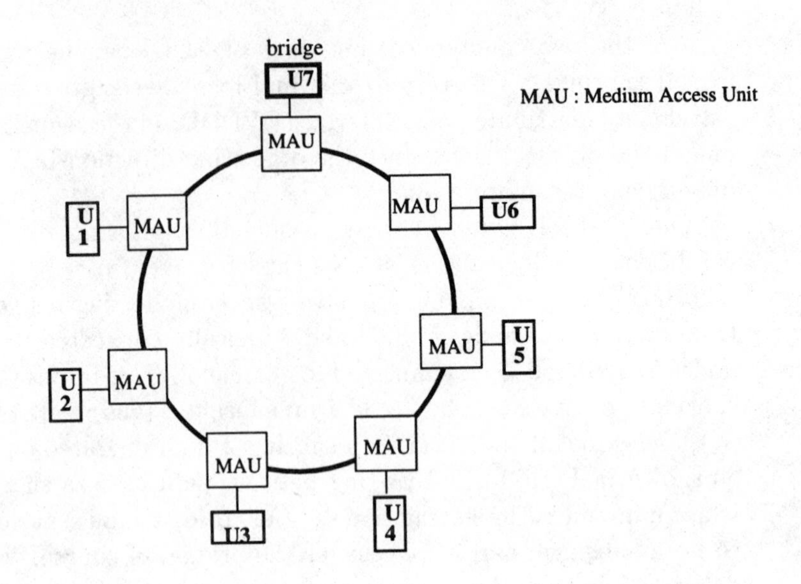

(a) Physical topology

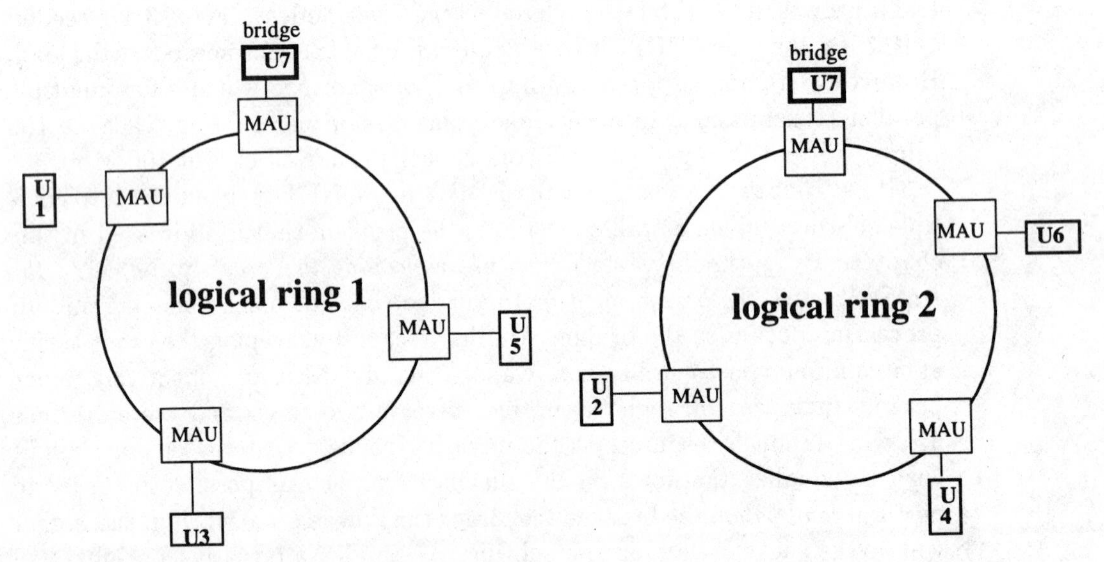

(b) Two logical subrings

Figure 13.1 The WDDI ring network.

Section 13.2 discusses the system architecture and states our assumptions. A few examples to further motivate the partitioning problem are illustrated in Section 13.3. Section 13.4 elaborates on various possible optimization criteria that may be employed for partitioning the nodes into subrings. Approaches based on traffic flow optimizations are examined in Section 13.5. A delay-based optimization approach is developed in Section 13.6, with simulated annealing used to perform the optimization. Section 13.7 compares the performance of different partitioning algorithms under various network conditions.

13.2 System Architecture and Assumptions (Model)

The following assumptions are made about the system architecture.

1. $N+1$ nodes are connected via point-to-point fibers to form a physical ring. The nodes are numbered $1, 2, \ldots, N+1$. The $(N+1)$-th node serves as a *bridge*.

2. Each point-to-point fiber connects two adjacent nodes, and it consists of W parallel (and unidirectional) WDM channels, numbered $1, 2, \ldots, W$.[3]

3. N nodes are each equipped with one tunable laser[4] and one tunable filter, while the $(N+1)$-th node (which will serve as the bridge node) is equipped with W fixed-tuned transmitters and W fixed-tuned receivers, one for each channel.[5] Also, if the traffic flows between nodes are well defined, and their intensities are not expected to change much, then the lasers and filters at the nodes could be pre-fixed to operate on particular wavelengths, based on the optimization studies conducted in this work.

[3] Note that if W equals N, we can have one node per subring, and the system equivalently becomes a star architecture. However, to operate like a star, the bridge node must be able to process packets N times faster than a nonbridge node.

[4] Standard FDDI nodes have wideband (50-nm linewidth) LEDs operating in the 1300-nm band, so upgraded FDDI nodes could be equipped with tunable lasers operating in the 800-nm or 1550-nm bands since not much spacing would be available in the 1300-nm band. Also, not all FDDI nodes need to be upgraded, although our model here assumes this to be the case.

[5] Alternatively, instead of overloading the bridge (as might happen above for $W > 2$), we can have multiple bridge nodes, with each bridge bridging anywhere from 2 to W channels. Unfortunately, this also leads to a large search space for optimization, and we don't consider it any further.

The system cost will now be lower, but it will lack the flexibility to reconfigure itself as the traffic demands change.

4. *Node interface at a nonbridge node* (see Fig. 13.2): Information on the incoming fiber is first demultiplexed into the W wavelengths. If this node has both its transmitter and receiver tuned to the same wavelength k, then it will let all of the other wavelengths pass through its interface untouched. For wavelength k, this node will act like an add-drop multiplexer [Elre93], as in any ring protocol. Such an interface could be implemented in several ways, with one possibility being to employ a channel-dropping filter. We assume that nodes perform *source removal* as in the FDDI ring and IEEE 802.5 token ring, i.e., after a node transmits a packet on the ring, it is responsible for removing that packet from the ring after the packet has traveled around the ring once. Local information as well as information on the untouched wavelengths are then multiplexed back on to the outgoing fiber link. Note that a node does not perform any wavelength conversion.[6]

5. *Node interface at a bridge node* (see Fig. 13.2): After demultiplexing the information from different wavelengths on the incoming fiber, each wavelength is treated with an add-drop multiplexer, and all wavelengths are then multiplexed back onto the outgoing fiber. (A similar strategy is used for creating SONET self-healing ring networks [Elre93].) The bridge should also be equipped with memory to buffer any packets that need to be forwarded by it from one subring to the others. We assume that the bridge maintains a separate queue for each subring, to prevent head-of-the-line blocking.

6. *Who runs the partitioning algorithm*: Our present approach is a "static" one which requires that, when the traffic matrix changes, the new partitioning should be determined. Since the bridge node has access to all of the traffic on all subrings, it can perform the repartitioning operation whenever required. The formulation of a dynamic reconfiguration approach in which all nodes participate in a distributed algorithm to perform repartitioning is an open problem.

7. *Role of the bridge node*: The bridge node handles traffic from different subrings. Essentially, the network can be thought of a $W \times W$

[6] A node performs wavelength conversion when it receives on one wavelength and transmits on another wavelength.

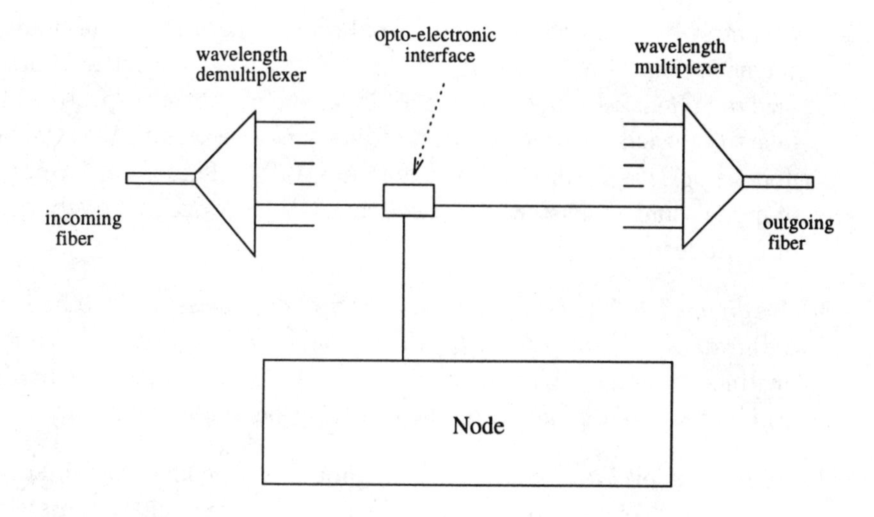

(a) Interface for nonbridge node

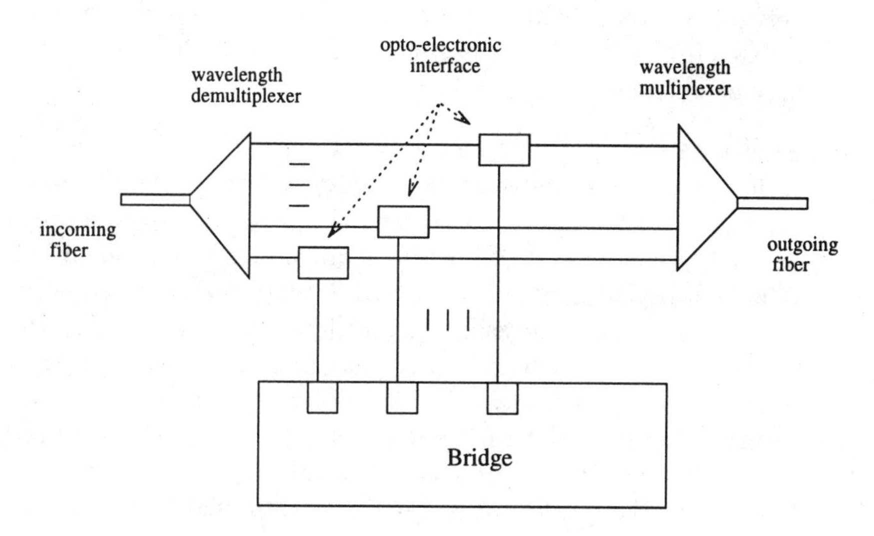

(b) Interface for bridge node

Opto-electronic interface has the necessary photodiode, optical source, and electronics

Figure 13.2 WDDI node and bridge interfaces.

packet switch, interconnecting W subrings, where W is the number of wavelengths, with the bridge node acting as the packet switch. The bridge node is assumed to have an independent physical-protocol interface to each subring. For the rest of this work, we assume that the bridge node is *passive*, i.e., it does not generate traffic. Therefore, for purposes of models and algorithms discussed here, the network effectively consists of N nodes.

8. *Traffic model*: The traffic arrival at any node i is assumed to be Poisson with rate λ_i. We denote by $l_{i,j}$ the probability that a packet at node i is destined to node j, where $i, j = 1, 2, \ldots, N$, and $l_{i,i} = 0$ for all i. Thus, traffic flow from node i to node j is also Poisson with rate $a_{ij} = \lambda_i l_{ij}$.

9. *Additional architectural issues*: Although not considered in our approach here, one should be mindful of the following characteristics (advantages) offered by a WDM-based ring network. Some of the nodes may be operated on an independent ring, without connection to any bridge, e.g., to provide security. Similarly, one may create multiple subnetworks operating with their own partitions and bridges, but otherwise isolated from one another. In general, the rates on different WDM rings may be different.

10. *Previous work on WDM ring networks*: In [Elre93], a new architecture is proposed for implementing unidirectional and bidirectional[7] self-healing ring networks[8] using WDM technology. The architecture of this network, which is a WDM-based upgrade of a conventional unidirectional self-healing ring utilizing SONET add-drop multiplexers [Bell91], consists of N nodes generating traffic flows, and these traffic flows are directed to a single hub node (which is essentially like our bridge node). Each of the N nonbridge nodes is assigned a unique wavelength for transmission to, and reception from, the hub node. In our approach, the network architecture is a WDM-based upgrade of the conventional FDDI network, with the number of wavelengths independent of the number of nodes. Additional work related to that in [Elre93] on unidirectional ring network architectures using WDM technology can be found in [WuKC89, WaCh92, Elre92]. However, these papers do not address

[7] A unidirectional ring requires one fiber to carry a duplex channel, while a bidirectional ring network requires two fibers to carry a duplex channel.

[8] A self-healing ring is a network architecture that connects a set of nodes in a physical ring topology with fault-tolerance capability to counter network component failures.

the problem of optimized partitioning, which is the main contribution of this chapter.

13.3 Illustrative Examples

Before outlining our partitioning algorithms, let us clarify the problem via some illustrations. Consider $W = 2$ and the traffic matrix, $\{a_{ij}\}$, in Table 13.1. (Assume that the bridge node, node 7, does not generate traffic, and therefore, is not included in the traffic matrix.)

from (i) ↓ / to (j) →	1	2	3	4	5	6
1	0	6	6	1	1	1
2	6	0	6	1	1	1
3	6	6	0	1	1	1
4	1	1	1	0	6	6
5	1	1	1	6	0	6
6	1	1	1	6	6	0

Table 13.1 Traffic matrix no. 1, showing two communities of interest.

from (i) ↓ / to (j) →	1	2	3	4	5	6
1	0	3	3	3	3	3
2	3	0	3	3	3	3
3	3	3	0	3	3	3
4	3	3	3	0	3	3
5	3	3	3	3	0	3
6	3	3	3	3	3	0

Table 13.2 Traffic matrix no. 2, showing balanced traffic flows.

An optimal (flow-based) partitioning will be to put nodes 1, 2, and 3 on one subring, and nodes 4, 5, and 6 on the second subring. Then, the traffic flow through each of the two subrings will be 54. And the amount of traffic forwarded through the bridge is 18. If we put all the nodes on one big ring, the total traffic flow through the ring would be 90. So, note that partitioning has reduced the maximum ring flow. In fact, if the capacity of each WDM channel (as well as that of the large FDDI network) was anywhere between 54 and 90, the single ring would not be able to support the loading, but the two-partition mode above will. In other words, if the ring network was FDDI, then each unit of traffic flow in this example could scale up to a maximum value of 100 Mbps/90 = 1.11 Mbps for the single-ring case, while this value would be 100 Mbps/54 = 1.85 Mbps for the two-partition case (for a total capacity of 166.5 Mbps). Note that the total capacity is not 200 Mbps even though that much is allocated in the fiber (two rings, each at 100 Mbps). This is because the cross-traffic packets "chew up" bandwidth on both rings.

Now, consider the balanced traffic matrix in Table 13.2. The same partitioning as above, viz., nodes 1, 2, and 3 on one subring, and nodes 4, 5, and 6 on the other subring will yield flows of 72 on each subring, and the amount of traffic forwarded through the bridge is 54. A full ring will have the same earlier flow of 90.

As expected, skewed traffic leads to more performance improvement with partitioning – (1) it causes lower flow on the subrings and (2) it reduces the amount of processing at the bridge. These could serve as our *optimization criteria* for flow-based algorithms.

13.4 Optimization Criteria

Our criteria for partitioning the nodes into subrings fall under two categories: (1) *traffic-flow*-based and (2) *delay*-based. Traffic-flow based criteria depend only the traffic flows (i.e., the traffic matrix) and do not take into account other parameters of the network, such as fiber lengths, bridge processing times, etc. The latter parameters are accommodated by the delay-based criteria. We examine two traffic-flow-based criteria, and analyze a delay-based model for the network. These criteria are examined below.

13.4.1 MIN-CROSS

In a WDDI network, the bridge can become a traffic bottleneck. Since the subrings connected to the bridge operate at high speeds, the bridge needs to be able to switch the packets from one ring to another very fast. There is an opto-electronic conversion at the bridge, followed by electronic routing and a conversion back from electronic to optical domain. This imposes a delay penalty for packets that are routed via the bridge. So, the objective of the MIN-CROSS algorithm is to minimize the cross-traffic that flows through the bridge.

Let S_i be the set of nodes in subring i. MIN-CROSS seeks to minimize the cross-traffic given by

$$\sum_{i,j} \sum_{x \in S_i, y \in S_j, i \neq j} a_{xy}$$

where we reiterate that the traffic matrix is $\{a_{xy}\}, 1 \leq x, y \leq N$.

13.4.2 MIN-DIFF

When we partition nodes into subrings, it might happen that some subring has a much higher traffic flowing on it than other subrings. As a result, packets on the heavily-loaded subring tend to have higher delays than packets on the other subrings. To avoid this situation, we can partition the nodes such that the traffic flows in subrings are balanced, i.e., as close to one another as possible. Flow on subring i, T_i, is given by

$$T_i = \text{flow within subring } i + \text{flow out of subring } i + \text{flow into subring } i$$

i.e.,

$$T_i = \sum_{x,y \in S_i} a_{xy} + \sum_{x \in S_i, y \notin S_i} a_{xy} + \sum_{y \in S_i, x \notin S_i} a_{xy}$$

MIN-DIFF seeks to minimize the following quantity:

$$\max_i \{T_i\} \; - \; \min_i \{T_i\}$$

13.4.3 MIN-DELAY

The MIN-DELAY criterion partitions the nodes such that the overall network-wide average packet delay is minimized. By employing a model for a set of interconnected ring networks [HoMu90], we analyze the throughput and delay characteristics for the WDDI network.

13.5 Flow-Based Algorithms

This section studies the complexities of, and algorithms for, the MIN-CROSS and the MIN-DIFF criteria for partitioning nodes of the WDDI network.

13.5.1 MIN-CROSS Algorithms

This subsection examines the complexity and algorithms for partitioning a WDDI network into W subrings, with the optimization criterion being minimization of the traffic flow across the bridge. To reiterate, MIN-CROSS seeks to minimize the cross-traffic given by:

$$\sum_{i,j} \sum_{x \in S_i, y \in S_j, i \neq j} a_{xy}$$

where S_i is the set of nodes in subring i, and $A = \{a_{xy}\}$, $1 \leq x, y \leq N$, is the traffic matrix.

This problem can be reduced to the *min-k-cut problem* in graph theory, and it can be solved *exactly.* We model the problem as follows. An undirected weighted graph $G = (V, E)$ represents the network, where V is the set of vertices (nodes), E is the set of edges (links), and the number of vertices $|V| = N$. Assume that there is an edge between each pair of vertices, i.e., the network is "fully connected." (This network is not the same as our WDDI network.) We assign weights to edges of this graph as follows. The weight on the edge between node u and node v is the sum of traffic from node u to node v and traffic from node v to node u. It is noteworthy that, if nodes u and v are on different subrings, then the traffic across the bridge due to nodes u and v is the sum of $u - to - v$ traffic and $v - to - u$ traffic, while if the nodes u and v are on the same subring, then they do not contribute to the bridge traffic at all. A $W - cut$ of G is a set of edges R in G, such that removing those edges from G splits G into W components. The weight of a $^{W-cut}$ is the sum of the weights on the edges comprising the set R. We deduce that the bridge traffic is minimized when the partition forms a *minimum weight* $W - cut$ of the graph G.

Lemma: The traffic forwarded by the bridge, for a W-partition of the nodes, is minimized, if the partition is the min-W-cut of the graph G.

Proof: Any partition into W groups of nodes in G determines a $W - cut$ of G. The traffic across the bridge is the weight of the $W - cut$ and therefore the *Lemma* follows. The problem of finding a *minimum-weight W-cut*, given arbitrary W, is NP-complete [GoHo88].

From the above observation, therefore, we examine algorithms for finding the minimum weight $W - cut$ of a (traffic) weighted graph G. To partition the nodes into two groups using the MIN-CROSS criterion, we can use standard *maximum-flow-finding algorithms* to locate the *minimum 2-cut* [EdKa72]. When the number of partitions W is greater than two, we can still perform the partitioning in polynomial time using a $O(N^{W^2})$ algorithm [GoHo88]. However, this algorithm is exponential in W, and is described below [GoHo88]:

1. Let $m = W - 2$ if W is even, else $m = W - 1$.
2. Repeat for every m-node source subset S[9] and $(W-1)$-node sink subset T out of N nodes.
   (a) Make source S as one node and the sink T as one node.
   (b) Find minimum cut with S as source and T as sink.
   (c) Recursively partition the sink set into $W - 1$ sets.
   (d) Record the weight of the resulting $W - cut$, if minimum.
3. Repeat for every L-node subset of N $(L < W - 2)$.
   (a) Make the L nodes as the first partition.
   (b) Recursively partition the remaining $N - L$ nodes into $W - 1$ sets.
   (c) Record weight of resulting $W - cut$ if minimum.

The partitioning produced by the exact algorithms described above has an undesirable characteristic that the number of nodes in any partition is unrestricted. This may lead to partitions where some subring may have too many nodes, while others may have too few. This is illustrated in the following example. Consider a nine-node network, with the traffic matrix in Table 13.3, where the traffic from node i to node j, a_{ij}, is a random value between 0 and 7. (Note that in this example, the bridge node – node 9 – does not generate any traffic, and therefore, is not included in the traffic matrix.)

The exact MIN-CROSS algorithm for $W = 2$ generates the following partition – subring 1 has one node {1}, subring 2 has seven nodes {2,3,4,5,6,7,8}, and the minimized cross-traffic has value 47. The fact that one of the subrings has only one node may not be desirable, even though this partitioning minimizes the cross-traffic.

So, additionally, we may like to constrain the number of nodes in any partition. This problem, of minimizing the weight of the $W - cut$, while at the same time restricting the number of nodes in any partition, has been

[9] S is a subset of $\{1, 2, \ldots, N\}$ and S has m elements.

from (i) ↓ / to (j) →	1	2	3	4	5	6	7	8
1	0	1	2	4	7	1	5	1
2	3	0	0	3	2	2	7	4
3	6	5	0	6	6	1	5	2
4	2	7	4	0	4	6	7	6
5	6	5	6	4	0	2	1	6
6	3	3	7	1	1	0	5	3
7	6	6	7	0	2	6	0	4
8	0	7	2	0	3	7	4	0

Table 13.3 Traffic matrix no. 3, with each nondiagonal entry being an uniformly distributed random number between 0 and 7.

extensively researched for the two-partition case [Sanc89, KeLi70], and has been shown to be NP-Complete. The Kernighan-Lin algorithm [KeLi70] is a classic heuristic algorithm to solve this problem. There have been several modifications and improvements to this algorithm. Here, we use a variant of the Kernighan-Lin heuristic adapted from [MeSe94].

Kernighan-Lin Algorithm for $W = 2$

We are given a set of costs for each edge of the network. The problem is to partition the graph of N nodes, not including the bridge, into two sets, S_1 and S_2, so as to minimize the total cost of the partition. The additional restriction is that the size of subring i is upper bounded by n_i for $i = 1,2$, such that $N \leq n_1 + n_2$.

The Kernighan-Lin heuristic uses *swap* as a basic operation. The *swap* operation exchanges nodes in different subrings, thereby giving rise to a new partition. The central idea behind the algorithm is to replace the search for one favorable swap by search for a favorable *sequence* of swaps, using the *cut-cost* to guide the search. Thus, a favorable sequence of k swaps is not found by examining a neighborhood containing all the sequences (of k swaps), but it is obtained sequentially as follows:

1. Add $n_1 + n_2 - N$ dummy vertices to the graph, with no edges incident on them.

2. Divide the vertices arbitrarily into sets S_1 and S_2 of sizes n_1 and n_2, respectively.

3. Repeatedly perform improvement *passes* over this initial partitioning as described below until a pass makes no improvement to the total weight $W(S_1, S_2)$ of the partition. In each pass, do the following:

   (a) Find vertices $v_1 \in S_1$ and $v_2 \in S_2$ such that swapping v_1 and v_2 reduces the cost of the partition, $W(S_1, S_2)$, by the largest possible amount.

   (b) Swap v_1 and v_2, and mark them as temporarily locked, preventing them from being considered for the rest of the pass. The new partition (S_1, S_2) is noted.

   (c) Repeat Steps (a) and (b) until all vertices in S_1 or S_2 are marked locked. In the sequence of (S_1, S_2) pairs which were generated, find the pair with the minimum cost $W(S_1, S_2)$, and reset S_1 and S_2 to that point. This ends the pass, and all vertices are marked unlocked.

4. Remove all dummy vertices from S_1 and S_2.

Kernighan and Lin [KeLi70] report that, for equal-sized partitions where $n_1 = n_2 = N/2$, locally optimal solutions for 0-1 matrices are globally optimal about 10% of the time, and they are within 1% or 2% of the globally optimal solution about 70% of the time. The running time of the algorithm has been reported empirically to be $O(N^{2.4})$.

Extensions for $W > 2$

The algorithm for multiway partitioning, i.e., for a given value of W where $W > 2$, is based on the two-partitioning algorithm and has been adapted from [MeSe94]. It uses *move* as a basic operation. The *move* operation moves a node from one subring to another. The algorithm examines a sequence of move operations for searching the neighborhood of its current configuration, and is described below:

1. Find an initial W-partition which can be arbitrary.

2. Repeat the following improvement passes until a pass does not improve the cross-traffic:

   (a) Mark all nodes as locked.

(b) Find the best feasible move, i.e., find a node i in partition p_0, and a partition p_1 such that moving node i from p_0 to p_1 reduces the cross-traffic by the greatest amount.

(c) Move node i to partition p_1. Note the current partition.

(d) Repeat steps (b) and (c) until no move is found. In the sequence of partitions in this pass, select the one with the lowest cross-traffic and reset the partition to it.

13.5.2 MIN-DIFF Algorithm

This section examines the complexity of the MIN-DIFF partitioning problem, and studies MIN-DIFF-based partitioning algorithms. First, we reduce the MIN-DIFF partitioning problem for two partitions to a well-known partitioning problem and consequently show that it is NP-Complete. We then use the reduction to provide a pseudo-polynomial exact algorithm[10] for the MIN-DIFF partitioning problem when $W = 2$. Finally, we outline a heuristic algorithm for multiway partitioning ($W > 2$), that employs the two-way partitioning algorithm as a subroutine.

MIN-DIFF for $W = 2$

We model the problem as follows. Let $A = \{a_{ij}\}$ be the $N \times N$ traffic matrix. (Recall that the $(N + 1)$-th node is the bridge, and nodes $1, 2, \ldots, N$ need to be placed in one of the two partitions.) Let $x_1, x_2, \ldots, x_N$ be binary variables corresponding to these N nodes. Let the set of nodes in each partition be denoted by S_k, k = 0,1. So, $x_i = k$ implies that $i \in S_k$, k = 0,1.

Then, traffic between all nodes in partition 1 = $(x_1, x_2, \ldots, x_N)A(x_1, x_2, \ldots, x_N)^T$, where 'T' denotes the transpose operator.

Similarly, traffic in partition 0 = $(1 - x_1, 1 - x_2, \ldots, 1 - x_N)A(1 - x_1, 1 - x_2, \ldots, 1 - x_N)^T$.

Therefore, the MIN-DIFF objective is:

$$\textbf{Minimize}: \quad \Big|(x_1, x_2, \ldots, x_N)A(x_1, x_2, \ldots, x_N)^T - (1 - x_1, 1 - x_2, \ldots, 1 - x_N)A(1 - x_1, 1 - x_2, \ldots, 1 - x_N)^T\Big| \quad (13.1)$$

[10] A pseudo-polynomial algorithm has a running time that is polynomial in its input size, but the running time also depends on the value of some parameter to the problem.

Expression (13.1) simplifies to

$$\textbf{Minimize}: \quad \Big|(1,1,\ldots,1)A(1,1,\ldots,1)^T - (1,1,\ldots,1)A(x_1,x_2,\ldots,x_N)^T - (x_1,x_2,\ldots,x_N)A(1,1,\ldots,1)^T\Big| \tag{13.2}$$

Now, the total traffic in the network equals $(1,1,\ldots,1)A(1,1,\ldots,1)^T = \sum_{i,j} a_{ij}$.

Let $a_k = \sum_{i=1}^N a_{ik} + \sum_{j=1}^N a_{kj}$. Note that $2\sum_{i,j} a_{ij} = \sum_{i=1}^N a_i$.

Then, $(1,1,\ldots,1)A(x_1,x_2,\ldots,x_N)^T + (x_1,x_2,\ldots,x_N)A(1,1,\ldots,1) = \sum_{i=1}^N a_i x_i$.

Therefore, Expression (13.2) reduces to

$$\textbf{Minimize}: \quad \left|\sum_{i,j} a_{ij} - \sum_{i=1}^{N} a_i x_i\right|$$

Thus, the MIN-DIFF partitioning problem is now reduced to finding a subset S of nodes, such that $\sum_{i\in S} x_i a_i$ is as close to $\sum_{i,j} a_{ij}$ as possible. We can show that the MIN-DIFF partitioning problem is NP-Complete by reducing it to the PARTITION problem in [GaJo79]. The PARTITION problem is defined as follows [GaJo79]:

Given a finite set A and a size $s(a)$ for each $a \in A$, is there a subset $A' \subseteq A$ such that $\sum_{a\in A'} s(a) = \sum_{a\in A-A'} s(a)$?

Given an instance of the MIN-DIFF partitioning problem, we can obtain an instance of the PARTITION problem by assuming a set A of N elements and assigning weight a_i to element i. The MIN-DIFF objective function will be zero if and only if there exists a partition of the elements of the set A.

Given an instance of the PARTITION problem, a set A of N elements and weight a_i assigned to element i, we can generate an instance of the MIN-DIFF partitioning problem. The traffic matrix is defined as follows:

$a_{ii} = 0$ for $i = 1, 2, \ldots, N$.
$a_{1i} = a_1/[2(N-1)]$ for $i = 2, 3, \ldots, N$.
$a_{i1} = a_1/[2(N-1)]$ for $i = 2, 3, \ldots, N$.
$a_{ik} = (a_k - \sum_{j=1}^{k} a_{jk})/2(N-k-1)$ for $i = k+1, k+2, \ldots, N$.
$a_{ki} = (a_k - \sum_{j=1}^{k} a_{kj})/2(N-k-1)$ for $i = k+1, k+2, \ldots, N$.

Given this traffic matrix, the MIN-DIFF partition function is: **Minimize** $\left| \sum_i a_i/2 - \sum_{i=1}^{N} a_i x_i \right|$. The MIN-DIFF partition function is zero if and only if there exists a partition of the elements of set A.

Though the PARTITION problem is NP-Complete, there exists a pseudo-polynomial time algorithm to solve the problem. We shall use the pseudo-polynomial algorithm to solve the MIN-DIFF problem. The MIN-DIFF instance is mapped to an instance of the PARTITION problem, and a well-known psuedo-polynomial algorithm described in [GaJo79] is used to solve this problem.

Given the traffic matrix we assign a weight a_i to element i. Let the total traffic be B. The algorithm maintains a table $t(i,j)$, $1 \leq i \leq N$, $0 \leq j \leq 2B$, such that $t(i,j)$ is 1 if and only if there is a subset of $\{a_1, a_2, \ldots, a_N\}$ for which the sum of the element sizes is exactly j. The t array can be filled with a simple procedure, e.g., row by row. For $i = 1$, $t(1,j) = 1$ if and only if either $j = 0$ or $j = s(a)$. Each subsequent row is filled using the entries in the previous row. For $1 < i \leq N$, $0 \leq j \leq 2B$, the entry $t(i,j) = 1$ if and only if either $t(i-1,j) = 1$ or $s(a_i) \leq j$ and $t(i-1, j - s(a_i)) = 1$. We solve the instance of MIN-DIFF by finding k such that $t(N,k) = 1$ and there is no j, $B - k < j < B + k$, such that $t(N,j) = 1$.

The running time of the algorithm is $O(NB)$ and depends on the total traffic flow in the network. We can scale the traffic flows to values that lie in a small range and make the algorithm run faster. A fast and simple $O(N^2)$ heuristic algorithm for the two-partitioning problem is described below.

1. Find the pair of nodes (i,j) such that $a_{ij} + a_{ji}$ is minimum and place them in different subrings.

2. Repeat the following until all nodes are put in either subring.

   (a) Choose node i (which has not yet been placed in any subring) and subring k ($k = 1$ or 2), such that the difference of the traffic in the subrings is minimized by the greatest amount.

   (b) Place node i in subring k.

MIN-DIFF for $W > 2$

We use the two-partition algorithm as a subroutine in the multiway MIN-DIFF partitioning algorithm. A description of the multiway MIN-DIFF partitioning algorithm, for a given W, follows.

1. Find an initial feasible W-partition. The initial partition can be arbitrary.

2. Repeat while there is an improvement found in a pass.

   (a) Choose any two partitions.

   (b) Merge the nodes in the two partitions.

   (c) Run the MIN-DIFF two-way partitioning algorithm on the set of nodes.

   (d) If the MIN-DIFF value of the new partitioning is less than that of the old partitioning, set improvement to true.

   (e) If there is no improvement for any choice of two partitions [among $N(N-1)$ possible choices], exit.

13.6 Delay-Based Algorithms

In the algorithms presented in the previous section, the *traffic flows* in the subrings led to the optimization criteria. This section studies partitioning algorithms that try to minimize the network-wide *average delay*. Furthermore, the algorithm will be general enough to create an arbitrary number of partitions. This enables us to study the effect of the number of rings on the delay-throughput characteristics of the network, for arbitrary traffic matrices.

13.6.1 Performance Analysis

We adopt the analysis for a system of interconnected ring networks from [HoMu90]. The network has N nodes, each of which has an infinite input packet buffer. The nodes are numbered $1, 2, \ldots, N$. The packet arrival at any node s is assumed to be Poisson with rate λ_s. The packet transmission times are arbitrarily distributed with mean b and second moment $b^{(2)}$. When the ring is operating in the W-partition mode, the subrings are assumed to be numbered $1, 2, \ldots, W$ in order, and subring i is assumed to have Q_i nodes. The time required for a packet to reach the bridge from node j is r_j. Each packet encounters an extra processing delay of P at the bridge. The probability that a packet originating at source node s with destination d is l_{sd}, where s, $d = 1, 2, \ldots, N$. S_j is the set of all nodes in subring j. The probability $q_s^{(j)}$ that a packet originating at node s will be destined to a node on subring j

is $q_s^{(j)} = \sum_{d \in S_j} l_{sd}$. The intensity of the traffic going into subring j from all other subrings (through the bridge node) is $\gamma^{(j)} = \sum_{s \notin S_j} \lambda_s q_s^{(j)}$.

Throughput Analysis

The maximum throughput that can be supported by the multiple-partition ring network is obtained from stability considerations of each ring in the network. The network is stable if every subring on the network, as well as the bridge node, is stable. A subring is stable if every nonbridge node on the subring is stable. A node on a subring is stable if it receives at most one new packet arrival between two visits of the free token, on the average. From the above reasoning [HoMu90], we find that subring j is stable if the following conditions hold:

$$\lambda_s \sum_{d \in S_j} 2r_d + \sum_{d \in S_j} \min(\lambda_s, \lambda_d)b \; + \; \min(\lambda_s, \gamma^{(j)})b < 1 \tag{13.3}$$

where $s \in S_j$, and

$$\gamma^{(j)} \sum_{d \in S_j} 2r_d + \sum_{d \in S_j} \min(\gamma^{(j)}, \lambda_d)b \; + \; \gamma^{(j)}b < 1 \tag{13.4}$$

The above two equations correspond to the stability of node s and the stability of the bridge node on subring j, respectively.

Delay Analysis

Let us denote the expected waiting time for a packet, measured from the instant of its arrival at a node until its transmission starts, by $E[W_s]$ for all $s \in S_j$. Also, let $E[W_{0j}]$ be the expected waiting time for a packet buffered at the bridge and destined for a node on subring j, measured from the instant the packet arrives at the bridge until its transmission is initiated by the bridge on subring j. The expressions for $E[W_s]$ and $E[W_{0j}]$ may be found in [HoMu90]. The mean packet delay for a packet originating from a node on subring i to any node on the same subring is the averaged sum of the queueing delay at a node, transmission time at a node, and the propagation delay on subring i, and is given by:

$$\overline{D_1^{(i)}} = \frac{1}{Q_i} \sum_{s \in S_i} \left(E[W_s] + b + \sum_{d \in S_i} r_d \right) \tag{13.5}$$

If the packet is destined for a node on subring j, $j \neq i$, it will encounter a delay due to bridging, whose value is the averaged sum of the expected waiting time at the bridge, transmission time at bridge, and the propagation time on subring j, and is given by:

$$\overline{D_2^{(i)}} = E[W_{0j}] + b + \sum_{d \in S_j} r_d \quad (13.6)$$

Thus, the average delay for packets originating from subring i is given by:

$$\overline{D^{(i)}} = \overline{D_1^{(i)}} + \frac{1}{Q_i} \sum_{s \in S_i} \sum_{j \neq i} (P + \overline{D_2^{(i)}}) q_s^{(j)} \quad (13.7)$$

The overall packet delay for the network is then given by:

$$\overline{D} = \frac{1}{N} \sum_{i=1}^{W} Q_i \overline{D^{(i)}} \quad (13.8)$$

13.6.2 Partitioning Algorithm

The simulated annealing algorithm [KiGV83] was used to perform the partitioning of nodes into the rings with minimum overall network packet delay as the optimization criterion. The inputs to the algorithm were the number of partitions needed and the traffic matrix. The objective function was the average delay in the network, as calculated by Eqn. (13.8). The simulated annealing algorithm starts off from a random initial configuration, where a configuration refers to a partitioning of nodes. At each step of the algorithm, a random neighboring configuration is considered for further search (optimization), where a neighboring configuration is the one obtained from the current configuration by swapping two nodes belonging to different partitions. If the neighbor reduces the objective function (optimality criterion), then the neighbor is chosen. Otherwise, if the neighbor does not reduce the objective function, then also it may be chosen with a certain probability. The latter property provides "backtracking" which may enable the solution to escape from "local minima" in the optimization search space. The simulated annealing algorithm "freezes" when no improvement takes place over several consecutive iterations, i.e., if from a particular configuration $\mathcal{X}$, a number n of consecutive iterations do not lead to a better configuration, then $\mathcal{X}$ is the final solution. A description of the simulated annealing algorithm follows:

1. Set current configuration to be a random initial configuration.

2. Repeat the following steps while configuration is not *frozen*:
   (a) Set new configuration to be a neighbor of current configuration.
   (b) Set δ to the difference between the objective function value corresponding to the new configuration and the objective function value of the current configuration.
   (c) If $\delta > 0$, set current configuration to be the new configuration; otherwise set current configuration to be the new configuration with probability $\exp^{-\delta/\beta}$, where β is a control parameter.

The control parameter β is chosen to be 0.9 in our numerical examples. This number has been found in the literature to perform satisfactorily.

13.7 Illustrative Examples

This section compares the performance of different partitioning algorithms under various network conditions.

13.7.1 Network Description

We make the following assumptions about the network's characteristics:

- *Traffic characteristics:* We examine typical traffic patterns that are observed on operational networks. The traffic types considered are the following.
  - *Clustered traffic:* In clustered traffic, the nodes are divided into c clusters. The packet arrival probability between nodes on the same cluster is k times the probability of packet arrival between nodes in different clusters. For the numerical examples presented below, we choose $c = 2$ and $k = 5$.
  - *Server traffic:* In server-type traffic, we designate certain nodes in the network to be servers. The packet arrival probability for packets generated by a nonserver node and destined to a server is s times the packet arrival probability between nonserver nodes. For our numerical examples, we employ $s = 5$.
  - *Pseudo-random traffic:* In pseudo-random traffic, traffic between any pair of nodes is a uniformly distributed random variable.

- *The bridge does not generate any traffic.*

- *Network parameters:* The parameters for the network are normalized with respect to the average packet transmission time, i.e., $b = 1$ time unit. The second moment of packet transmission time $b^{(2)}$ is assumed to be 1 (i.e., all packets are of fixed length of duration 1 time unit), and time r_j, taken to pass a packet between a node j and the bridge is assumed to be a constant = 0.01 time units.

- *Bridge processing:* In order to study the effect of bridge processing time on network performance, we consider this parameter, denoted by P, to take on a range of values between 0.1 (low) and 10.1 (high) time units.

- *Algorithms:* We use the Kernighan-Lin (KL) algorithm (for $W = 2$) with different parameters for maximum partition sizes. By KL1, KL2, and KL3, we denote the KL algorithm with maximum partition sizes $N/2$, $3N/4$, and $N - 1$, respectively. Heuristic algorithms for multiway partitioning were used for MIN-DIFF and MIN-CROSS. Simulated annealing was used for MIN-DELAY.

- For a given traffic matrix, a given flow-based algorithm from Section 13.5 provides a partitioning of the nodes. Then, the delay analysis of Section 13.6.1 is employed to obtain the corresponding average network-wide packet delay.

- Statistics were collected over 10 runs, each with a different traffic matrix of the same traffic type. Results are reported by averaging over the ten runs.

13.7.2 Delay vs. N Characteristics (Two Partitions)

First, we study the effect of various network parameters by considering the two-partition network. For each traffic type, we examine the impact of arrival rate and bridge processing times on the average packet delay for different partitioning algorithms. In order to examine the interplay of different network parameters, we study the network under light load (0.1 packets per unit time aggregated over all nodes) and heavy load (0.8 packets per unit time over all nodes).

Figure 13.3 shows average packet delay plotted against number of nodes, for partitions produced by different partitioning algorithms for server-type

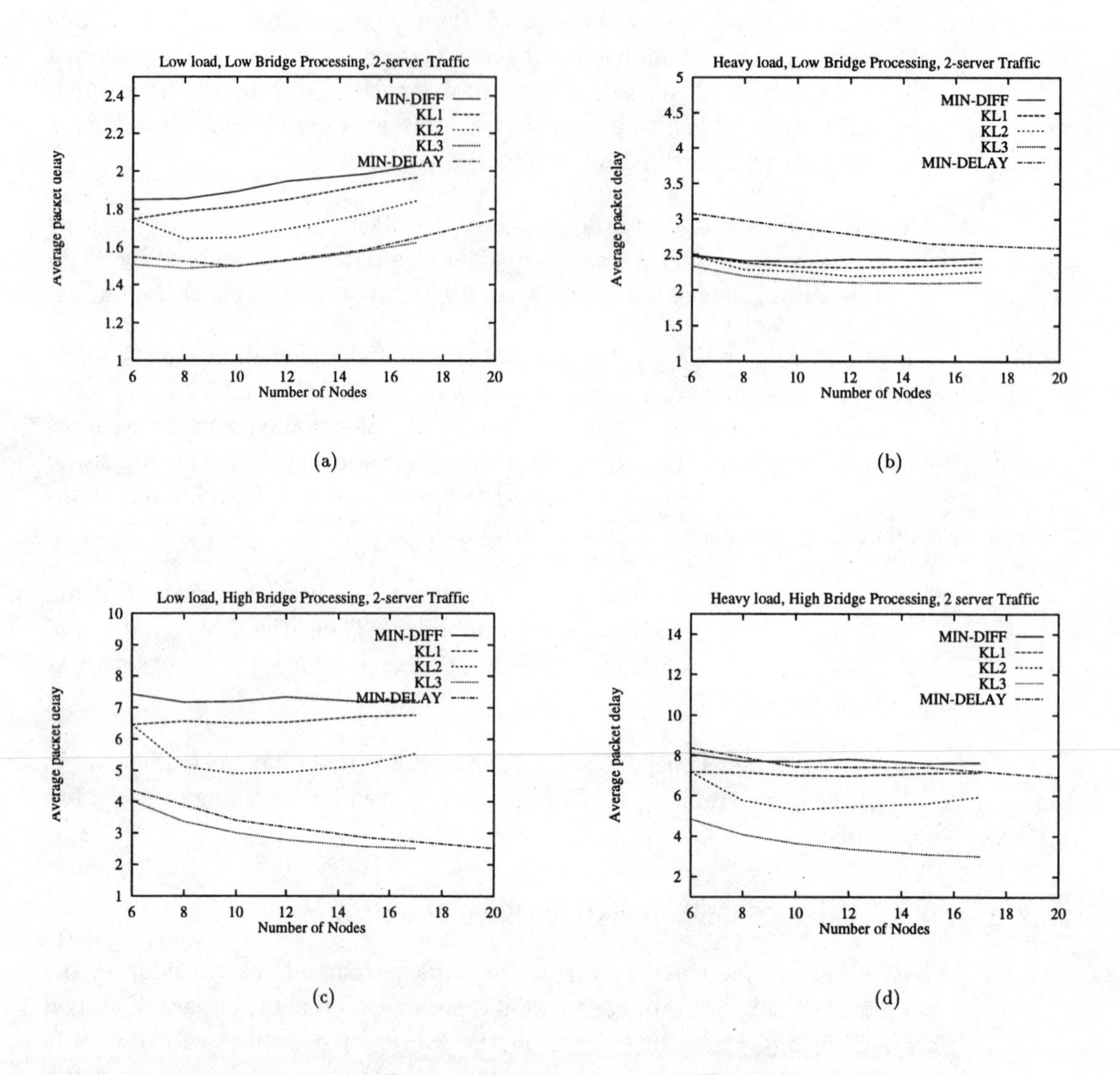

Figure 13.3 Delay versus N characteristics for two-server traffic ($W = 2$).

traffic. Note that algorithm KL3 has the best delay characteristics and MIN-DIFF usually has the worst (independent of arrival rates and bridge processing times). Note, however, that at heavy load and low bridge processing time (Fig. 13.3(b)), the difference in delays between different algorithms is minimal. This may be due to the fact that, under these conditions, the queueing delay at nodes dominates the delay, and that the partitioning of nodes into clusters has little effect.

For clustered traffic shown in Fig. 13.4, algorithm KL1 generally does the best. It is interesting to observe that the difference in delay characteristics of different algorithms is small. This may be because all of the algorithms tend to properly identify clusters in the traffic matrix. Delay vs. N characteristics for pseudo-random traffic shown in Fig. 13.5 are similar to those for server-type traffic.

For all traffic types, Figs. 13.3(a), 13.4(a), and 13.5(a) demonstrate that, at low load and for low bridge processing time, the delay characteristic decreases initially with increasing N and then shows a slow rising trend. This may be because when the number of nodes is small, the dominant delay component is the queueing delay component at a node, whereas, as we increase the number of nodes, the dominant delay component is the time spent waiting for a token. When the arrival rate is high or the bridge processing is large, the delay decreases initially and then is almost insensitive to the number of nodes.

Delay vs. Throughput Characteristics (Two Partitions)

In Fig. 13.6, we plot the delay-throughput characteristics for different partitioning algorithms for a 17-node network for low ($P = 0.1$) and high ($P = 10.1$) bridge processing times.

Note that, for two-server traffic and low bridge processing (Fig. 13.6(a)), algorithms KL3 and MIN-DELAY perform comparably, and have the best delay under low loads. However, algorithms KL3 and MIN-DELAY become unstable at an offered load of approximately 0.85. MIN-DIFF partitioning, on the other hand, performs worse than the KL algorithms at low loads, but supports higher network loads of up to 1.2. Similar characteristics are observed for pseudo-Random traffic.

For clustered traffic, MIN-DELAY and MIN-DIFF perform comparably, and have the best delay characteristics under low as well as heavy loads, when bridge processing time is low. When bridge processing time is high, MIN-DELAY has the best delay characteristics for low as well as heavy loads.

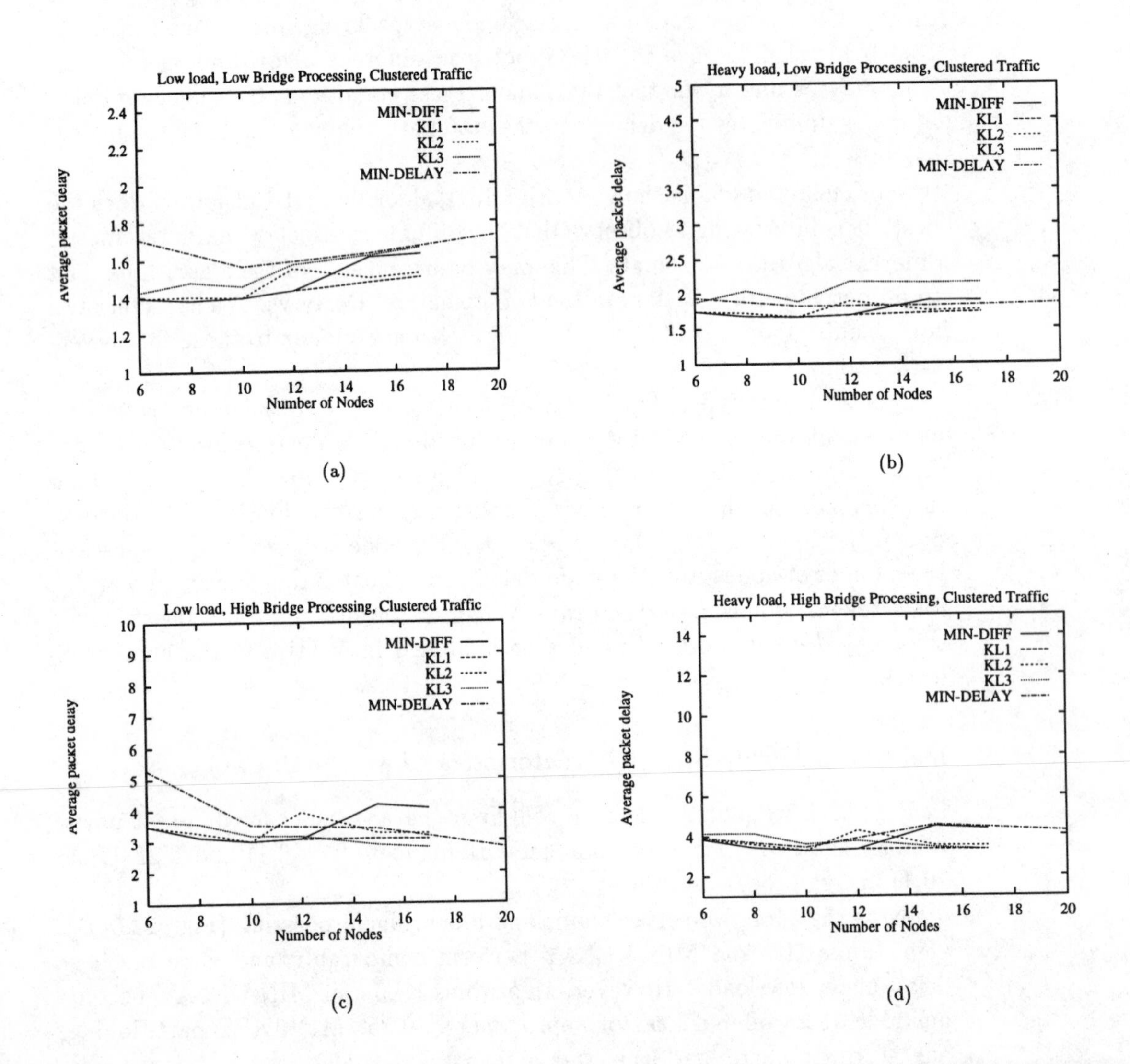

Figure 13.4 Delay versus N characteristics for clustered traffic, $c = 2$, $k = 5$ ($W = 2$).

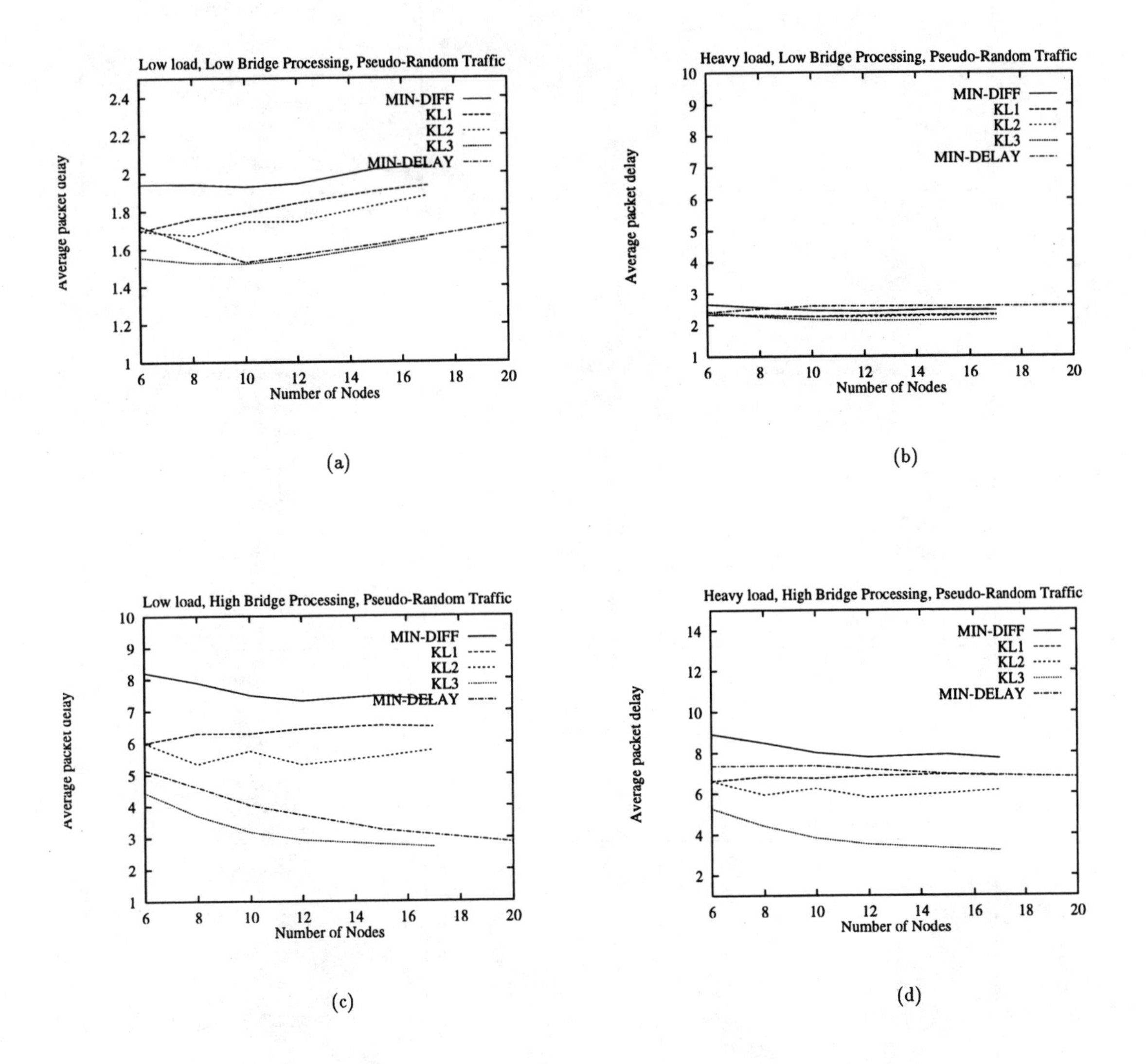

Figure 13.5 Delay versus N characteristics for pseudo-random traffic ($W = 2$).

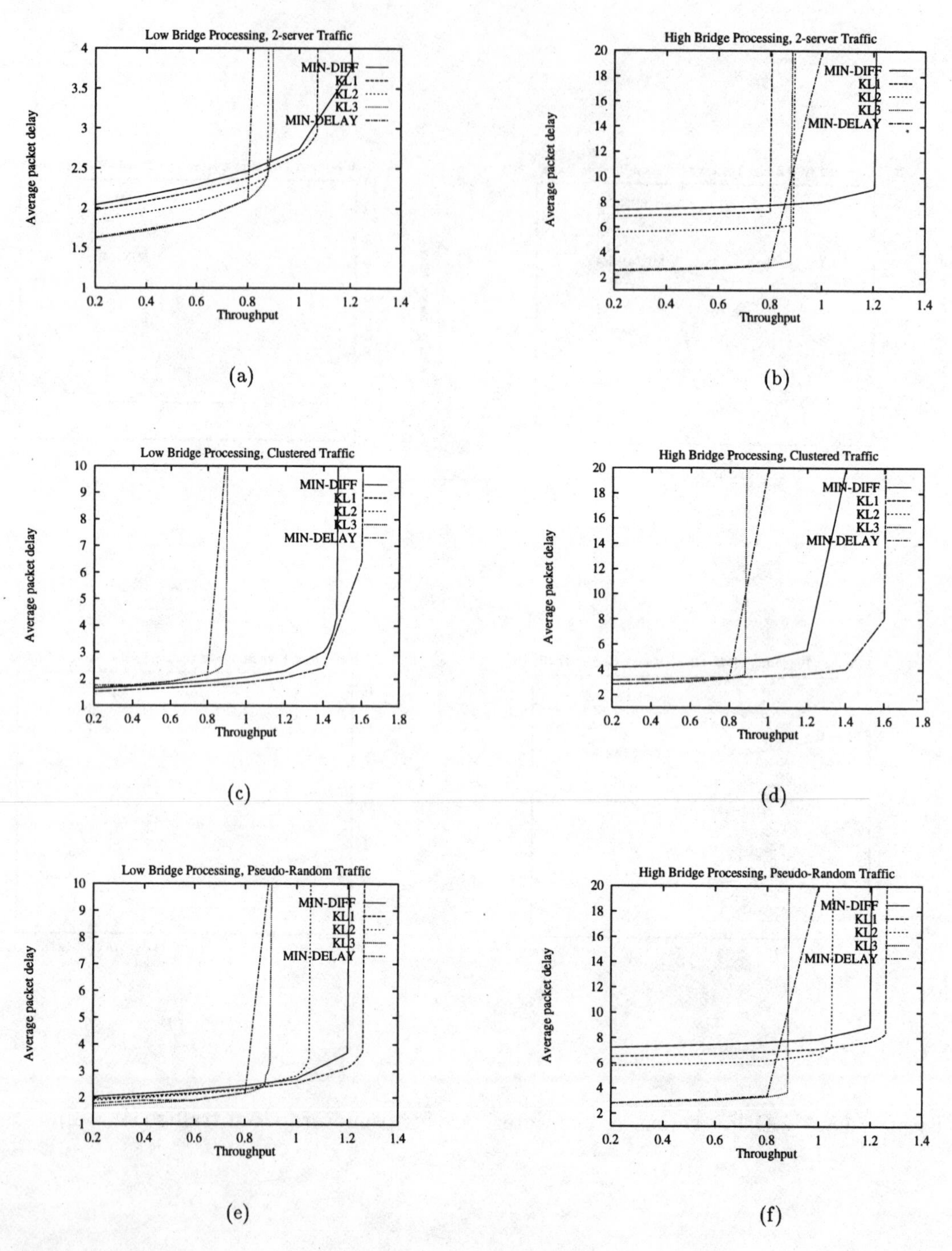

Figure 13.6 Delay versus throughput characteristics ($W = 2$).

When bridge processing time is large ($P = 10.1$), Fig. 13.6(b), (d), and (f) indicates that algorithms KL3 and MIN-DELAY perform the best under all traffic conditions and for low loads.

13.7.3 Two or Greater Partitions

In this section, we study a multiple-partition network. In particular, we examine the performance of MIN-DIFF-based and MIN-CROSS-based partitioning algorithms (producing two or more partitions) for a 17-node network under different traffic conditions and low and high values of bridge processing time.

MIN-DIFF Algorithm

Figure 13.7(a) shows delay characteristics for the heuristic multiway MIN-DIFF partitioning algorithm, for partition sizes ranging between two and eight, server-type traffic, and low bridge processing time. Note that the maximum throughput supported by the network increases with the number of partitions. A similar characteristic is observed when bridge processing time is high as shown in Fig. 13.7(b). Similar delay vs. throughput characteristics are also observed for clustered and pseudo-random traffic.

Figure 13.8 plots delay vs. the number of partitions for different values of arrival rates and bridge processing times. Observe that, when the bridge processing time is high, the delay initially increases with the number of partitions and then tends to remain almost constant beyond a certain number of partitions. Under high bridge processing times, a small number of partitions seems to be advantageous at low to moderate loads.

When bridge processing time is low, network delay tends to behave differently for different loads. For a lightly loaded network, the delay increases initially with the number of partitions and then becomes almost insensitive to the number of partitions. For a moderate to heavily loaded network, the delay decreases initially with the number of partitions and then remains almost constant.

MIN-CROSS Algorithm

Figure 13.9(a) shows delay characteristics for the heuristic multiway MIN-CROSS partitioning algorithm, for partition sizes ranging between two and eight, server-type traffic, and low bridge processing time. Note that the maximum throughput supported by the network *does not* increase significantly

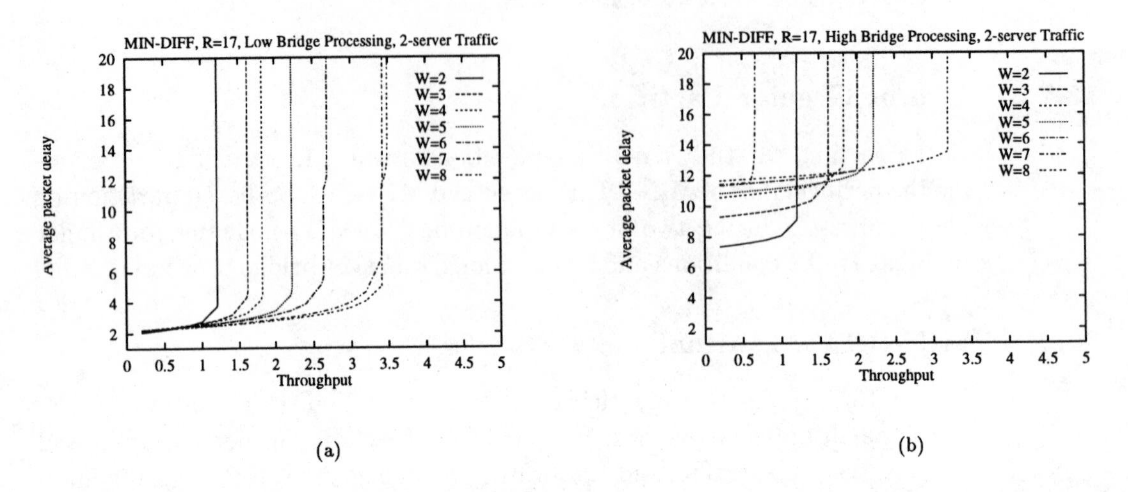

Figure 13.7 Delay vs. throughput characteristics for multiple partitions for MIN-DIFF-based algorithms.

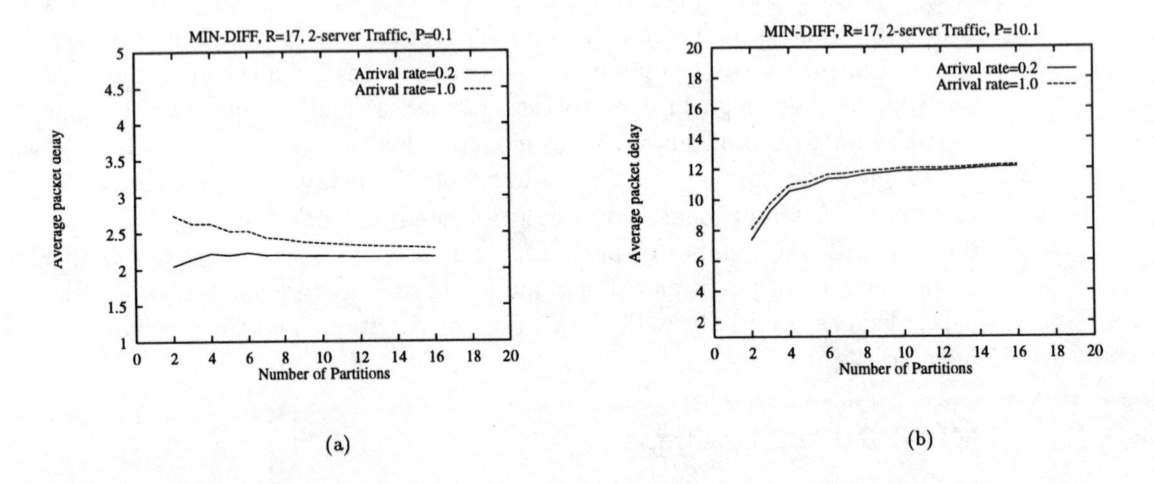

Figure 13.8 Delay vs. number of partitions for MIN-DIFF-based algorithms.

with the number of partitions. A similar characteristic is observed when bridge processing time is high as shown in Fig. 13.9(b). Similar delay vs. throughput characteristics are also observed for clustered and pseudo-random traffic.

Figure 13.10 plots delay vs. the number of partitions for different values of arrival rates and bridge processing times. We observe that, when the bridge processing time is high, the delay increases almost linearly with the number of partitions. Under high bridge processing times, a small number of partitions seems to be advantageous at low to moderate loads. When bridge processing time is low, the delay tends to increase very gently with the number of partitions.

13.8 Summary

This chapter examined how a fiber optic ring network, such as a SONET ring or FDDI, can be operated over multiple wavelengths on its existing fiber plant consisting of point-to-point fiber links. Under our new approach, called wavelength distributed data interface (WDDI), the network nodes, by using WDM, can be partitioned to operate over multiple subnetworks, with each subnetwork operating independently on a different wavelength, and intersubnetwork traffic forwarding performed by a bridge. We examined the architecture of WDDI nodes and bridges. We investigated how, based on prevailing traffic conditions, partitioning of nodes into subnetworks can be performed in an optimized fashion. In particular, we developed and examined the performance of two flow-based algorithms and a delay-based method. The MIN-CROSS algorithms partition the nodes into subrings such that the total traffic flow in the subrings is minimized. The MIN-DIFF algorithms tend to balance the traffic flows in all subrings. Simulated annealing was used to minimize the network-wide average packet delay.

The MIN-CROSS-based algorithms were found to perform well under low network loads, whereas MIN-DIFF-based algorithms supported a higher maximum network throughput. MIN-DIFF-based algorithms showed better delay characteristics for clustered traffic conditions, while MIN-CROSS-based algorithms performed better, delay-wise, for server and random traffic. The maximum throughput supported by the network increased with multiple partitions (as might be expected), while the average packet delay did not necessarily decrease by increasing the number of partitions. The bridge processing

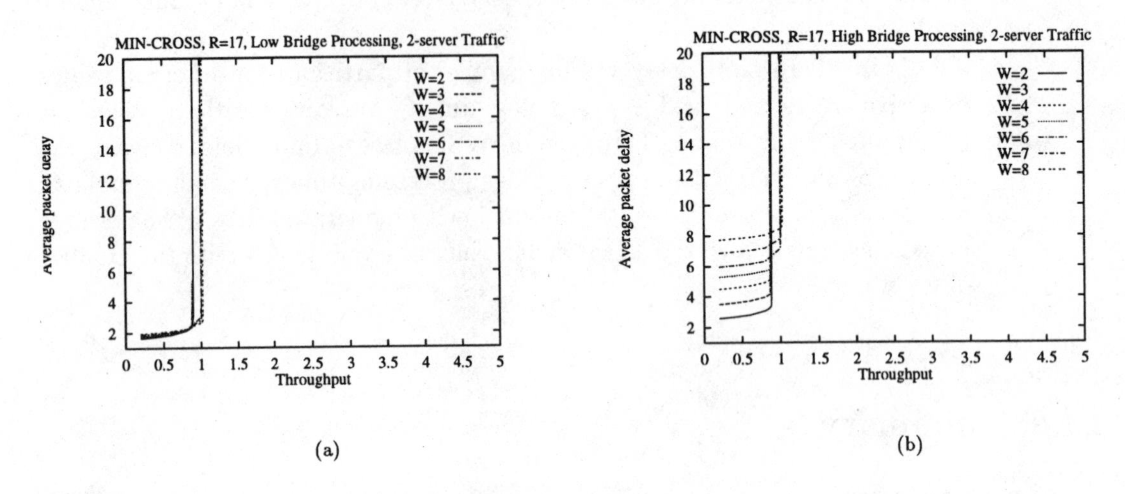

Figure 13.9 Delay vs. throughput characteristics for multiple partitions for MIN-CROSS-based algorithms.

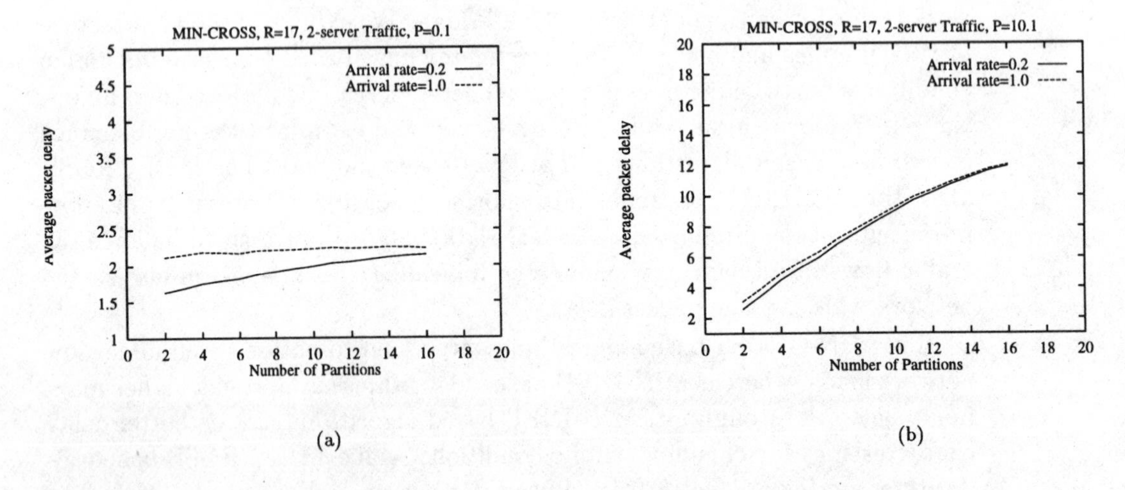

Figure 13.10 Delay vs. number of partitions for MIN-CROSS-based algorithms.

time is a significant factor in determining the optimal number of partitions that might be employed under different traffic conditions.

A number of open problems still need to be addressed. While this work assumes only one bridge, it would be interesting to model networks with multiple subnetworks and multiple bridges. In addition, we have assumed that the traffic matrix is static. But traffic intensities can change (perhaps over longer periods of time). Approaches that can detect such changes and perform dynamic reconfiguration will need to be examined.

Exercises

13.1. For the following traffic matrix and $W = 2$, find the partitioning which minimizes the cross traffic across a bridge. The bridge node does not generate or receive any traffic. Traffic matrix:

$$A = \begin{bmatrix} 0 & 1 & 3 & 2 & 1 \\ 4 & 0 & 2 & 5 & 3 \\ 1 & 6 & 0 & 1 & 2 \\ 3 & 2 & 4 & 0 & 3 \\ 1 & 2 & 1 & 4 & 0 \end{bmatrix}$$

13.2. For the same traffic matrix as above and $W = 2$, find the partitioning which results in the minimum difference in traffic between the two rings.

13.3. Consider the following argument. The goal of MIN-CROSS, with $W = 2$, is to reduce the traffic that crosses the bridge node. However, if one of the two partitioned rings is empty, except for the bridge node, then no traffic at all will cross the bridge, and the problem is solved optimally (and trivially). If you agree, give reasons, and if you think the argument is erroneous, explain why.

13.4. MIN-CROSS and MIN-DIFF optimization criteria can lead to different results for the same traffic characteristics. Verify that this is true. Propose a hybrid optimization criterion that would capture the properties that MIN-DIFF and MIN-CROSS are trying to optimize. Also, propose an algorithm that performs the necessary partitioning, per the above optimization criterion.

13.5. Consider a WDM ring network with N stations and W wavelengths. Assume an $N \times N$ traffic matrix $a_{ij}, 1 \leq i, j \leq N$. Assume that the brige processing time per unit traffic flowing across bridge is p, and assume that the network must be operated such that the total processing at the bridge P is bounded as $P_{low} \leq P \leq P_{high}$. How may partitions must be formed and what is the partitioning of the nodes (for a feasible operating point)? Make other assumptions as appropriate.

13.6. Explain the results in Figs. 13.7 and 13.9.

13.7. Imagine that you are the manager of an FDDI network with N user workstations. You are responsible for end-users getting the desired delay-throughput response from the network.

(a) You want to study how the network would scale with increasing user traffic. How would you perform traffic measurements to obtain an understanding of the traffic loading and delay characteristics in the network?

(b) Given that you have obtained an $N \times N$ traffic matrix (each matrix entry is in bytes/hour), how would you normalize the traffic matrix for purposes of further analysis?

(c) You have decided to upgrade your FDDI network to a WDDI network with W wavelengths. How would you decide on W?

(d) Given the measured traffic and W = total number of wavelengths available, you want to operate with k subrings, where k is the optimal operating point. How would you choose k?

(e) Given that you have chosen k, how would you partition the nodes into k subrings?

13.8. Imagine that you now want to engineer the network such that it can dynamically reconfigure based on prevailing traffic conditions.

(a) How can you dynamically measure traffic?

(b) Give an outline of a centralized algorithm to dynamically compute the optimal operating point.

(c) How frequently is it feasible for the network to reconfigure? How does the network handle disruptions due to a reconfiguration?

(d) What is the effect of reconfiguration on higher-layer protocols?

Chapter 14

All-Optical Cycle Elimination

A transparent (wide-area) wavelength-routed optical network may be constructed by using wavelength crossconnect switches – also referred to as wavelength-routing switches (WRSs) – connected together by fiber to form an arbitrary mesh structure. The network is accessed through electronic stations that are attached to some of these crossconnects.

These wavelength crossconnect switches have the property that they may configure themselves into unspecified states. Each input port of a switch is always connected to some output port of the switch whether or not such a connection is required for the purpose of information transfer. Due to the presence of these unspecified states, there exists the possibility of setting up unintended *all-optical cycles* in the network (viz., *a loop with no terminating electronics in it*). If such a cycle contains amplifiers [e.g., erbium-doped fiber amplifiers (EDFAs)], there exists the possibility that the net loop gain is greater than the net loop loss. The amplified spontaneous emission (ASE) noise from amplifiers can build up in such a feedback loop to saturate the amplifiers and result in *oscillations* of the ASE noise in the loop. Such all-optical cycles as defined above (and hereafter referred to as "white" cycles) must be eliminated from an optical network in order for the network to perform any useful operation. Furthermore, for the realistic case in which the wavelength crossconnects result in signal crosstalk, there is a possibility of having closed cycles with oscillating crosstalk signals.

This chapter examines algorithms that set up new transparent optical connections upon request while avoiding the creation of such cycles in the network. These algorithms attempt to find a route for a connection and then (in a postprocessing fashion) configure switches such that white cycles that might get created would automatically get eliminated. In addition, our call-setup algorithms can avoid the possibility of crosstalk cycles.

14.1 Introduction

Wavelength-routed optical (wide-area) networks using wavelength crossconnects have been proposed for the purpose of providing transparent interconnection and wavelength reuse [Alex93, Brac93]. The wavelength crossconnect switches may be implemented using acoustooptic tunable filters (AOTFs) or WDM multiplexer/demultiplexer-based technologies. The wavelength crossconnects may or may not include a wavelength-translation functionality.

However, as a direct consequence of transparency, these networks can form closed cycles that contain amplifiers, e.g., EDFAs. Such closed cycles create the possibility of oscillations and amplifier saturation due to the recirculation of amplified spontaneous emission (ASE) noise from the EDFAs. Hence, searching for and eliminating such closed cycles becomes very important; otherwise, the network will be nonfunctional.

The wavelength crossconnects considered here are $K \times K$ reconfigurable optical switching elements which can achieve any permutation of input to output ports, independently for each of W wavebands,[1] $\lambda_1, \lambda_2, \ldots, \lambda_W$. That is, each input port of a crossconnect switch is always connected to some output port of the switch whether or not such a connection is required for the purpose of information transfer. As a result, there exists the possibility of setting up unintended all-optical cycles in the network (viz., a loop with no terminating electronics in it). If such a cycle contains amplifiers, (e.g., EDFAs), there is a possibility that the net loop gain is greater than the net loop loss.

End-users communicate with one another through electronic access stations which are attached to some of these crossconnects. The access stations exchange information using signals on wavelengths within the wavebands λ_1, $\lambda_2, \ldots, \lambda_W$. Consider the EDFA gain curve shown in Fig. 14.1.

---

[1] A waveband may contain just one wavelength or a band of several wavelengths.

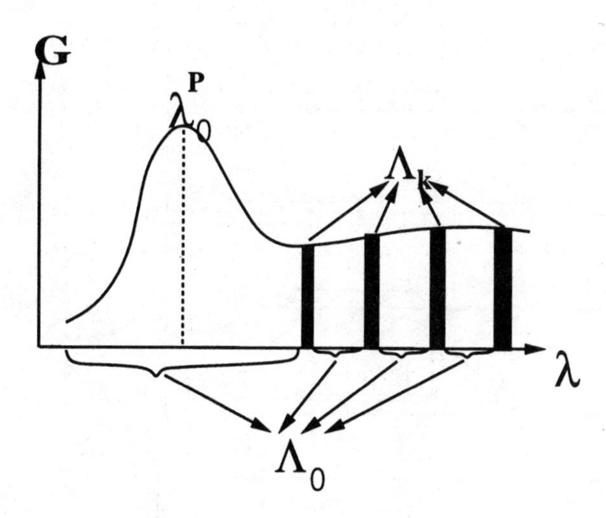

Figure 14.1 EDFA gain curve.

We define Λ_k to be the desired set of wavebands used for communication between the access stations.

We define Λ_0 as the remaining set of wavelengths in the amplifier spectrum, excluding the set Λ_k.

We will examine acoustooptic tunable filter (AOTF) crossconnects [Smit90] and other types of wavelength crossconnects for the way they treat the build up of ASE from the EDFAs.

14.1.1 Wavelength Crossconnect Switches

The transmission properties of the AOTF crossconnect are depicted in Fig. 14.2(a) and (b). As a first approximation, let us assume that this device impresses (on its incident signal spectrum) piecewise flat, transparent passbands for wavebands λ_1 through λ_W, separated by opaque stopbands. Figure 14.2(a) illustrates the device's transmission spectra when operated in the BAR state for all wavebands, i.e., when no RF control signals are applied. The spectrum above output port 1 shows that, in the BAR state, the device is perfectly transparent from input 1 to output 1 at all wavelengths; the spectrum shown at output port 2 shows that, in this BAR state, the device is opaque at all wavelengths from input 1 to output 2. Figure 14.2(b) shows transmission spectra when the device is in the CROSS state for wavebands 1 and 2 (physically achieved by applying the appropriate RF drive frequencies), and in the

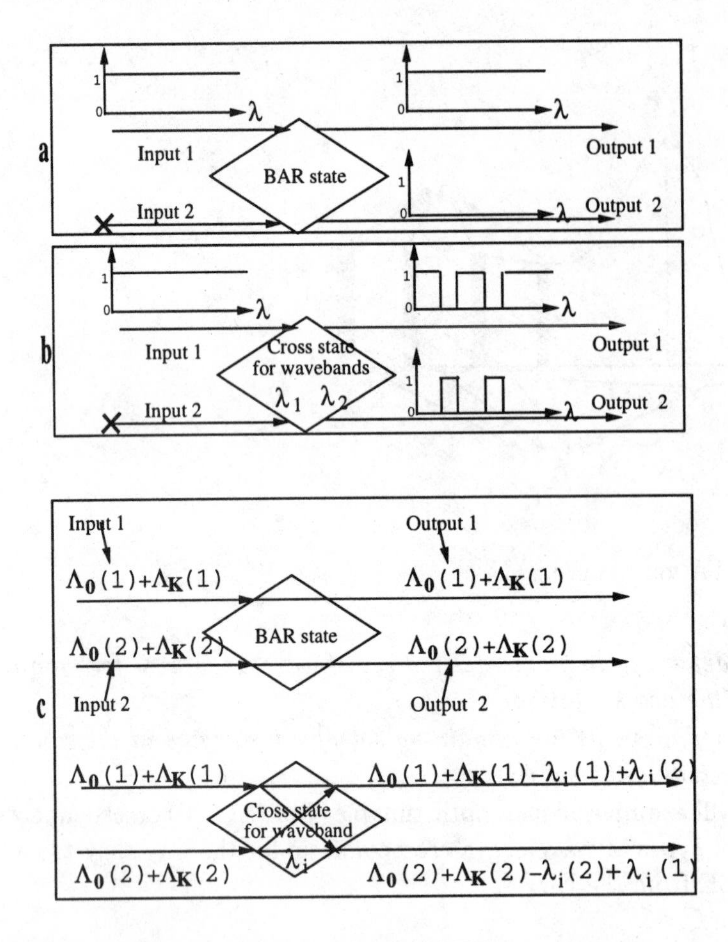

Figure 14.2 Wavelength routing using AOTF.

BAR state for all the other wavebands. In this case, the transmission from input 1 to output 2 is characterized by flat, transparent passbands occupying λ_1 and λ_2, separated by opaque stopbands. The transmission from input 1 to output 1 is then simply the complement of the lower spectrum. The device's transmission properties when illuminated at port 2 are directly obtainable from the above by symmetry. The above description summarizes the physical behavior of the AOTF under the idealized assumption that its passbands are piecewise flat. Multiple RF control signals can be applied simultaneously to the AOTF with the result that many wavebands may be switched independently and simultaneously. Note that, by default, the AOTF is always in the BAR state for all wavelengths (entire set Λ_0 and set Λ_k), but it can be put in the CROSS state only for wavelengths in the set Λ_k. Figure 14.2(c) shows an example of a 2×2 AOTF in CROSS state for the waveband Λ_i and in the BAR state otherwise (the notation is self-explanatory). Larger-size AOTF crossconnects have the same properties as the 2×2 shown here. Other types of wavelength crossconnects (mux/demux-based) also possess characteristics similar to the ones described above.

There are two important observations that we can make from the above discussion:

1. Wavelength crossconnects can result in unspecified states for some wavelengths, i.e., *wavelengths may pass through the crossconnects, uninterrupted, in the BAR state, without being set up for this purpose.*

2. Setting up a connection on a waveband between a single input port and a single output port may result in an unspecified connection between another pair of ports. For example, in Fig. 14.2(b), say we wish to create a connection only from input 1 to output 2 on waveband 5. Since the crossconnect is symmetric, setting up a connection from input 1 to output 2 on waveband 5 automatically results in an undesired and unspecified connection between input 2 and output 1 on waveband 5.

Based on the above observations, by considering multiple crossconnects connected to one another via fiber to form an arbitrary physical topology, we can show that the corresponding optical network can result in closed optical cycles due to unspecified states in the crossconnects.

When the recirculating wavelengths include the set Λ_0, we call the cycle a Λ_0 cycle.

When the recirculating wavelengths exclude the set Λ_0 but include the set Λ_k, we call the cycle a Λ_k cycle.

In the presence of ASE noise from the optical amplifiers, such feedback cycles can saturate the amplifiers and can cause the ASE noise to oscillate. An example of a wavelength-routed network using AOTFs and EDFAs is shown in Fig. 14.3. The ASE from the EDFAs passes through the AOTFs in the BAR state and forms closed feedback loops. Hence, such cycles must be eliminated as a first principle.

This chapter builds up on previous work [BaBr94], and proposes network-layer solutions to this problem. Furthermore, the presence of crosstalk in the crossconnects can exacerbate the problem. Our algorithms can eliminate crosstalk as well.

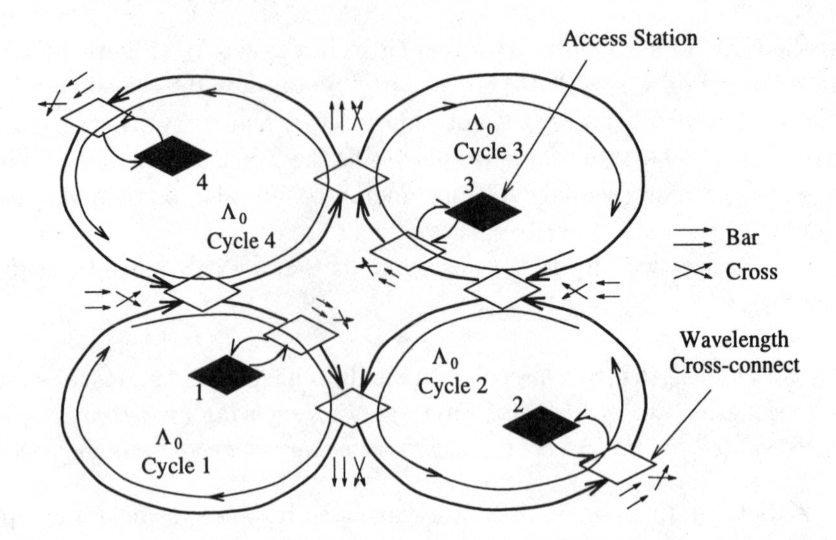

Figure 14.3 Wavelength-routed network with Λ_0 cycles.

This chapter examines the problems that arise in the simple scenario where only a single waveband is present, amplifiers are perfect with flat gain curves, etc. It then develops a corresponding solution. Issues related to multiple wavelengths and precise device models have not yet been examined, and are open research problems.

Section 14.2 lists our assumptions for this study. Section 14.3 describes a number of different modules which comprise our solution, while Section 14.4 provides the details of the corresponding algorithms used in each module. Some illustrative examples are presented in Section 14.5.

14.2 Network Assumptions

Our assumptions for this study are listed below:

1. The physical network consists of (electronic) access stations, wavelength crossconnects (switches), and physical fiber links. A fiber link connects either an access station to a switch or a switch to another switch to form a fiber network, e.g., see Figs. 14.4 and 14.5. Figure 14.4, which is a subgraph of the NSFNET T3 backbone network, employs five switches and five access stations, with one access station per switch, while Fig. 14.5, which is a random network generated by one of our algorithms,

consists of eight switches, but only four of them have an access station attached.

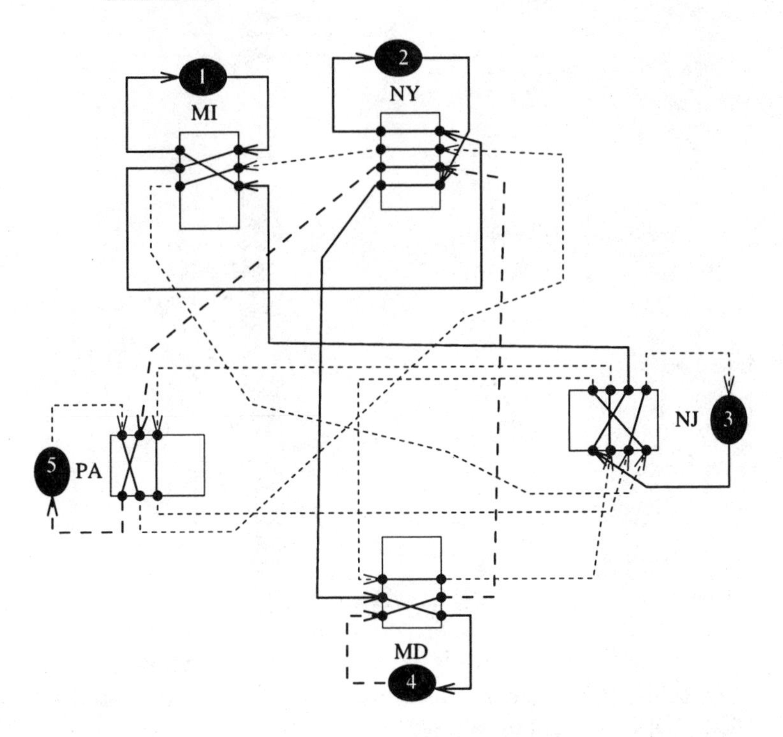

Figure 14.4 An example five-station, five-switch subgraph of the NSFNET T3 backbone. This network is used for the example "static" analysis results.

2. Presently, we consider only a single waveband in the network.

3. The in-degree (number of input ports) of any switch is equal to its out-degree (number of output ports). The same is true for an access station. Since we are considering a single waveband, an access station's in-degree and out-degree will be further constrained to unity.

4. So, if switch i is attached to $D(i)$ other switches in the physical graph, then switch i must be of size $[D(i) + m(i)] \times [D(i) + m(i)]$, where $m(i)$ is the number of access stations attached to switch i. Note that a switch may possibly have no access stations attached to itself, i.e., $m(i) = 0$, as in switches 5, 6, 7, and 8 in Fig. 14.5.

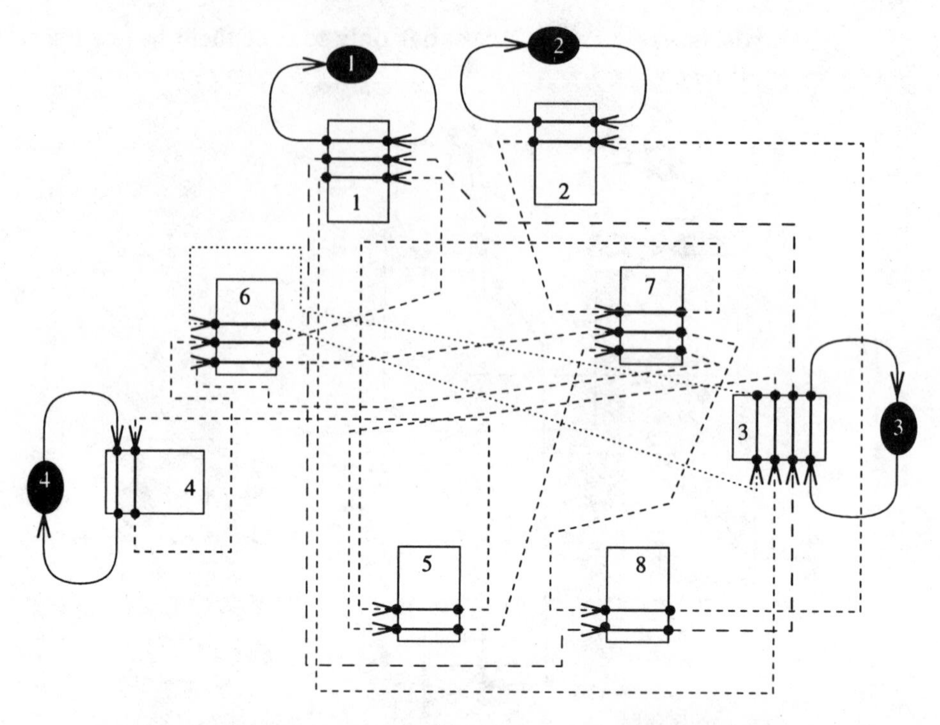

Figure 14.5 A (random) four-station, eight-switch network generated by Module 1. Note that this network contains Λ_0 cycles (as indicated by dashed and dotted lines) when all switches are in BAR state.

5. Each input port of a switch *always* connects to one output port of the switch, independent of whether such a port-to-port connection is intended or not.

6. At this point, we are assuming nearly perfect filters. Crosstalk problems are addressed by our algorithms, but we have not addressed the problems created by switching multiple wavelengths on a fiber.

7. We may design the network's physical connections such that "Λ_0 cycles" are eliminated when switches are in default (BAR) state [BaBr94]. Alternatively, the network must have been previously designed this way.

14.3 Overview of Solution and Algorithms

This section provides an outline of our network-layer solution approach to eliminate all-optical cycles in a wavelength-routed optical network (see Fig. 14.6). The solution approach consists of six modules. High-level descriptions of these modules are provided below, and details of the specific algorithms that they employ are provided in Section 14.4.

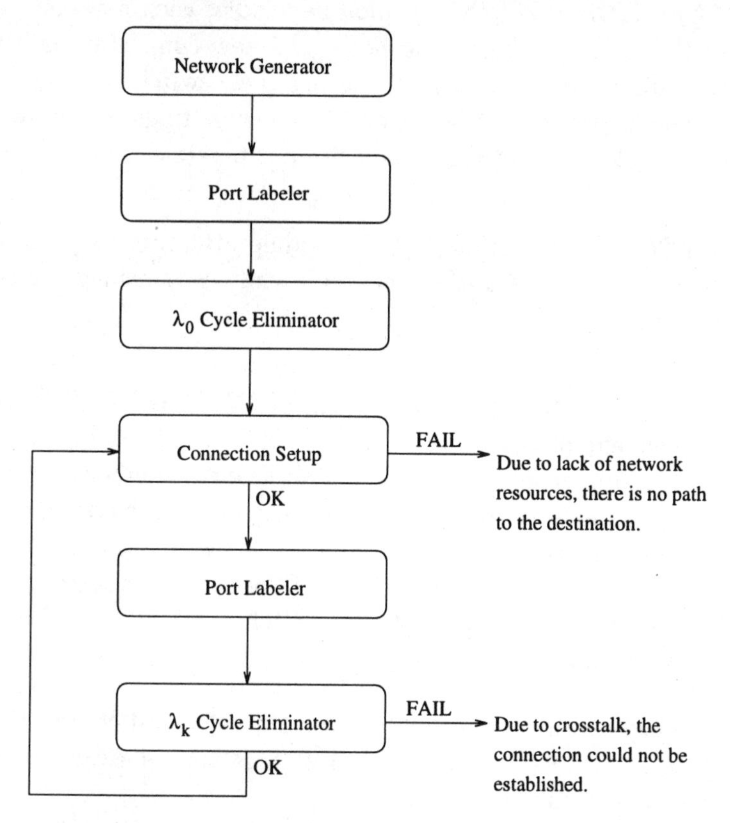

Figure 14.6 Flow chart of modules.

1. *NETWORK GENERATOR*: This module generates a random network consisting of access stations and switches. It is not needed when a physical description of the actual network to be studied is already available.

2. *SWITCH PORT LABELER*: This module labels the input and output ports of each switch according to whether the port carries an active connection or not.

3. Λ_0 *CYCLE ELIMINATOR*: This module removes any Λ_0 cycles in the default state of the network (when all switches are in BAR state). When a random network is generated, it may contain Λ_0 cycles when the switches are in BAR state. These cycles need to be removed only once, i.e., when designing the network.

4. *CONNECTION SETUP*: This module handles each new connection request that needs to be established. The new connection is attempted without disturbing the current routing (i.e., switch settings) of existing connections. This module outputs new settings of the switches to accommodate the requested connection, if possible.

5. Λ_k *CYCLE ELIMINATOR*: This module attempts to eliminate both Λ_k cycles and crosstalk in the network while establishing the requested connection.

6. *(STATIC) NETWORK CONFIGURATION ANALYZER*: This module is independent of the other modules. It is used off-line to statically assess the properties of a given network, e.g., number of connections achievable at any one time in the network without the occurrence of Λ_0 and Λ_k cycles and crosstalk. In particular, given that a $d \times d$ switch can be in one of $d!$ states, how many of the network's states are good or bad, given that each of the network switches can be in any state.

Figure 14.6 gives an overview of how the different modules in this solution approach interact with one another. All of these modules are described in detail in the following section.

14.4 Details of Algorithms

For each of the modules outlined in the previous section, we now provide its goal, required inputs, the output that it generates, method/algorithm used, and possible enhancements, wherever applicable. Readers who are interested in only the overview of these modules/algorithms, but not their details, may skip this section, and go directly to Section 14.5.

14.4.1 NETWORK GENERATOR (Module 1)

Goal: Generate a random network for testing various algorithms. (This module is not needed if the physical network's connectivity pattern is known.)

Input: Number of access stations; number of switches; minimum and maximum size of each switch (switch size currently is uniformly distributed between this minimum and maximum); and Boolean indicating possibility of more than one access station per switch.

Output: File containing number of access stations; number of switches; links connecting the stations and switches of the network; and switch settings. Switch settings are output from this module with switches in BAR state. (If a known physical graph is to be analyzed, its connectivity information should be provided according to the format output by this module.)

Methods/Algorithms Used: The program first randomly determines the size of each switch, which is currently based on an uniform distribution between the minimum and maximum size of switches. This switch size calculation is easily modifiable to accommodate other distributions (including a different specification for each switch). The program then randomly connects access stations to switches, and switches to other switches.

The program then "corrects" the following three problems that may occur in a randomly generated network:

self-loop (i.e., a switch connected to itself),
a switch that is connected to another switch more than once, and
a disconnected network (i.e., inability of a station to reach all other stations or switches).

The corrections are performed in a random fashion in order to maintain the true random nature of the network.

Possible Enhancements: Placements may be based on input parameters to examine propagation delay of a link (currently, the program just indicates which physical links exist, not the distance of a link; physical location of a link; etc.). The addition of more input parameters will allow flexibility in the types of random networks generated (e.g., server stations, connectivity requirements, etc.).

14.4.2 SWITCH PORT LABELER (Module 2)

Goal: To analyze the current network configuration and label all the ports on each switch as either *white*, *black_used*, or *black_unused*. A switch port, either input or output, is referred to as *black* if it is *not* part of an all-optical cycle; it is referred to as *black_used* if it is carrying an (active, intended) connection; otherwise, it is called *black_unused*.

To count number of white cycles in the network and list switches which each white cycle passes through.

To compare the intended and the established interconnection patterns of stations and report any discrepancy. This information is later utilized by the algorithm which eliminates these white cycles in Modules 3 and 5.

Input: Number of access stations; number of switches; current settings of each switch; intended interconnection pattern of stations; and physical links in the network.

Output: (1) Number of white cycles; (2) list of switches in the path of each white cycle; (3) for each switch in the network, label of each port (*white*, *black_used*, *black_unused*), number of the white cycle, if any, passing through each port, settings information; and (4) physical links in the network.

Method Used: Each switch and each access station is broken into as many nodes as there are input and output ports in it. An adjacency matrix is formed for the graph consisting of these "port" nodes. The physical link information and the settings information is used to form edges in this graph. Starting from any node in this graph, we traverse a path, which is guaranteed to return to the starting node. (Since this graph is Eulerian , it can be broken into edge-disjoint cycles.) We then determine if this cycle is *white* (goes through only switch ports) or *black* (goes through at least one access station). This is done until all the edges in the graph have been traversed. The interconnection pattern of stations is found out by starting from the node corresponding to the output port of each access station and keeping note of the destination. A comparison is made between the active connections intended and those established; any discrepancy results in an error message.

14.4.3 Λ_0 CYCLE ELIMINATOR (Module 3)

Goal: To create a network that contains no Λ_0 cycles when the switches of the network are in BAR state.

Input: Number of access stations; number of switches; size and settings (BAR state by default) of each switch; physical links in the network; and labels of the switch ports – *white*, *black_used*, or *black_unused*. (At this point – the physical network design stage – there will be no port labeled *black_used* since no connections have been set up yet.)

Output: File containing full network specifications [number of access stations, number of switches, size and settings (BAR state) of each switch, physical links in the network] for the input network converted such that it contains no Λ_0 cycles.

Methods/Algorithms Used: The algorithm "visits" each switch (in order) that currently contains an input port that is labeled *white*. It then attempts to find another input port that is labeled *black_unused* at that switch. If one is found, it relabels all ports that were part of that *white* cycle to *black_unused* and "swaps" the switch settings of those two input ports such that the *white* labeled input port now goes to where the *black* labeled input port used to go, and the *black* labeled input port now goes to where the *white* labeled input port used to go. This combines the two cycles (one *black*, one *white*) into one larger *black* cycle (*black* since this larger cycle still contains at least one access station that caused the smaller original *black* cycle to be *black*). If no *black_unused* port at that switch is found, then go to the next switch that contains a *white* port and repeat the process.

If, after going through all switches containing *white* cycles, some switches still contain *white* cycles, repeat the process until no switch contains any *white* cycle. If there is any way to eliminate *white* cycles, this method is guaranteed to break them. Due to the Eulerian property of this network, we are guaranteed that there is always a way to break *white* cycles at this point, so we can say that this program always succeeds in breaking *white* cycles.

After the cycle breaking is complete, the switches will be in a specified state. We want to achieve this network configuration but with the switches in BAR state. Therefore, move the physical output port connections to achieve this network when the switches are in the BAR state.

As an example, if for a particular switch, input port 0 is connected to output port 3 of that switch, connect output port 0 to what output port 3 used to be connected to. This will finally result in a network that has no Λ_0 cycles when the switches are in the BAR state.

Possible Enhancements: Make the algorithm more efficient. There are a few ways that the algorithm can be made to run in time O(number of white ports), but these improvements have not been implemented yet.

14.4.4 CONNECTION SETUP (Module 4)

Goal: To establish a new connection in the network by creating new settings for the switches, without changing the paths of the existing connections.

Input: A file containing the current state of the network as given by: number of access stations; number of switches; current settings of each switch; intended interconnection pattern of stations; physical links in the network; and a list of new connections to be added, given by <Source station, Destination station> pair.

Output: SUCCESS, if a switch setting can be found to satisfy the new connection, or
FAILURE, if no such path can be found to establish the new connection because of unavailability of network resources such as fiber links and switch ports.

In the case of SUCCESS, the algorithm returns: (1) number of white cycles; (2) list of switches in the path of each white cycle; (3) for each switch in the network, label of each port (white, black_used, black_unused), number of the white cycle, if any, passing through each port, settings information; and (4) physical links in the network.

Method Used: The algorithm used for setting up the new connection is Dijkstra's shortest-path algorithm. First, we eliminate the active links (links in the path of existing connections) from the graph. Then, we add new edges from all idle input ports of a switch to all its idle output ports. All edges in this new graph are given equal weight (value = 1). If no edge exists between two stations, the corresponding entry in the weight array is set to ∞. We run Dijkstra's algorithm on this graph. The resulting path, if any, is used to change the switch settings.

The labeling of ports in this graph is performed as described in Section 14.4.2.

Possible Enhancements: The shortest-path algorithm may not always be the best one. By taking the shortest path for this connection, we might decrease the chances for setting up a new connection later, without white cycles and crosstalk. (See the example discussed later in Section 14.5.)

We must allow teardown of existing connections. We need to study the blocking probability of any connection after running through a random sequence of connection setups and teardowns.

14.4.5 Λ_k CYCLE ELIMINATOR (Module 5)

Goal: To configure the switches to achieve a network that contains no Λ_k cycles after certain network connections are set up.

Input: Number of access stations; number of switches; size and current settings of each switch; physical links in the network; and labels of the switch ports – *white*, *black_used*, or *black_unused*. For a connection that was just requested, the path it uses will appear as a *black_used* path to this program.

Output: File containing full network specifications [number of access stations, number of switches, size and settings (BAR state) of each switch, physical links in the network] for the input network converted such that it contains no Λ_k cycles.

Methods/Algorithms Used: The algorithm "visits" each switch (in order) that currently contains an input port that is labeled *white*. It then attempts to find another input port that is labeled *black_unused* at that switch. (Note: Here, a *white* cycle cannot be combined with a *black_used* port because, if that happened, a node would be transmitting traffic that entered this switch twice and we would have a problem with crosstalk.) If one is found, the program relabels all ports that were part of that *white* cycle to *black_unused* and "swaps" the switch settings of those two input ports such that the *white* labeled input port now goes to where the *black* labeled input port used to go, and the *black* labeled input port now goes to where the *white* labeled input port used to go. This combines the two cycles (one *black*, one *white*) into one larger *black* cycle (*black* since this larger cycle still contains at least one access station that caused the smaller original *black* cycle to be *black*). If no *black_unused* port at that switch was found, then go to the next switch that contained a *white* port and repeat the process.

If, after going through all switches containing *white* cycles, some switches still contain *white* cycles, repeat the process until no switches contain *white* cycles. If there is any way to eliminate *white* cycles, this method is guaranteed to break them. This procedure is not guaranteed to succeed at this point due to the crosstalk limitations. If crosstalk were not a concern, we could combine the *white* ports with either *black_unused* or *black_used* ports and this procedure could be guaranteed to succeed.

Possible Enhancements: Make the algorithm more efficient. There are a few ways that the algorithm can be made to run in time O(number of white ports), but these improvements have not been implemented yet.

14.4.6 (STATIC) NETWORK CONFIGURATION ANALYZER (Module 6)

Goal: To place all the switches in the network in all possible settings and determine: (1) total number of states of the network, (2) number of GOOD states (no white cycles and no crosstalk in any active connections), and (3) number of permutation connection patterns of access stations which are achievable in this network.

Input: Number of access stations; number of switches; size of switches; and physical links in the network.

Output: Total number of states in the network.
- Total number of permutation connections in which none of the access stations is idle.
- Number of settings which establish these patterns:
  - with possible white cycles and crosstalk,
  - without white cycles but with possibility of crosstalk, or
  - without both white cycles and crosstalk.
- Number of such connections which are impossible without white cycles or crosstalk.
- A list of these impossible connections.
- A file with as many lines as there are states giving the information for each state such as:

interconnection pattern established,
number of white cycles, and
presence/absence of crosstalk.

Method Used: A $d \times d$ switch can be placed in any of $d!$ settings. Suppose there are m switches, each of size $d \times d$. Then, the total number of states of the network is $(d!)^m$. We exhaustively place the network in each of these states and then find out the interconnection pattern that has been established. The number of white cycles in the network is also counted. An active path is one between an access station and another station. All self-connections are assumed to be idle connections. In each of the active paths, we check to see if we encounter the same switch more than once. If so, then there is crosstalk in this path. For each state, this information is output in a single line in a specific format.

14.5 Illustrative Examples

This section discusses some illustrative examples which demonstrate how our algorithms (modules) presented in the previous sections may be employed to operate, or to learn about the properties of, a wavelength-routed optical network. In Section 14.5.1, we consider the "dynamic" case in which, given an optical network, calls need to be set up without the possibility of white cycles and crosstalk. In Section 14.5.2, we study the "static" properties of a wavelength-routed optical network, viz., its proportion of good states, etc. Modules 1 through 5 from Section 14.4 are used for the "dynamic" study, while Module 6 is employed for the "static" analysis.

14.5.1 Dynamic Analysis

Application of Algorithms on Example Physical Networks

To study the application of our algorithms on some example physical networks, we first consider a random network generated by Module 1, with switch ports labeled by Module 2, as shown in Fig. 14.5. Note that this network does contain Λ_0 cycles in the initial state of the network when all the switches are in the BAR state. These Λ_0 cycles are indicated by the dashed lines in Fig. 14.5. The result of eliminating these Λ_0 cycles by Module 3 is shown in Fig. 14.7.

Then, the following two connections are set up: (A) from station 2 to station 4; and (B) from station 4 to station 3. The state of the network after

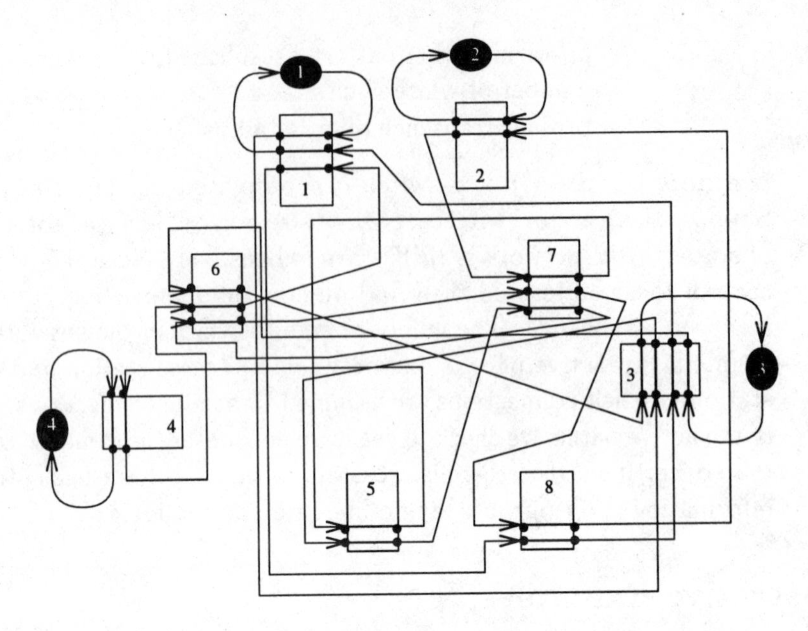

Figure 14.7 Network after elimination of Λ_0 cycles using Module 3.

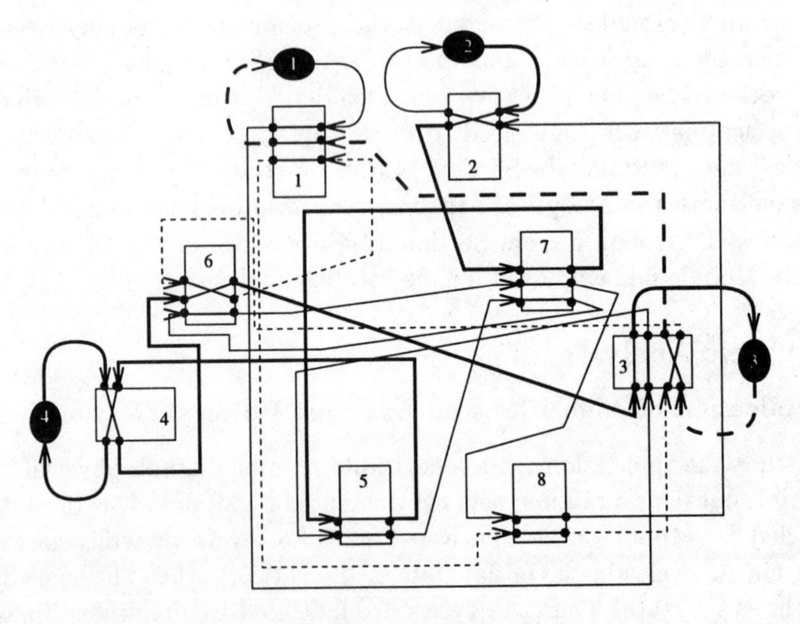

Figure 14.8 Network after establishing two connections – heavy lines – (from station 2 to station 4 and from station 4 to station 3) using Module 4. However, a new connection – dashed heavy line – (from station 3 to station 1) is causing a Λ_k cycle – dashed light line.

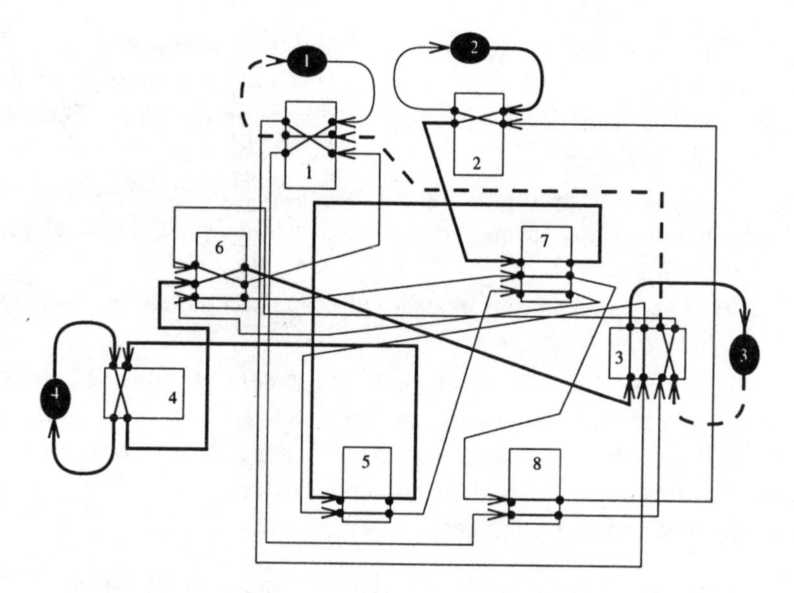

Figure 14.9 Network after elimination of Λ_k cycles using Module 5.

establishing these connections is shown in Fig. 14.8. The links carrying traffic are indicated by thick continuous lines.

Now, we would like to set up a new connection (C) from station 3 to station 1. The path suggested by the shortest-path algorithm is indicated by the thick dashed lines in Fig. 14.8. Notice, however, that this routing choice results in a Λ_k cycle, which is indicated by thin dashed lines. To avoid this Λ_k cycle, the Λ_k Cycle Eliminator Algorithm (Module 5 in Section 14.4.5) is employed. It changes a few switch settings, and the resulting network is shown in Fig. 14.9. Notice that none of the paths for the existing connections is altered.

Performance Analysis of Dynamic Algorithms

To study the performance of the dynamic algorithms, we assume the following: (1) Calls (i.e., connection requests, also referred to as messages) arrive at each station according to a Poisson process, with an intensity (load) of α per unit time at each station, and the holding time for each call is exponentially distributed with a mean of 100 time units. (2) A call arriving at each station is equally likely to be destined to any of the other stations. (3) Each station can source at most one call at a time, i.e., calls are not buffered, or a call arrival is allowed only if the station's transmitter is idle.

A call may be blocked due to unavailability of resources, viz., (1) if the destination is busy, or (2) if there is no free path (through fiber links and switches) to the destination even if the destination is available. This is referred to as *resource blocking.*

A call that is not resource-blocked may still be considered to be blocked if its establishment would have introduced crosstalk. This is referred to as *crosstalk blocking.*

Calls that are blocked due to either of the above blocking phenomenon are *immediately lost.*

Now, consider Fig. 14.4, which has 5 access stations, 17 links, and 5 switches, and which is a subset of the NSFNET T3 backbone. Calls arrive to this network according to the above model. For this network, we would like to study its throughput and call blocking performance when the call-setup algorithms outlined in earlier sections are used.

We would also like to study, for the same number of access stations and physical links, the effect of reducing/expanding the number of switches. In one extreme, we consider a network of two switches, since a network of just a single switch is not very meaningful with our assumptions – our assumption that an access station has in-degree = out-degree = 1 would mean that, if the network has N access stations and L links, then the single giant switch must have $N - L$ external (and useless) fiber links connecting one of its output ports to one of its input ports. In the other extreme, we consider the network to consist of as many 2×2 switches as possible; thus, the largest number of switches turns out to be $\lfloor L/2 \rfloor$. For a particular number of switches, we find that the network demonstrates the best performance if all switches are as equally sized as possible. Thus, in our network construction, we maintain the difference between the largest-sized and the smallest-sized switches to be unity; also, whenever possible, access stations are attached to the larger-sized switches in order to achieve better blocking performance.[2]

Expanding the number of switches would mean reducing the sizes of (some or all of) the switches (for the same number of fiber links). Smaller switches, in turn, would reduce the number of possible network states (as determined by all possible combinations of switch settings),[3] so that resource blocking would increase. When the number of switches is reduced, the sizes of the switches

---

[2]We use these principles for generating the numerical results in Figs. 14.10 through 14.15; however, these principles are not followed in the random graph in Fig. 14.5.

[3]This phenomenon will be quantified in the static analysis in Section 14.5.2.

will increase, in general, for the same number of links. However, now, a larger number of network states will also be unusable due to crosstalk, as we shall observe in the static analysis in Section 14.5.2.

The dynamic performance of the 5-station, 17-link network in Fig. 14.4 has been studied, along with that of its derivatives with an expanded/contracted number of switches M as outlined in earlier paragraphs. The loading of calls is varied from $\alpha = 0.01$ to $\alpha = 2$, and the corresponding performance results for different numbers of switches are shown in Figs. 14.10 through 14.12. These results were obtained by simulating calls for 100,000 time units, collecting the pertinent statistics, and then averaging the statistics over five runs (i.e., five different graphs) for each switch size. In general, as expected, the higher the offered load, the higher is the probability of call blocking due to unavailable resources and crosstalk, and the lower is the fraction of unblocked calls. One anomaly we would like to comment on is the value of crosstalk blocking for $M = 6$ and $M = 8$. Due to the small number of access stations, it is fairly easy to get all of the stations communicating with other stations. For $M \leq 6$, five stations would be prevented from talking to one another, at once, due to crosstalk. In the eight-switch case, five stations would usually not be able to talk to one another simultaneously due to lack of a path. So, in effect, the probability of crosstalk blocking is going down. Also, note that a call is considered resource-blocked first and crosstalk-blocked second, and resource blocking is the dominant phenomenon (the absolute value for resource blocking is generally over an order of magnitude higher than crosstalk blocking).

The effect of number of switches M is interesting. At light loads, when M is increased, the probability of resource blocking increases (and the fraction of unblocked calls decreases) because, now, a call has to find its way through several small switches. When the loading is heavy, each station's receiver is busy most of the time, and this dominates the probability of resource blocking, which becomes nearly independent of M. Correspondingly, the fraction of unblocked messages is nearly independent of M for heavy loading.

When M is small, crosstalk blocking increases with increasing load. When M is increased, for light loading, crosstalk blocking increases, since most calls are not resource-blocked.

We repeated these experiments on a larger network, viz., the 14-station, 56-link NSFNET T1 network that existed until early 1992, before it was upgraded to the NSFNET T3 network. For this example, we studied not only the capabilities of the 14-switch network, but also those of its related constructions employing $M = 2$ through $M = 26$ switches. The corresponding

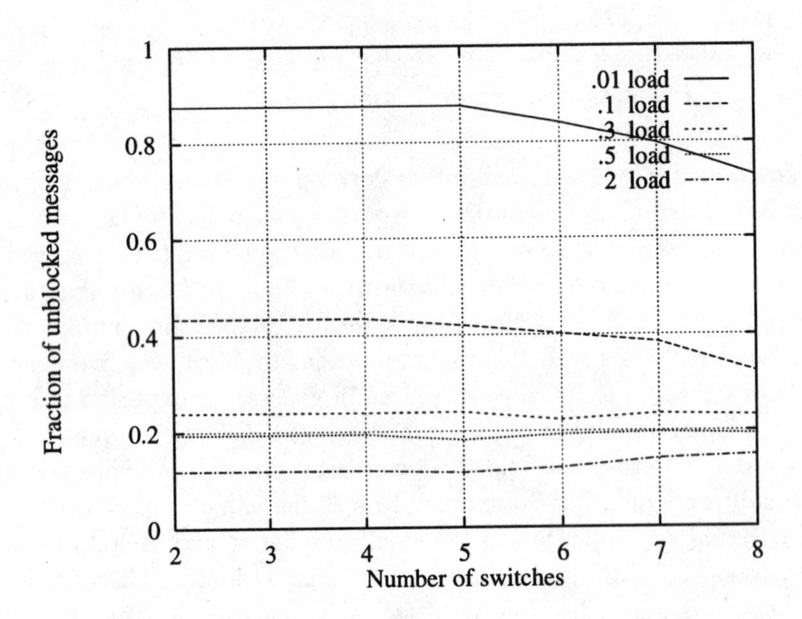

Figure 14.10 Fraction of unblocked calls vs. M for the 5-station, 17-link network.

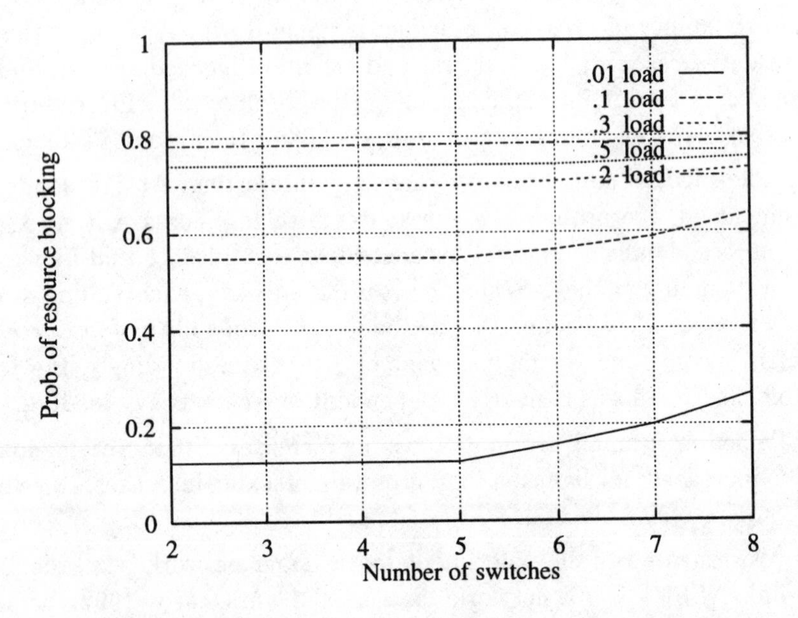

Figure 14.11 Probability of resource blocking vs. M for the 5-station, 17-link network.

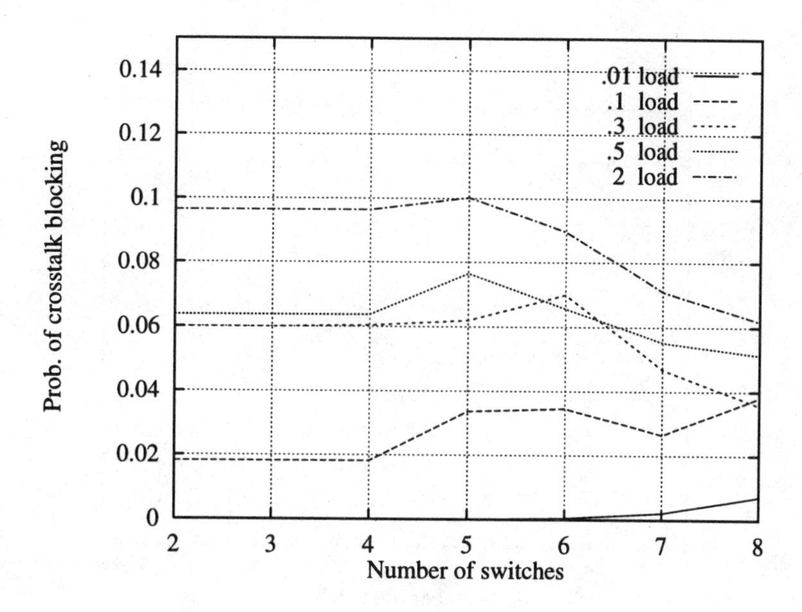

Figure 14.12 Probability of crosstalk blocking vs. M for the 5-station, 17-link network.

performance results are shown in Figs. 14.13 through 14.15. Now, each simulation was run for 2,000 time units, which is not very long but which seemed to provide reasonably accurate results (also, each run took quite a long amount of computer time due to the larger network size). As before, statistics were compiled and averaged over five different graphs for each value of M. The results in Figs. 14.13 through 14.15 reinforce our earlier observations on the smaller five-station network. However, now, the value of M seems to play a more dramatic role with respect to increasing blocking (and decreasing the fraction of unblocked calls) for higher values of M.

14.5.2 Static Analysis

Consider Fig. 14.4, which has five access stations and five switches. In one extreme, when all five stations are busy, we have a complete permutation connection pattern. Notice that the receiving stations form a permutation of the sending stations. The number of such permutation connection patterns formed from N stations is given by $\chi(N)$, e.g., $\chi(5) = 44$. Now, among all the network states which support such a permutation connection pattern, we keep track of how many are unaffected by cycles and crosstalk. In fact, we keep

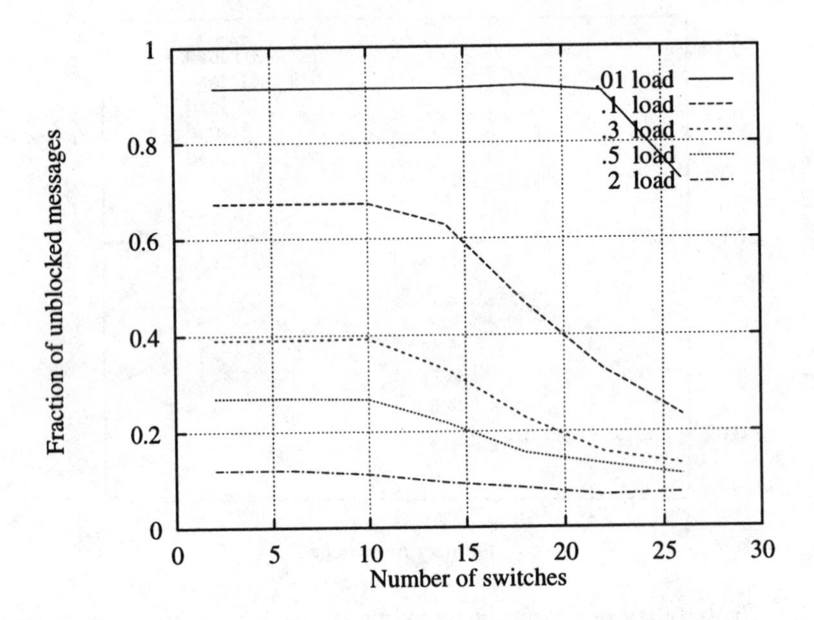

Figure 14.13 Fraction of unblocked calls vs. M for the 14-station, 56-link network.

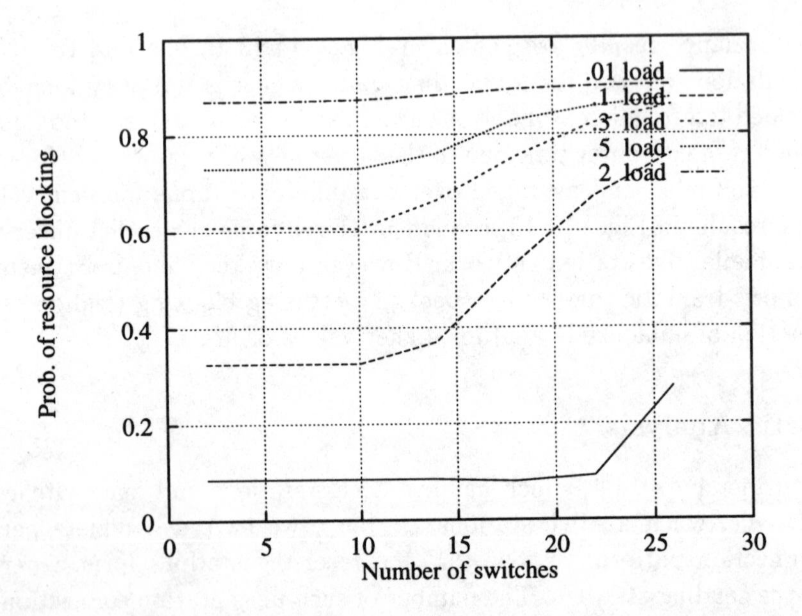

Figure 14.14 Probability of resource blocking vs. M for the 14-station, 56-link network.

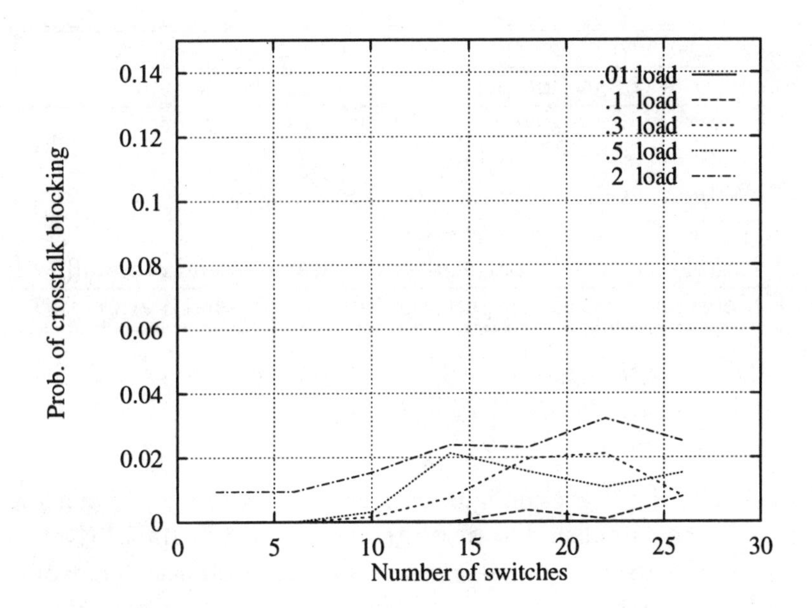

Figure 14.15 Probability of crosstalk blocking vs. M for the 14-station, 56-link network.

count of the number of states which achieve *each* such pattern without cycles and crosstalk. If we do not find even a single state to support a pattern, then that particular connection pattern is *impossible* to achieve in the network with the given constraints. At the other extreme, when all five stations are idle, the network could be in any of its states with all of its transmitters being idle. Between these two extremes, we have the case of exactly k stations being idle, where $1 \leq k \leq 4$. We are interested in finding out if each of these connection patterns is achievable on this network. In other words, does there exist at least one network state which supports each of these connection patterns?

Assume that a permutation routing pattern has to be set up without white cycles or crosstalk, i.e., we need to set up five connections, with each access station being the source of a connection, as well as the sink of a (another) connection, i.e., there are no self-loops. The results in Table 14.1 are obtained after an exhaustive search through all possible switch settings of the network shown in Fig. 14.4.

Total number of switch settings (network states)	124416
Total number of connection permutations (with five stations)	44
No. of network states allowing the permutations of connections	
With cycles and crosstalk	9696
Without cycles but with crosstalk	6528
Without cycles and crosstalk	84
No. of possible connection patterns (without cycles and crosstalk)	36
No. of impossible connection patterns (due to cycles and crosstalk)	8

Table 14.1 Static analysis via exhaustive search on the network with five access stations, 5 switches, and 22 links.

The following eight configurations were the set of connections which were impossible to establish due to white cycles and crosstalk. (Notation: Each group represents the complete set of five connections. The sources of the connections are 1 through 5 from left to right. The respective destinations are given below.)

```
2  3  1  5  4          3  1  2  5  4
4  3  2  5  1          4  3  5  1  2
4  5  2  1  3          5  3  2  1  4
5  3  4  2  1          5  4  2  3  1
```

Now, consider all possible connection patterns with five stations. The number of connection patterns of five stations with different number of idle stations are the following:

- No idle stations = 44
- One idle station = 265
- Two idle stations = 320
- Three idle stations = 130
- Four idle stations = 20
- Five idle stations = 1
- Total = 780

We would like to study the capability of the network in Fig. 14.4, as well as its derivatives with an expanded/contracted number of switches (as was studied in Section 14.5.1), to support the above connection patterns without white cycles and crosstalk. In this regard, we could study networks only for $M \geq 5$, since for smaller values of M, the number of network states was too large for our computing facilities to handle. In Table 14.2, for each value of M between 5 and 8 corresponding to an example graph, we show the total number of states, number (and percentage) of states without cycles, and number (and percentage) of states without crosstalk. We also show the number of connection patterns (out of 780 total) that could not be supported because of resource blocking, cycles, and crosstalk.

M	*Settings*			*Impossible Connections*		
	Total	*Without Cycles*	*Without Crosstalk*	*No Route*	*Cycles*	*Crosstalk*
5	124416	62208 (50.00%)	11146 (8.95%)	0	0	8
6	15552	10016 (64.40%)	2742 (17.63%)	6	6	8
7	3456	2456 (71.06%)	1036 (29.97%)	51	51	52
8	768	310 (40.36%)	83 (10.80%)	663	663	696

Table 14.2 Static analysis with an expanded/contracted number of switches.

The above results indicate the following:

1. The total number of states in the network, as well as the number of good states (i.e., states without white cycles and crosstalk), decreases as the number of switches increases.

2. However, the percentage of good states increases as we go from $M = 5$ to $M = 7$, and then decreases for $M = 8$. We have observed similar phenomenon for other graphs with the same values of M. This may partially explain the anomalous results in crosstalk blocking that we alluded to earlier in Fig. 14.12 while discussing the dynamic network performance.

3. The number of impossible connection patterns increases drastically when we go from $M = 6$ to 7 to 8, since the corresponding number of good states also reduces drastically.

We have not been able to perform static analysis of the 14-station network because the state space explodes. However, we believe that a sampling technique may be applicable to get a reasonable idea of the fraction of network states that are good (or bad). That is, we need not enumerate all states, but pick each state with a certain probability in order to obtain reasonable statistics in a reasonable amount of time. However, this is still an open problem.

14.6 Summary

This chapter considered the problem of setting up *unintended* all-optical cycles in a transparent, wavelength-routed, optical network employing wavelength crossconnect switches. If such a cycle contains amplifiers so that the net loop gain is greater than the net loop loss, the ASE noise from the amplifiers can build up, oscillate in the loop, and saturate the amplifiers. We examined algorithms that can set up new transparent optical connections upon request while avoiding the creation of such cycles in the network. These algorithms attempted to find a route for a connection and then (in a postprocessing fashion) configured the switches such that white cycles that might get created automatically got eliminated. In addition, our call-setup algorithms avoided the possibility of crosstalk. We also developed mechanisms for statically analyzing the capabilities of a network, viz., the absolute value and the fraction of good (and bad) states, and the corresponding connection patterns the network can (and cannot) support.

A large number of open research problems still exist. First, algorithms need to be developed to allow multiple wavebands. Also, problems arising out of interactions between physical layer issues and network routing problems need to be examined closely, such as precise characteristics for amplifiers and crossconnects.

Exercises

14.1. Consider the network in Fig. 14.16. The 2×2 switch configurations are the same for all wavelengths and are as depicted in the figure.

Determine the all-optical cycles in the network for each of the following cases. (When we say that the configuration of a switch is reversed, we mean that the switch will be in bar state if it was in cross state previously, and that it will be in cross state if it was in bar state previously.)

(a) Switch configurations are as shown in the figure.

(b) Configuration of switch 8 is reversed.

(c) Configuration of switch 6 is reversed.

(d) Configuration of switch 3 is reversed.

(e) Configuration of switch 2 is reversed.

(f) Configurations of switches 3 and 6 are reversed.

(g) Configurations of switches 4 and 6 are reversed.

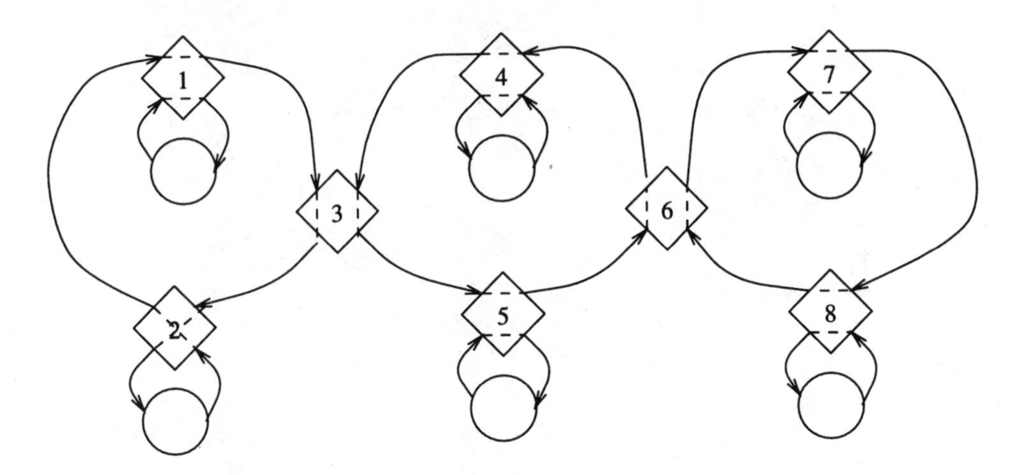

Figure 14.16 Network for Problem 14.1.

14.2. Consider the network in Fig. 14.17 with two wavelengths. Each switch state is indicated for each wavelength separately. For each of the two wavelengths, determine the all-optical cycles in the network for each of the following cases. (Parts (b) and (c) have five subparts each.)

(a) Switch configurations are as shown in the figure.

(b) Switch configurations are reversed for wavelength λ_1 only, as in Problem 14.1, parts (b) through (f).

(c) Switch configurations are reversed for wavelength λ_2 only, as in Problem 14.1, parts (b) through (f).

14.3. Consider again the network in Fig. 14.17. Let us assume that node 1 wants to send a transmission to node 2. Node 1 could accomplish this by utilizing λ_2 and configuring switch 1 and switch 2 into the cross states for λ_2, and switch 3 into the bar state for λ_2.

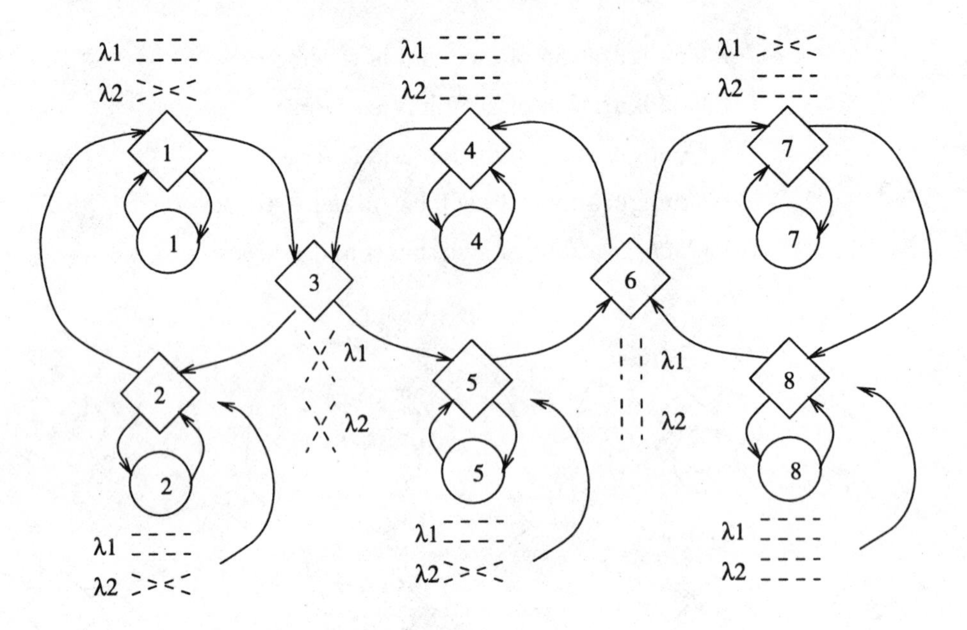

Figure 14.17 Network for Problems 14.2 and 14.3.

(a) Indicate any all-optical cycles that develop if the switches were configured in this fashion.

(b) If there are no other active connections in the network at the time of above transmission, how could the cycle(s) be broken?

14.4. Oftentimes, optical switches can suffer from a significant amount of crosstalk for a particular wavelength. For this reason, it is sometimes desirable to not route a signal through the same switch more than one time. Figure 14.18 shows the network configuration for a particular wavelength. Existing connections using particular links and switch settings are highlighted using thick lines. (These connections are $1 \rightarrow 3$, $3 \rightarrow 2$, $4 \rightarrow 5$, and $5 \rightarrow 4$.) Without reconfiguring the existing connections, is it possible to route a connection between node 2 and node 1 without (a) sending the signal through the same switch more than once, and (b) without creating an all-optical cycle?

14.5. Perform static analysis, and show your results as in Table 14.1 for the following networks:

(a) 14.16, and

(b) 14.18.

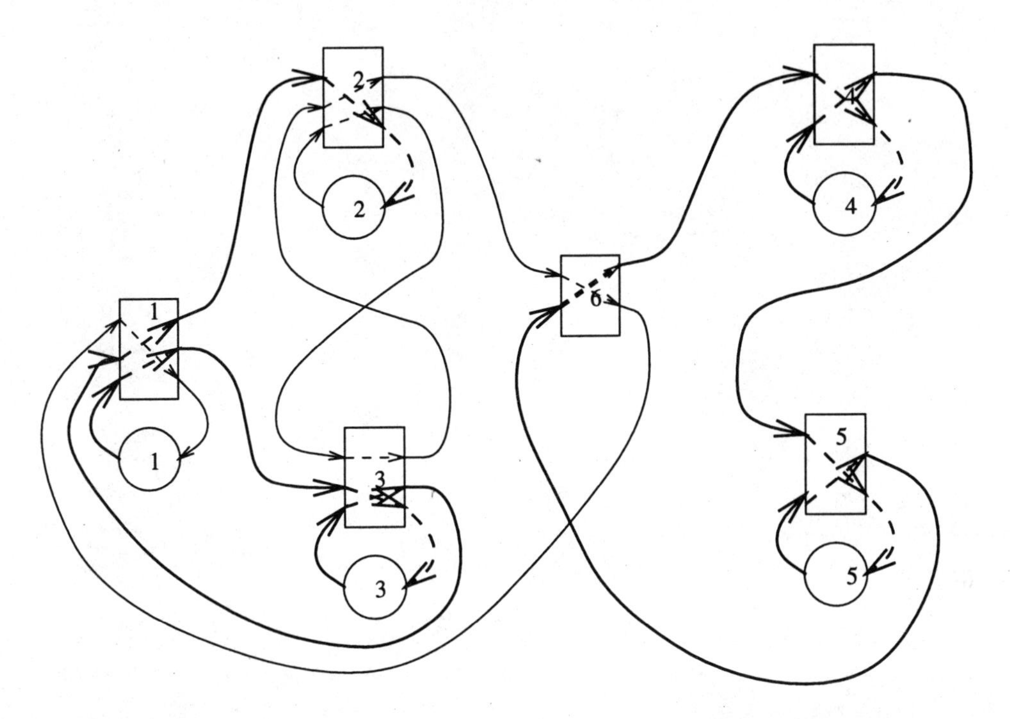

Figure 14.18 Network for Problem 14.4.

Chapter 15

Optimizing Amplifier Placements in an Optical LAN/MAN

Optical networks based on passive star couplers and employing wavelength-division multiplexing (WDM) have been proposed for deployment in local and metropolitan areas. These networks suffer from splitting, coupling, and attenuation losses. Since there is an upper bound on transmitter power and a lower bound on receiver sensitivity, optical amplifiers are usually required to compensate for the power losses mentioned above. Since amplifiers are costly, it is desirable to minimize their total number in the network. However, an optical amplifier has constraints on the maximum gain and the maximum output power it can supply; thus optical amplifier placement becomes a challenging problem. In fact, the general problem of minimizing the total amplifier count is a mixed-integer *nonlinear* problem. Previous studies have attacked the amplifier-placement problem by adding the "artificial" constraint that all wavelengths, which are present at a particular point in a fiber, be at the same power level. This constraint simplifies the problem into a solvable mixed-integer *linear* program. Unfortunately, this artificial constraint can miss feasible solutions that have a lower amplifier count but do not have the constraint

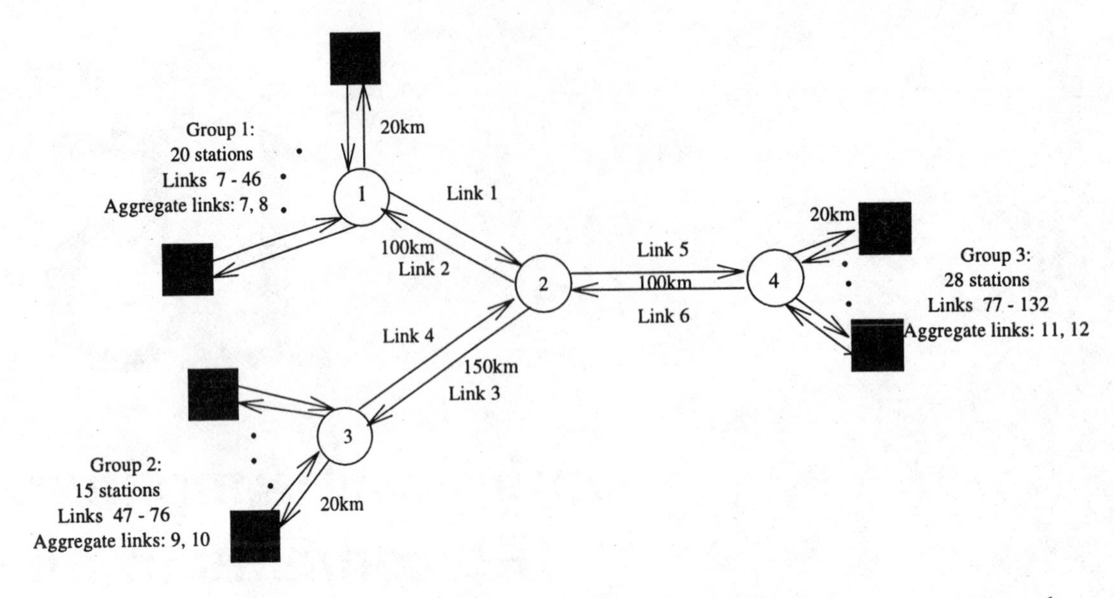

Figure 15.1 Example of a passive-star-based optical metropolitan-area network (slightly modified version of the one used in [LTGC94]).

on equally powered wavelengths. This chapter presents a method to solve the *minimum-amplifier-placement problem* while avoiding the equally-powered-wavelength constraint. By allowing signals to operate at different power levels, our method can reduce the number of amplifiers required in several small to medium-sized networks.

15.1 Introduction

15.1.1 Network Environment

The focus of this study is on a class of the next-generation optical local/metropolitan-area networks (LAN/MAN) which span distances from fewer than a kilometer to a few tens of kilometers and which provide loop-free communication paths between all source-destination pairs.[1] A large-distance version of such a network is depicted in Fig. 15.1, and it consists of $N = 63$ stations and $M = 4$ passive optical star couplers ("stars"), such that each star is connected to other stars and/or stations via two unidirectional fiber links. The

[1] Such networks have been referred to in the literature as access networks, passive optical networks (PONs) [TaWB95], etc.

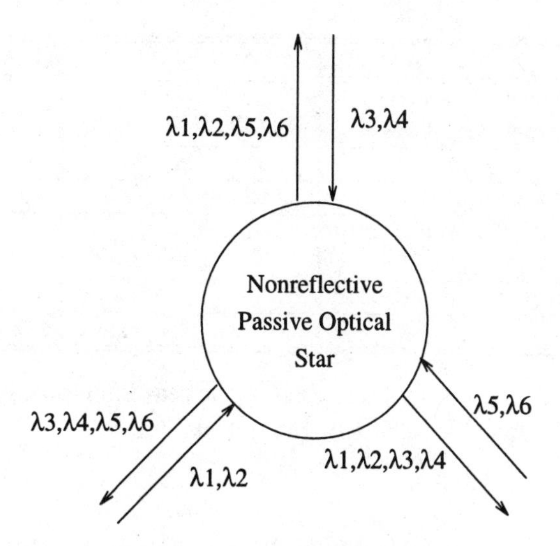

Figure 15.2 Example of a nonreflective star.

passive star coupler provides a broadcast facility, but it must also be of the "nonreflective" type (to be elaborated below) in order to prevent loops in the network.

Our study will consider the case where each station in the network has a fixed-wavelength transmitter and is set to operate on its own unique wavelength channel. Each station either has a tunable receiver or a receiver array in order to receive signals from all of the other stations. The objective is to ensure that a station's transmission can be received by every other station after being subject to losses and gains as the signal traverses through different parts of the network.

The network consists of optical stars that are nonreflective. A nonreflective star consists of pairs of inputs and outputs, and each output carries all of the wavelengths that were incident on all of the inputs except for the wavelengths that were carried on its own paired input (see Fig. 15.2 for an example). Such stars have been employed in the Level-0 all-optical network (AON) [Alex93]. Nonreflective stars are needed in order to avoid interference due to loops ("echoes") in the network. A star in the network with k input fibers and k output fibers operates such that the power on each wavelength on an input fiber is divided evenly among the $k - 1$ output fibers. This is referred to as the *splitting loss* at a star. (Note that the splitting loss can be different for different-sized stars in the network.)

As the sample network in Fig. 15.1 shows, these networks can be deployed

Parameter	*Description*	*Range*	*Value Used*
p_{sen}	Minimum signal power at receiver and the amplifier sensitivity level	−30dBm at 1 Gbps	−30 dBm
G_{max}	Maximum small-signal gain	≤25 dB (MQW)	20 dB
$P_{NONLIN,max}$	Maximum total power in fiber	10 to 50mW	10 mW
P_{max}	Maximum total output power of amp and transmitter		0 dBm
P_{sat}	Internal saturation power of the amp		1.298 mW
α	Fiber attenuation		0.2 dB/km

Table 15.1 Important parameters and their values used in the amplifier-placement algorithms. (See [MaOn91] for some of the details.)

as part of a metropolitan-area network (MAN). We require that each transmitted signal/wavelength be received at all of the other receivers at a power level greater than a station's receiver sensitivity level, denoted by p_{sen}.[2] However, apart from the splitting loss due to the stars mentioned above, there is signal attenuation on the fibers given by the parameter α dB/km. Even though attenuation losses for fiber are relatively low (approximately 0.2-dB/km loss[3]) compared to other transmission media, larger networks (MANs) and networks with numerous splitting/coupling losses will require amplification to allow a transmitted signal to reach the receivers at a detectable level. The constraints on the system are shown in Table 15.1, along with typical values for each parameter. $P_{NONLIN,max}$ defines the power level, in a fiber, above which a signal encounters significant nonlinear effects. However, the total power at any point in the network is usually bounded by a lower value P_{max}, which is the maximum output power of an amplifier and a transmitter. P_{sat} is the internal saturation power of the optical amplifier. G_{max} is the maximum small-signal gain of the optical amplifier. These parameter values (last column of Table 15.1) will be used in our illustrative numerical examples in Section 15.3. We do not consider other system factors that might be relevant in determining the actual system performance, such as amplifier noise and crosstalk at the receivers.

[2] The signal-to-noise ratio (SNR) of the wavelengths is another important parameter (see [ChNT92]) that may also have an important role to play in amplifier placements.

[3] The 0.2-dB loss per kilometer of fiber is close to the absolute minimum due to the fundamental limits of Rayleigh scattering loss and infrared material loss.

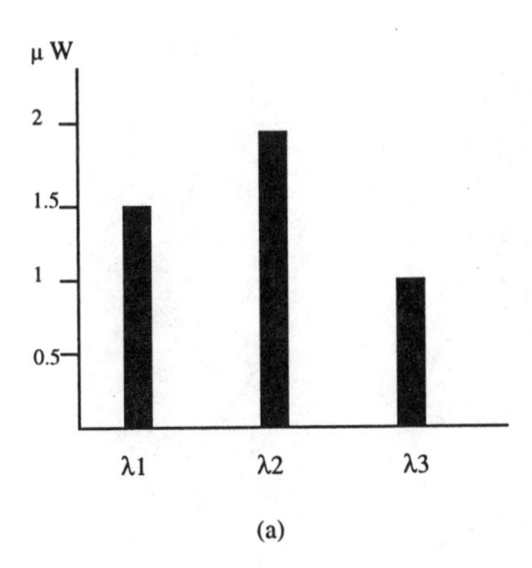

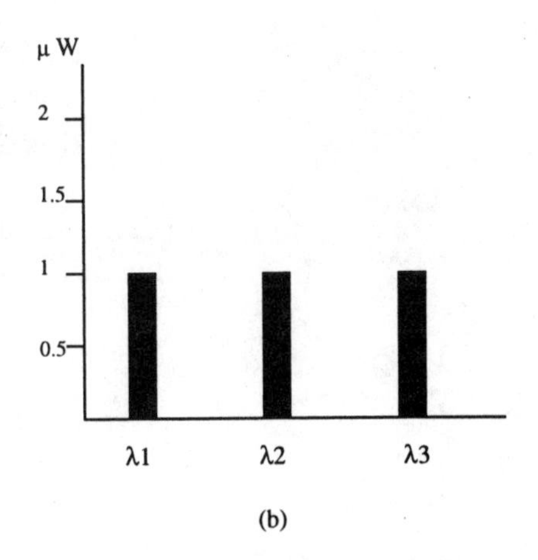

Figure 15.3 Two examples of powers on three wavelengths passing through a fiber.

15.1.2 Problem Definition

In the network setting described above, it is important to quantify the *minimum number of amplifiers* required to operate the network and to *determine their exact placements* in the network. In such a network, when signals on different wavelengths originating from different locations in the network arrive at an amplifier, their power levels could be very different. This phenomenon is known as the *near-far effect*, and it results in inefficient utilization of the individual amplifier. The difference in power levels of the input wavelengths can significantly limit the amount of amplification available since the higher-powered wavelengths could saturate the amplifier and limit the gain seen by the lower-powered wavelengths. Figure 15.3 shows, at some location on a fiber link, a case where three wavelengths have different power levels and a case where the three wavelengths have the same power level. In Fig. 15.3(a), the total power is 4.5 μW, and in Fig. 15.3(b), it is 3 μW. Since the per-wavelength amplifier sensitivity is 1 μW ($= -30$ dBm), in both cases an amplifier will be required before the signals suffer any more attenuation. However, since an amplifier has a limited total output power, the amount of achievable gain is greater when the total input power is less. This would allow the wavelengths in Fig. 15.3(b) to receive a higher gain than the wavelengths in Fig. 15.3(a). Also, allowing wavelengths in the same fiber to be at different power levels changes the minimal-amplifier-placement problem from a mixed-integer linear program (MILP) [RaIM96] into a mixed-integer nonlinear program, as we shall show later in this chapter.

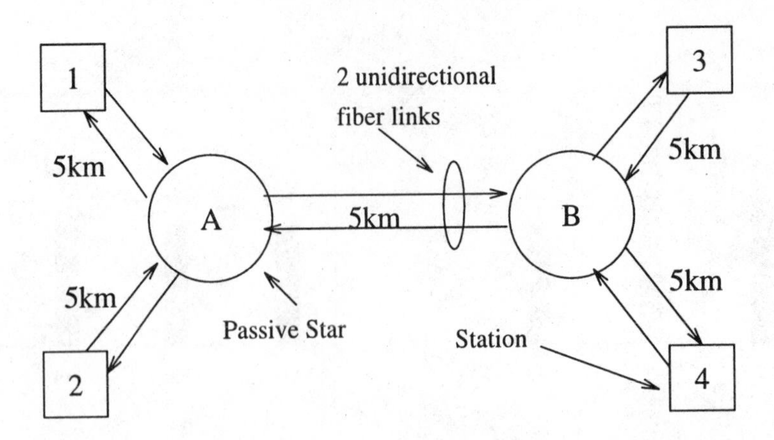

Figure 15.4 Simple two-star network that needs no amplifiers to operate.

Previous optical amplifier-placement schemes [LTGC94, RaIM96] bypassed these problems by restricting all of the wavelengths at any given point in a fiber to be at the same power level. Unfortunately, requiring wavelengths to be at the same power level often forces the designer to add more amplifiers than the minimum necessary in order for the receivers to receive signals at or above the receiver sensitivity level. Since each optical amplifier costs around $25,000, every attempt should be made to minimize their number in the network. It is also desirable to reduce the number of amplifiers used in the network based on noise, maintenance, and fault-tolerance considerations.

Our study was motivated by the network in Fig. 15.4. For reasonable network parameters, this network can operate without using any amplifiers. However, if the power levels for all wavelengths must be equal on any given link, as required by the MILP approach in [RaIM96], then an amplifier (on one of the links between stars A and B) will have to be added to the network. This is because, if we fix the output power of star A to be some value x, then the signals from stations 3 and 4 must reach star B with an output power higher than x. Without an amplifier, signals from stations 1 and 2 reach star B at a power less than x, which means that wavelengths on the link from star B to station 3 (and similarly on the link from star B to station 4) will have unequal powers. Therefore, requiring equal power on all wavelengths adds an unnecessary amplifier to this network. As we shall soon see, allowing wavelengths to be at unequal powers eliminates the need for any amplifiers in this network.

In this chapter, we examine a scheme that minimizes the number of amplifiers for the network setting in Fig. 15.1 without the restriction that wavelengths in the same fiber be at the same power level. The method works as follows: (1) determine whether or not it is possible to design the network taking into consideration the limitations of the devices (e.g., the power budget of the amplifiers), (2) generate a set of constraints to closely describe the problem setting, which turns out to be a nonlinear program, (3) pass the set of constraints to a nonlinear solver, such as C code for Feasible Sequential Quadratic Programming (*CFSQP*) [PaTi93], in order to solve for the minimum number of amplifiers needed for the entire network, and (4) determine the exact placements of the optical amplifiers. Numerical examples will show that this network-wide optimization method without the equal-power constraint often results in solutions that require fewer amplifiers than the solutions in [LTGC94, RaIM96].

15.1.3 Amplifier Gain Model

Currently, we employ a simplified model for the gain of a generic optical amplifier. The simplifying assumptions are that the amplifier has a flat gain over the wavelengths being amplified and that the amplifier gain is homogeneously broadened.[4] A flat gain can be achieved through various techniques discussed in Chapter 2 such as (1) notch filters, (2) different pump laser powers, (3) Mach-Zehnder filters, and (4) demultiplexers and attenuators. However, assuming that optical amplifiers are homogeneous is an approximation. For each specific amplifier [erbium-doped fiber amplifier (EDFA), semiconductor optical amplifier (SOA), etc.], we need to develop a gain model depending on its degree of homogeneity in order to accurately solve the amplifier-placement problem.

Based on the above assumptions, the gain model for our amplifiers is given by (from [Sieg86])

$$\frac{P_{\text{in}}}{P_{\text{sat}}} = \frac{1}{G-1} \ln\left(\frac{G_0}{G}\right) \tag{15.1}$$

where P_{in} is the total input power (across all wavelengths) to the amplifier in mW, P_{sat} is the internal saturation power in mW, G is the actual gain achieved (in absolute scale, *not* dB), and G_0 is the small-signal gain (which is the gain achievable for small values of input power when the amplifier does

[4]By homogeneous broadening, we mean that a single high-powered wavelength, which saturates the amplifier, can bring down the gain available for all of the wavelengths uniformly.

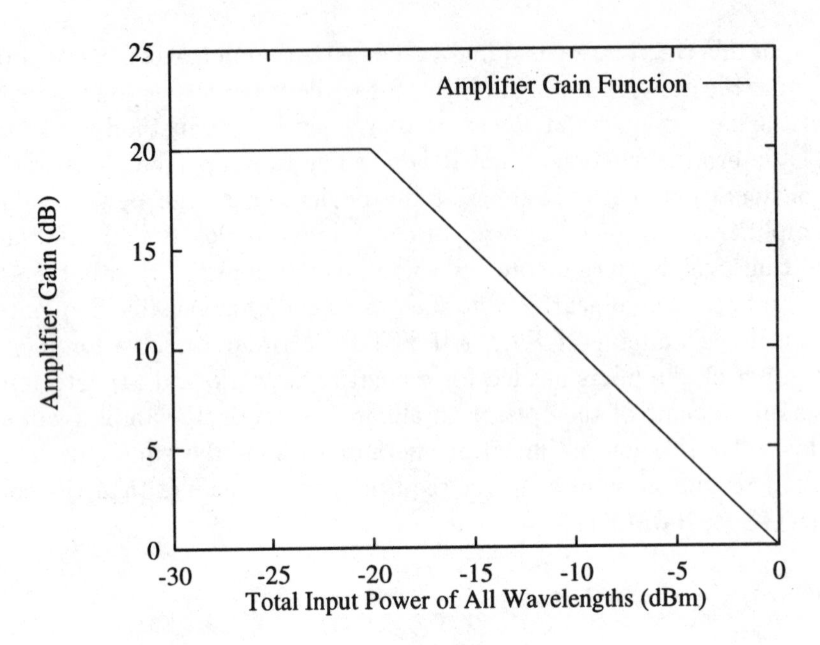

Figure 15.5 Original amplifier gain model approximations used in previous studies [LTGC94].

not saturate, again in absolute scale). Since the formula for G is not an explicit formulation, we use an iterative method to solve for the value of G. Our amplifier model has been designed into our solution as a generic gain module that can be easily replaced when a more accurate model for a specific amplifier is used.

Previous studies [LTGC94] used the gain model in Fig. 15.5. In this model, it is assumed that the full small-signal gain of the amplifier is realizable until the point at which the amplifier output becomes power-limited. At this point, the amplifier is assumed to enter saturation and the gain starts to drop. This "point" of saturation occurs in the example of Fig. 15.5 at a total input power of -20 dBm. At lower input powers, the amplifier is assumed to be able to supply the full small-signal gain of $G_{\max} = 20$ dB. The more-accurate model (Eqn. (15.1)), which is used in this study, is plotted in Fig. 15.6 and shows how saturation does not happen at a specific point but is really a continuous effect. In fact, note that, even for small input powers, the amplifier is not able to supply the full small-signal gain of $G_{\max} = 20$ dB. The numerical differences between the models are not huge, but are significant enough so that a network designer may have thought a design was feasible (based on the

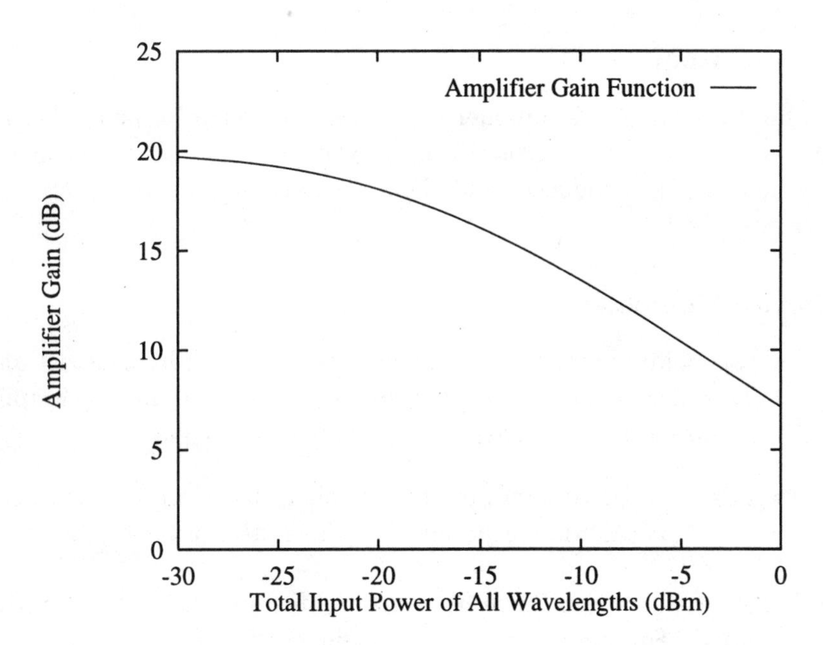

Figure 15.6 More-accurate amplifier gain model used in this study.

model in Fig. 15.5), when in fact it may not satisfy the design specifications (based on the more-accurate model in Fig. 15.6).

15.2 Solution Approach

Given a network as in Fig. 15.1, we would like to minimize the number of amplifiers used in the network without violating the device capabilities and constraints. Throughout this study, we assume that the stars are connected together in the form of a tree and that all neighbors have two unidirectional links connecting each other. A mathematical formulation of the problem is provided in Section 15.2.1. Unfortunately, the resulting mixed-integer nonlinear optimization problem is extremely difficult to solve. Hence, we carefully avoid the integral constraints by modifying the formulation, specifically the objective function, and solve the resulting nonlinear optimization problem. The description of the solution strategy is provided in Section 15.2.2. The output from the solver is fed to an Amplifier-Placement Module which outputs the exact positions and gains of the amplifiers. The functionality of the Amplifier-Placement Module is described in Section 15.2.3.

15.2.1 Formulation

In this subsection, the amplifier-placement problem is formulated as an Integer Nonlinear Constrained Optimization Problem. First, the notation used in the formulation is introduced, and then the constraints and objective functions are described.

Device Parameters

- p_{sen} = Minimum power required on a wavelength for detection in dBm. This represents both the receiver sensitivity level and the amplifier sensitivity level, which have been assumed to be equal.

- P_{max} = Maximum power available from an amplifier in mW
  = Maximum power of a transmitter in mW

It is not necessary that the maximum amplifier output and transmitter powers be identical. For simplicity, we have assumed them to be equal.

- G_{max} = Maximum (small-signal) amplifier gain in dB

- α = Signal attenuation in dB/km

Problem Variables

This section introduces the variables used in the problem formulation. Note that, among the variables representing the power levels, those beginning with lowercase ($p_l^{\text{min,beg}}$, $p_{x,i}$, p_i^{xmit}) are measured in dBm and those with uppercase (P_l^{beg}, P_l^{min}) in mW. Also, the variables in lowercase represent the per-wavelength power levels, whereas the ones in uppercase represent the aggregate power over all the wavelengths on the respective link.

- N = number of access stations in the network
  = number of wavelengths in the network.

- M = number of stars in the network.

- L = number of links in the network = $2(N + M - 1)$.

Note that stars are identified by the indices $1, 2, \ldots, M$ and stations by the indices $M{+}1, M{+}2, \ldots, M{+}N$. As we shall soon see, this provides notational convenience when we refer to the source/destination of a link, irrespective of whether it is a station or a star. Also, the wavelengths in the network are identified by the indices $M+1, M+2, \ldots, M+N$ of the source stations. We associate the following parameters with each link l, $1 \leq l \leq L$:

- s_l = source of link l, $1 \leq s_l \leq (M+N)$.
- d_l = destination of link l, $1 \leq d_l \leq (M+N)$.
- Λ_l = set of powered wavelengths carried by link l.
- n_l = number of amplifiers on link l.
- L_l = length of link l in km.
- SG_l = actual total *Supplied Gain* on link l in dB.
- $p_l^{\text{min,beg}}$ = power level of the least-powered wavelength arriving at link l, in dBm.
- P_l^{beg} = total power at the beginning of link l, in mW.
- P_l^{min} = total power on link l when all signals are $\geq p_{\text{sen}}$ and at least one wavelength is equal to p_{sen}, in mW.
- $gmax_l$ = maximum gain available from an amplifier on link l, in dB.

Consider the star i, $1 \leq i \leq M$.

- D_i = in-degree of star i = out-degree of star i.
- $p_{x,i}$ = power of wavelength x at the output of star i, in dBm.

Consider the station i, $(M+1) \leq i \leq (M+N)$.

- p_i^{xmit} = transmitted power of wavelength i at station i, in dBm.

Useful Functions

The following functions allow conversion between the milliwatt (regular) and dBm (log) scales.

$$
\begin{aligned}
ToDB(\xi) &= 10\log_{10}(\xi) \\
ToMW(\xi) &= 10^{\xi/10}
\end{aligned}
$$

They are used to express the constraints conveniently in the appropriate scale.

Basic and Nonbasic Variables

Given a network, the values of the topology-specific variables N, M, L, s_l, d_l, Λ_l, L_l, and D_i are fixed, irrespective of the amplifier-placement algorithm chosen. The only basic variables used in the formulation are p_i^{xmit}, SG_l, and n_l. Note that the variables P_l^{beg}, $p_l^{\text{min,beg}}$, P_l^{min}, $p_{x,i}$, and $gmax_l$ are nonbasic variables and can be expressed in terms of the basic variables as follows:

For link l, whose source is a star, i.e., $1 \leq s_l \leq M$, we have

$$p_l^{\text{min,beg}} = \underset{x \in \Lambda_l}{\text{Min}} \; p_{x,s_l} \tag{15.2}$$

and we also have

$$P_l^{\text{beg}} = \sum_{x \in \Lambda_l} ToMW(p_{x,s_l}) \tag{15.3}$$

For link l, whose source is a station, i.e., $(M+1) \leq s_l \leq (M+N)$, we have

$$p_l^{\text{min,beg}} = p_{s_l}^{\text{xmit}} \tag{15.4}$$

and we also have

$$P_l^{\text{beg}} = ToMW(p_{s_l}^{\text{xmit}}) \tag{15.5}$$

For any link l, the total power drops to its minimum level when at least one of the wavelengths is equal to the sensitivity level (p_{sen}). Hence, on link l, starting with an aggregate power level P_l^{beg}, when the weakest signal is at a power level $p_l^{\text{min,beg}}$, after appropriate scale changes, we have

$$P_l^{\text{min}} = ToMW[ToDB(P_l^{\text{beg}}) - (p_l^{\text{min,beg}} - p_{\text{sen}})] \tag{15.6}$$

The equation above is best explained with an example. Consider a link containing three wavelengths λ_1, λ_2, and λ_3. Suppose the power levels on these wavelengths at the beginning of the link were $2\,\mu$W (-26.99 dBm), $3\,\mu$W (-25.23 dBm), and $5\,\mu$W (-23.01 dBm), respectively. Now, the weakest signal is on wavelength λ_1, and from Eqn. (15.2), we have $p_l^{\text{min,beg}} = -26.99$ dBm. Also, from Eqn. (15.3), we have $P_l^{\text{beg}} = 2\,\mu\text{W} + 3\,\mu\text{W} + 5\,\mu\text{W} = 10\,\mu\text{W}$. Now, with a link attenuation (α) of 0.2 dB/km, and a sensitivity level (p_{sen}) of -30 dBm, this group of wavelengths can travel ($p_l^{\text{min,beg}}$ –

$p_{\text{sen}})/\alpha = (-26.99 + 30)/0.2 = 15.05$ km before the power of wavelength λ_1 drops below p_{sen}. At this point, the powers on the three wavelengths are $1\,\mu$W (-30 dBm), $1.5\,\mu$W (-28.24 dBm), and $2.5\,\mu$W (-26.02 dBm), respectively. Hence, the aggregate "minimum" power ($P_l^{\min}$) is $1 + 1.5 + 2.5\,\mu\text{W} = 5\,\mu$W. This value can be derived from the Eqn. 15.6 since

$$ToMW[ToDB(10\,\mu\text{W}) - (-26.99 + 30)] = ToMW(-20 - 3.01) = 5\,\mu\text{W}$$

For links from stations to stars, i.e., $(M+1) \leq s_l \leq (M+N)$ and $1 \leq d_l \leq M$, we have

$$p_{s_l,d_l} = p_{s_l}^{\text{xmit}} + SG_l - \alpha L_l - ToDB(D_{d_l} - 1) \quad (15.7)$$

For links between stars, i.e., $1 \leq s_l, d_l \leq M$, we have

$$\forall_{x \in \Lambda_l} \; p_{x,d_l} = p_{x,s_l} + SG_l - \alpha L_l - ToDB(D_{d_l} - 1) \quad (15.8)$$

For any link l,

$$gmax_l = G(P_l^{\min}, G_{\max}, P_{\text{sat}}) \quad (15.9)$$

We note that various amplifier gain models can be used to obtain this function G.

Constraints

Inequalities. Consider the link l, $1 \leq l \leq L$. The powers on each of the wavelengths at the beginning of the link l should be at least the sensitivity level, p_{sen}. This can be ensured by requiring that the *weakest* signal has a power level of at least p_{sen} as follows.

$$p_l^{\text{min,beg}} \geq p_{\text{sen}} \quad (15.10)$$

The powers on each of the wavelengths at the end of each link l should be at least p_{sen}. This is to enable the receivers to detect the signals correctly. Thus,

$$p_l^{\text{min,beg}} + SG_l - \alpha L_l \geq p_{\text{sen}} \quad (15.11)$$

The above inequalities (Eqns. (15.10) and (15.11)) ensure that the signal powers remain at or above p_{sen} everywhere along the fiber links and throughout the network.

There are upper limits on the maximum power carried by all the signals in a link. This value P_{max} is the same for transmitters and amplifiers, and hence at the beginning of link l, we have

$$P_l^{\text{beg}} \leq P_{\text{max}} \tag{15.12}$$

Similarly, at the end of the link l, we have

$$ToDB(P_l^{\text{beg}}) + SG_l - \alpha L_l \leq ToDB(P_{\text{max}}) \tag{15.13}$$

Since we need to divide the total supplied gain SG_l among the n_l amplifiers on link l, we have

$$SG_l \leq gmax_l \, n_l \tag{15.14}$$

However, the gain SG_l should require no fewer than n_l amplifiers; thus,

$$SG_l > gmax_l \, (n_l - 1) \tag{15.15}$$

Integrality Constraints. Consider the link l, $1 \leq l \leq L$. The number of amplifiers, n_l, on any link l, is an integral value. Hence, we require that

$$n_l \quad \text{is an integer.} \tag{15.16}$$

Objective function

$$\textbf{Minimize}: \quad \sum_{l=1}^{L} n_l \tag{15.17}$$

Complexity

The only basic variables used in the formulation are p_i^{xmit}, SG_l, and n_l. The others can be computed either beforehand from the topology or at run-time as a function of the basic variables. Hence, we have

- number of variables $= 2L + N$,
- number of integer constraints $= L$, and
- number of nonlinear inequalities $= 6L$.

Reasons for Nonlinearities

The approach presented in our current study differs from the one in [RaIM96] in that it allows the different wavelengths on a link to be at different power levels. Whereas the method in [RaIM96] needed to place amplifiers whenever all the wavelengths on the link were at their lowest power level, now the placement of the amplifier is constrained by the *weakest* signal on the link. Hence, on each link, we need to identify the wavelength coming in with the lowest power level ($p_l^{\text{min,beg}}$). This introduces a nonlinear term in the formulation (Eqn. (15.2)). Moreover, the maximum gain ($gmax_l$) available at an amplifier on a link is dependent on the precise mix of the power levels on its incoming wavelengths. This computation cannot be performed off-line and results in nonlinear constraints (see Eqns. (15.14) and (15.15)).

15.2.2 Solver Strategies

The mixed-integer nonlinear optimization problem resulting from Section 15.2.1 is an extremely difficult one to solve and is highly computation-intensive. In order to reduce the computation complexity, it is possible to eliminate the integral constraints altogether. This can be done by removing the variables n_l from the formulation, and hence the constraints in Eqns. (15.14) and (15.15) disappear. We define a *new objective function*:

$$\textbf{Minimize}: \quad \sum_{l=1}^{L} SG_l/gmax_l \tag{15.18}$$

which is close to the original one, since $n_l = \lceil SG_l/gmax_l \rceil$. The *starting point* of the problem space is especially important for this nonlinear search. We initialize the basic variables of the problem, namely, SG_l and p_i^{xmit} such that

$$\begin{aligned} SG_l &= 0 \\ p_i^{\text{xmit}} &= ToDB(P_{\text{max}}) \end{aligned}$$

i.e., the network is initialized to a state when all the transmitters are operating at their highest powers and all of the links have zero gain. However, we could also use the solution from [RaIM96] as a feasible starting point. Since the new objective function is not identical to the one in the integral case, the solver might end up minimizing the exact function $SG_l/gmax_l$ and not the number of amplifiers in the network. To handle this situation, we adopt a *nonintrusive measurement* approach, where, at every feasible point along the search path to the optimum solution taken by the nonlinear program solver, we evaluate the *exact* objective function and remember the point in the search space which resulted in the minimum value for the *exact* objective function thus far.

The ensuing heuristic search has the following interesting properties:

1. It contains significantly fewer variables and constraints. In fact, it has only
   - $L + N$ variables,
   - $4L$ inequalities, and
   - *zero* integer constraints.
2. All the constraints and the objective function are easily differentiable. Hence, the gradients can be fed to the nonlinear program solver to aid it in its search for the optimum solution.

The nonlinear program solver, *CFSQP*, which was used for this study achieves the minimization of the smooth objective function subject to general smooth constraints through the generation of feasible iterates. If the starting point is infeasible, it generates a point satisfying the constraints by solving a strictly convex quadratic program (QP). It then uses a nonmonotone line search [GrLL86] forcing a decrease of the objective function within at most three iterations. There are, however, limitations to this approach and they are discussed below:

1. *Local minima:* The nonlinear program solver might terminate at a point corresponding to a local minimum for the objective function. This happens, for example, when the starting point corresponds to the Linear Program solution (see Table 15.2 and the examples in Figs. 15.1 and 15.9).
2. *Feasible point generation:* When the starting point is infeasible, subject to the constraints, the solver may not be able to locate a feasible point in the problem space. With *CFSQP*, this problem can be fixed by using a different quadratic programming solver to generate the feasible point. However, finding a feasible point becomes increasingly difficult as the number of network elements grow (i.e., more network elements means more variables).

3. *Integer variables:* The nonlinear program solver (*CFSQP*), which we used in this study, is not well-suited to handle integer variables. Hence, its results for this problem could be improved upon by using specialized mixed-integer nonlinear program solvers.

The output of the nonlinear program solver is fed to the Amplifier-Placement Module which is described next.

15.2.3 Amplifier-Placement Module

This module uses the values of SG_l and p_i^{xmit} output by the nonlinear program solver to determine the exact location and gain of the amplifiers in the network. It operates on a link-by-link basis as follows. It computes the maximum value of the gain available from each amplifier on a link l ($gmax_l$) using Eqn. (15.9) and, hence, the number of amplifiers (n_l) required on that link. It also computes the power levels of the different wavelengths at the output of the stars ($p_{x,i}$). Then, it follows the *As Soon As Possible (ASAP)* method for the amplifier placement, which operates as follows. For all but the last amplifier on a link, this method places an amplifier on a link as soon as the input power is low enough to allow the *maximum* gain; and for the last amplifier on a link, it places the amplifier as soon as the input power is low enough to allow the *remaining* gain. Several other methods of splitting the gain (SG_l) along the link l, including uniform distribution among the n_l amplifiers, are possible. The ASAP method was chosen to maintain the power levels of the signals as high as possible. Further discussions on various approaches to gain splitting can be found in [Lin90].

15.3 Illustrative Examples

The link-by-link method in [LTGC94] was designed to equalize the powers of the wavelengths in the network, as opposed to trying to minimize the number of amplifiers in the network. By forcing the powers of all wavelengths to be equal to p_{sen} at the beginning of most links, the algorithm placed amplifiers simply by knowing how many wavelengths were on a link. If the number of wavelengths on a link was precomputed, this allowed the algorithm to operate on each link individually (locally) without knowing what was happening on other links. This led to a very simple amplifier-placement algorithm. Unfortunately, as was shown in [RaIM96] and can also be seen in Table 15.2, this approach does not minimize the number of amplifiers needed in the network.

Network	*Link-by-link method [LTGC94]*	*Equally-powered wavelengths (LP) [RaIM96]*	*Unequally-powered wavelengths (NLP) (this chapter)*
Lower bound	$N+2(M-1)$	$M-1$	0
Simple 2 star (Fig. 15.4)	6	1	0
Tree (Fig. 15.7)	44	14	0
MAN (Fig. 15.8)	38	6	4
Scaled-up MAN (Fig. 15.9)	48	16	16*
Scaled-down MAN (Fig. 15.10)	38	4	0
Previous study (Fig. 15.1)	79	77	77*

Table 15.2 Number of amplifiers needed for the various amplifier-placement schemes. (Note that $N =$ number of stations and $M =$ number of stars for the lower bound computation. A "*" in column 4 indicates that the NLP solver could not perform better than the LP solution, even when it was given multiple feasible starting points, including the solutions found in [LTGC94] and [RaIM96].)

The transmitter powers can be adjusted to avoid placing amplifiers on the links which originate at a station. However, since signals on all other links start off with the minimum power (p_{sen} on each wavelength), we know that the algorithm will place an amplifier on every single link not originating at a station in the network. Note that there are $L-N$ such links in the network which originate at a star (recall that $L =$ number of links, $N =$ number of stations, and $M =$ number of stars); thus we obtain the lower bound of $L-N=2(N+M-1)-N=N+2(M-1)$ on the number of amplifiers used by the method in [LTGC94]. This algorithm performs the poorest, in comparison to other placement schemes, on networks that have short links because the other algorithms can usually avoid placing an amplifier on a short link simply by exiting the originating star with enough power to traverse the short link. We show the results of this algorithm for various networks in column 2 of Table 15.2.

The global method in [RaIM96] allowed wavelengths at the beginning of the links to be above the absolute minimum allowed, p_{sen}. However, the powers on all of the wavelengths at any given point in the network were required to be equal; this equally-powered-wavelengths constraint enabled the computation of the maximum gain ($gmax_l$) available on a link, by knowing just the number of wavelengths on the link. The amplifier-placement problem can be formulated as a mixed-integer linear program and solved exactly.

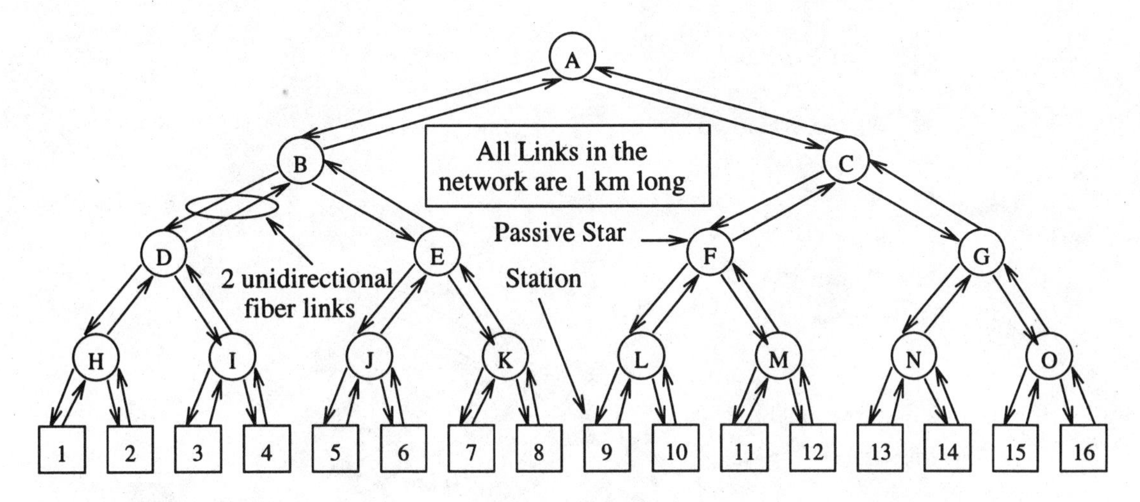

Figure 15.7 Mid-sized tree-based network needing no amplifiers to function.

Consider a pair of adjacent stars in the network. Taking into account the attenuation loss along the links connecting the stars and the splitting losses at the stars, we require that there be at least one amplifier on either of these links. The lower bound on the number of amplifiers required using the Linear-Program (LP) method in [RaIM96] is thus $M - 1$, where M is the number of stars in the network. (See [RaIM96] for details.)

The method described in this chapter (see Section 15.2) is a global one too; however, unlike the LP method in [RaIM96], it allows the wavelengths at any point in the network to operate at unequal powers. The solution obtained to the amplifier-placement problem is not guaranteed to be the optimum because of the presence of local minima. Moreover, the only available lower bound on the number of amplifiers required by this Nonlinear Program (NLP) method is the trivial one (i.e., not needing any amplifier). Next, we compare the results of these three approaches to amplifier placement on certain sample networks (see Table 15.2).

As mentioned earlier, the network in Fig. 15.4 motivated this study. While both the earlier approaches (the link-by-link method and the LP method) required a few amplifiers to operate the network, the NLP method described in this chapter does not require any.

The network in Fig. 15.7 is the motivating network, described above, taken to the extreme. This network has many stars and yet it needs no amplifiers to function. Table 15.2 reveals that the new method was indeed able to come up with the solution of not needing any amplifiers. This is the type of network

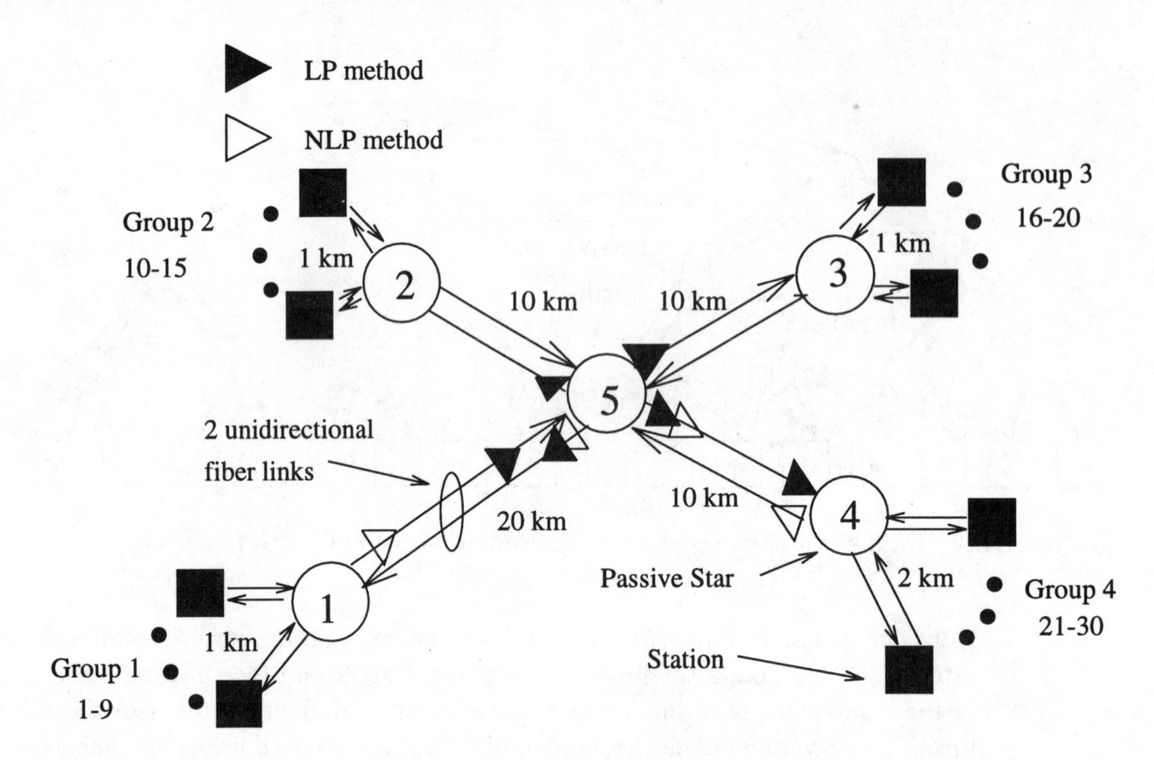

Figure 15.8 A possible MAN network.

where the unequally-powered-wavelengths solution is clearly superior to the previous two amplifier-placement methods. Although it is arguable whether this network is realistic or not, we have presented it here in order to give the reader some insight as to the conditions in which the new method performs best.

The network in Fig. 15.8 is meant to be a realistic design of a MAN. This network was designed in a semi-random fashion with some heuristics to guide the design. Table 15.2 shows that the new method was able to find a solution which required fewer amplifiers than the methods in [LTGC94] and [RaIM96]. Figure 15.8 also provides an insight into how the actual placements of amplifiers differ between the equally-powered-wavelengths method (LP) and the unequally-powered-wavelengths method (NLP). The amplifiers that are filled black are the locations at which the equally-powered-wavelengths method placed the six amplifiers it deemed necessary to operate. The empty, or filled white, amplifiers are the locations where the unequally-powered-wavelengths method placed the four amplifiers it deemed necessary. The numerical in-

Link (Star→Star)	*LP (gain and distance from start of link)*	*NLP (gain and distance from start of link)*
1→5	Gain 5.41 dB at 17.90 km	Gain 0.46 dB at 1.31 km
5→1	Gain 16.90 dB at 1.09 km	Gain 0.45 dB at 0 km
2→5		
5→2	Gain 16.55 dB at 1.09 km	
3→5		
5→3	Gain 15.76 dB at 1.09 km	
4→5		Gain 0.46 dB at 0.30 km
5→4	Gain 17.03 dB at 1.09 km Gain 1.74 dB at 10.0 km	Gain 0.45 dB at 0 km

Table 15.3 Exact amplifier placements for the network depicted in Fig. 15.8.

formation on exact gains and exact placements of the amplifiers can be seen in Table 15.3. The power levels of the signals at the transmitters and receivers can be found in Table 15.4. Note that the equally-powered-wavelengths constraint results in more amplifiers and a higher overall gain in the network. Note also that the transmitters are unable to operate at their maximum power for the same reason. However, when wavelengths are allowed to operate at different power levels, the NLP solution requires just the minimum overall gain to operate the network. As mentioned in Section 15.2.2, because of the presence of local minima introduced by the new objective function, the solver (i.e., *CFSQP*) is unable reduce the number of amplifiers further by combining the low gains at the amplifiers on adjacent links into a higher gain at a single amplifier.

This network serves as the reference point for a study into the effects of scaling network distances up and scaling network distances down, which will be discussed below.

As previously noted in Section 15.1, an amplifier becomes less efficient when multiple wavelengths passing through it are operated at different power levels. If a link were long enough, we would expect that this inefficiency would start to require the addition of more amplifiers. On the other hand, we would expect that, if links were short, then wavelengths at different power levels might not require the addition of more amplifiers and might allow us to potentially save even more amplifiers at critical points in the network. The network in Fig. 15.9 is meant to study the effects on the solution when we have links that span longer distances, and the network in Fig. 15.10 is

Stations	*LP method*		*NLP method*	
	Transmitter power	*Receiver power*	*Transmitter power*	*Receiver power*
1–9 (Group 1)	−16.68 dBm	−26.62 dBm	0.00 dBm	From G1: −9.94 dBm
				From G2: −28.04 dBm
				From G3: −27.25 dBm
				From G4: −30.00 dBm
10–15 (Group 2)	−15.03 dBm	−23.21 dBm	0.00 dBm	From G1: −28.03 dBm
				From G2: −8.18 dBm
				From G3: −23.94 dBm
				From G4: −26.70 dBm
16–20 (Group 3)	−15.82 dBm	−23.21 dBm	0.00 dBm	From G1: −27.24 dBm
				From G2: −23.94 dBm
				From G3: −7.39 dBm
				From G4: −25.90 dBm
21–30 (Group 4)	−12.61 dBm	−23.41 dBm	0.00 dBm	From G1: −30.00 dBm
				From G2: −26.70 dBm
				From G3: −25.91 dBm
				From G4: −10.80 dBm

Table 15.4 Transmitter and receiver powers for the network depicted in Fig. 15.8.

meant to study the effects on the solution when a network has shorter links. Both of these networks are the same as the network in Fig. 15.8 except that the distances have been scaled up and down, respectively, by a factor of 10. As we observe in Table 15.2, the results seem to verify our earlier predictions. The new method is not able to find a better solution than the equally-powered-wavelengths solution for the larger network in Fig. 15.9, even when it was given multiple feasible starting points (including the solutions found in [LTGC94] and [RaIM96]). We cannot be certain that a better solution does not exist, but our new method was not able to find one. Our method's solution is not guaranteed to be the best because it could have become stuck at a local minimum. If our new method is stuck at a local minimum, we potentially can miss the global minimum solution. This differs from the LP solution which does find the global minimum solution (subject to the equally-powered-wavelengths constraint). On the other hand, the new NLP method is able to come up with a better solution for the smaller network (Fig. 15.10). In fact, as we predicted, our new method was able to take advantage of the

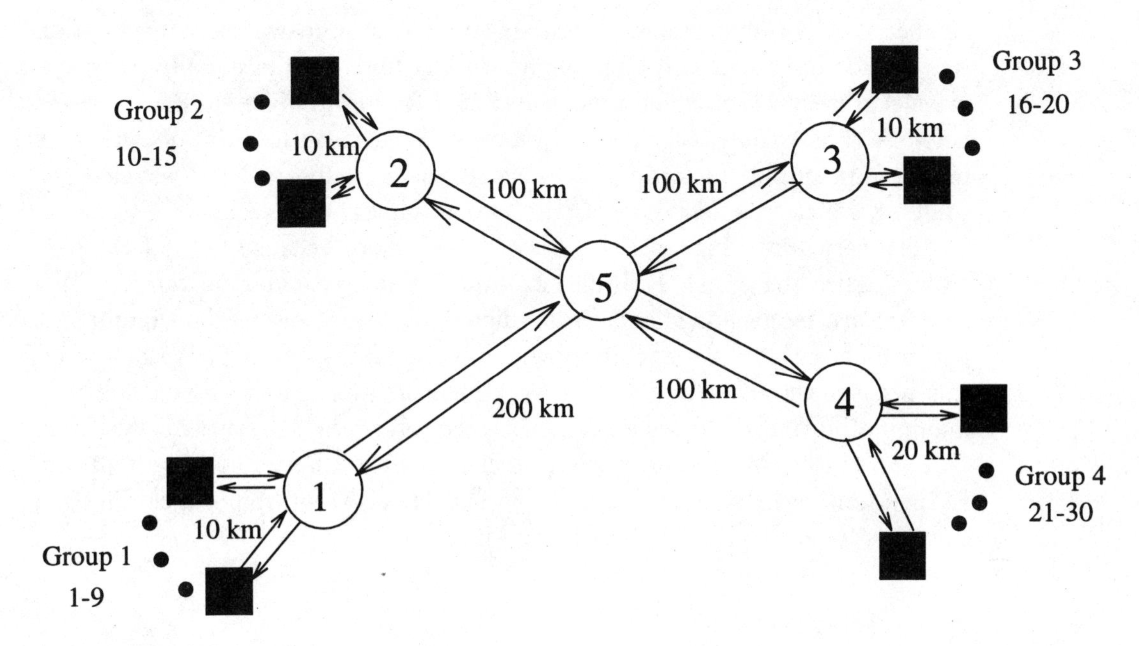

Figure 15.9 A scaled-up version of the MAN network in Fig. 15.8.

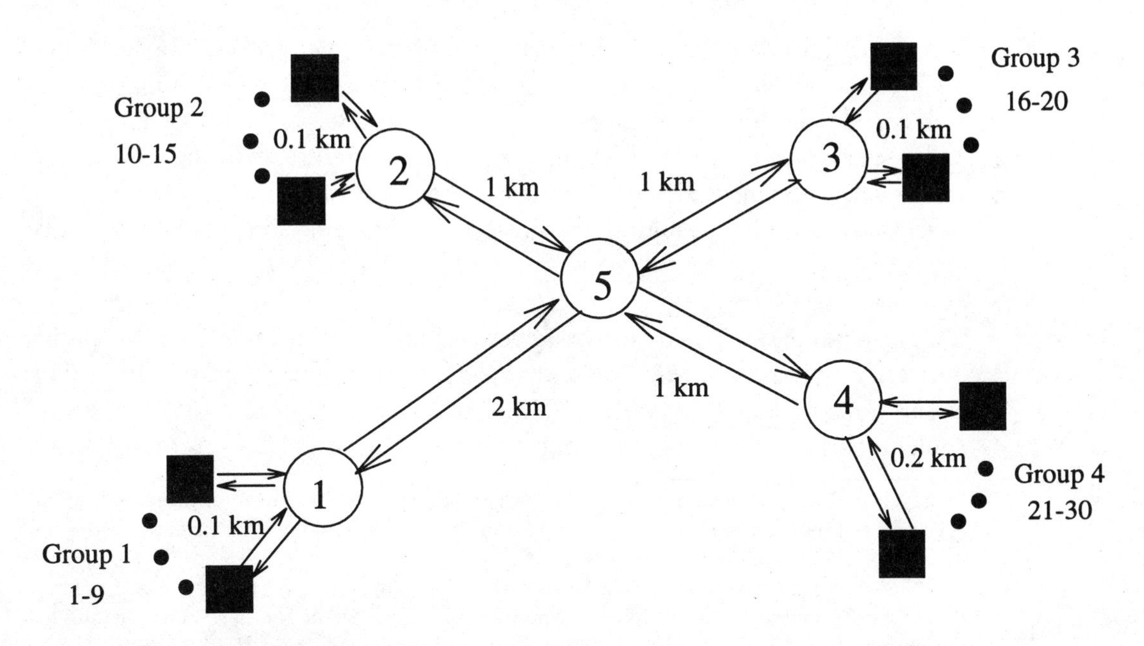

Figure 15.10 A scaled-down version of the MAN network in Fig. 15.8.

smaller network environment. The unequally-powered-wavelengths solution was able to use zero amplifiers compared to four for the equally-powered-wavelengths solution, which was a savings of four amplifiers. In the reference network (Fig. 15.8), the unequally-powered-wavelengths solution was able to use four amplifiers compared to six for the equally-powered-wavelengths solution, which was a savings of only two amplifiers.

The network in Fig. 15.1 is also examined here because both of the previous studies [LTGC94, RaIM96] examined this particular network.[5] This network has many nodes and we predicted that our new method might not perform better than the equally-powered-wavelengths solution. We predicted this because the more nodes a network has, the more variables the solver is manipulating and the more local minima the solver can get stuck at. As Table 15.2 shows, the solver was unable to come up with a better solution than the LP solution, even when given multiple feasible starting points including the solutions found in [LTGC94] and [RaIM96].

15.4 Open Problems

15.4.1 Switched Networks

The algorithms described in this chapter were designed to operate on "loopless" networks where there is only one path from a source to a destination. In a switched network, there can potentially be multiple paths between a source and a destination. Since the above algorithms operate knowing how many wavelengths are on a given link, they assume that all wavelengths that can possibly reach a link could all be present on that link simultaneously. This approach has the potential to place more amplifiers in the network than is absolutely necessary.

The example switched network given in Fig. 15.11 includes multiple paths between any source-destination pair. When examining the "permutation"[6] of connections that use the WR1-PS3 link, notice that all eight of the stations could each have a connection setup that use this half of L1. Actually, there

[5] The number of nodes for group 3 was reduced from 35 to 28 nodes because the original network, as proposed in [LTGC94], was infeasible because signals exited the star of degree 35 with power below $p_{sen} = -30$ dBm.

[6] "Permutation" refers to the source-to-destination pairings when written as a permutation with the source station implicitly known by its position in the list. As an example, the permutation "3 1 2" means station 1 (source) is connected to station 3 (destination), station 2 is connected to station 1, and station 3 is connected to station 2. A station not connected to anyone is indicated as connected to itself.

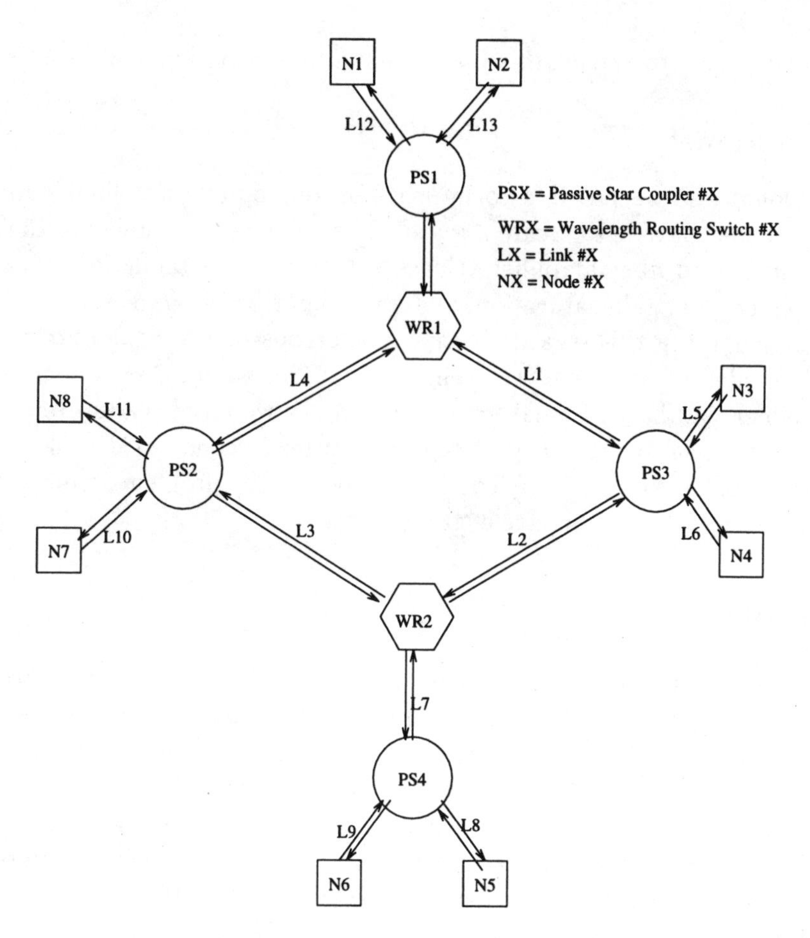

Figure 15.11 A sample switched network.

is always at least one permutation of connections that would cause any of the "halves" of link L1, L2, L3, and L4 to carry eight connections. Now, if amplifiers were placed in this network to allow any possible configuration of connections, all "halves" of links L1 through L4 would have to be designed with enough amplifiers to carry eight connections in the worst case. Now, it is fairly easy to see that, if the connections are set up in a "smart" fashion, a link never has to carry eight connections. In fact, a link should never have to carry more than two connections in this network. Designing links to carry two connections instead of eight, since the network will then potentially need only one-fourth the power on these links, can result in a significant savings in the number of amplifiers. It should be possible to extend our current algorithms to allow them to exploit this phenomenon that occurs in switched networks.

15.4.2 Gain Model

It should also be possible to improve on our optical amplifier gain model. One should be able to create a reasonably accurate gain model of the popular erbium-doped fiber amplifier (EDFA). Analytical methods for modeling the amplifier gain, gain saturation, and noise described in [GiDe91] are expected to be helpful in this regard. It may also be possible to expand our amplifier gain model to handle per-wavelength gain. This would allow one to model an amplifier that has a nonflat gain spectrum. It would also allow one to model the small gain for wavelengths that are normally considered to lie outside of the "amplifier bandwidth." The formulation of the problem would have to be changed to handle per-wavelength gain too.

15.5 Summary

This chapter considered the problem of minimizing the number of optical amplifiers in an optical LAN/MAN. This study departed from previous studies by allowing the signal powers of different wavelengths on the same fiber to be at different levels. Although this increases the complexity of the amplifier-placement algorithm, numerical results show that certain networks do benefit from this method by requiring fewer amplifiers. Our results demonstrated that smaller networks (in terms of distance) benefited the most from this new method. Larger networks tended not to benefit as much because (1) using unequally powered wavelengths hurts the efficiency of the amplifiers too much if long links have to be traversed, and (2) larger networks have more local minima causing our solver to sometimes miss the global "optimal" solution.

Exercises

15.1. Consider the set of wavelengths shown in Fig. 15.3(a) and (b).

(a) Using the amplifier gain model in Fig. 15.5, compute the maximum per-wavelength gain available and the maximum per-wavelength output power achievable when the set of wavelengths are input to a single amplifier.

(b) Repeat the above using the gain model in Fig. 15.6.

15.2. Consider the cascade of amplifiers on a single link as shown in Fig. 15.12. Let the set of wavelengths shown in Fig. 15.3(a) and (b) be at the input to amplifier A. Throughout this problem, use the amplifier gain model in Fig. 15.5.

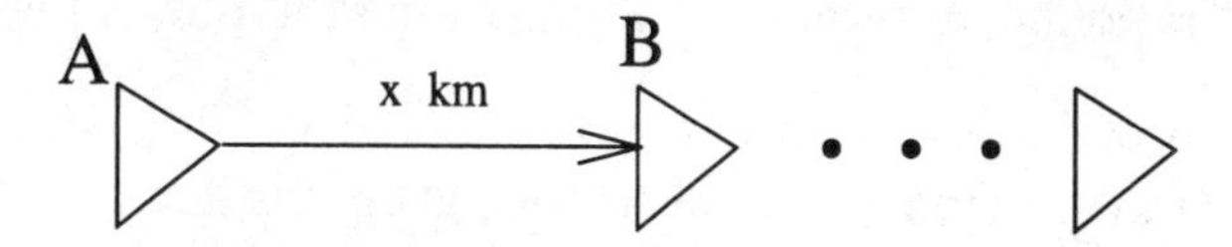

Figure 15.12 A cascade of amplifiers along a link.

(a) First, consider the case when there are just two amplifiers in the cascade: A and B. For each of the two sets of input wavelengths, compute the distance x which the set of wavelengths can traverse along the link before a second amplifier B needs to be placed. Assume that the amplifier A is operating at the maximum gain possible.

(b) Now, consider a cascade of ($n > 2$) amplifiers. Assume that the amplifiers are placed only when needed and are operated at their maximum possible gain. Compute the minimum span length (distance between the first and last amplifiers) for which one set of input wavelengths, from Fig. 15.3, at the first amplifier results in one additional amplifier over the other.

(c) Which set of input wavelengths results in better utilization of the amplifiers? Why?

15.3. Derive the lower bound of $(M - 1)$ amplifiers required to operate a network with M stars when the power on all of the wavelengths is the same (the equal-powered-wavelengths case).

15.4. Consider a point-to-point 440-km fiber optic link employing four optical amplifiers, each with a small-signal gain of 20 dB and an internal saturation power of 1.2 mW. Let the transmitted power be 0 dBm and fiber attenuation = 0.2 dB/km. Use the gain model in Fig. 15.6.

Consider the following amplifier placements along the link [for parts (a) and (b) below], and calculate the received power in both the cases.

(a) All the optical amplifiers are equispaced over the first 200 km from the transmitter end.

(b) Distance between the first amplifier and the transmitter is 200 km and the remaining three amplifiers are equispaced between the first amplifier and the receiver end.

(c) Comment on the feasibility of these two placements when the signal-to-noise ratio at the receiver will determine the power budget.

15.5. Consider the broadcast optical network depicted in Fig. 15.13. Each of the nodes transmits on a unique wavelength. Assume that (1) fiber loss = 0.2 dB/km, (2) star couplers are echoless (i.e., signals will not "bounce" between star couplers, (3) splitting loss at a size $X + 1$ star is X, and (4) gain is the same for all wavelengths passing through an amplifier.

(a) Determine the received power levels for all possible source-destination groups.

(b) Assuming that the receiver sensitivity is -30 dBm, does the above network ensure that signals between all pairs of source-destination groups are received at a power above the sensitivity level?

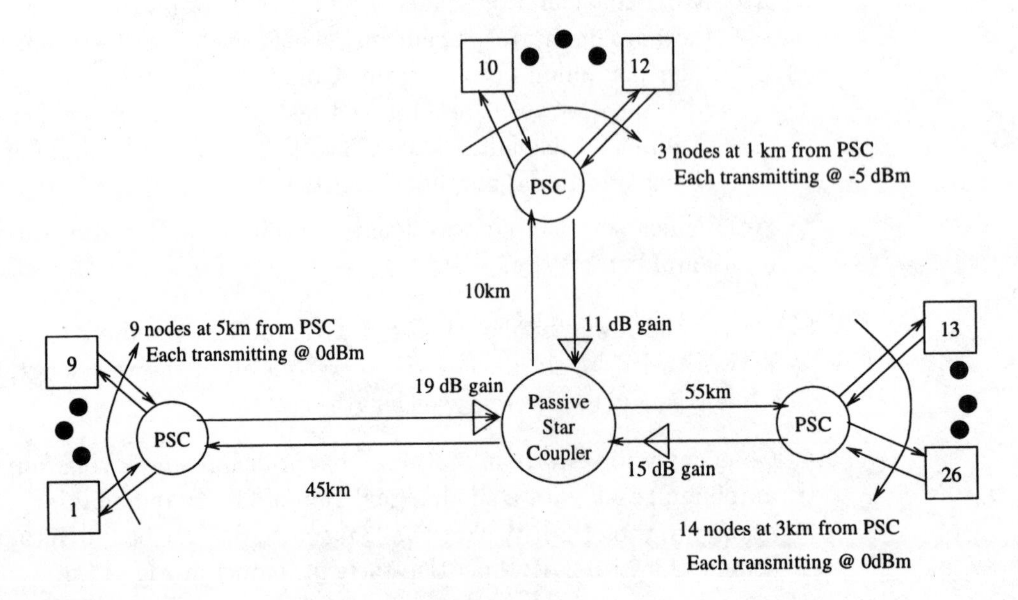

Figure 15.13 Network for Problems 15.5 and 15.6.

15.6. Often a good lower bound for the number of amplifiers a network requires is to assume that an amplifier placed on any link is able to supply the maximum gain possible. The maximum gain of an amplifier can be calculated by assuming that all wavelengths signals reach the amplifier at its sensitivity level and determining the maximum output power.

(a) For the network in Fig. 15.13, use the following formula to calculate the maximum gain that can be supplied on all of the internal links between passive star couplers (6 unidirectional links total):

$$P_{in}/1.298mW = 1/(1-G)\ln(100/G)$$

where G is the absolute gain of an amplifier (not dB), and P_{in} is the total input power to an amplifier in mW.

(b) Using the numbers calculated in part (a), determine the lower bound on the number of amplifiers needed in the network depicted in Fig. 15.13.

15.7. Consider the portion of a network shown in Fig. 15.14. The three stations A, B, and C are located at distances of 5 km, 10 km and 15 km respectively from star 1. Let l denote the link from star 1 to star 2 of length 200 km. Assume that the transmitters at the stations are operating at the maximum possible power levels.

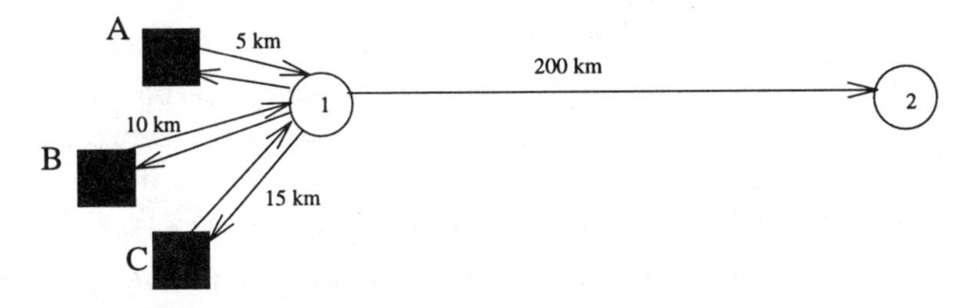

Figure 15.14 A portion of a network.

(a) What is the splitting loss suffered by any of the wavelengths incident on star 1 (i) in the dB scale and (ii) in the absolute scale?

(b) Calculate $p_l^{\min,\text{beg}}$, P_l^{beg}, and $P_l^{\min}$ in the appropriate scales.

15.8. Throughout this chapter, we focused on broadcast networks based on echoless passive stars. In this problem, we will examine a distribution network which carries unidirectional traffic from a set of transmitters to a set of receivers. Consider the network in Fig. 15.15. It consists of

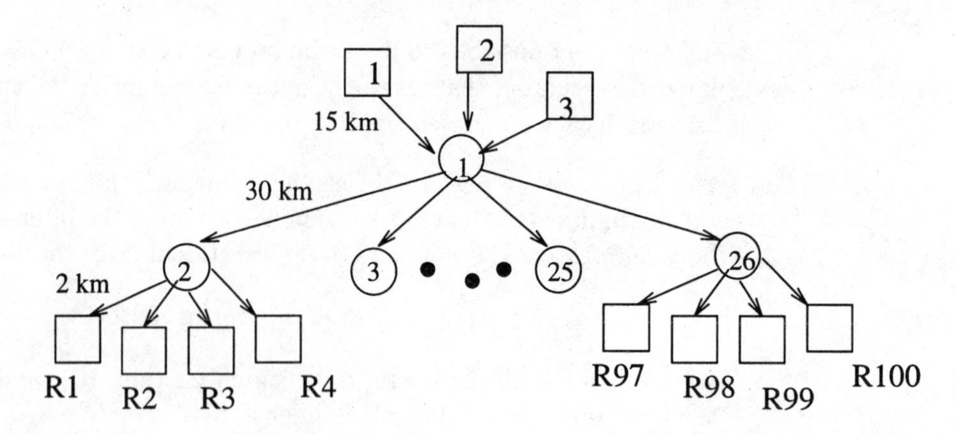

Figure 15.15 A distribution network.

3 transmitters, 100 receivers, and 26 passive stars, each with a fan-in of 1 and fan-out of 4. For this problem, use the amplifier gain model in Fig. 15.5.

(a) Calculate the number of amplifiers required to operate this network, given that all of the 3 transmitters are transmitting at their maximum power levels and that the amplifiers are placed in the network only when needed.

(b) Now, suppose you are allowed to reduce the power levels at the transmitters. Show that you can operate the network with fewer amplifiers than in the above case, without violating the constraint on equally-powered wavelengths.

(c) Suppose, further, that you can operate the 3 wavelengths at unequal power levels. Does this help you save additional amplifiers in the network? Explain.

Chapter 16

Optical TDM and CDM Networks

As we have already discussed, the recent years have witnessed an outstanding growth in the area of telecommunication and computer networking, leading to an acute need for very-high-bandwidth transmission systems. Optical fibers can indeed supply the required bandwidth, provided that one can tap this bandwidth through appropriate opto-electronic interfaces. Wavelength-division multiplexing (WDM) offers one possible technique to exploit the huge bandwidth of optical fibers through concurrent transmissions from the network nodes at nonoverlapping wavelengths. However, WDM networks with a large number of closely spaced wavelengths, propagating with high launched power, over long span of dispersion-shifted optical fibers, can lead to significant system degradation arising from fiber nonlinearities. Furthermore, in order to utilize the fiber bandwidth most efficiently by WDM, one needs to employ coherent optical transmission techniques along with heterodyne optical receivers, which are costlier than their noncoherent counterparts.

Alternative techniques to access the huge bandwidth have been and are being explored for high-speed networking applications. One such longer-term approach to ultra-fast transmission is time-division multiplexing (TDM) in optical networks leading to the optical time-division multiplexed (OTDM) networks [BCHK96, BGDB96, SeBP96]. While this chapter will review the basics of OTDM and some related efforts, the interested reader may wish to consult several OTDM papers in [JLT96, JSAC96] for detailed information.

Code-division multiplexing (CDM), also known as spread spectrum, is another technique that has been and is being explored as another alternative [Sale89, SaBr89]. The technologies for both of these techniques are less mature than WDM, but a discourse on optical networks would be incomplete without a discussion of these approaches. This chapter will briefly explore both optical TDM and CDM, and the interested reader may wish to consult the references for further details.

16.1 Optical TDM Networks

16.1.1 Basics of TDM

Time-division multiplexing (TDM) is a well-known technique that has been widely used in point-to-point digital telecommunication systems and networks, with high-speed TDM switches connecting such point-to-point links and providing time-slot switching between them. TDM's broadcast network version, known as time-division multiple access (TDMA), has also been extensively applied in satellite communication systems.

In a TDM-based system, a number of nodes, say N, share a common broadcast channel (i.e., the transmission medium) in round-robin fashion. Sharing is usually *equal*, i.e., when a node's turn comes up, it transmits for T units of time before relinquishing the channel. (However, *unequal* sharing can be effected by having node i transmit for k_iT time units, where $i = 1, 2, \ldots, N$, and the k_i are arbitrary integers that are known to all the nodes. Unless otherwise stated, we will assume equal sharing with all of the k_i equaling unity.)

Nodes in a TDM system must also be perfectly *synchronized*, so that nodal transmissions do not collide with one another (as in a random, multiaccess communication network), and they do not "spill over" into their adjacent nodes' transmission windows.

In a *synchronous TDM system*, a node gets its allotted time even if it is idle. Thus, time on such a synchronous TDM channel consists of a sequence of equal-sized, nonoverlapping *frames*, with each node getting exactly one turn to transmit in a frame. (Such a frame is also referred to as a *cycle*, *period*, or a *TDM frame* in the literature.) Also, note that two consecutive transmission instants at any node are separated by exactly one frame length.

To assist a receiver in a TDM system to maintain frame synchronization, the beginning of each frame is "marked" by a special bit pattern. Since all frames are of the same length, and since each node's relative transmission instant in a frame and its transmission duration are known to all nodes, a nodal transmitter need not identify itself explicitly – this information is implicit. For a point-to-point TDM link (as in Fig. 16.1), the TDM multiplexer provides this synchronization information, while in a distributed, broadcast, multiaccess TDMA network, the synchronization bits may be transmitted by any designated node, which is generally the one that is first to transmit data in a frame.

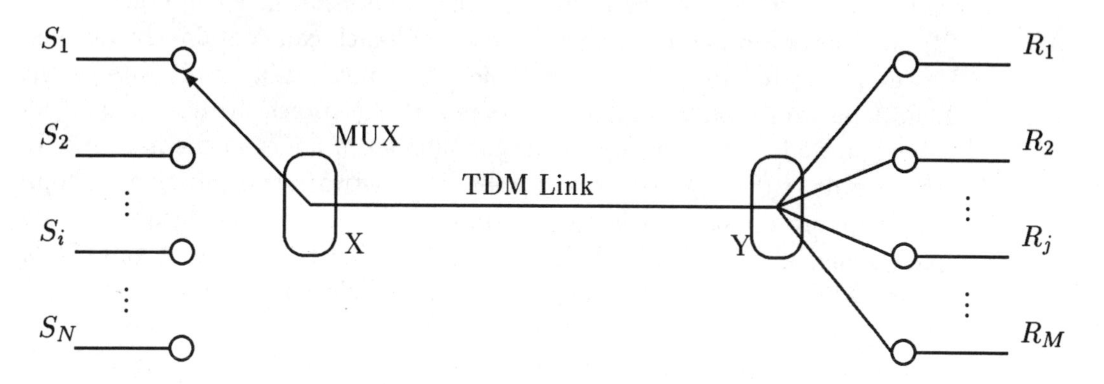

Figure 16.1 A TDM link and multiplexer.

An *asynchronous TDM system*, on the other hand, does not waste any time on idle nodes; thus, while all nodes are still "scanned" in round-robin fashion, only the active nodes transmit on the channel. Consequently, the frame lengths are not fixed any longer, and a nodal transmitter must also identify itself explicitly in its transmission, thereby increasing the amount of overhead information per frame.

Trade-offs to be considered when designing or comparing a synchronous to an asynchronous TDM system include simplicity of design or operation, channel utilization (and, hence, access delay), and amount of overhead.

Optical TDM systems, to be discussed below, are generally synchronous, especially for local applications, because their high data rate allows very little processing time.

Network architectures have been proposed that employ switches such that a TDM packet is routed all-optically from one of the incoming fiber ports of a switch to its desired output port. There is generally no notion of a synchronous TDM frame on any link, so according to our above classification, this is an *asynchronous TDM system.* Frequently, such a network is referred to as a *photonic packet network* in the literature.

16.1.2 Optical TDM

Consider the passive-star-based, broadcast-and-select local optical network in Fig. 1.7. Instead of WDM, nodes can use a synchronous TDM system for multiaccess communication, so that we now have a local optical TDMA network. The nodal transmissions can be implemented in one of two possible ways: (1) multiplexing the data from the sources using bit-interleaving or (2) multiplexing based on time slots, each of which can contain several bits. However, regardless of the above choice, the physical-layer bit rate of the TDMA network would need to be enhanced to N times the data rate of an individual node, where N represents the number of nodes in the network. In other words, if the user data rate is r bps, the network transmission medium must support a data transmission rate of Nr bps over the entire network span. Thus, in an optical TDMA network (hereafter referred to as an OTDM network), if the individual nodes have outbound electrical data stream at 10 Gbps with 10 such nodes connected to the network, then each node should be able to transmit at a rate of 100 Gbps on the optical fiber along with the global synchronization with the other nodes in the network. This is indeed a nontrivial task as opto-electronic interfaces would find it difficult to support the transmission rates of this order along with the requirement to preserve the optical pulse shapes against fiber dispersion and timing jitter.

Implementation of the functional units that would constitute an OTDM network require special considerations for the capability of handling the ultra-fast optical signals, which include the generation of narrow optical pulses with high repetition rates and their multiplexing as well as demultiplexing with appropriate time synchronization, transmission of the OTDM signal over the optical fibers, and processing of the en-route optical signals at the network nodes using optical buffers and other processing elements.

In the following subsections, we describe the salient features and operating principles of the basic subsystems in OTDM networks, viz.,

1. optical sources,

2. modulation and multiplexing,

3. transmission of high-speed optical signals, i.e., solitons,

4. demultiplexing and clock recovery,
5. optical processing, and
6. optical TDM network architectures and proposals.

16.1.3 Optical Sources

As mentioned before, for high-speed OTDM transmission, an optical pulse should be very narrow. For example, an OTDM network having 10 nodes with each generating data at 10 Gbps would require a composite optical transmission of 100 Gbps. This will, in turn, imply a bit interval of about 10 picoseconds (ps), and hence the optical pulses to be transmitted should have widths no more than 5 ps (i.e., 50% duty cycle). This represents a typical return-to-zero (RZ) optical transmission, which is a necessity to accommodate further pulse stretching due to fiber dispersion and timing jitter during the course of signal propagation along the fiber. The transmitted optical pulses in the OTDM network should also be transform-limited, which implies that the time-bandwidth product of the optical pulses should be limited to 0.44 [BGDB96]. For a given pulse width, the latter constraint would also put an upper limit on the spectral width of the optical signal, which, in turn, would significantly reduce the fiber dispersion and help in preserving the optical pulse shape during the course of propagation in the fiber.

Possible techniques for generation of such an optical pulse include gain-switched semiconductor lasers and mode-locked fiber lasers. Gain switching is a versatile technique which requires the driving of a conventional Distributed FeedBack (DFB) laser with a high-quality electrical sine wave, producing short pulses at Gbps repetition rates. It is therefore straightforward to synchronize this pulse source with a specific data rate as well. However, the optical pulse trains produced by gain switching are highly chirped (i.e., they have large spectral spread) and they may possess nonnegligible levels of timing jitter. These problems can be alleviated by limiting the spectral spread due to chirping by optical filtering followed by pulse compression using fibers with appropriate dispersion compensation.

The other alternative for generating a narrow optical pulse train is to use a mode-locked laser (MLL). The external-cavity semiconductor MLL is a development of the commercially available grating-tuned external-cavity semiconductor laser, which provides a narrow linewidth source. The major new requirement is for the provision of a high-quality RF bias connection to the laser chip, which allows the laser to be driven with a single frequency RF signal. The length of the laser cavity is adjusted so that the optical round-trip time is equal to the period (or harmonics) of the RF drive signal. The net laser output would consist of a train of short optical pulses at a well-defined repetition frequency. Synchronization to a particular data rate would require an adjustment to match the cavity length to the RF frequency.

16.1.4 Modulation and Multiplexing

As mentioned earlier, in OTDM networks, one needs to employ bit or slot interleaving depending on the network under consideration. For packet-switched networks, packets need to be interleaved optically, i.e., "packet compression" needs to be performed, leading to the composite OTDM signal for high-speed transmission. The generation of the OTDM signal can be viewed as a three-step process, as shown in Fig. 16.2, and as outlined below.

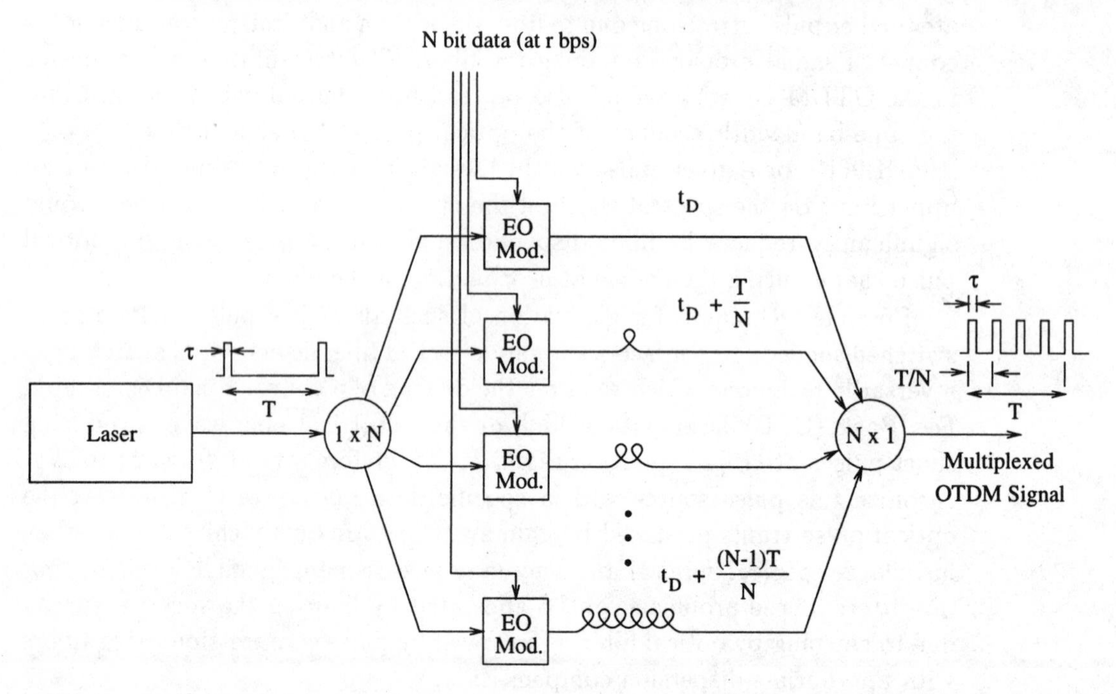

Figure 16.2 Generation of the OTDM signal: packet compression.

A laser generates a periodic stream of narrow optical pulses (transform-limited) at the electrical data rate (r bps). The optical pulse stream is then subjected to three steps of processing.

First, the pulse stream is split into N separate channels, where N represents the compression factor or the ratio by which the individual channel rate (r) gets enhanced through optical multiplexing. The width (τ) of each optical pulse is chosen such that $\tau < T/N$, where $T = 1/r$ represents the bit interval of the electrical data stream.

In the next step, optical pulses in all of the N channels are modulated by N data bits in parallel, which are derived from the original electrical data stream through serial-to-parallel conversion. The modulation is carried out in each channel by an external modulator which is a high-speed electro-optic on-off switch. The modulator output is passed through all of the N channels, each equipped with an optical delay line that provides a progressively increasing delay equal to $t_D + nT/N$, where $n = 0, 1, 2, \ldots, (N-1)$, and t_D represents a fixed-delay overhead in each channel.

Finally, the optical pulse outputs from the N channels are combined together (i.e., superimposed) leading to a passively multiplexed *compressed* OTDM signal at Nr bps. It is important to note that, since $\tau < T/N$, the compressed optical signal assumes a return-to-zero (RZ) format. For example, with $r = 10$ Gbps and $N = 10$, we get $T/N = 10$ ps, and hence optical pulses should have the width $\tau < 10$ ps for RZ signaling. This constraint would ensure that, even after compression, an optical pulse would not "encroach" into one its adjacent pulses. In practice, with the above set of system parameters, τ should be typically within 5 ps (i.e., RZ signal with 50% duty cycle), which would provide a clearance of 5 ps which will, in turn, help in absorbing the fiber dispersion to that extent. The requirements for the optical delay lines would therefore be t_D, t_D + 10 ps, t_D + 20 ps, ..., t_D + 90 ps, respectively for the 10 channels.

Thus, in general, with an electrical bit rate of r bps and N nodes, multiplexing in an OTDM node can be implemented by splitting the modulated optical pulse train into N streams and by subsequently combining them after subjecting the N streams to progressively increasing incremental fiber delays equal to $t_D, t_D + 1/Nr, t_D + 2/Nr, \ldots, t_D + (N-1)/Nr$.

16.1.5 Transmission of Ultrafast OTDM Signal Using Soliton

In a TDM system, one needs to increase the transmission rate of the lightwave communication systems. For this purpose, one needs to resort to a nonlinear phenomenon in optical fibers, called optical *soliton*, which enables optical transmission at very high rates over very long distances. Below, we first explain the basic principle of soliton transmission followed by the relevant issues while applying soliton transmission to an OTDM network.

The basic concept of soliton transmission is the compensation of the group velocity dispersion (hereafter, referred to as dispersion for the sake of brevity) by the action of a specific nonlinear property of the silica, namely, the Kerr effect. The optical fiber is inherently a dispersive medium which makes the various frequency components in the propagating optical pulses travel with different velocities, thereby resulting in *broadening* of the optical pulses . Under normal conditions, the higher-frequency components (i.e., shorter wavelength components) will travel slower than the lower-frequency components (viz., longer wavelength components). In silica, this situation is reversed at wavelengths longer than the zero-dispersion wavelength, λ_0. This phenomenon is known as anomalous dispersion, where the lower-frequency components of a pulse travel slower than the higher-frequency components. As a result, the pulse acquires negative chirp, where the instantaneous frequency decreases across the pulse from its leading edge to its trailing edge.

Interestingly, the pulse broadening due to (anomalous) dispersion in optical fibers can be compensated by utilizing the Kerr nonlinearity, leading to distortionless transmission of optical pulses in optical fibers. Due to the Kerr effect, an optical fiber exhibits a dependence of the refractive index of its constituent material (silica) on the light intensity as an optical pulse stream passes through it. The intense peak of an optical pulse experiences a higher refractive index than its tail-ends, resulting in a modulation of the phase across the pulse as it propagates through the fiber. This effect is known as self-phase modulation (SPM). This phase-shift, in turn, causes a positive chirp, resulting in an instantaneous increase in the frequency across the pulse from its leading edge to its trailing edge.

In the presence of both SPM and anomalous dispersion, the positive chirp acquired by the optical pulse due to SPM will cancel the negative chirp acquired by the same pulse due to anomalous dispersion. If the envelope of the initial optical pulse has the ideal hyperbolic secant shape and its intensity has a proper value, then according to the nonlinear Schrödinger equation (NLSE), SPM will completely compensate for the broadening caused by the effect of anomalous dispersion on that pulse. This results in the formation of remarkably stable nonlinear pulses, called solitons. For comparison, Figs. 16.3 and 16.4 show the propagation of a nonhyperbolic secant pulse (super-Gaussian pulse) and a hyperbolic secant pulse, respectively, along a fiber in the anomalous dispersion regime [CCLA95]. Notice how the pulse is broadened in the former, while it is virtually distortionless in the latter. With the use of soliton control techniques, transmission at 10 Gbps over essentially infinite distances has been demonstrated [Naka93].

Features of Solitons

Soliton-based systems employ essentially intensity-modulated direct-detection (IM-DD) transmission systems, and hence the detection process is as simple as in noncoherent optical systems. However, the central issue in soliton-based systems becomes the generation of a train of high-quality soliton pulses and maintaining the optical power level above a threshold during the course of propagation through fiber. As discussed earlier, the narrow pulses are generated using gain-switched or mode-locked lasers along with the necessary band-limiting and pulse-compression techniques. The generated pulse streams are then multiplexed/compressed and preamplified for transmission over dispersion-shifted optical fibers along with periodic in-line amplification using erbium-doped fiber ampliers (EDFA) over the entire transmission span. Important considerations in the transmission system design include estimating the optimum number of EDFAs that would help preserve the optical pulse shape and keep the EDFA-induced timing jitters under tolerable limits over the maximum transmission distance of the OTDM network.

16.1.6 Demultiplexing and Clock Recovery

In contrast with the multiplexing operation, which can be implemented by passive technique, the demultiplexing operation needs to employ an active technique. In general, demultiplexing requires us to "gate" the received OTDM signal to take out the relevant slots or bits at every N^{th} interval; thus, a reliable demultiplexing operation necessitates an appropriate framing of the incoming optical pulses (e.g., forming the correct N-bit word for a bit-interleaved TDM system). This, in turn, needs an accurate estimate of the clock timing information in the incoming optical signal. Incidentally, the task of the optical clock recovery becomes easier for OTDM signals, as the OTDM signals are essentially transmitted in RZ format (to accommodate fiber dispersion), and the baseband spectrum of the RZ optical signals has a spectral line at the clock frequency ($= Nr$ Hz).

In one of the possible schemes for optical clock recovery, a local optical clock whose repetition rate is controlled by some RF drive source is locked to the incoming OTDM pulse stream [BCHK96]. Such schemes have been employed using four-wave-mixing (FWM) and gain switching in diode amp-

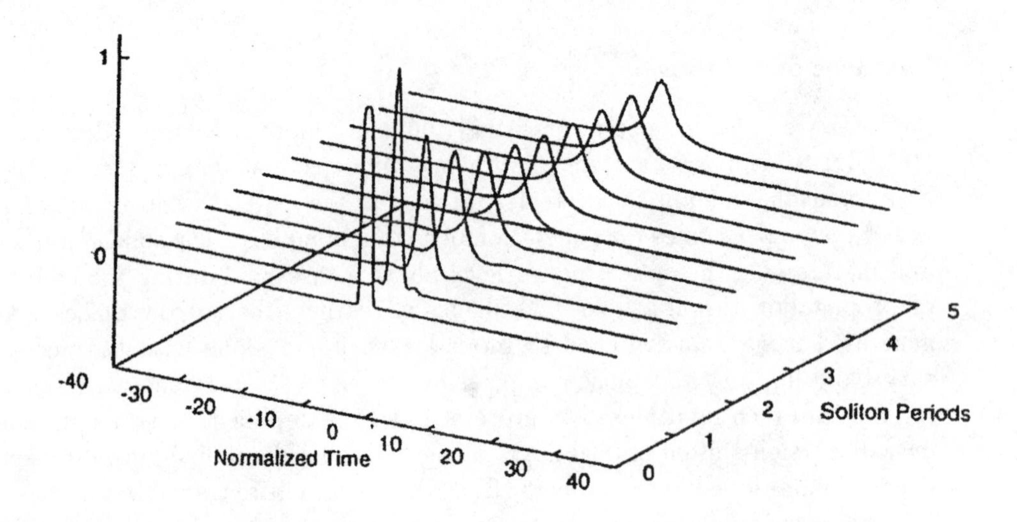

Figure 16.3 Evolution of a nonhyperbolic secant pulse in a fiber.

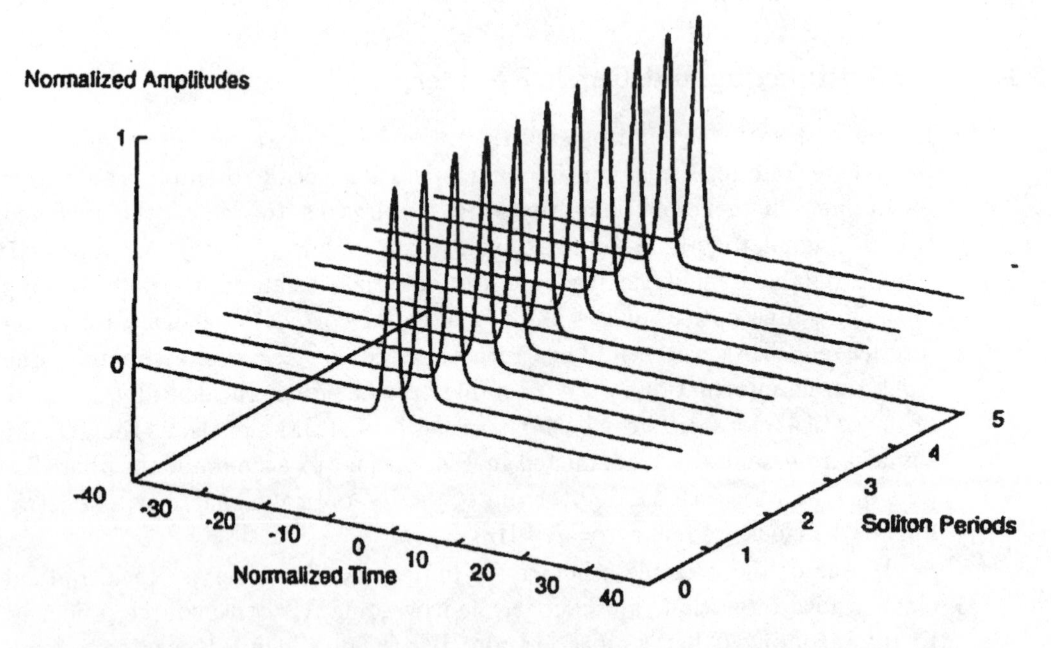

Figure 16.4 Evolution of a hyperbolic secant pulse in a fiber.

lifiers and second harmonic generation as the all-optical bit phase sensor. In another method, a nonlinear optical loop mirror (NOLM) is used as an optical bit phase sensor [BCHK96, SeBP96, Smit95, SPGK93]. The NOLM consists of a 3-dB fiber coupler with two of the ports joined through a length of optical fiber. Optical clock pulses are split at the coupler into two counter-propagating pulses which acquire identical phase shifts as they traverse the loop. These pulses interfere constructively at the coupler and are reflected back through their input port. The loop can be imbalanced by introducing a high-power data-pulse stream that co-propagates with one of the clock pulses, nonlinearly shifting its phase. The two clock pulses are no longer in phase when they arrive back at the 3-dB coupler, and some fraction of the input signal is transmitted through the output port. The NOLM output power is employed as an error signal to synchronize the local clock with the incoming optical signal.

Figure 16.5 presents the schematic of an all-optical clock recovery subsystem employing NOLM with a figure-eight configuration [BCHK96]. One

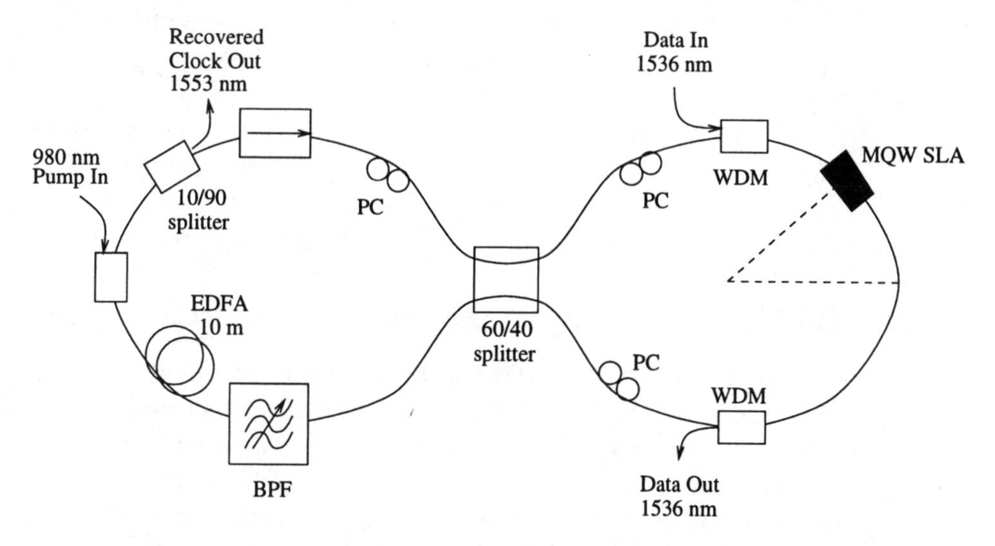

Figure 16.5 All-optical clock recovery system (BPF = band-pass filter, PC = polarization controller) [BCHK96].

of the two loops in the figure-eight is an NOLM containing a semiconductor laser amplifier (SLA) located off-center in the NOLM. The other loop contains gain and output coupling for the fiber laser, an isolator to force unidirectional operation, and a bandpass filter. The data pulses acting on the SLA induce a time-dependent phase shift which, in turn, modifies the transmissivity of the NOLM. The error signal generated from the NOLM helps the fiber laser on the other loop to mode-lock to the data rate.

Having recovered the optical clock, the demultiplexer needs to buffer the desired slot or header of the incoming optical signal using an optical storage, and subsequently it needs to slow down the data rate (by a factor of $1/N$) in order to interface the desired signal with the receiver for subsequent data processing.

16.1.7 Optical Processing

As mentioned in the previous section, the OTDM receiver requires buffering which needs to be implemented with the optical storage devices until the received data rate is slowed down. The necessary rate conversion can be achieved with an optical storage loop in conjunction with an optical switch. For example, for a 100-Gbps OTDM network receiver with a packet size of 10 kbits, the storage requirement for a packet would be 100 ns. Based upon the anticipated tolerable latencies, storage time requirements are likely to be no more than hundreds of microseconds for such applications [BGDB96], which can be implemented by the above storage loop arrangements.

It should also be noted that there does not exist any general-purpose random-access optical memory. To circumvent this problem to some extent, optical delay lines (or loops) may be used as "buffers." Their use will be examined in some of the photonic packet-switching architectures below.

16.1.8 Optical TDM Network Architectures and Proposals

A synchronous optical TDMA local network can be built using a passive-star broadcast medium [SeBP96]. An alternative architecture for an optical TDM LAN is based on a helical bus, and it will be discussed below. Also to be discussed in this subsection are two alternative architectures for photonic packet switches for wide-area optical TDM applications. Some additional photonic switching projects that a reader may find interesting but are not covered here include photonic slot routing [CEFS97] and ATM optical switching (ATMOS) [MBBG96].

An Example Optical TDM LAN: The HLAN Architecture

The HLAN architecture, described in [BCHK96], avoids channel collisions by using a slotted system with empty slot markers that indicate when a node can write data into a slot. HLAN is implemented on a helical unidirectional bus in which the physical fiber wraps around twice to visit each node three times, and the fiber medium is considered to be divided into three nonoverlapping "segments" (see Fig. 16.6). A headend periodically generates equal-sized frames of empty slots and puts them on the bus. The Guaranteed-Bandwidth traffic is transmitted on the GBW segment, the Bandwidth-On-Demand traffic is transmitted on the BOD segment, and data is received on the RCV segment. All users receive traffic on the third segment, i.e., all receivers are downstream of the transmitters. Each node in the network is equipped with a header/slot processor, protocol logic units, clock recovery mechanisms, and opto-electronic buffers. Although the network has been designed to operate at 100 Gbps, it is scalable in principle to faster media rates.

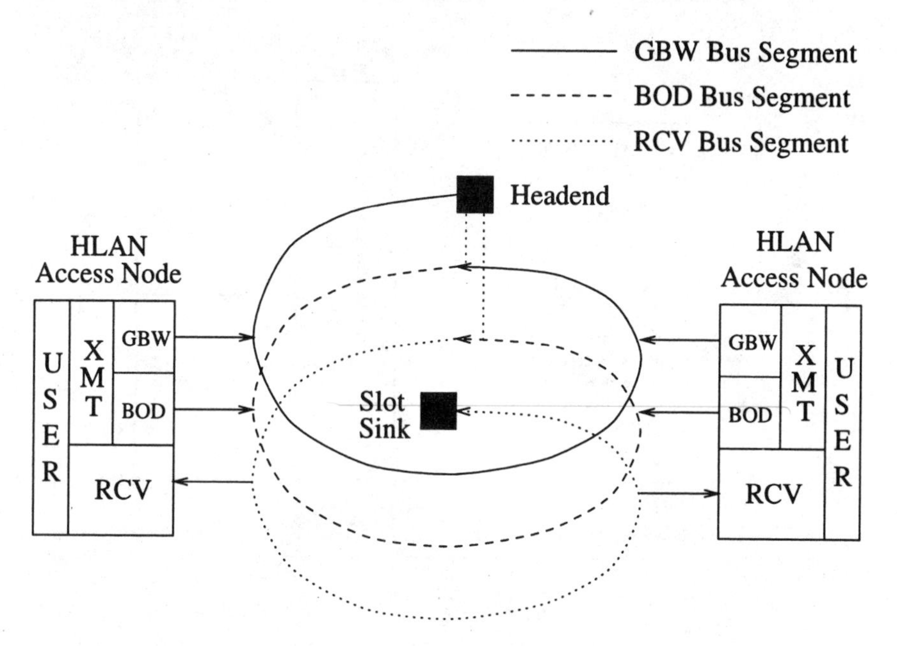

Figure 16.6 The HLAN architecture.

An Example Photonic Packet Switch: The Staggering Switch

The *staggering switch*, which is an "almost-all-optical" packet switch has been proposed in [Haas93]. In an "almost-all-optical" network, the data path is fully optical, but the control of the switching operation is performed electronically. One of the advantages of such switching over its electronic counterpart is that it is transparent, i.e., except for the control information, the payload may be encoded in an arbitrary format or at an arbitrary bit rate. The main problem in the implementation of packet-switched optical networks is the lack of random-access optical memory.

The staggering switch architecture employs an output-collision-resolution scheme that is controlled by a set of delay lines with unequal delays. The architecture is based on two rearrangeably nonblocking stages interconnected by optical delay lines with different amounts of delay. The work in [Haas93] investigates the probability of packet loss and the switch latency as a function of link utilization and switch size. In general, with proper setting of the number of delay lines, the switch can achieve an arbitrary low probability of packet loss. Figure 16.7 gives a simple overview of the switch architecture.

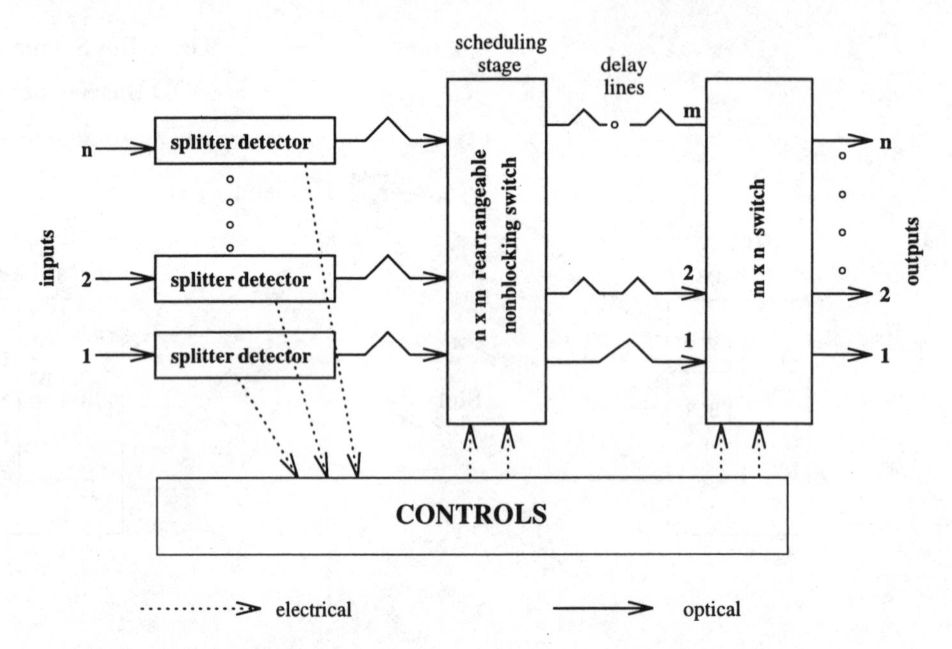

Figure 16.7 The staggering switch architecture.

Another Example Photonic Packet Switch: Contention Resolution by Delay Lines (CORD)

Another architecture which deals with contention in a packet-switched optical network is the Contention Resolution by Delay Lines (CORD) architecture [CFKM96]. The CORD architecture consists of a number of 2×2 crossconnect elements and delay lines (see Fig. 16.8). Each delay line functions as a buffer for a single packet. If two packets contend for the same output port, one packet may be switched to a delay line while the other packet is switched to the proper output. The packet which was delayed can then be switched to the same output after the first packet has been transmitted.

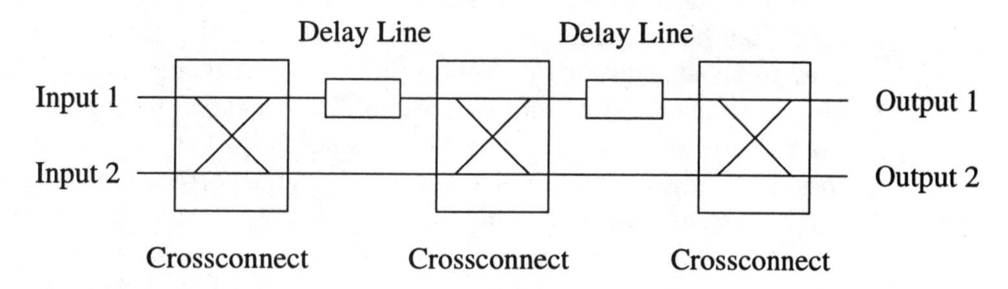

Figure 16.8 The CORD architecture.

16.2 Optical CDMA Networks

16.2.1 Basics of CDMA

Unlike WDM and TDM networks which can be used in a broadcast as well as a switched environment, code-division multiplexing (CDM) mainly finds usefulness in a broadcast environment as a multiple-access scheme. Hence, the corresponding networks are referred to as code-division multiple access (CDMA) networks.

A communication channel provides a fixed amount of bandwidth. In order to allow multiple users to share a channel, a multiple-access technique must be employed. In WDMA/FDMA, the bandwidth of the channel is subdivided into a number of smaller wavelength/frequency bands. In TDMA, time is divided into slots, and users take turns accessing the channel. In CDMA, all users are allowed to access the entire bandwidth of the channel simultaneously. In order to distinguish individual transmissions in a CDMA system, each user is assigned a unique code which is used to encode the user's data. A receiver which is tuned to the correct code is able to selectively receive the desired transmission.

Let us consider a very simple example to explain the CDMA principle. CDMA could be interpreted as different people in the same room talking to one another simultaneously using different languages. However, since you understand your own language only, the interference generated by other languages you do not speak or understand is marginal.

One advantage of CDMA is that there is no fixed limit to the number of users. In FDMA, there are a fixed number of channels, and once the number of users reaches the number of channels, no additional users can be accommodated. In TDMA, when all of the time slots are occupied, no further traffic can be accommodated. In CDMA, additional code sequences may be used to allow additional users into the system. Although adding more users degrades the quality of service in terms of lower signal-to-noise ratio, the degradation is gradual.[1] Another advantage of CDMA is that it provides some form of security. Only receivers which have the correct code may receive a given message. There are other advantages of CDMA, e.g., its bit rate is not fixed, as in TDMA or WDMA, and the user can use a bit rate which is below the maximum bit rate allowed in the system, and thus proportionally limit the interference it generates with respect to other transmitting nodes. In optics, CDMA was originally considered interesting since it does not require wavelength stability of transmitter and receiver (which is a nontrivial problem), and since it does not require synchronization in the system.

In CDMA each user is assigned an "almost orthogonal" pseudo-random code sequence which is multiplied by each data bit. This allows the users to access the same frequency band at the same time. The data rate of the new sequence is called the chipping rate f_c. If the baseband data rate is f_b, then the processing gain is $G_P = f_c/f_b$. For any single user, the other simultaneous transmissions will act as interference. Thus the more users that are simultaneously transmitting, the more interference there is and the higher P_e will be. The limit on the number of users L is determined by the P_e requirement and the equation:

$$P_e = \frac{1}{2} \operatorname{erfc} \sqrt{2 \frac{1}{L-1} \frac{f_c}{f_b}}$$

An advantage of CDMA is that it could be overlaid onto frequency bands already occupied by narrowband users. The narrowband signals will collapse when multiplied by the chipping function in the receiver, and will be considered as added interference.

[1]One can design a code for a CDMA system with a large number of codewords, and assign one codeword to each user in a statistical manner. Then, the channel allows only a subset of users to transmit simultaneously without performance degradation below a certain threshold, but the subset can be any subset of users in the system. In the other two multiplexing techniques (WDMA and TDMA), this is obviously not true.

One drawback to CDMA is that the power from closer transmitters will be higher than the power from more distant transmitters, resulting in high degradation for the distant transmitters. A proposed solution for this problem is the use of power control schemes. In a typical power control scheme, the receiver tells the senders whether to increase or decrease signal power so that the power received from each station is almost equal. This problem can also be dealt with using "hard limiters."

Another drawback is that if the original bit rate is already high, physical components won't be able to handle a much higher chip rate. (In optical CDMA, this drawback can be somewhat overcome if the coding is done in optics.)

16.2.2 Spread Spectrum

CDMA is based on spreading techniques which spread the original data signal over the entire frequency spectrum, and in some cases over the time domain. Some of these techniques are direct-sequence spread spectrum, frequency hopping, and time hopping.

Direct-Sequence Spread Spectrum

In direct-sequence spread spectrum, each bit of data is represented by a sequence of coded bits called chips. If the code sequence has length L, then each data bit will be represented by L chips. Each sequence of L chips is transmitted in the same time duration as an original data bit; thus the effective transmission rate is increased by a factor of L. The increase in transmission rate has the effect of spreading the signal over a larger frequency band.

The processing gain is defined as $10 \log(Bss/B)$, which is a measure of the difference in the signal-to-noise ratio of a signal before and after spreading. So, if the coded stream is transmitted over a channel, at the receiver, after the signal is decoded or despreaded, the signal-to-noise ratio goes up by the processing gain.

Frequency Hopping

In frequency hopping, users spread their signals over the spectrum by switching frequencies during transmission. The bandwidth is divided into N frequency channels, and the transmitted signals hop from frequency to frequency. Each user has a different hopping pattern which corresponds to that user's code.

In slow frequency hopping, one or more bits are transmitted per hop. For example, a user will transmit 5 data bits on frequency channel f1, hop to frequency f2, transmit another 5 bits, etc. In fast frequency hopping, each bit is transmitted over a number of hops. For example, the first bit is transmitted over frequencies f1, f5, f2, while the next bit is transmitted over frequencies f3, f6, f4. Fast frequency hopping provides higher reliability, since each symbol is transmitted more than once. In this case, if certain channels are jammed, or have high interference, there are several other channels on which the symbol may be received.

Time Hopping

In time hopping, a message is transmitted in bursts over a long period. The time between bursts is determined by a hopping pattern. For example, a message is broken up into n subpackets [LaSa90, RaSa84]. The subpackets are transmitted over n time slots at times $t_0, t_1, t_2, \ldots, t_{n-1}$, where $t_0 < t_1 < t_2 < \ldots < t_{n-1}$. The intervals between t_i and t_{i+1} for $i = 0, 1, 2, \ldots, n-2$ are determined by the hopping pattern.

16.2.3 Code Sequences

Another important aspect of CDMA is the code sequence. The code sequences used in radio-frequency CDMA are pseudo-noise (PN) or pseudo-random sequences, viz., sequences of zeros and ones which resemble a random data pattern. The codes, however, are not random. They are generated in a deterministic way and have specific properties. Some properties of maximal-length PN sequences include:

- Balance property: The number of zeros and ones is balanced. Actually, each code contains $2^{(N-1)} - 1$ zeros and $2^{(N-1)}$ ones.

- Run property: 1/2 runs of length 1, 1/4 runs of length 2, 1/8 runs of length 3, ... , etc.

- Autocorrelation property: Autocorrelation is the measure of similarity between a signal $f(t)$ and a time-shifted version of itself $f(t-\tau)$. The autocorrelation of a code sequence can be obtained by looking at the number of agreements and disagreements. Autocorrelation between the sequence and a shifted version of the sequence is always -1. This property allows ease of synchronization.

- Cross-correlation property: Cross-correlation is the measure of similarity between two codewords, and its value is always -1 for two different codewords.

There are a number of approaches to generate PN codes, and the preferred approach to generate these codes is shown in Fig. 16.9. This hardware implementation of a PN sequence generator consists of N flip-flops and some additional logic, and it generates a maximal-length PN sequence. The length of the sequence is given by $L = 2^N - 1$.

The number of independent maximal-length sequences that can be generated is bounded by $S \leq L - 1/N$. Depending on how the logic is configured, sequences with small cross-correlation may be generated.

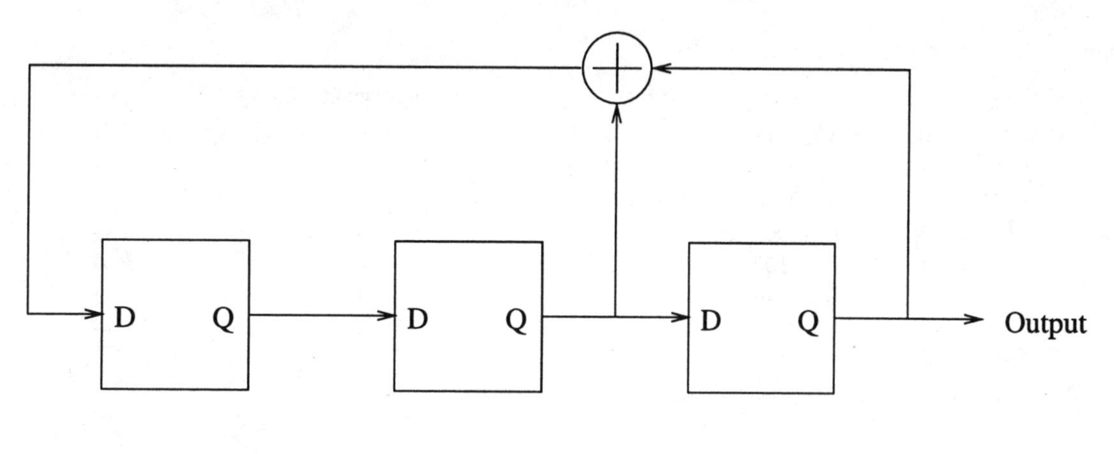

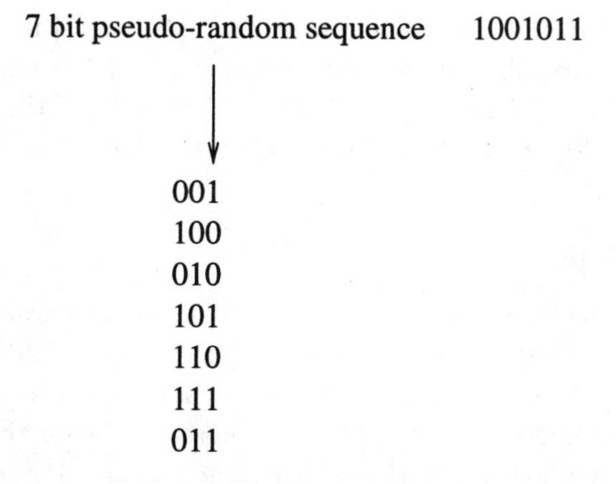

Figure 16.9 A pseudo-random sequence generator.

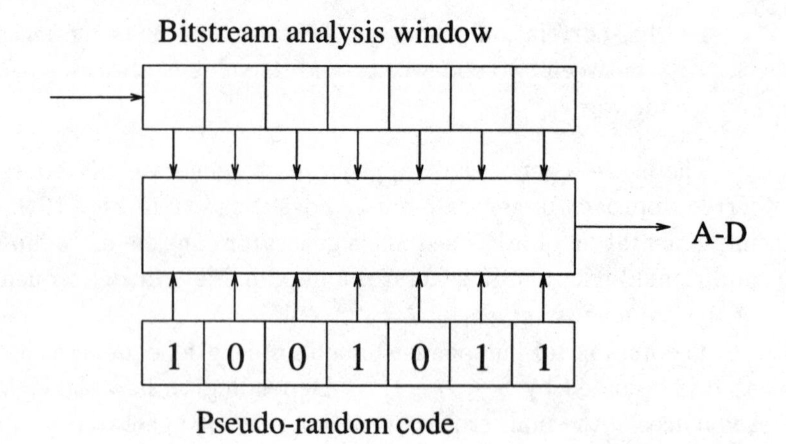

Reference sequence: 1001011

Shift	Sequence	Agreements (A)	Disagreements (D)	A-D
0	1001011	7	0	7
1	1100101	3	4	-1
2	1110010	3	4	-1
3	0111001	3	4	-1
4	1011100	3	4	-1
5	0101110	3	4	-1
6	0010111	3	4	-1

Figure 16.10 A CDMA receiver.

An implementation of a CDMA receiver is shown in Fig. 16.10. The input sequence is compared with the code sequence, and the number of disagreements are subtracted from the number of agreements. If the resulting value is above a certain threshold, then the signal is decoded to its appropriate value.

16.2.4 CDMA Example

In Fig. 16.11, two data sequences are shown, along with their corresponding codewords. (This example assumes chip as well as bit synchronization.) By performing a bit-wise multiplication on each chip, the coded sequence is generated. This coded sequence is then transmitted over the channel.

Figure 16.12 illustrates how the two coded streams overlap and interfere with each other on the data channel.

In Fig. 16.13, the combined signal is multiplied by the original coded

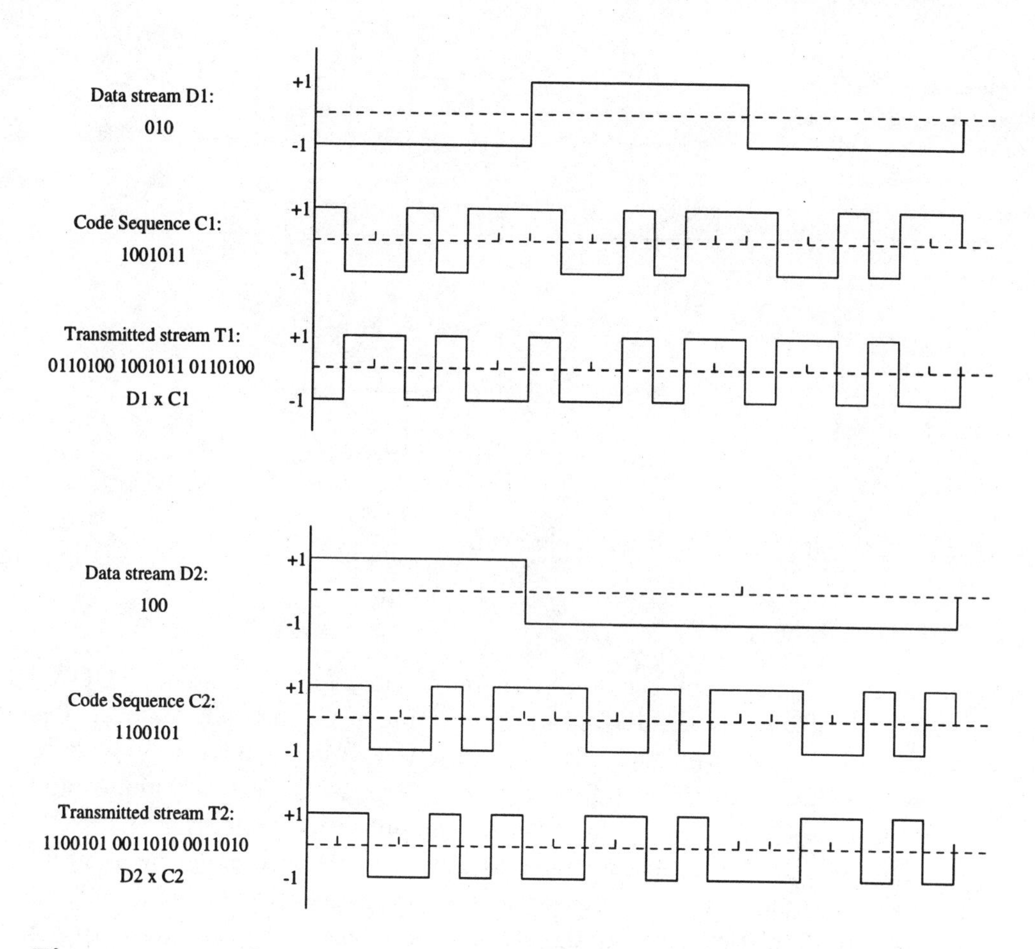

Figure 16.11 Original data streams and coded, transmitted streams.

sequence of one of the sources. The resulting signal can then be integrated over each bit duration to obtain the original data bits. Since the multiplication of the coded sequence with the interfering signal results in a pseudo-random sequence with an (almost) equal number of $+1$s and -1s, the interfering signal can be treated as random noise.

Figure 16.14 illustrates the decoded signal in the presence of multiple interfering signals. It can be seen that, even with a large number of interfering signals, the original data stream is still identifiable.[2]

[2]A bit is declared to be a 1 (or a 0) if the received signal, after integration over a bit duration, turns out to be higher (or lower) than a certain threshold. See Fig. 16.14.

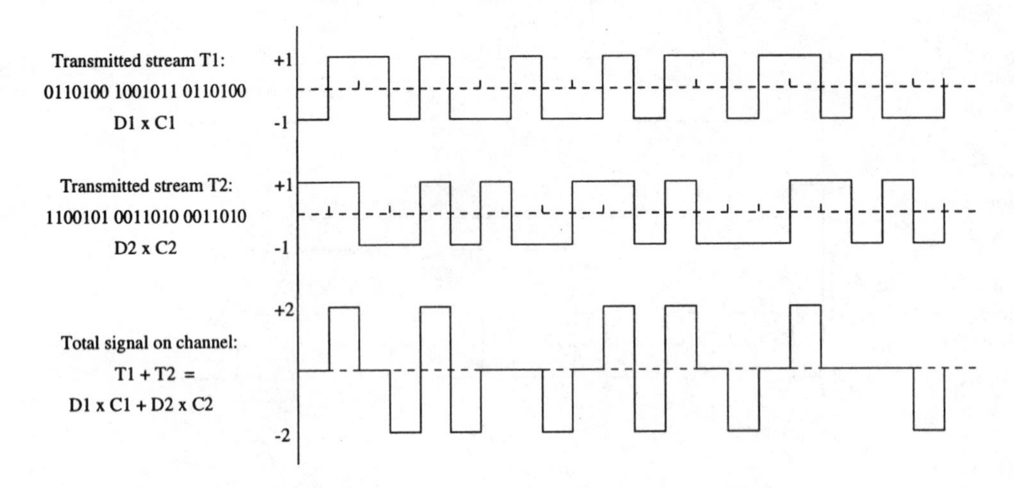

Figure 16.12 Combined signals on data channel.

16.2.5 Optical CDMA

In optical systems which use amplitude modulation [or on-off keying (OOK)] with direct detection, the above implementation of CDMA won't work. The reason is that laser amplitudes can't be negative. Also, it is very difficult to process optical signals. Therefore, different types of CDMA techniques and different codes must be used for optical communication systems.

Optical CDMA can be implemented using either noncoherent or coherent detection techniques.[3] In noncoherent detection, the receiver looks only at the amplitude of the received signal, not the phase. An example is direct detection. In coherent detection, the receiver looks at both the amplitude and phase of the signal. In general, coherent detection gives better performance, but is much more difficult to implement, requiring tight control over the phase of the signal. Details on coherent optical CDMA systems may be found in [WeSa93]. Noncoherent optical CDMA techniques include time spreading, frequency hopping, and spatial spreading.

The time-spreading technique makes use of temporal spreading codes. Each bit of information is represented by a number of light pulses, and the coding is specified by the spacing of the light pulses. When a user sends data, a '0' bit is represented by no light pulses, while a '1' bit is represented by the

[3]Here is a simple illustrative logic behind the problem. Suppose you have a low (l) and a high (h) signal. In a coherent system, we will have (1) $l \times l = h$, while in a noncoherent system, we will have (2) $l \times l = l$. (Electronic) CDMA (i.e., all previous examples on CDMA) follow Logic (1), while this subsection on optical CDMA will follow Logic (2).

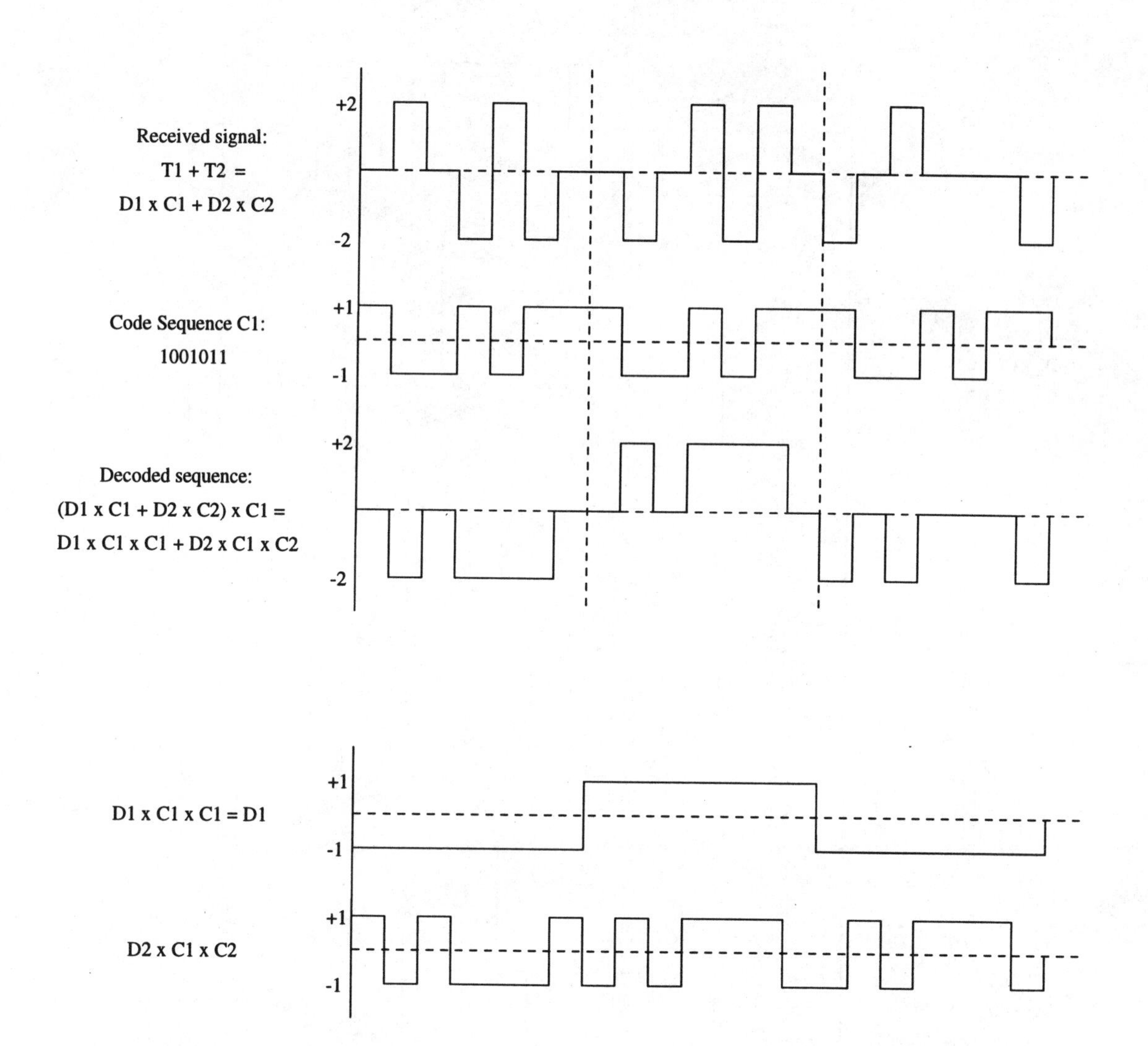

Figure 16.13 Decoded sequence consisting of original signal and pseudo-noise.

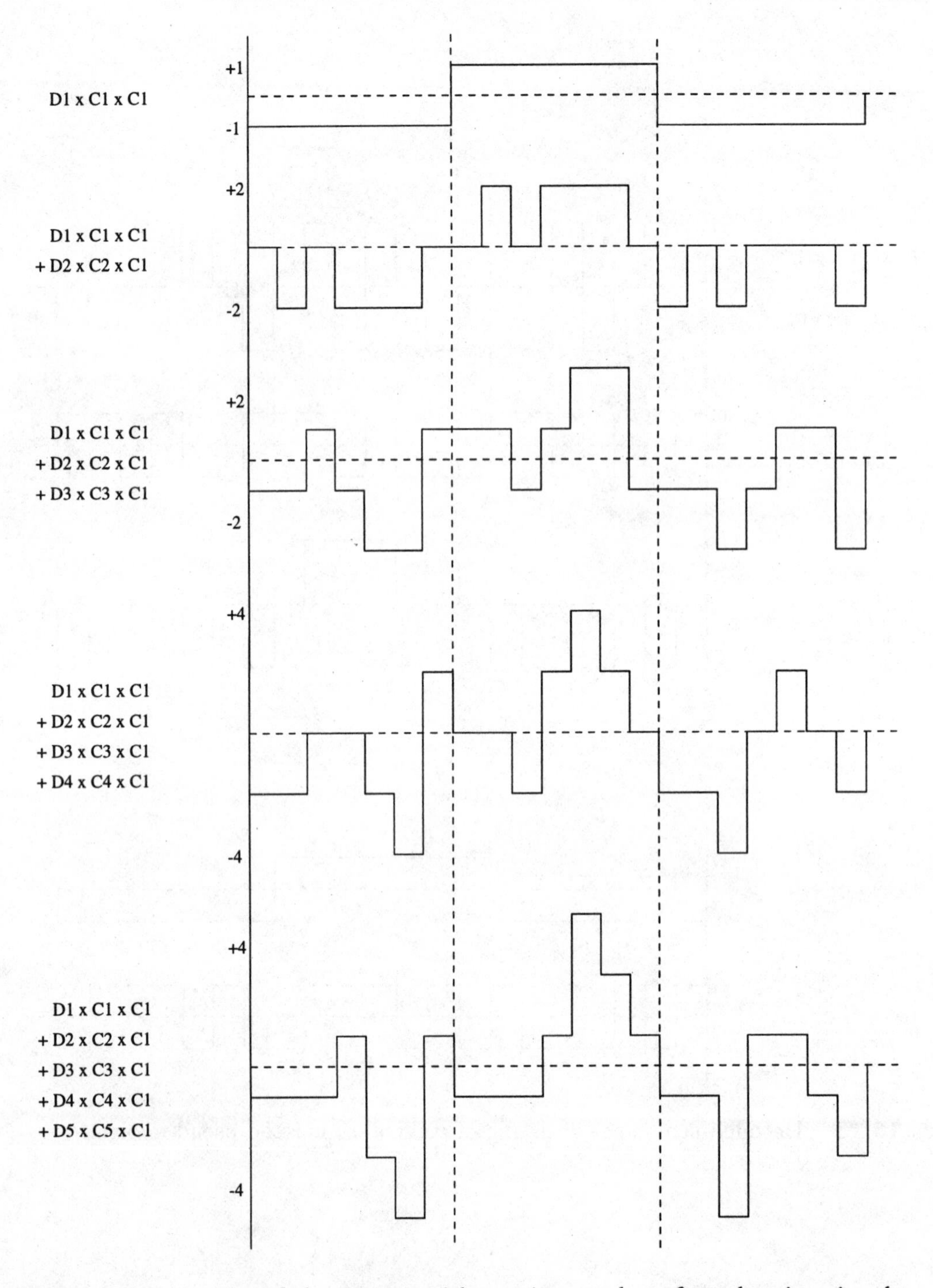

Figure 16.14 Decoded sequence with varying number of overlapping signals.

transmission of a codeword with light pulses in the proper locations within the word.

At the receiver, in the absence of noise, the decoder checks each of the positions for the given code, and counts the number of light pulses received. If the number of light pulses received is greater than a certain threshold, then the bit is decoded as a '1'; otherwise it is decoded as a '0'. Note that, if a '1' is sent, it will always be decoded as a '1,' since interfering signals will not cause any of the transmitted pulses to be reduced in power. However, if a '0' is sent, it is possible to get errors caused by interfering users. If other users happen to transmit pulses in the specified slots, then a '1' may be detected.

One implementation of a time-spreading CDMA system is shown in Fig. 16.15. At the transmitter, each data bit is spread into a number of pulses. The code, which determines the spacing of the smaller pulses, is specified by the delay lines following an optical splitter. The power levels of the resulting coded pulses are below the normal detection threshold, making it difficult for other nodes to detect or intercept the transmissions. At the receiver, an optical splitter/combiner with appropriate delay lines is used to reconstruct the original data bit. Since the splitter at the receiver also reduces the power of the signals, some amplification may be needed before detecting the original signal.

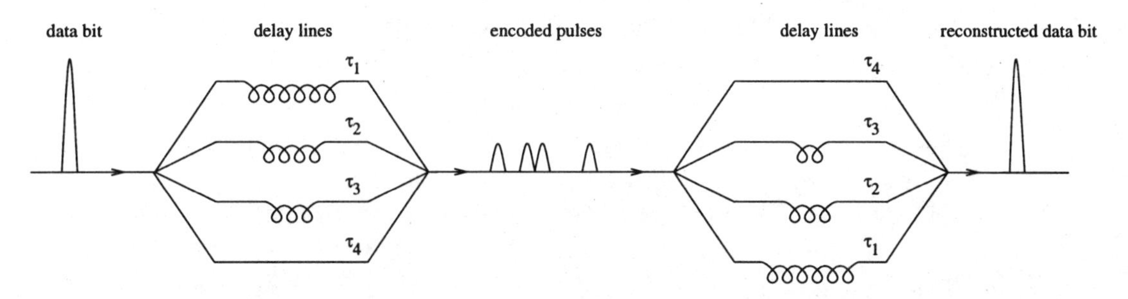

Figure 16.15 Implementation of a CDMA coder and decoder based on optical splitters and combiners.

In frequency-hopping optical CDMA, a node transmits to another node while rapidly changing the channel on which it is transmitting. The sequence of channels is determined by a pseudo-random code which is known only to the source and the destination. In this scheme, an eavesdropper may be able to intercept portions of the message, but is unlikely to intercept a significant amount of the message.

Another approach to implementing optical CDMA is to combine time spreading with frequency hopping. Each pulse in the codeword is transmitted at a different wavelength. This approach to CDMA allows for more codewords and less cross-correlation.

Yet another approach to optical CDMA is called multivariate CDMA which uses spatial spreading. This approach involves the use of several fibers in parallel, each fiber with its own broadcast network. The coding determines which combination of fibers to use.

In optical CDMA, the codes used are different from the PN codes used in radio-frequency CDMA approaches, since noncoherent optical signals are unable to take on negative values. One family of optical codes is called optical orthogonal codes (OOC). Some requirements of OOCs include low autocorrelation and low cross-correlation.

The primary difference between OOC and PN codes is that PN codes have an equal number of zeros and ones, while OOCs have more zeros. The sparseness of ones in OOCs is necessary in order to ensure that different codes don't overlap in many positions. If codes begin to overlap in too many positions, then interference among different users will result in a higher number of errors. The number of simultaneous users in an OOC system is bounded by $(L-1)/K(K-1)$ where L is the length of the codeword, and K is the number of pulses per codeword [Sale89]. Alternative codes for optical CDMA systems have been considered. These codes include prime sequence codes [PrSF86] and generalized temporal codes [AzSa92].

An example of optical time-spreading CDMA is shown in Fig. 16.16. There are two users in the system who are sharing an optical channel using optical CDMA. User 1 has a code $C1 = 1001010$, while User 2 has a code $C2 = 0011001$. Each bit in each user's data stream is multiplied by the user's code sequence to produce the appropriate stream of transmitted light pulses. The transmitted light pulses of User 1 and User 2 are superimposed on the optical channel, resulting in a received stream of pulses R. In this example, we assume that the receivers use a "hard-limiter" (which bounds the signal value). Each bit duration in the received pulse stream is then compared to the appropriate code sequence, and the number of matches are counted. If the number of matches is greater than or equal to the threshold, then the bit is decoded to a "1", otherwise the bit is decoded to a "0". In Fig. 16.16, the received stream R is compared to the code sequence $C1$. In the first bit duration, there is one match, while in the second bit duration, there are three matches. Thus, the decoded sequence is 01, which matches the original data stream $D1$.

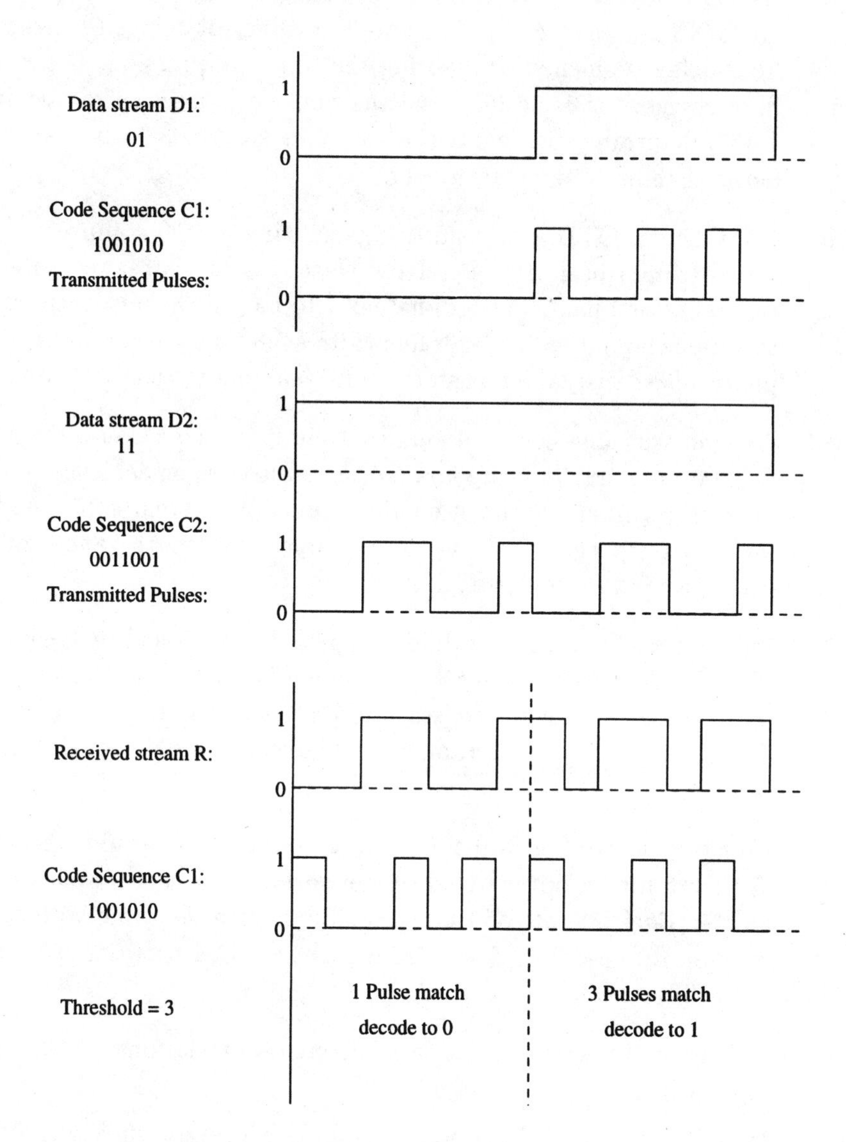

Figure 16.16 Optical time-spreading CDMA.

Exercises

16.1. An OTDM transmitter employs a 10-times compression over a basic 1-Gbps NRZ data stream. Calculate the values of delays required for the fiber-delay segments for realizing the time compression. If the residual fiber segment is 1 cm long and the fiber core has a refractive index of 1.457, determine the lengths of the fiber segments to be used in each delay channel of the transmitter.

16.2. Consider an OTDM transmitter generating a 1:10 compressed optical signal from a basic 10-Gbps NRZ electrical data stream. If the optical pulses in the final OTDM signal need to have 40% duty cycle, calculate the upper bound on the laser linewidth such that the compressed optical pulses also satisfy the transform limit enabling the soliton transmission.

16.3. Explain why one needs to employ soliton transmission in an ultrafast OTDM network. Indicate how soliton transmission achieves such transmission on fiber and mention the important system components used for this purpose. What is the suitable wavelength range for soliton propagation through optical fiber?

16.4. Indicate the fundamental differences between optical orthogonal codes (OOCs) used for optical CDMA networks and the PN sequence codes used for conventional wireless CDMA networks. Explain with suitable examples and block diagrams how these codes are detected in a CDMA receiver.

16.5. Using the upper bound for the number of users, determine the transmission rate for an optical CDMA network for 15 users, each working at an electrical bit rate of 1 Gbps, and compare the same with the transmission rate required for an equivalent OTDM network. Comment on their difference.

16.6. Calculate the autocorrelations and cross-correlations for the codes in Fig. 16.17.

Which codes meet the properties of a pseudo-random sequence?

16.7. Consider a direct-sequence spread spectrum system in which three users share a single channel. Show the received waveform and the decoded waveform (i.e., as in Fig. 16.13) at each receiver for the following situation:

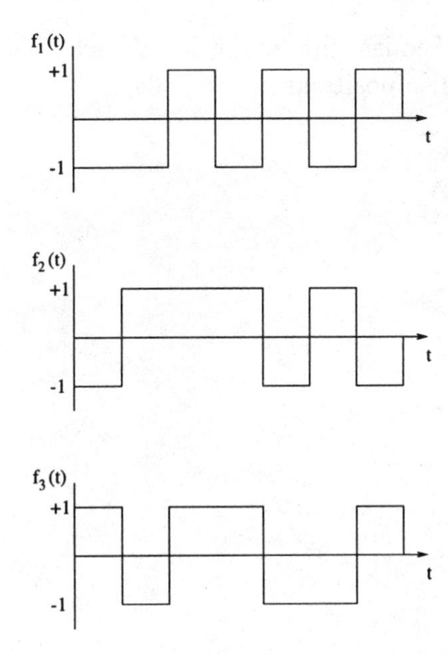

Figure 16.17 CDMA codes.

(a) Source S1 transmits data stream D1 = 101 using code sequence C1 = 0111010,

(b) Source S2 transmits data stream D2 = 001 using code sequence C2 = 1001110, and

(c) Source S3 transmits data stream D3 = 110 using code sequence C3 = 1010011.

16.8. Find a family of optical orthogonal codes with length 32 and with three pulses per code.

16.9. Suppose three users access an optical fiber medium using optical CDMA. Optical orthogonal codes of length 8, and having two pulses per code are used. The codes for each user are as follows:

- Code S_1 has pulses in positions 0 and 2,
- Code S_2 has pulses in positions 0 and 3, and
- Code S_3 has pulses in positions 0 and 7.

Find the bit-error rate (BER) caused by interference from other users. Assume that there is no other source of signal degradation.

16.10. Consider the following sets of codes with length 24. Which set is better? Why? (Numbers indicate pulse positions in the codes.)

(a) Set I

- S_1: 0, 7, 23
- S_2: 0, 4, 6
- S_3: 0, 9, 14

(b) Set II

- S_1: 0, 7, 23
- S_2: 0, 5, 8
- S_3: 0, 9, 14

Appendix A

Further Reading

For further reading, one can refer to a large number of resources on the subject matter. For in-depth coverage on optical WDM device technology, please see [Gree92]. While the book's coverage of optical device technology is quite broad, its material on network architectures represents the state of the art in 1992, and is outdated.

Valuable tutorial/review/survey articles are the following: [Brac90], [Rama93], [Mukh92a], and [Mukh92b], although these articles are also quite outdated by now since they were written before wavelength routing had become mature.

Most of the latest breakthroughs in optical network device technology are reported at the conferences [OFC] and [ON], and in the journal [JLT]. Papers reporting most of the latest developments on optical network architectures can be found in the conference [Info]; the journals [ToN], [TComm], and [JSAC]; and the magazines [Comm] and [Net]. (Papers on optical networking are published in other journals and conference proceedings also; please refer to the Bibliography for an extensive listing.)

Several specific journal special issues which deal with both device and network aspects and which are "collector's items" for anyone working in the field are [JSAC90], [JLT93], [JHSN95], [JSAC96], and [JLT96]. An additional special issue [JSAC98] addressing related topics is also in its preparation stage.

A.1 General Resources and Publications

[Comm.] *IEEE Communications Magazine*, published monthly.

[Info.] *IEEE Infocom Conference*, held annually during Spring.

[JHSN95.] *Journal of High-Speed Networks*, Special issue on Optical Networks, vol. 3, nos. 1-2, January-April 1995.

[JLT.] *IEEE/OSA Journal of Lightwave Technology*, published monthly.

[JLT93.] *IEEE/OSA Journal of Lightwave Technology*, Special issue on Broadband Optical Networks, vol. 11, no. 5/6, May/June 1993.

[JLT96.] *IEEE/OSA Journal of Lightwave Technology*, Special issue on Multiwavelength Optical Networks, vol. 14, no. 6, June 1996.

[JSAC90.] *IEEE Journal on Selected Areas in Communications*, Special issue on Dense Wavelength Division Multiplexing Techniques for High Capacity and Multiple Access Communication Systems, vol. 8, no. 6, Aug. 1990.

[JSAC96.] *IEEE Journal on Selected Areas in Communications*, Special issue on Optical Networks, vol. 14, no. 5, June 1996.

[JSAC98.] *IEEE Journal on Selected Areas in Communications*, Special issue on High-Capacity Optical Transport Networks, anticipated publication date: late 1998.

[Net.] *IEEE Network Magazine*, published every other month.

[OFC.] *Optical Fiber Communications (OFC) Conference*, held annually in February.

[ON.] *IEEE/LEOS Summer Topical Meeting on Optical Networks*, held during July of every even-numbered year.

[TComm.] *IEEE Transactions on Communications*, published monthly.

[ToN.] *IEEE/ACM Transactions on Networking*, published every even-numbered month.

A.2 Enabling Technologies

[Brac90.] C. A. Brackett, "Dense wavelength division multiplexing networks: Principles and applications," *IEEE Journal on Selected Areas in Communications*, vol. 8, pp. 948-964, Aug. 1990.

[Gree92.] P. E. Green, Jr., *Fiber Optic Networks*, Prentice-Hall, 1992.

A.3 Tutorials/Surveys/Reviews

[**Mukh92a.**] B. Mukherjee, "WDM-Based Local Lightwave Networks – Part I: Single-Hop Systems," *IEEE Network Magazine*, vol. 6, no. 3, pp. 12-27, May 1992.

[**Mukh92b.**] B. Mukherjee, "WDM-Based Local Lightwave Networks – Part II: Multihop Systems," *IEEE Network Magazine*, vol. 6, no. 4, pp. 20-32, July 1992.

[**Rama93.**] R. Ramaswami, "Multi-wavelength lightwave networks," *IEEE Communications Magazine*, vol. 31, no. 2, pp. 78-88, Feb. 1993.

Appendix B

Glossary of Important Terms

all-optical cycle. In a wavelength-routed network, each input port of a wavelength-routing switch may always remain connected to some output port regardless of whether such connection is required or not. Due to such unspecified and uncontrolled connections, unintended optical paths may be set up in the network. Optical amplifiers in such paths may repeatedly amplify their own noise thereby saturating the gain of the amplifiers.

all-optical network. A wavelength-routed optical network wherein the information path between the source and the destination remains entirely optical. Such an all-optical path offers protocol transparency to the network.

broadcast-and-select network. A network in which information transmitted by any node passes by the interface of all other nodes, and each node independently selects which part of the information it wants to receive.

channel sharing. A technique used when the number of channels needed exceeds the number of wavelengths available so that some of the (WDM) channels must be shared to allow network operation (e.g., each WDM channel may be shared using TDM).

circuit switching. A type of transmission in which a dedicated path is established between two nodes (with zero or more intermediate switching nodes) in order to carry information. Most efficient for a connection that will remain up for a fairly long time duration.

code-division multiple access (CDMA). Multiple channels on the same wavelength, but separated by the way they encode data. Usually has high overhead to encode data.

control channel. A channel used to conduct pretransmission coordination (e.g., on which data channel will data transmission occur, at what time will the transmission start, etc.)

crosstalk. The undesirable effect of a transmission on one channel interfering with the transmitted signal on another channel.

dark fiber. Fiber optic cable that is currently not being used.

deflection routing. A method of misrouting ("deflecting") packet(s) at a switching node at which two or more packets are competing for the same output channel. Packets that lose the contention are "deflected" to other parts of the network instead of being locally buffered at the switching node.

electronic speed (bandwidth). The maximum speed at which electronic components can operate.

electrooptic conversion. Conversion technique employed when moving information from the electronic domain to the optical domain or vice versa. Typically employed at each hop in a multihop network.

embedding. A technique used to create a virtual (logical) topology on a fixed physical topology by properly configuring the channels in the network (viz., which node transmits to which other node(s) over which channel(s)).

erbium-doped fiber amplifier. A amplifier that is able to amplify signals in the wavelength range of 1540 nm to 1570 nm which is within one of the low-loss operating regions of a fiber optic cable. However, gain is not uniform over the amplified region.

(wavelength-agile, tunable) filter (receiver). An optical filter that can be used to select a particular wavelength channel to receive information on, and, if tunable, it can be reconfigured to allow a multitude of possible network configurations.

graph coloring. The algorithms that are used to ensure that, if two or more lightpaths share a common physical fiber link on their paths, they must necessarily be operated on different wavelengths on that link. Also referred to as a "color clash" constraint.

hop distance. The number of hops it takes for a packet to travel from its source to its final destination. A network's diameter is the maximum number of hops it takes for a packet to get from any node to any other node (assuming that shortest paths are always used).

(wavelength-agile, tunable) laser (transmitter). An electromagnetic transmission device that can be modulated to carry information over an optical fiber, and, if the laser is tunable, a multitude of possible virtual network configurations can be created.

laser array. A set of fixed-tuned lasers which are integrated into a single component, with each laser operating at a different wavelength so that, if each of the wavelengths in the array is modulated independently, then multiple transmissions may take place simultaneously.

lightpath. The all-optical path through which the information flows in a wavelength-routed optical network. A lightpath provides "single-hop" connectivity between its end points. A lightpath may be composed of a single wavelength (in a network without wavelength converters), or it may consist of multitude of wavelengths (in the presence of wavelength converters).

lightwave network. A network in which multiple channels, each capable of operating at peak electronic speed, are multiplexed onto a fiber optic cable. These networks have a theoretical capacity of tens of terabits per second (Tbps).

logical topology (a.k.a. called virtual topology). The set of lightpaths that are embedded on a physical fiber network are said to form a multihop logical topology.

medium-access protocol. Method to determine when a node has the right to use (transmit on and/or receive from) a given channel (e.g., random access, reservation, etc.)

multicasting. The ability to transmit from a single source to multiple destinations.

multihop network. A network in which a packet may hop through zero or more intermediate nodes before it reaches its final destination.

optical bandwidth. The large information-carrying capacity, measured in tens of Tbps, that occurs in the low-loss regions of an optical fiber.

optical power loss. Losses that occur in a fiber optic network due to connections (splitting, coupling, and insertion loss) and propagation (loss due to signal attenuation). These losses must be taken into account while designing the network to ensure that it meets its power budget.

packet delay. The entire delay experienced by a packet, usually measured from the instant of its generation at the source node until it is completely delivered at its destination, including propagation delay(s) in the route to its destination, and queuing delays at the source and intermediate nodes.

packet switching. A technique in which each packet contains enough information in its header so that it can be independently switched at intermediate nodes in the network and routed to its final destination. At each intermediate node, the packet may be queued up while waiting for an output channel to clear up.

passive (wavelength) router. An N-input, N-output device that can separately route each of several wavelengths incident on an input fiber to the same wavelength on separate output fibers, based on a *fixed* routing matrix.

passive-star coupler. An N-input, N-output passive broadcast device that can divide the power from each of its N input fibers and direct them to each of its N output fibers simultaneously.

photonic packet switching. An all-optical network in which packets (time slots) are switched all-optically from an input to an output at a photonic switch. If all packets are of the same size and all inputs at the switch are synchronized, then the corresponding ability of the switch to route packets is called photonic slot routing.

physical topology. The physical layout of the network (e.g., bus, star, multistar, mesh, tree, etc.)

pretransmission coordination. A technique in which some information is transmitted on a control channel before the data is transmitted on a data channel.

receiver collision. The loss of a transmission due to the fact that the receiver was occupied receiving another transmission on another channel.

regular structure. A structure where each node in the network has the same number of transmitters and the same number of receivers, and the entire network has well-defined (structured) nodal-interconnection pattern.

routing mechanism. The method by which a packet travels from the source to the destination including what happens if there is contention for the output links (e.g., shortest path routing, adaptive routing, deflection routing, etc.)

scalability/modularity. The ability to grow (scale) a network to a larger size, but be able to do so in small increments (modules).

single-hop network. A network in which a packet travels from its source to its destination directly (in one hop). The packet does not encounter an electrooptic conversion before reaching its final destination.

soliton. Soliton transmission is a technique to transmit ultrafast optical signals while maintaining the shape of the optical pulse even in the presence of fiber dispersion by utilizing an optical fiber nonlinearity.

splitter. A device that splits the power input from one optical fiber onto two or more output fibers.

time-division multiplexing (TDM). A medium-access technique in which a given wavelength is carved up into a number of different channels based on periodic time slices.

tuning range. The range of wavelengths over which a tunable device (transmitter or receiver) can operate.

tuning time. The time it takes for a (tunable) transmitter or receiver to switch from one channel to another. Reciprocal of tuning speed.

wavelength assignment. See graph coloring.

wavelength channel (a.k.a. WDM channel). A channel which is centered at a specific wavelength within the low-loss region of an optical fiber.

wavelength-continuity property. In the absence of any wavelength conversion device, a lightpath is required to be on the same wavelength channel throughout its path in the network; this requirement is referred to as the *wavelength continuity* property of the lightpath.

wavelength converter (a.k.a. wavelength changer, wavelength shifter).] A device that can *convert* the optical data arriving on one wavelength along a fiber link into another wavelength and forward it along the next fiber link.

wavelength-convertible switch [a.k.a. wavelength interchanging crossconnect (WIXC)].] A wavelength-routing switch that is also equipped with a wavelength-conversion facility.

wavelength-division multiplexing (WDM). A medium-access technique in which a single fiber can have multiple channels by having each channel operate at a different center wavelength (or frequency).

wavelength-routing switch (WRS) [a.k.a. wavelength crossconnect, wavelength-selective crossconnect (WSXC)].] A space/wavelength division multiplexed optical switch used to route and add/drop multi-wavelength optical signals at the nodes of all-optical networks.

WDM network control and management. The mechanisms to set up and tear down lightpaths, possibly in a distributed fashion, in a wavelength-routed network.

Bibliography

[ABCR94] A. Aggrawal, A. Bar-Noy, D. Coppersmith, R. Ramaswami, S. Schieber, and M. Sudan, "Efficient routing and scheduling algorithms for optical networks," *Proceedings, Fifth Annual ACM-SIAM Symposium on Discrete Algorithms*, pp. 412–423, 1994.

[ACGK86] E. Arthurs, J. M. Cooper, M. S. Goodman, H. Kobrinski, M. Tur, and M. P. Vecchi, "Multiwavelength optical crossconnect for parallel-processing computers," *Electronic Letters*, vol. 24, pp. 119–120, 1986.

[AGKV88] E. Arthurs, M. S. Goodman, H. Kobrinski, and M. P. Vecchi, "Hypass: An optoelectronic hybrid packet-switching system," *IEEE Journal on Selected Areas in Communications*, vol. 6, pp. 1500–1510, Dec. 1988.

[ASWC96] V. Arya, D. W. Sherrer, A. Wang, R. O. Claus, and M. Jones, "Temperature compensation scheme for refractive index grating-based optical fiber devices," *Proceedings, SPIE*, vol. 2594, pp. 52–59, 1996.

[AaKo89] A. Aarts and J. Korst, *Simulated Annealing and Boltzmann Machines*, New York: John Wiley & Sons, 1989.

[AcKa89] A. S. Acampora and M. J. Karol, "An overview of lightwave packet networks," *IEEE Network Magazine*, vol. 3, pp. 29–41, Jan. 1989.

[AcSh91] A. S. Acampora and S. I. A. Shah, "Multihop lightwave networks: A comparison of store-and-forward and hot-potato routing," *Proceedings, IEEE INFOCOM '91*, Bal Harbour, FL, pp. 10–19, April 1991.

[Acam87] A. S. Acampora, "A multichannel multihop local lightwave network," *Proceedings, IEEE Globecom '87*, Tokyo, Japan, Nov. 1987.

[Agra89] G. P. Agrawal, *Nonlinear Fiber Optics*, Academic Press, Inc., 1989.

[Alba83] A. Albanese, "Star network with collision avoidance circuits," *Bell Systems Technical Journal*, vol. 62, pp. 631–638, March 1983.

[Alex93] S. B. Alexander et al., "A precompetitive consortium on wide-band all-optical networks," *IEEE/OSA Journal of Lightwave Technology*, vol. 11, pp. 714–735, May/June 1993.

[Alfe88] R. C. Alferness, "Titanium-diffused lithium niobate waveguide devices," in *Guided-Wave Optoelectronics* (T. Tamir, ed.), Chapter 4, New York: Springer-Verlag, 1988.

[Alfe93] R. C. Alferness, "Widely tunable semiconductor lasers," *OFC/IOOC '93 Technical Digest*, San Jose, CA, pp. 11–12, 1993.

[Ande95] D. Anderson, "Low-cost mechanical fiber-optic switch," *OFC '95 Technical Digest*, vol. 8, San Diego, CA, pp. 185–186, Feb. 1995.

[Ayan89] E. Ayanoglu, "Signal flow graphs for path enumeration and deflection routing analysis in multihop networks," *Proceedings, IEEE Globecom '89*, Dallas, TX, pp. 1022–1029, Nov. 1989.

[AzBM96] M. Azizoglu, R. A. Barry, and A. Mokhtar, "Impact of tuning delay on the performance of bandwidth-limited optical broadcast networks with uniform traffic," *IEEE Journal on Selected Areas in Communications*, vol. 14, pp. 935–944, June 1996.

[AzSa92] M. Azizoglu, J. A. Salehi, and Y. Li, "Optical CDMA via temporal codes," *IEEE Transactions on Communications*, vol. 40, no. 7, pp. 1162–1170, July 1992.

[Aziz91] M. Azizoglu, *Phase Noise in Coherent Optical Communications*, Ph.D. Dissertation, Department of Electrical Engineering and Computer Science, Massachusetts Institute of Technology, 1991.

[BBCG86] E.-J. Bachus, R.-P. Braun, C. Casper, and E. Grossman, "Ten-channel coherent optical fiber transmission," *Electronic Letters*, vol. 22, pp. 1002–1003, 1986.

[BCHK96] R. A. Barry, V. W. S. Chan, K. L. Hall, E. S. Kintzer, et al., "All-optical network consortium – Ultrafast TDM networks," *IEEE Journal on Selected Areas in Communications*, vol. 14, no. 5, pp. 999–1013, June 1996.

[BESO92] A. Budman, E. Eichen, J. Schlafer, R. Olshansky, and F. McAleavey, "Multigigabit optical packet switch for self-routing networks with subcarrier addressing," *Technical Digest, Optical Fiber Communications Conference (OFC '92)*, San Jose, CA, pp. 90–91, Feb. 1992.

[BGDB96] L. P. Barry, P. Guignard, J. Debeau, R. Boittin, and M. Bernard, "A high-speed optical star network using TDMA and all-optical demultiplexing techniques," *IEEE Journal on Selected Areas in Communications*, vol. 14, no. 5, pp. 1030–1038, June 1996.

[BMSW94] E. Bradley, E. Miles, R. Stone, and E. Wooten, "High speed tunable optical filter with variable channel spacing," *Proceedings, IEEE/LEOS '94 Summer Topical Meetings on Optical Networks and their Enabling Technologies*, Lake Tahoe, NV, pp. 57–58, July 1994.

[BaBr94] K. Bala and C. A. Brackett, "Cycles in wavelength-routed optical networks," *IEEE/LEOS Summer Topical Meeting Digest, Optical Networks and Their Enabling Technologies*, Lake Tahoe, NV, pp. 7-8, July 1994.

[BaFG90] J. A. Bannister, L. Fratta, and M. Gerla, "Topological design of the wavelength-division optical network," *Proceedings, IEEE INFOCOM '90*, San Francisco, CA, pp. 1005–1013, June 1990.

[BaHu94] R. A. Barry and P. A. Humblet, "On the number of wavelengths and switches in all-optical networks," *IEEE Transactions on Communications*, vol. 42, pp. 583–591, Feb.-April 1994.

[BaHu96] R. A. Barry and P. A. Humblet, "Models of blocking probability in all-optical networks with and without wavelength changers," *IEEE Journal on Selected Areas in Communications*, vol. 14, pp. 858–867, June 1996.

[BaMS94a] S. Banerjee, B. Mukherjee, and D. Sarkar, "Heuristic algorithms for constructing optimized structures of linear multihop lightwave networks," *IEEE Transactions on Communications*, vol. 42, pp. 1811–1826, Feb.–Apr. 1994.

[BaMS94b] D. Banerjee, B. Mukherjee, and S. Ramamurthy, "The multidimensional torus: Analysis of average hop distance and application as a multihop lightwave network," *Proceedings, IEEE International Conference on Communications (ICC '94)*, New Orleans, pp. 1675–1680, May 1994.

[BaMu93a] S. Banerjee and B. Mukherjee, "The photonic ring: Algorithms for optimized node arrangements," *Journal of Fiber and Integrated Optics, Special issue on Networking with Optical Technology*, vol. 12, pp. 133–171, 1993.

[BaMu93b] S. Banerjee and B. Mukherjee, "Algorithms for optimized node arrangements in shufflenet based multihop lightwave networks," *Proceedings, IEEE INFOCOM '93*, San Francisco, CA, pp. 557–564, March 1993.

[BaMu93c] S. Banerjee and B. Mukherjee, "Fairnet: A WDM-based multiple channel lightwave network with adaptive and fair scheduling policy," *IEEE/OSA Journal of Lightwave Technology*, vol. 11, pp. 1104–1112, May/June 1993.

[BaMu96] D. Banerjee and B. Mukherjee, "Practical approaches for routing and wavelength assignment in large all-optical wavelength-routed networks," *IEEE Journal on Selected Areas in Communications*, vol. 14, pp. 903–908, June 1996.

[BaSB91] K. Bala, T. E. Stern, and K. Bala, "Algorithms for routing in a linear lightwave network," *Proceedings, IEEE INFOCOM '91*, Bal Harbour, FL, Apr. 1991.

[Bala96] K. Bala et al., "WDM network economics," *Proceedings, NFOEC '96*, Denver, CO, 1996.

[Bane92] S. Banerjee, *Optimally-Configured High-Speed Networks*, Ph.D. Dissertation, University of California, Davis, Department of Computer Science, Dec. 1992.

[Barr93] R. A. Barry, *Wavelength Routing in All-Optical Networks*, Ph.D. Dissertation, Massachusetts Institute of Technology, Sept. 1993.

[Bell91] Bellcore Technical Advisory, TR-TSY-000496, Issue 2, Supplement 1, Sept. 1991.

[Berk94] M. Berkelaar, "lpsolve: Readme file," Documentation for the lpsolve program, 1994.

[BiGu92] D. Bienstock and O. Gunluk, "A degree sequence problem related to network design," *Networks*, vol. 24, pp. 195–205, July 1994.

[Birm96] A. Birman, "Computing approximate blocking probabilities for a class of all-optical networks," *IEEE Journal on Selected Areas in Communications*, vol. 14, pp. 852–857, June 1996.

[BoDo91] K. Bogineni and P. W. Dowd, "Collisionless media access protocols for high speed communication in optically interconnected parallel computers," *Proceedings, SPIE*, vol. 1577, pp. 276–287, Sept. 1991.

[BoFB94] F. Borgonovo, L. Fratta, and J. A. Bannister, "On the design of optical deflection-routing networks," *Proceedings, IEEE INFOCOM '94*, Toronto, Canada, pp. 120–129, June 1994.

[BoMu95b] M. S. Borella and B. Mukherjee, "Limits of multicasting in a packet-switched WDM single-hop local lightwave network," *Journal of High Speed Networks*, vol. 4, no. 2, pp. 156–167, 1995.

[BoMu95c] M. S. Borella and B. Mukherjee, "A reservation-based multicasting protocol for WDM local lightwave networks," *Proceedings, IEEE International Conference on Communications (ICC '95)*, Seattle, WA, pp. 1277–1281, June 1995.

[BoMu96] M. S. Borella and B. Mukherjee, "Efficient scheduling of nonuniform packet traffic in a WDM/TDM local lightwave network with arbitrary transceiver tuning latencies," *IEEE Journal on Selected Areas in Communications*, vol. 14, pp. 923–934, June 1996.

[BoSD93] K. Bogineni, K. M. Sivalingham, and P. W. Dowd, "Low-complexity multiple access protocols for wavelength-division multiplexed photonic networks," *IEEE Journal on Selected Areas in Communications*, vol. 11, pp. 590–604, May 1993.

[Brac90] C. A. Brackett, "Dense wavelength division multiplexing networks: Principles and applications," *IEEE Journal on Selected Areas in Communications*, vol. 8, pp. 948–964, Aug. 1990.

[Brac93] C. A. Brackett et al., "A scalable multiwavelength multihop optical network: A proposal for research on all-optical networks," *IEEE/OSA Journal of Lightwave Technology*, vol. 11, pp. 736–753, May/June 1993.

[CAMS96] T.-K. Chiang, S. K. Agrawal, D. T. Mayweather, D. Sadot, C. F. Barry, M. Hickey, and L. G. Kazovsky, "Implementation of STARNET: A WDM computer communications network," *IEEE Journal on Selected Areas in Communications*, vol. 14, pp. 824–839, June 1996.

[CCLA95] K. C. Chan, H. H. Y. Cheng, H. F. Liu, and K. A. Ahmed, "Application of optical solitons in high-speed and wavelength-division-multiplexed lightwave communication systems," *Journal of High Speed Networks*, vol. 4, pp. 41-59, 1995.

[CEFS97] I. Chlamtac, V. Elek, A. Fumagalli, and C. Szabo, "Scalable WDM network architecture based on photonic slot routing and switched delay lines," *Proceedings, IEEE INFOCOM '97*, Kobe, Japan, April 1997.

[CEGI96] G.-K. Chang, G. Ellinas, J. K. Gamelin, M. Z. Iqbal, and C. A. Brackett, "Multiwavelength reconfigurable WDM/ATM/SONET network testbed," *IEEE/OSA Journal of Lightwave Technology*, vol. 14, pp. 1320–1340, June 1996.

[CFKM96] I. Chlamtac, A. Fumagalli, L. G. Kazovsky, P. Melman, et al., "CORD: Contention resolution by delay lines," *IEEE Journal on Selected Areas in Communications*, vol. 14, pp. 1014–1029, June 1996.

[CSBH89] K.-W. Cheung, D. A. Smith, J. E. Baron, and B. L. Heffner, "Multiple channel operation of an integrated acousto-optic tunable filter," *Electronics Letters*, vol. 25, pp. 375–376, 1989.

[ChBa95] C. Chen and S. Banerjee, "Optical switch configuration and lightpath assignment in wavelength routing multihop lightwave networks," *Proceedings, IEEE INFOCOM '95*, Boston, MA, pp. 1300–1307, June 1995.

[ChBa96] C. Chen and S. Banerjee, "A new model for optimal routing and wavelength assignment in wavelength division multiplexed optical networks," *Proceedings, IEEE INFOCOM '96*, pp. 164–171, 1996.

[ChDR90] M.-S. Chen, N. R. Dono, and R. Ramaswami, "A media access protocol for packet-switched wavelength division multiaccess metropolitan area networks," *IEEE Journal on Selected Areas in Communications*, vol. 8, pp. 1048–1057, Aug. 1990.

[ChFZ96] I. Chlamtac, A. Faragó, and T. Zhang, "Lightpath (wavelength) routing in large WDM networks," *IEEE Journal on Selected Areas in Communications*, vol. 14, pp. 909–913, June 1996.

[ChFu91] I. Chlamtac and A. Fumagalli, "Quadro-stars: High performance optical WDM star networks," *Proceedings, IEEE Globecom '91*, Phoenix, AZ, pp. 1224–1229, Dec. 1991.

[ChFu94] I. Chlamtac and A. Fumagalli, "Quadro-Star: A high performance optical WDM star network," *IEEE Transactions on Communications*, vol. 42, pp. 2582–2591, Aug. 1994.

[ChGK92] I. Chlamtac, A. Ganz, and G. Karmi, "Lightpath communications: An approach to high bandwidth optical WANs," *IEEE Transactions on Communications*, vol. 40, pp. 1171–1182, July 1992.

[ChGK93] I. Chlamtac, A. Ganz, and G. Karmi, "Lightnets: Topologies for high speed optical networks," *IEEE/OSA Journal of Lightwave Technology*, vol. 11, pp. 951–961, May/June 1993.

[ChGa88a] I. Chlamtac and A. Ganz, "Channel allocation protocols in frequency-time controlled high speed networks," *IEEE Transactions on Communications*, vol. 36, pp. 430–440, April 1988.

[ChGa88b] I. Chlamtac and A. Ganz, "A multibus train communication architecture for high-speed fiber optic networks," *IEEE Journal on Selected Areas in Communications*, vol. SAC-6, pp. 903–912, July 1988.

[ChGa90] I. Chlamtac and A. Ganz, "Towards alternative high speed network concepts: The swift architecture," *IEEE Transactions on Communications*, vol. 38, pp. 431–439, April 1990.

[ChNT92] A. R. Chraplyvy, J. A. Nagel, and R. W. Tkach, "Equalization in amplified WDM lightwave transmission systems," *IEEE Photonics Technology Letters*, vol. 4, pp. 920–922, Aug. 1992.

[ChYu91] M. Chen and T.-S. Yum, "A conflictfree protocol for optical WDM networks," *Proceedings, IEEE Globecom '91*, Phoenix, AZ, pp. 1276–1291, Dec. 1991.

[ChYu94] K. M. Chan and T. S. Yum, "Analysis of least congested path routing in WDM lightwave networks," *Proceedings, IEEE INFOCOM '94*, Toronto, Canada, pp. 962–969, June 1994.

[ChZA92] R. Chipalkatti, Z. Zhang, and A. S. Acampora, "High-speed communication protocols for optical star networks using WDM," *Proceedings, IEEE INFOCOM '92*, Florence, Italy, pp. 2124–2133, May 1992.

[Chra84] A. R. Chraplyvy, "Optical power limits in multi-channel wavelength-division-multiplexed systems due to stimulated Raman scattering," *Electronics Letters*, vol. 20, no. 2, pp. 58–59, 1984.

[Chra90] A. R. Chraplyvy, "Limits on lightwave communications imposed by optical-fiber nonlinearities," *IEEE/OSA Journal of Lightwave Technology*, vol. 8, pp. 1548–1557, Oct. 1990.

[CiGS93] I. Cidon, I. Gopal, and A. Segall, "Connection establishment in high-speed networks," *IEEE/ACM Transactions on Networking*, vol. 1, pp. 469–482, Aug. 1993.

[Coch95] P. Cochrane, *Optical Network Technology – Foreword*, New York: Chapman & Hall, 1995.

[DGLR90] N. R. Dono, P. E. Green, K. Liu, R. Ramaswami, and F. F.-K. Tong, "A wavelength division multiple access network for computer communication," *IEEE Journal on Selected Areas in Communications*, vol. 8, pp. 983–993, Aug. 1990.

[DJMP94] T. Durhuus et al., "All optical wavelength conversion by SOA's in a Mach-Zender configuration," *IEEE Photonic Technology Letters*, vol. 6, pp. 53–55, Jan 1994.

[DMJD96] T. Durhuus et al., "All-optical wavelength conversion by semiconductor optical amplifiers," *IEEE/OSA Journal of Lightwave Technology*, vol. 14, pp. 942–954, June 1996.

[Darc87] T. E. Darcie, "Subcarrier multiplexing for multiple-access lightwave networks," *IEEE/OSA Journal of Lightwave Technology*, vol. LT-5, pp. 1103–1110, 1987.

[DeNa95] J.-M. P. Delavaux and J. A. Nagel, "Multi-stage erbium-doped fiber amplifier designs," *IEEE/OSA Journal of Lightwave Technology*, vol. 13, pp. 703–720, May 1995.

[DSGP93] F. Delorme, S. Slempkes, P. Gambini, and M. Puleo, "Fast tunable distributed-Bragg-reflector laser for optical switching," *OFC/IOOC '93 Technical Digest*, San Jose, CA, pp. 36–38, 1993.

[Dela94] D. Delacourt, "Integrated optics on lithium niobate," in *Advances in Integrated Optics* (S. Martellucci, A. N. Chester, and M. Bertolotti, eds.), Chapter 4, pp. 79–93, New York: Plenum Press, 1994.

[DoBo92] P. W. Dowd and K. Bogineni, "Simulation analysis of a collisionless multiple access protocol for a wavelength division multiplexed star-coupled configuration," *Proceedings, 25th Annual Simulation Symposium*, Orlando, FL, April 1992.

[Dowd91] P. W. Dowd, "Random access protocols for high speed interprocessor communication based on an optical passive star topology," *IEEE/OSA Journal of Lightwave Technology*, vol. 9, pp. 799–808, June 1991.

[Dowd92] P. W. Dowd, "Wavelength division multiple access channel hypercube processor interconnection," *IEEE Transactions on Computers*, 1992.

[DrEK91] C. Dragone, C. A. Edwards, and R. C. Kistler, "Integrated optics $n \times n$ multiplexer on silicon," *IEEE Photonics Technology Letters*, vol. 3, pp. 896–899, Oct. 1991.

[DHKK89] C. Dragone, C. H. Henry, I. P. Kaminow, and R. C. Kistler, "Efficient multichannel integrated optics star coupler on silicon," *IEEE Photonics Technology Letters*, vol. 1, pp. 241–243, Aug. 1989.

[EGZJ93] A. F. Elrefaie, E. L. Goldstein, S. Zaidi, and N. Jackman, "Fiber-amplifier cascades with gain equalization in multiwavelength unidirectional inter-office ring network," *IEEE Photonics Technology Letters*, vol. 5, pp. 1026–1031, Sept. 1993.

[EdKa72] J. Edmonds and R. M. Karp, "Theoretical improvements in the algorithmic efficiency for network flow problems," *Journal of the ACM*, vol. 19, pp. 248-264, 1972.

[EiMe88] M. Eisenberg and N. Mehravari, "Performance of the multichannel multihop lightwave network under nonuniform traffic," *IEEE Journal on Selected Areas in Communications*, vol. 6, pp. 1063–1078, Aug. 1988.

[EiPW93] M. Eiselt, W. Pieper, and H. G. Weber, "Decision gate for all-optical retiming using a semiconductor laser amplifier in a loop mirror configuration," *Electronic Letters*, vol. 29, pp. 107–109, Jan. 1993.

[Elre92] A. F. Elrefaie, "Self-healing WDM ring networks with all-optical protection path," *Technical Digest, OFC '92*, San Jose, CA, pp. 255–256, Feb. 1992.

[Elre93] A. F. Elrefaie, "Multiwavelength survivable ring network architectures," *Proceedings, IEEE International Conference on Communications (ICC '93)*, pp. 1245–1251, Geneva, May 1993.

[EvIS76] S. Even, A. Itai, and A. Shamir, "On the complexity of timetable and multicommodity flow problems," *SIAM Journal of Computing*, vol. 5, pp. 691–703, 1976.

[FDDI90] International Standards Organization, *Information Processing Systems – Fiber Distributed Data Interface*, Parts I, II, and III, ISO 9314-1-1989, ISO 9314-2-1989, ISO 9314-3-1990.

[FTCM94] F. Forghieri, R. W. Tkach, A. R. Chraplyvy, and D. Marcuse, "Reduction of four-wave mixing crosstalk in WDM systems using unequally spaced channels," *IEEE Photonics Technology Letters*, vol. 6, no. 6, pp. 754–756, 1994.

[FrGK73] L. Fratta, M. Gerla, and L. Kleinrock, "The flow deviation method: An approach to store-and-forward communication network design," *Networks*, vol. 3, pp. 97–133, 1973.

[FuTa83] A. Fukuda and S. Tasaka, "The equilibrium point analysis – a unified analytic tool for packet broadcast networks," *Proceedings, IEEE GLOBECOM '83*, San Diego, CA, pp. 33.4.1–33.4.8, Nov. 1983.

[Fuji88] M. Fujiwara et al., "A coherent photonic wavelength-division switching system for broadband networks," *Proceedings, European Conference on Communications (ECOC '88)*, pp. 139–142, 1988.

[GGKV90] M. S. Goodman, J. L. Gimlett, H. Kobrinski, M. P. Vecchi, and R. M. Bulley, "The LAMBDANET multiwavelength network: Architecture, applications, and demonstrations," *IEEE Journal on Selected Areas in Communications*, vol. 8, pp. 995–1004, Aug. 1990.

[GaGa92a] A. Ganz and Y. Gao, "A time-wavelength assignment algorithm for a WDM star network," *Proceedings, IEEE INFOCOM '92*, Florence, Italy, pp. 2144–2150, May 1992.

[GaGa92b] A. Ganz and Y. Gao, "Traffic scheduling in multiple WDM star systems," *Proceedings, IEEE International Conference on Communications (ICC '92)*, Chicago, IL, June 1992.

[GaJo79] M. R. Garey and D. S. Johnson, *Computers and Intractability: A Guide to the Theory of NP-Completeness*, New York: W. H. Freeman and Company, 1979.

[GaKo91] A. Ganz and Z. Koren, "WDM passive star protocols and performance analysis," *Proceedings, IEEE INFOCOM '91*, Bal Harbour, FL, pp. 991–1000, April 1991.

[GaLZ92] A. Ganz, B. Li, and L. Zenou, "Reconfigurability of multi-star based lightwave LANs," *Proceedings, IEEE International Conference on Communications (ICC '92)*, Chicago, IL, June 1992.

[GaLi92] A. Ganz and B. Li, "Broadcast-wavelength architecture for a WDM passive star-based local area network," *Proceedings, IEEE International Conference on Communications (ICC '92)*, Chicago, IL, pp. 837–842, June 1992.

[GiDe91] C. R. Giles and E. Desurvire, "Modeling erbium-doped fiber amplifiers," *IEEE/OSA Journal of Lightwave Technology*, vol. 9, no. 2, pp. 271–283, Feb. 1991.

[GiTe91] R. Gidron and A. Temple, "Teranet: A multihop multichannel ATM lightwave network," *Proceedings, IEEE International Conference on Communications (ICC '91)*, Denver, CO, pp. 602–608, June 1991.

[GlKW94] B. Glance, I. P. Kaminow, and R. W. Wilson, "Applications of the integrated waveguide grating router," *IEEE/OSA Journal of Lightwave Technology*, vol. 12, pp. 957–962, June 1994.

[GlSc90] B. S. Glance and O. Scaramucci, "High-performance dense FDM coherent optical network," *IEEE Journal on Selected Areas in Communications*, vol. 6, pp. 1043–1047, Aug. 1990.

[GoHo88] O. Goldschmidt and D. S. Hochbaum, "Polynomial algorithm for the k-cut problem," *Proceedings, 29th Annual Symposium on the Foundations of Computer Science*, pp. 444–451, 1988.

[GrLL86] L. Grippo, F. Lampariello, and S. Lucidi, "A nonmonotone line search technique for Newton's method," *SIAM Journal on Numerical Analysis*, vol. 23, pp. 707-716, 1986.

[Gree92] P. E. Green, "Misunderstood issues in lightwave networking," *OMAN Summer Topicals*, Santa Barbara, CA, pp. 47–48, Aug. 1992.

[Gree93] P. E. Green, *Fiber Optic Networks*, Englewood Cliffs, NJ: Prentice-Hall, 1993.

[Gree96] P. E. Green, "Optical networking update," *IEEE Journal on Selected Areas in Communications*, vol. 14, pp. 764–779, June 1996.

[GLTJ92] M. Gustavsson, B. Lagerstrom, L. Thylen, M. Janson, L. Lundgren, A.-C. Morner, M. Rask, and B. Stoltz, "Monolithically integrated 4 × 4 InGaAsP/InP laser amplifier gate switch arrays," *Electronic Letters*, vol. 28, pp. 2224–2225, Nov. 1992.

[HaDL92] S. L. Hansen, K. Dybdal, and L. C. Larsen, "Gain limit in erbuim-doped fiber amplifiers due to internal Rayleigh backscattering," *IEEE Photonics Technology Letters*, vol. 4, pp. 559–561, June 1992.

[HaKR96] E. Hall, J. Kravitz, R. Ramaswami, et al., "The Rainbow-II gigabit optical network," *IEEE Journal on Selected Areas in Communications*, vol. 14, pp. 814–823, June 1996.

[HaKS87] I. M. I. Habbab, M. Kavehrad, and C.-E. W. Sundberg, "Protocols for very high speed optical fiber local area networks using a passive star topology," *IEEE/OSA Journal of Lightwave Technology*, vol. LT-5, pp. 1782–1794, Dec. 1987.

[HaSo96] S. F. Habiby and M. J. Soulliere, "WDM standards: A first impression," *SPIE Photonics West*, vol. 2690, San Jose, CA, pp. 97–108, Jan. 1996.

[Haas93] Z. Haas, "The 'staggering switch': An electronically controlled optical packet switch," *IEEE/OSA Journal of Lightwave Technology*, vol. 11, pp. 925–936, May/June 1993.

[Hech92] J. Hecht, *Understanding Lasers: An Entry-Level Guide*, New York: IEEE Press, 1992.

[Hech93] J. Hecht, *Understanding Fiber Optics*, vol. 2, Indianapolis, IN: Sams Publishing, 1993.

[Henr85] P. S. Henry, "Lightwave primer," *IEEE Journal of Quantum Electronics*, vol. QE-21, pp. 1862–1879, Dec. 1985.

[Hill93] G. Hill et al., "A transport network layer based on optical network elements," *IEEE/OSA Journal of Lightwave Technology*, vol. 11, pp. 667–679, May/June 1993.

[Hint90] H. S. Hinton, "Photonic switching fabrics," *IEEE Communications Magazine*, vol. 28, pp. 71–89, Apr. 1990.

[HlKa91] M. G. Hluchyj and M. J. Karol, "ShuffleNet: An application of generalized perfect shuffles to multihop lightwave networks," *IEEE/OSA Journal of Lightwave Technology*, vol. 9, pp. 1386–1397, Oct. 1991.

[HoMu90] W. W. Ho and B. Mukherjee, "A multiple-partition token ring network," *Computer Communications*, vol. 14, no. 3, pp. 133-142, April 1991.

[HuHa90] P. A. Humblet and W. M. Hamdy, "Crosstalk analysis and filter optimization of single- and double-cavity Fabry-Perot filters," *IEEE Journal on Selected Areas in Communications*, vol. 8, pp. 1095–1107, Aug. 1990.

[HuRS93] P. A. Humblet, R. Ramaswami, and K. N. Sivarajan, "An efficient communication protocol for high-speed packet-switched multichannel networks," *IEEE Journal on Selected Areas in Communications*, vol. 11, pp. 568–578, May 1993.

[InBB94] J. Iness, S. Banerjee, and B. Mukherjee, "GEMNET: A generalized, shuffle-exchange-based, regular, scalable, and modular multihop network based on WDM lightwave technology," Technical Report CSE-94-8, Department of Computer Science, University of California, Davis, June 1994.

[InMu96] J. Iness and B. Mukherjee, "Simulation-based case study of three degrees of sparse wavelength conversion in wavelength routed optical networks," Technical Report CSE-96-16, Department of Computer Science, University of California, Davis, Dec. 1996.

[ISII95] A. Inoue, M. Shigehara, M. Ito, M. Inai, Y. Hattori, and T. Mizunami, "Fabrication and application of fiber Bragg grating – A review," *Optoelectronics - Devices and Technologies*, vol. 10, pp. 119–130, Mar. 1995.

[Ines97] J. Iness, *Efficient Use of Optical Components in WDM-Based Optical Networks*, Ph.D. Dissertation, University of California, Davis, Department of Computer Science, 1997.

[Ishi91] A. Ishimaru, *Electromagnetic Wave Propagation, Radiation, and Scattering*, Englewood Cliffs, NJ: Prentice Hall, 1991.

[JHSN95] "Special Issue on WDM Networks," *Journal of High-Speed Networks*, vol. 4, no. 1/2, 1995.

[JLT93] "Special Issue on Broad-Band Optical Networks," *IEEE/OSA Journal of Lightwave Technology*, vol. 11, May/June 1993.

[JLT96] "Special Issue on Multiwavelength Optical Technology and Networks," *IEEE/OSA Journal of Lightwave Technology*, vol. 14, June 1996.

[JSAC90] "Special Issue on Dense Wavelength Division Multiplexing Techniques for High Capacity and Multiple Access Communication Systems," *IEEE Journal on Selected Areas in Communications*, vol. 8, Aug. 1990.

[JSAC96] "Special Issue on Optical Networks," *IEEE Journal on Selected Areas in Communications*, vol. 14, June 1996.

[JaRS93] F. J. Janneillo, R. Ramaswami, and D. G. Steinberg, "A prototype circuit-switched multi-wavelength optical metropolitan-area network," *IEEE/OSA Journal of Lightwave Technology*, vol. 11, pp. 777–782, May/June 1993.

[JeAy96] G. Jeong and E. Ayanoglu, "Comparison of wavelength-interchanging and wavelength-selective cross-connects in multi-wavelength all-optical networks," *Proceedings, IEEE INFOCOM '96*, San Francisco, CA, pp. 156–163, March 1996.

[JiMI95] F. Jia, B. Mukherjee, and J. Iness, "Scheduling variable-length messages in a single-hop multichannel local lightwave network," *IEEE/ACM Transactions on Networking*, vol. 3, pp. 477–488, Aug. 1995.

[JiMu92a] F. Jia and B. Mukherjee, "Performance analysis of a generalized receiver collision avoidance (RCA) protocol for single-hop WDM lightwave networks," *Proceedings, SPIE '92*, Boston, MA, Sept. 1992.

[JiMu92b] F. Jia and B. Mukherjee, "Bimodal throughput, nonmonotonic delay, optimal bandwidth dimensioning, and analysis of receiver collisions in a single-hop WDM local lightwave network," *Proceedings, IEEE Globecom '92*, Orlando, FL, pp. 1896–1900, Dec. 1992.

[JiMu93a] F. Jia and B. Mukherjee, "The receiver collision avoidance (RCA) protocol for a single-hop WDM lightwave network," *IEEE/OSA Journal of Lightwave Technology*, vol. 11, pp. 1053–1065, May/June 1993.

[JiMu93b] F. Jia and B. Mukherjee, "A high-capacity, packet switched, single-hop local lightwave network," *Proceedings, IEEE GLOBECOM '93*, Houston, TX, pp. 1110–1114, Dec. 1993.

[Jia93] F. Jia, *Architectures and Protocols for High-Speed Multichannel Networks*, Ph.D. Dissertation, University of California, Davis, Department of Computer Science, Sept. 1993.

[Joha96] S. Johansson, "Transport network involving a reconfigurable WDM network layer – An European demonstration," *IEEE/OSA Journal of Lightwave Technology*, vol. 14, pp. 1341–1348, June 1996.

[KOYM87] H. Kawaguchi et al., "Tunable optical wavelength conversion using a multielectrode distributed-feedback laser diode with a saturable absorber," *Electronic Letters*, vol. 23, no. 20, pp. 1088–1090, 1987.

[KaPo93] L. G. Kazovsky and P. T. Poggiolini, "STARNET: A multi-gigabit-per-second optical LAN using a passive WDM star," *IEEE/OSA Journal of Lightwave Technology*, vol. 11, pp. 1009–1027, May/June 1993.

[KaSh91] M. J. Karol and S. Z. Shaikh, "A simple adaptive routing scheme for congestion control in ShuffleNet multihop lightwave networks," *IEEE Journal on Selected Areas in Communications*, vol. 9, pp. 1040–1051, Sept. 1991.

[Kami90] I. P. Kaminow, "FSK with direct detection in optical multiple-access FDM networks," *IEEE Journal on Selected Areas in Communications*, vol. 6, pp. 1005–1014, Aug. 1990.

[Karp72] R. M. Karp, "Reducibility among combinatorial problems," in *Complexity of Computer Computations*, New York: Plenum Press, pp. 85–104, 1972.

[KeLi70] B. W. Kernighan and S. Lin, "An efficient heuristic procedure for partitioning graphs," *Bell Systems Technical Journal*, vol. 49, pp. 291–307, 1970.

[KiGV83] S. Kirkpatrick, C. D. Gelatt, and M. P. Vecchi, "Optimization by simulated annealing," *Science*, pp. 671–680, May 1983.

[KoAc96a] M. Kovačević and A. S. Acampora, "Benefits of wavelength translation in all-optical clear-channel networks," *IEEE Journal on Selected Areas in Communications*, vol. 14, pp. 868–880, June 1996.

[KoAc96b] M. Kovačević and A. S. Acampora, "Electronic wavelength translation in optical networks," *IEEE/OSA Journal of Lightwave Technology*, vol. 14, pp. 1161–1169, June 1996.

[KoCh89] H. Kobrinski and K.-W. Cheung, "Wavelength-tunable optical filters: Applications and technologies," *IEEE Communications Magazine*, vol. 27, pp. 53–63, Oct. 1989.

[KoGB93] M. Kovacevic, M. Gerla, and J. A. Bannister, "Time and wavelength division multiaccess with acoustooptic tunable filters," *Journal of Fiber and Integrated Optics*, vol. 12, pp. 113–132, Aug. 1993.

[KoGB95] M. Kovacevic, M. Gerla, and J. A. Bannister, "On the performance of shared-channel multihop lightwave networks," *Proceedings, IEEE INFOCOM '95*, Boston, MA, pp. 544–551, Apr. 1995.

[KrHa90] A. Krishna and B. Hajek, "Performance of shuffle-like switching networks with deflection," *Proceedings, IEEE INFOCOM '90*, San Francisco, CA, June 1990.

[Krup82] R. S. Krup, "Stabilization of alternate routing networks," *Proceedings, IEEE International Conference on Communications (ICC '82)*, Philadelphia, PA, June 1982.

[LTGC94] C.-S. Li, F. F.-K. Tong, C. J. Georgiou, and M. Chen, "Gain equalization in metropolitan and wide area optical networks using optical amplifiers," *Proceedings, IEEE INFOCOM '94*, Toronto, Canada, pp. 130–137, June 1994.

[LaAc90b] J.-F. P. Labourdette and A. S. Acampora, "Partially reconfigurable multihop lightwave networks," *Proceedings, IEEE Globecom '90*, San Diego, CA, pp. 34–40, Dec. 1990.

[LaAc91] J.-F. P. Labourdette and A. S. Acampora, "Logically rearrangeable multihop lightwave networks," *IEEE Transactions on Communications*, vol. 39, pp. 1223–1230, Aug. 1991.

[LaAc94] J.-F. P. Labourdette and A. S. Acampora, "Reconfiguration algorithms for rearrangeable lightwave networks," *IEEE Transactions on Communications*, vol. 39, pp. 1223–1230, Aug. 1994.

[LaHA94] J. F. P. Labourdette, G. W. Hart, and A. S. Acampora, "Branch-exchange sequences for reconfiguration of lightwave networks," *IEEE Transactions on Communications*, vol. 42, pp. 2822–2832, Oct. 1994.

[LaPT96] J. P. R. Lacey, G. J. Pendock, and R. S. Tucker, "Gigabit-per-second all-optical 1300-nm to 1550-nm wavelength conversion using cross-phase modulation in a semiconductor optical amplifier.," *Proceedings, Optical Fiber Communication (OFC '96)*, San Jose, CA, vol. 2, pp. 125–126, 1996.

[LaSa90] A. Lam and D. Sarwate, "Time-hopping and frequency-hopping multiple-access packet communications," *IEEE Transactions on Communications*, vol. 38, pp. 875–888, June 1990.

[LeCh96] T.-P. Lee et al., "Multiwavelength DFB laser array transmitters for ONTC reconfigurable optical network testbed," *IEEE/OSA Journal of Lightwave Technology*, vol. 14, pp. 967–976, June 1996.

[LeLi93] K.-C. Lee and V. O. K. Li, "A wavelength convertible optical network," *IEEE/OSA Journal of Lightwave Technology*, vol. 11, pp. 962–970, May/June 1993.

[LeLi94a] K. C. Lee and V. O. K. Li, "A circuit rerouting algorithm for all-optical wide-area networks," *Proceedings, IEEE INFOCOM '94*, Toronto, Canada, pp. 954–961, June 1994.

[LeLi94b] K. C. Lee, V. O. K. Li, et al., "Routing for all-optical networks using wavelengths outside erbium-doped fiber amplifier bandwidth," *IEEE/OSA Journal of Lightwave Technology*, vol. 13, pp. 791–801, May 1995.

[LeLi95] K. C. Lee and V. O. K. Li, "Optimization of a WDM packet switch with wavelength converters," *Proceedings, IEEE INFOCOM '95*, Boston, MA, pp. 423–430, June 1995.

[LeSu94] C. Lee and T. Su, "2*2 single-mode zero-gap directional-coupler thermo-optic waveguide switch on glass," *Applied Optics*, vol. 33, pp. 7016–7022, Oct. 1994.

[LeZa89] T.-P. Lee and C.-E. Zah, "Wavelength-tunable and single-frequency lasers for photonic communications networks," *IEEE Communications Magazine*, vol. 27, pp. 42–52, Oct. 1989.

[LiGa92] B. Li and A. Ganz, "Virtual topologies for WDM star LANs: The regular structure approach," *Proceedings, IEEE INFOCOM '92*, Florence, Italy, pp. 2134–2143, May 1992.

[Lin90] H.-D. Lin, "Gain splitting and placement of distributed amplifiers," IBM Research Report RC 16216 (#72010), Oct. 1990.

[LuRa94] R. Ludwig and G. Raybon, "BER measurements of frequency converted signals using four-wave mixing in a semiconductor laser amplifier at 1, 2.5, 5, and 10 Gbit/s," *Electronic Letters*, vol. 30, pp. 338–339, Jan 1994.

[MBBG96] F. Masetti, J. Benoit, F. Brillouet, J. M. Gabriagues, et al., "High speed high capacity ATM optical switches for future telecommunication transport networks," *IEEE Journal on Selected Areas in Communications*, vol. 14, no. 5, pp. 979–998, June 1996.

[MDJD96] B. Mikkelsen et al., "Wavelength conversion devices," *Proceedings, Optical Fiber Communication (OFC '96)*, San Jose, CA, vol. 2, pp. 121–122, 1996.

[MDJP94] B. Mikkelsen et al., "Polarization insensitive wavelength conversion of 10 Gbit/s signals with SOAs in a Michelson interferometer," *Electronic Letters*, vol. 30, pp. 260–261, Feb 1994.

[MRBM94] B. Mukherjee, S. Ramamurthy, D. Banerjee, and A. Mukherjee, "Some principles for designing a wide-area optical network," *Proceedings, IEEE INFOCOM '94*, Toronto, Canada, pp. 110–119, June 1994.

[MWWP92] M. W. Maeda, J. R. Wullert II, A. E. Willner, J. Patel, and M. Allersma, "Wavelength-division multiple access network based on centralized common-wavelength control," *Technical Digest, Optical Fiber Communications Conference (OFC '92)*, San Jose, CA, p. 85, Feb. 1992.

[MaMI72] D. W. Matula, G. Marble, and J. D. Isaacson, "Graph coloring algorithms," in *Graph Theory and Computing* (R. C. Read, ed.), ch. 10, pp. 109–122, New York and London: Academic Press, 1972.

[MaON91] K. Magari, M. Okamoto, and Y. Noguchi, "1.55 μm polarization insensitive high gain tensile strained barrier MQW optical amplifier," *IEEE Photonics Technology Letters*, vol. 3, no. 11, pp. 998-1000, Nov. 1991.

[Maho93] M. J. O'Mahony, "Optical amplifiers," in *Photonics in Switching* (J. E. Midwinter, ed.), vol. 1, pp. 147–167, San Diego, CA: Academic Press, 1993.

[Maho95] M. J. O'Mahony et al., "The design of a European optical network," *IEEE/OSA Journal of Lightwave Technology*, vol. 13, pp. 817–828, May 1995.

[Matu72] D. W. Matula, "k-components, clusters and slicings in graphs," *SIAM Journal of Applied Mathematics*, vol. 22, 1972.

[Maxe85] N. F. Maxemchuk, "Regular mesh topologies in local and metropolitan area networks," *AT&T Technical Journal*, vol. 64, pp. 1659–1686, Sept. 1985.

[Maxe87] N. F. Maxemchuk, "Routing in the Manhattan street network," *IEEE Transactions on Communications*, vol. 35, pp. 503–512, May 1987.

[MeBo96] P. M. Melliar-Smith and J. E. Bowers, "Thunder and lightning," *ARPA Networking PI Meeting*, Charleston, SC, Feb. 1996.

[MePD95] S. Melle, C. P. Pfistner, and F. Diner, "Amplifier and multiplexing technologies expand network capacity," *Lightwave Magazine*, pp. 42–46, Dec. 1995.

[MeSe94] A. Merchant and B. Sengupta, "Multiway graph partitioning with applications to PCS networks," *Proceedings, IEEE INFOCOM '94*, Toronto, Canada, pp. 593–600, June 1994.

[Mehr90] N. Mehravari, "Performance and protocol improvements for very high speed optical fiber local area networks using a passive star topology," *IEEE/OSA Journal of Lightwave Technology*, vol. 8, pp. 520–530, April 1990.

[Mest95] D. J. G. Mestdagh, *Fundamentals of Multiaccess Optical Fiber Networks*, The Artech House Optoelectronics Library, Artech House, 1995.

[MoAz95a] A. Mokhtar and M. Azizoglu, "Hybrid multiacess for all-optical LANs with nonzero tuning delays," *Proceedings, IEEE International Conference on Communications (ICC '95)*, Seattle, WA, pp. 1272–1276, June 1995.

[MoAz95b] A. Mokhtar and M. Azizoglu, "Multiacess in all-optical networks with wavelength and code concurrency," *Journal of Fiber and Integrated Optics*, vol. 14, no. 1, pp. 37–51, 1995.

[MoAz96a] A. Mokhtar and M. Azizoglu, "Adaptive routing algorithms for wavelength-routed all-optical networks," *Proceedings, 1996 IEEE Midwest Symposium on Circuits and Systems*, Ames, IA, Aug. 1996.

[MoAz96b] A. Mokhtar and M. Azizoglu, "Adaptive wavelength routing in all-optical networks," *IEEE/ACM Transactions on Networking*, 1996 (submitted).

[MoTo93] J. B. Moore and D. E. Todd, "Recent developments in distributed feedback and distributed Bragg reflector lasers for wideband long-haul fiberoptic communication systems," *Proceedings, IEEE Southeastcon '93*, Charlotte, NC, p. 9, April 1993.

[Mukh92a] B. Mukherjee, "WDM-based local lightwave networks – Part I: Single-hop systems," *IEEE Network Magazine*, vol. 6, no. 3, pp. 12–27, May 1992.

[Mukh92b] B. Mukherjee, "WDM-based local lightwave networks – Part II: Multihop systems," *IEEE Network Magazine*, vol. 6, no. 4, pp. 20–32, July 1992.

[Naka93] M. Nakazawa et al., "Experimental demonstration of soliton data transmission over unlimited distances with soliton control in time and frequency domain," *Electronic Letters*, vol. 29, pp. 729-730, 1993.

[NewF94a] "New Focus, Inc., 1994 Product Catalog," 1994.

[NewF94b] Personal communication with representatives from New Focus, Inc., April 1994.

[NoTo95] C. A. Noronha and F. A. Tobagi, "Routing of multimedia streams in reconfigurable WDM optical networks," *Journal of High Speed Networks*, vol. 4, pp. 133–153, March 1995.

[OHLJ85] N. A. Olsson, J. Hegarty, R. A. Logan, L. F. Johnson, K. L. Walker, L. G. Cohen, B. L. Kasper, and J. C. Campbell, "68.3 km transmission with 1.37 Tbit km/s capacity using wavelength division multiplexing of ten single-frequency lasers at 1.5 μm," *Electronic Letters*, vol. 21, pp. 105–106, 1985.

[OkMS95] K. Okamoto, K. Moriwaki, and S. Suzuki, "Fabrication of 64$\times$64 arrayed-waveguide grating multiplexer on silicon," *Electronic Letters*, vol. 31, pp. 184–186, Feb. 1995.

[OkSu96] K. Okamoto and A. Sugita, "Flat spectral response arrayed-waveguide grating multiplexer with parabolic waveguide horns," *Electronic Letters*, vol. 32, pp. 1661–1662, Aug. 1996.

[OkTa91] K. Okamoto, H. Takahashi, et al., "Design and fabrication of integrated-optic 8$\times$8 star coupler," *Electronic Letters*, vol. 27, pp. 774–775, April 1991.

[Opti95] Optivision, Inc. home page, 1995, http://www.optivision.com/.

[PaSe95] G. Panchapakesan and A. Sengupta, "On multihop optical network topology using Kautz digraph," *Proceedings, IEEE INFOCOM '95*, Boston, MA, pp. 675–682, April 1995.

[PaSt86] D. B. Payne and J. R. Stern, "Transparent single mode fiber optical networks," *IEEE/OSA Journal of Lightwave Technology*, vol. LT-4, pp. 864–869, 1986.

[PaTi93] E. R. Panier and A. L. Tits, "On combining feasibility, descent and superlinear convergence in inequality constrained optimization," *Mathematical Programming*, vol. 59, pp. 261-276, 1993.

[Pank92] R. K. Pankaj, *Architectures for Linear Lightwave Networks*, Ph.D. Dissertation, Department of Electrical Engineering and Computer Science, MIT, Cambridge, MA, Sept. 1992.

[PiSa94] G. R. Pieris and G. H. Sasaki, "Scheduling transmissions in WDM broadcast-and-select networks," *IEEE/ACM Transactions on Networking*, vol. 2, pp. 105–110, April 1994.

[Powe93] J. P. Powers, *An Introduction to Fiber Optic Systems*, Homewood, IL: Irwin, 1993.

[PrSF86] P. R. Prucnal, M. A. Santoro, and T. R. Fan, "Spread spectrum fiber optic local area network using optical processing," *IEEE/OSA Journal of Lightwave Technology*, vol. 4, pp. 547–554, May 1986.

[RaIM96] B. Ramamurthy, J. Iness, and B. Mukherjee, "Optimizing amplifier placements in a multi-wavelength optical network," Technical Report CSE-96-17, Department of Computer Science, University of California, Davis, Dec. 1996.

[RaIM97] B. Ramamurthy, J. Iness, and B. Mukherjee, "Minimizing the number of optical amplifiers needed to support a multiwavelength optical MAN," *Proceedings, IEEE INFOCOM '97*, Kobe, Japan, April 1997.

[RaSa84] P. Rasky and D. Sarwate, "Performance analysis of a time-hopping multiple-access communication scheme," *Proceedings, Eighteenth Annual Conference on Information Science and Systems*, Princeton University, NJ, pp. 131–132, March 1984.

[RaSa97] R. Ramaswami and G. H. Sasaki, "Multiwavelength optical networks with limited wavelength conversion," *Proceedings, IEEE INFOCOM '97*, Kobe, Japan, April 1997.

[RaSe96] R. Ramaswami and A. Segall, "Distributed network control for wavelength routed optical networks," *Proceedings, IEEE INFOCOM '96*, San Francisco, CA, pp. 138–147, March 1996.

[RaSi94] R. Ramaswami and K. Sivarajan, "Optimal routing and wavelength assignment in all-optical networks," *IEEE/ACM Transactions on Networking*, vol. 3, pp. 489–500, Oct. 1995.

[RaSi95] R. Ramaswami and K. Sivarajan, "Design of logical topologies for wavelength-routed all-optical networks," *Proceedings, IEEE INFOCOM '95*, Boston, MA, pp. 1316–1325, April 1995.

[RaSi96] R. Ramaswami and K. N. Sivarajan, "Design of logical topologies for wavelength-routed optical networks," *IEEE Journal on Selected Areas in Communications*, vol. 14, pp. 840–851, June 1996.

[RaTh87] P. Raghavan and C. D. Thompson, "Randomized rounding: A technique for provably good algorithms and algorithmic proofs," *Combinatorica*, vol. 7, no. 4, pp. 365–374, 1987.

[RoAm94] G. N. Rouskas and M. H. Ammar, "Multi-destination communication over single-hop lightwave WDM networks," *Proceedings, IEEE INFOCOM '94*, Toronto, Canada, pp. 1520–1527, June 1994.

[RoAm94b] G. N. Rouskas and M. H. Ammar, "Dynamic reconfiguration in multihop WDM networks," *Journal of High Speed Networks*, vol. 4, no. 3, pp. 221–238, 1995.

[SBJS93] T. E. Stern, K. Bala, S. Jiang, and J. Sharony, "Linear lightwave networks: Performance issues," *IEEE/OSA Journal of Lightwave Technology*, vol. 11, pp. 937–950, May/June 1993.

[SPGK93] J. P. Sokoloff, P. R. Prucnal, I. Glesk, and M. Kane, "A terahertz optical asymmetric demultiplexer (TOAD)," *IEEE Photonic Technology Letters*, vol. 5, no. 7, pp. 787–790, 1993.

[SaBr89] J. A. Salehi and C. A. Brackett, "Code division multiple-access techniques in optical fiber networks–Part II: Systems performance analysis," *IEEE Transactions on Communications*, vol. 37, no. 8, pp. 834–842, Aug. 1989.

[SaIa96] R. Sabella and E. Iannone, "Wavelength conversion in optical transport networks," *Journal of Fiber and Integrated Optics*, vol. 15, no. 3, pp. 167–191, 1996.

[Sale89] J. A. Salehi, "Code division multiple-access techniques in optical fiber networks–Part I: Fundamental principles," *IEEE Transactions on Communications*, vol. 37, no. 8, pp. 824–833, Aug. 1989.

[Sams97] "Single forward pumping EDFA," EDFA Datasheet, Samsung Electronics Co. Ltd., 1997.

[Sanc89] L. A. Sanchis, "Multiple-way network partitioning," *IEEE Transactions on Computers*, vol. 38, no. 1, pp. 62–81, Jan. 1989.

[ScAl90] R. V. Schmidt and R. C. Alferness, "Directional coupler switches, modulators, and filters using alternating $\delta\beta$ techniques," *Photonic Switching* (H. S. Hinton and J. E. Midwinter, eds.), pp. 71–80, New York: IEEE Press, 1990.

[ScSp96] M. A. Scobey and D. E. Spock, "Passive DWDM components using MicroPlasma optical interference filters," *OFC '96 Technical Digest*, San Jose, CA, pp. 242–243, 1996.

[Schn94] R. Schnabel et al., "Polarization insensitive frequency conversion of a 10-channel OFDM signal using four-wave mixing in a semiconductor laser amplifier," *IEEE Photonic Technology Letters*, vol. 6, pp. 56–58, Jan 1994.

[SeBP96] S.-W. Seo, K. Bergman, and P. R. Prucnal, "Transparent optical networks with time division multiplexing," *IEEE Journal on Selected Areas in Communications*, vol. 14, no. 5, pp. 1039–1051, June 1996.

[ShCS93] J. Sharony, K. Cheung, and T. E. Stern, "The wavelength dilation concept in lightwave networks: Implementation and system considerations," *IEEE/OSA Journal of Lightwave Technology*, vol. 11, pp. 900–907, May/June 1993.

[SiRa91] K. Sivarajan and R. Ramaswami, "Multihop networks based on de bruijn graphs," *Proceedings, IEEE INFOCOM '91*, Bal Harbour, FL, pp. 1001–1011, April 1991.

[SiRa94] K. Sivarajan and R. Ramaswami, "Lightwave networks based on de bruijn graphs," *IEEE/ACM Transactions on Networking*, vol. 2, no. 1, pp. 70–79, 1994.

[Sieg86] A. E. Siegman, *Lasers*, University Science Books, pp. 298-301, 1986.

[Smit90] D. A. Smith et al., "Integrated-optic acoustically-tunable filters for WDM networks: System issues and network applications," *IEEE Journal on Selected Areas in Communications*, vol. 8, pp. 1151-1159, Aug. 1990.

[Smit95] D. W. Smith, ed., *Optical Network Technology*, BT Telecommunications Series, New York: Chapman & Hall, 1995.

[SnJo94] A. Sneh and K. M. Johnson, "High-speed tunable liquid crystal optical filter for WDM systems," *Proceedings, IEEE/LEOS '94 Summer Topical Meetings on Optical Networks and their Enabling Technologies*, Lake Tahoe, NV, pp. 59–60, July 1994.

[SoAz97] A. Somani and M. Azizoglu, "All-optical LAN interconnection with a wavelength selective router," *Proceedings, IEEE INFOCOM '97*, Kobe, Japan, April 1997.

[SuAS96] S. Subramaniam, M. Azizoglu, and A. K. Somani, "All-optical networks with sparse wavelength conversion," *IEEE/ACM Transactions on Networking*, vol. 4, pp. 544–557, Aug. 1996.

[SuGK91a] G. N. M. Sudhakar, N. Georganas, and M. Kavehrad, "Slotted ALOHA and reservation ALOHA protocols for very high-speed optical fiber local area networks using passive star topology," *IEEE/OSA Journal of Lightwave Technology*, vol. 9, pp. 1411–1422, Oct. 1991.

[SuGK91b] G. N. M. Sudhakar, M. Kavehrad, and N. Georganas, "Multi-control channel very high-speed optical fiber local area networks and their interconnections using passive star topology," *Proceedings, IEEE Globecom '91*, Phoenix, AZ, pp. 624–628, Dec. 1991.

[SuGK92a] G. N. M. Sudhakar, N. Georganas, and M. Kavehrad, "A multichannel optical star LAN and its application as a broadband switch," *Proceedings, IEEE International Conference on Communications (ICC '92)*, Chicago, IL, pp. 843–847, June 1992.

[SuGK92b] G. N. M. Sudhakar, M. Kavehrad, and N. Georganas, "A simple contention-based reservation scheme for optical star LAN using a passive bus topology," *Proceedings, Photonics '92*, 1992.

[SuMo89] T. Suda and S. Morris, "Tree LANs with collision avoidance: Station and switch protocols," *Computer Networks and ISDN Systems*, vol. 17, pp. 101–110, 1989.

[TCFG95] R. W. Tkach et al., "Four-photon mixing and high-speed WDM systems," *IEEE/OSA Journal of Lightwave Technology*, vol. 13, pp. 841–849, May 1995.

[TKBS91] T. D. Todd, Z. Khurshid, A. M. Bignell, and S. Sivakumaran, "Photonic multihop bus networks," *Proceedings, IEEE INFOCOM '91*, Bal Harbour, FL, pp. 981–990, April 1991.

[TONT90] H. Toba, K. Oda, K. Nosu, and N. Takato, "Factors affecting the design of optical FDM information distribution systems," *IEEE Journal on Selected Areas in Communications*, vol. 6, pp. 965–972, Aug. 1990.

[TaLa91] M. Tachibana, R. I. Laming, et al., "Erbium-doped fiber amplifier with flattened gain spectrum," *IEEE Photonics Technology Letters*, vol. 3, pp. 118–120, Feb. 1991.

[TaWB95] V. Tandon, M. Wilby, and F. Burton, "A novel upgrade path for transparent optical networks based on wavelength reuse," *Proceedings, IEEE INFOCOM '95*, Boston, MA, pp. 1308–1315, April 1995.

[Taka86] H. Takagi, *Analysis of Polling Systems*, Cambridge, MA: MIT Press, 1986.

[Tang94] K. W. Tang, "CayletNet: A multihop WDM-based lightwave network," *Proceedings, IEEE INFOCOM '94*, Toronto, Canada, June 1994.

[ToDV95] S.-R. Tong, D. H. C. Du, , and R. J. Vetter, "Design principles for multihop wavelength and time division multiplexed optical passive star networks," *Journal of High-Speed Networks*, vol. 4, no. 2, pp. 189–200, 1995.

[ToHa92] T. D. Todd and E. L. Hahne, "Local and metropolitan multimesh networks," *Proceedings, IEEE International Conference on Communications (ICC '92)*, Chicago, IL, pp. 900–904, June 1992.

[Todd92] T. D. Todd, "The token grid: Multidimensional media access for local and metropolitan networks," *Proceedings, IEEE INFOCOM '92*, Florence, Italy, pp. 2415–2424, May 1992.

[TrBM96] S. B. Tridandapani, M. S. Borella, and B. Mukherjee, "A lower bound on the expected cost of minimum-delay multicast traffic," *Proceedings, International Conference on Computer Communications and Networks (IC^3N '96)*, Washington, DC, pp. 287–292, Oct. 1996.

[TrBe97] D. Trouchet, A. Beguin, et al., "Passband flattening of PHASAR WDM using input and output star couplers designed with two focal points," *OFC '97 Technical Digest*, Dallas, TX, pp. 302–303, Feb. 1997.

[TrMe95] S. B. Tridandapani and J. S. Meditch, "Supporting multipoint connections in multihop WDM optical networks," *Journal of High-Speed Networks*, vol. 4, no. 2, pp. 169–188, 1995.

[TrMu95] S. B. Tridandapani and B. Mukherjee, "Channel sharing in multi-hop lightwave networks (Do we really need more channels?)," *Proceedings, IEEE Globecom '95*, Singapore, pp. 1773–1778, Nov. 1995.

[TrMu96] S. B. Tridandapani and B. Mukherjee, "Multicast traffic in multihop lightwave networks: Performance analysis and an argument for channel sharing," *Proceedings, IEEE INFOCOM '96*, San Francisco, CA, pp. 345–352, March 1996.

[TrMu97] S. B. Tridandapani and B. Mukherjee, "Channel sharing in multi-hop WDM lightwave networks: Realization, and performance of multicast traffic," *IEEE Journal on Selected Areas in Communications*, vol. 15, pp. 488–500, April 1997.

[VoGa97] R. Vodhanel, L. D. Garrett, et al., "National-scale WDM networking demonstration by the MONET consortium," *Supplement to Technical Digest – Postdeadline Papers, OFC '97*, Dallas, TX, p. PD27, Feb. 1997.

[WaAl96] R. E. Wagner, R. C. Alferness, et al., "MONET: Multiwavelength optical networking," *IEEE/OSA Journal of Lightwave Technology*, vol. 14, pp. 1349–1355, June 1996.

[WaCh92] S. S. Wagner and T. E. Chapuran, "Multiwavelength Ring Networks for Switch Consolidation and Interconnection," *Proceedings, IEEE International Conference on Communications (ICC '92)*, Chicago, IL, pp. 1173-1179, June 1992.

[Waxm88] B. M. Waxman, "Routing of multipoint connections," *IEEE Journal on Selected Areas in Communications*, vol. 6, pp. 1617–1622, Dec. 1988.

[WeSa93] A. M. Weiner and J. A. Salehi, "Optical code-division multiple access," *Photonics in Switching* (J. E. Midwinter, ed.), vol. 2, pp. 73–118, San Diego, CA: Academic Press, 1993.

[Whit95] T. J. Whitley, "A review of recent system demonstrations incorporating 1.3-μm praseodymium-doped fluoride fiber amplifiers," *IEEE/OSA Journal of Lightwave Technology*, vol. 13, pp. 744–760, May 1995.

[WiHw93] A. E. Wilner and S. M. Hwang, "Passive equalization of nonuniform EDFA gain by optical filtering for megameter transmission of 20 WDMA channels through a cascade of EDFA's," *IEEE Photonics Technology Letters*, vol. 5, pp. 1023–1026, Sept. 1993.

[Wies96] J. M. Wiesenfeld, "Wavelength conversion techniques," *Proceedings, Optical Fiber Communication (OFC '96)*, San Jose, CA, vol. Tutorial TuP 1, pp. 71–72, 1996.

[WuKC89] T. H. Wu, D. J. Kolar, and R. H. Cardwell, "High-speed self-healing ring architectures for future interoffice networks," *Proceedings, IEEE Global Communications (GLOBECOM '89)*, Dallas, TX, pp. 801–807, Nov. 1989.

[YCBK95] S. J. B. Yoo, C. Caneau, R. Bhat, and M. A. Koza, "Wavelength conversion by quasi-phase-matched difference frequency generation in AlGaAs waveguides," *Proceedings, Optical Fiber Communication (OFC '95)*, San Diego, CA, vol. 8, pp. 377–380, Feb. 1995.

[YCBK96] S. J. B. Yoo, C. Caneau, R. Bhat, and M. A. Koza, "Transparent wavelength conversion by difference frequency generation in AlGaAs waveguides," *Proceedings, Optical Fiber Communication (OFC '96)*, San Jose, CA, vol. 2, pp. 129–131, 1996.

[YLES96] J. Yates, J. Lacey, D. Everitt, and M. Summerfield, "Limited-range wavelength translation in all-optical networks," *Proceedings, IEEE INFOCOM '96*, San Francisco, CA, pp. 954–961, March 1996.

[YSIY96] H. Yasaka et al., "Finely tunable 10-Gb/s signal wavelength conversion from 1530 to 1560-nm region using a super structure grating distributed Bragg reflector laser," *IEEE Photonic Technology Letters*, vol. 8, pp. 764–766, June 1996.

[YaSh95] M. Yamada, M. Shimizu, et al., "Low-noise and high-power Pr^{3+}-doped fluoride fiber amplifier," *IEEE Photonics Technology Letters*, vol. 7, pp. 869–871, Aug. 1995.

[Yoo96] S. J. B. Yoo, "Wavelength conversion technologies for WDM network applications," *IEEE/OSA Journal of Lightwave Technology*, vol. 14, pp. 955–966, June 1996.

[ZPVN94] J. Zhou et al., "Four-wave mixing wavelength conversion efficiency in semiconductor traveling-wave amplifiers measured to 65 nm of wavelength shift," *IEEE Photonic Technology Letters*, vol. 6, no. 8, pp. 984–987, 1994.

[ZaFa92] C.-E. Zah, F. J. Favire, et al., "Monolithic integration of a multiwavelength compressive-strained multiquantum-well distributed-feedback laser array with star coupler and optical amplifiers," *Electronic Letters*, vol. 28, pp. 2361–2362, Dec. 1992.

[ZhAc90] Z. Zhang and A. S. Acampora, "Analysis of multihop lightwave networks," *Proceedings, IEEE Globecom '90*, San Diego, CA, pp. 1873–1879, Dec. 1990.

[ZhAc91] Z. Zhang and A. S. Acampora, "Performance analysis of multihop lightwave networks with hot-potato routing and distance-age-priorities," *Proceedings, IEEE INFOCOM '91*, Bal Harbour, FL, pp. 1012–1021, April 1991.

[ZhAc94] Z. Zhang and A. Acampora, "A heuristic wavelength assignment algorithm for multihop WDM networks with wavelength routing and wavelength reuse," *IEEE/ACM Transactions on Networking*, vol. 3, pp. 281–288, June 1995.

[ZiDJ92] M. Zirngibl, C. Dragone, and C. H. Joyner, "Demonstration of a 15×15 arrayed waveguide multiplexer on InP," *IEEE Photonics Technology Letters*, vol. 4, pp. 1250–1253, Nov. 1992.

Index

Λ_0 cycle eliminator, 443
Λ_k cycle eliminator, 445
3-dB coupler, 503

access station, 341, 432
acoustooptic tunable filter (AOTF), 432
active switch, 13, 15
adaptive routing, 180
all-optical channel, 7
all-optical cycle, 431
ALOHA/ALOHA protocol, 124
ALOHA/CSMA protocol, 126
amplified spontaneous emission (ASE) noise, 386, 431
amplifier bandwidth, 488
amplifier gain model, 469
amplifier placement, 84
amplifier sensitivity level, 472
amplifier-placement module, 479
amplifier-placement problem, 463
amplitude modulation, 514
AMTRAC, 139
anomalous dispersion, 500
AOTF crossconnect, 433
approximation algorithms, 324
arrayed waveguide grating (AWG) multiplexer, 63
As Soon As Possible (ASAP), 479
asynchronous TDM, 495
ATM, 3, 263
ATM optical switching (ATMOS), 504
attempt-and-defer, 138
attenuation, 3, 24, 31–32, 83, 385, 463, 466, 467, 481
    low-attenuation region, 24, 35, 48
auxiliary graph, 378
average packet delay, 282, 369, 397, 417
average propagation delay, 282
average queueing delay, 282

balanced bridge interferometric switch, 59
BAR state, 433, 443, 445
Bellman-Ford algorithm, 278
bimodal throughput characteristic, 126
binary hypercube, 187
black cycle, 442, 443, 446
black_unused cycle, 442
black_used cycle, 442
blocking probability, 358, 368, 445
branch-exchange, 276, 378
bridge, 397, 402
bridge traffic, 408
broadcast, 10, 465
broadcast network, 23
broadcast-and-select network, 14, 110, 496

call-setup algorithm, 432
carrier sense multiple access (CSMA), 123
CayleyNet, 188
CDMA

autocorrelation property, 510
balance property, 510
cross-correlation property, 511
CFSQP, 469, 478, 483
channel-sharing, 192, 219, 222
chipping rate, 508
chips, 509
circuit switching, 7, 58
circuit-switched network, 69, 342, 369
classification of WDM systems, 113
clear channel, 266
clock recovery, 497
code-division multiple access (CDMA), 507
code-division multiplexing (CDM), 4, 494, 507
coherent detection, 514
coherent optical CDMA, 514
color interchange, 337
combiner, 35
compressed OTDM signal, 499
conflict graph, 333
congestion, 330, 368, 375
connection, 381
connection arrival, 368
connection diagram, 378
connection holding-time, 368
connection management, 382
connection setup, 444
connection teardown, 445
control channel, 88, 113, 123
controller, 381
convex quadratic program (QP), 478
CORD, 506
counter-rotating rings, 399
coupler, 23, 35
CROSS state, 433
cross-channel interference, 32
cross-traffic, 399
crosstalk, 85, 386, 431, 446, 455, 466
crosstalk blocking, 450, 457
crosstalk cycle, 432
CSMA/N-server protocol, 126
CSMA/ALOHA protocol, 126
cycle breaking, 443

data transparency, 57
de Bruijn graph, 173, 181, 206
deflection routing, 173, 180, 377
delay characteristics, 282
demand assignment, 118
Destination Allocation (DA) protocol, 120
diameter, 173
diffraction grating, 51
Dijkstra's algorithm, 278, 444
dilation, 368, 375
direct detection, 514
direct-sequence spread spectrum, 509
directional coupler, 59
disconnected network, 441
dispersion, 31, 85
chromatic, 31
intermodal, 27, 31
material, 31
waveguide, 32
dispersion-shifted fiber, 387, 493
distance matrix, 271
distortion, 3, 32, 500
Distributed FeedBack (DFB) laser, 497
distributed protocols, 381
distributed queue dual bus (DQDB), 188
DT-WDMA protocol, 134

duty cycle, 497
Dynamic Allocation Scheme (DAS), 136
dynamic lightpath establishment (DLE), 325, 336

EDFA gain curve, 432
edge-disjoint cycle, 442
electromagnetic interference, 25
electronic packet switching, 259
embedding, 370
equally powered wavelengths, 464
equally-powered-wavelengths method (LP), 482
equilibrium point analysis (EPA), 151
erbium-doped fiber amplifier (EDFA), 431, 469
    gain model, 488
Eulerian graph, 442
Eulerian property, 443
exhaustive search, 455
extended breadth-first, 327

Fabry-Perot filter, 150
fast frequency hopping, 510
fault-tolerance, 404
feasible point generation, 478
fiber crossconnect (FXC), 393
fiber distributed data interface (FDDI), 118, 397
fiber length matrix, 295
fiber nonlinearities, 493
Fiber-Optic Crossconnect (FOX), 117
fixed assignment, 118, 119
fixed receiver, 113
fixed routing, 63
fixed transmitter, 113, 150
fixed-tuned receiver, 401
fixed-tuned tranceiver, 171
fixed-tuned transmitter, 401
flat gain curve, 436
flow and wavelength assignment (FWA), 174
flow deviation, 260, 261
flow deviation algorithm, 278
four-photon mixing, 74
four-wave mixing, 34, 74, 387, 501
frame synchronization, 495
frequency hopping, 509
frequency-hopping optical CDMA, 517

GEMNET, 203, 219, 225
    average hop distance
        bounds, 209
    diameter, 206
    hop distance, 207
    interconnection pattern, 206
    routing, 206
    routing algorithms, 208
    scalability, 213
        retunings, 215
generalized hypercube, 187, 194
global method, 480
good state, 446, 457
gradient, 478
graph, 410, 442
    adjacency matrix, 442
    indegree, 264, 437
    out-degree, 264, 437
graph coloring, 321, 330
grating array, 64
grating demultiplexer, 66
group velocity dispersion, 500
growth capability, 399

hard limiter, 509
helical bus, 504
high-performance parallel interface (HIPPI), 117
HLAN, 504
hop distance, 173, 227
hopping pattern, 509, 510
Hybrid TDM (HDTM), 137
HYPASS, 118
hyperbolic secant pulse, 500
hypercube, 173, 186

in-band polling protocol, 150
integer linear program (ILP), 324, 326
integer variables, 479
integral constraints, 477
Integrated Opto-Electronic-Integrated Chip (OEIC), 389
integrated WDM laser, 389
integrated-optics amplifier gate switch, 61
intensity-modulated direct-detection (IM-DD), 501
internal saturation power, 469
Internet, 3, 4
intersecting waveguide switch, 61
intersubnetwork traffic, 397

Kautz graph, 188
Kernighan-Lin Algorithm, 410
Kerr effect, 500

LAMBDANET, 116
LAN/MAN, 464
Latin Router, 373
Latin Square, 373
Least Congested Path (LCP) algorithm, 336
lightnet, 276
lightpath, 7, 16, 69, 89, 259, 322
lightpath length bound, 296
linear bus, 138
linear divider and combiner (LDC), 368, 369
linear dual bus, 173, 188
linear lightwave network, 368, 369
linear program (LP), 292, 324
linear programming, 291
link-by-link method, 479
load balancing, 399
local lightwave network, 109, 219
local minimum, 478, 484
logical subring, 399
logical topology, 222
loop, 431
loopless network, 486
LP output, 292
LP solution, 292
lpsolve, 333

Mach-Zender interferometer, 76
Manhattan Street Network, 173, 183, 193
maximal-length PN sequence, 511
maximum-flow-finding algorithm, 409
MaxMultiHop, 304
MaxSingleHop, 303
mechanical fiber-optic switch, 61
metropolitan-area network (MAN), 466
Michelson interferometer, 76
mid-sized tree-based network, 481
min-k-cut problem, 408
minimum 2-cut, 409
minimum-weight W-cut, 408
mixed-integer linear program (MILP), 376, 463, 467

mixed-integer nonlinear program, 463, 467
mode-locked laser (MLL), 497
Moore bound, 232
Moore graph, 232
multicast, 14
multicasting, 219, 224, 370
multichannel p(i)-persistent protocol, 139
multicommodity flow, 298, 321, 324
multihop, 14, 87, 109, 112, 220
  regular structures, 176
multihop network, 171
multihop topology, 87
multistage interconnection network (MIN), 307
multistar-based physical topology, 379
multiwavelength ring network, 397
multiway MIN-DIFF partitioning algorithm, 414

near-far effect, 467
negative chirp, 500
net loop gain, 431, 432
net loop loss, 431, 432
network configuration, 442
network configuration analyzer, 446
network cost model, 305
network generator, 441
network lag, 4
nodal degree, 285
node, 16
nonblocking network (NBN), 374
noncoherent optical CDMA, 514
nonflat gain spectrum, 488
nonhyperbolic secant pulse, 500
nonintegral multicommodity flow problem, 328
nonintrusive measurement, 478
nonlinear constraints, 477
nonlinear optical loop mirror (NOLM), 76, 503
Nonlinear Program (NLP), 481
nonlinear program solver, 478
nonlinear Schrödinger equation (NLSE), 500
nonmonotone line search, 478
nonreflective star, 465
NSFNET, 261, 310, 358

oblivious routing scheme, 374
off-line scheduling, 120
on-off keying (OOK), 514
optical amplifier, 23, 52–56, 83, 385, 463
  1R, 2R, 3R, 52
  amplified spontaneous emission (ASE) noise, 53
  doped-fiber amplifiers, 55
    pump wavelength, 55
  erbium-doped fiber amplifier (EDFA), 55
  Fabry-Perot amplifier, 54
  gain, 53
  gain bandwidth, 53
  gain equalization, 56, 385
  gain saturation, 53
  gain spectrum, 385
  in-line amplifier, 83
  internal saturation power, 466
  maximum output power, 466
  maximum small-signal gain, 466
  multistage EDFAs, 84
  nonuniform gain, 385
  polarization sensitivity, 53
  praseodymium-doped fluoride fiber amplifier (PDFFA), 55

receiver preamplifier, 83
semiconductor laser amplifier, 54
transmitter power booster, 83
traveling-wave amplifier, 54
unequal gain spectrum, 56
optical amplifier placement, 463
optical bandwidth, 4
optical CDM
chip rate, 4
optical CDMA, 514
optical circuit switching, 259
optical clock recovery, 501
optical delay lines, 504
optical fiber, 24
attenuation, 30
cladding, 25
core, 25
graded-index, 27
mode, 25, 28
multimode, 25, 28
nonlinearities, 32–35
Four-Wave Mixing, 34
nonlinear refraction, 32
Stimulated Brillouin Scattering, 34
Stimulated Raman Scattering, 33
numerical aperture, 28
single-mode, 25, 28
step-index, 25
total internal reflection, 25
Snell's Law, 25
optical orthogonal codes (OOC), 518
optical pulse broadening, 500
optical pulse stream, 498
optical receiver, 23, 44–52
coherent detection
difference signal, 45
detection mechanism
coherent detection, 45
direct detection, 45
filter, 23
acoustooptic, 49
electrooptic, 50
etalon, 47
Fabry-Perot filter, 47
fiber Bragg grating, 51
finesse, 46
fixed filter, 50
free spectral range, 46
grating filter, 51
liquid-crystal Fabry-Perot, 50
Mach-Zehnder chain, 48
multicavity, 47
multipass, 47
thin-film interference filter, 52
transfer function, 46
tunable, 46
tuning range, 46
tuning time, 46
photodetection, 45
photodetector, 45
PIN photodiode, 45
PN photodiode, 45
threshold device, 45
optical storage, 504
optical switch, 23, 66
packet switch, 66
CORD, 68
delay line, 67
HLAN, 69
staggering switch, 67
reconfigurable, 66
space division, 66
optical TDM

bit rate, 4
optical time-spreading CDMA, 518
optical transmission, 25
optical transmitter, 23, 37–44
laser, 37
acoustooptically tuned, 42
band gap, 40
cavity, 38
chirp, 44
Distributed Bragg Reflector), 43
Distributed FeedBack (DFB), 42
doped, 40
electrooptically tuned, 42
excitation device, 38
Fabry-Perot, 42
frequency instability, 41
injection current tuned, 42
laser array, 43
lasing medium, 38
linewidth, 41
longitudinal modes, 41
Mach-Zehnder interferometer, 44
mechanically tuned, 42
multi-quantum well (MQW), 40
on-off keying (OOK), 44
optical modulation, 43
photon, 38
population inversion, 37
quasi-stable substance, 38
stimulated emission, 37
tunable vs. fixed, 40
tuning range, 41
tuning time, 41
semiconductor diode laser, 39
optimal partition, 399, 405
optimization, 174, 188
channel-sharing, 220
GEMNET, 211
linear bus
delay-based heuristics, 192
flow-based heuristics, 189
minimum-amplifier-placement problem, 467, 471–479
mutihop
delay-based, 175
flow-based, 174
Rainbow, 163
reconfiguration
Markov decision process (MDP), 379
RWA, 326
virtual topology
reconfiguration, 309
virtual topology design, 259, 271
linear formulation, 291, 297
WDDI, 399, 406–407, 417–418
optimized partitioning, 405
oscillations, 431
OTDM, 493

PAC circuit, 123
packet compression, 498
packet switching, 58
packet-switched network, 369, 498
partially oblivious routing scheme, 374
PARTITION problem, 413
passive photonic loop (PLL), 118
passive router, 11
passive star, 10
passive-star coupler, 36, 64, 85, 110, 379, 463, 465
path stripping, 328

perfect filter, 438
perfect sharing, 227
permutation connection pattern, 446, 453
permutation network, 374
permutation routing, 375
phase difference, 65
photon, 38
photonic bus network (PBNet), 189
photonic packet network, 496
photonic packet switch, 504
photonic packet-switching architectures, 504
photonic slot routing, 504
physical topology, 15, 263
physical topology route, 272, 297
PN codes, 511
PN sequence generator, 511
polling, 152
polynomial-time algorithm, 378, 409
positive chirp, 500
possible MAN network, 482
power control scheme, 509
pretransmission coordination, 88, 113
processing gain, 508, 509
Protection-Against-Collision (PAC) switch, 123
pseudo-noise (PN), 510
pseudo-polynomial time algorithm, 414
pseudo-random code, 508
pseudo-random sequence, 510
pulse broadening, 500

Rainbow, 48, 116
  analysis, 155
  architecture, 150
  blocking probability, 159
  deadlock, 152, 163
  delay, 158
  optimal timeout duration, 163, 167
  state diagram, 153
  state space vector, 155
  throughput, 158
  timeout probability, 159
  timing for connection setup, 154
Rainbow-I, 117, 149
Rainbow-II, 117, 149
random access protocol, 121
random network, 441
random TDM protocol, 122
randomized path selection, 328
randomized rounding, 321, 324, 328
rearrangeably NBN, 374
receiver
  wavelength tunable, 16
receiver collision, 125
Receiver Collision Avoidance (RCA) protocol, 132
receiver sensitivity, 31, 463, 472
reconfiguration, 378
  circuit rerouting algorithms, 380
    move-to-vacant (MTV), 380
    parallel MTV-WR, 380
    wavelength-retuning (WR), 380
reconfiguration algorithm, 309
reconfiguration penalty, 379
refractive index, 25, 65, 500
Reservation-ALOHA, 129
resolvable channel, 83
resource blocking, 450
resource budgeting trade-offs, 304
return-to-zero (RZ), 497
reversed delta-beta coupler, 59
round-robin, 495
router, 23

routing and wavelength assignment (RWA), 8, 321
routing matrix, 11

satellite communication systems, 494
SC_GEMNET (Shared-Channel GEMNET), 226
scalability, 368
scheduling algorithm, 119
security, 404, 508
self-healing ring network, 404
self-loop, 441
self-phase modulation (SPM), 500
semiconductor laser amplifier (SLA), 503
semiconductor optical amplifier (SOA), 74, 469
sequential coloring algorithms, 325
sequential graph coloring, 331
shortest path, 173
shortest-path algorithm, 444
shortest-path delay matrix, 296
shortest-path flows, 278
ShuffleNet, 173, 176, 193, 206, 222
signal-to-noise ratio (SNR), 83, 386, 466
simulated annealing, 260, 261, 277, 417
single-hop, 14, 87, 109, 112
skewed traffic, 406
slotted-ALOHA, 121
slotted-ALOHA/ALOHA protocol, 121, 126
slotted-ALOHA/delayed-ALOHA protocol, 126
slow frequency hopping, 510
small-signal gain, 470
smallest-last (SL) vertex ordering, 332
soliton, 499, 500
soliton transmission, 500
SONET add-drop multiplexer, 404
SONET ring network, 397
Source Allocation (SA) protocol, 121
space-division switch, 15, 66
splitter, 35
    splitting ratio, 35
spread spectrum, 509
stability, 416
staggering switch, 505
STAR-TRACK, 118
STARNET, 117
static analysis, 453
static lightpath establishment (SLE), 325, 332
Stimulated Brillouin Scattering, 387
strict-sense NBN, 374
subcarrier multiplexing, 195
subrings, 397
switch
    confinement factor, 59
    coupling strength, 59
    crosspoint element, 58
        directive switch, 58, 59
        gate switch, 58, 61
    fiber crossconnect, 58
    logic device, 58
    relational device, 57, 66
switch port labeler, 442
switching elements, 57–69
Synchronous Digital Hierarchy (SDH), 391
synchronous TDM, 494

TDM multiplexer, 495
TDMA/N-Server Protocol, 137
temporal spreading code, 514
Testbeds, 91–97, 387–393

AON, 95, 387
LAMBDANET, 91
MONET, 93, 391
ONTC, 94, 389
RACE-MWTN, 92, 391
Rainbow-I, 91
Rainbow-II, 92
thermo-optic switch, 61
third-order nonlinearity, 74
throughput, 369
Thunder and Lightning, 118
time hopping, 509, 510
time-division multiple access (TDMA), 494
time-division multiplexing (TDM), 4, 219, 493
time-slot switching, 494
time-spreading CDMA, 517
time-wavelength assignment algorithm, 119
token grid, 186
toroid, 173
traffic matrix, 264, 272, 296, 405, 413
traffic model, 369, 404
clustered traffic, 418
pseudo-random traffic, 418
server traffic, 418
traffic routing, 272, 297
transceiver utilization, 313
transform-limited, 498
transmitter
wavelength tunable, 16
transmitter power, 30, 463
transparent optical connection, 432
tunable receiver, 87, 113, 123
tunable transmitter, 87, 113, 123

ultra-fast transmission, 493
unequally-powered-wavelengths method (NLP), 482
unspecified state, 431, 435

vertex-disjoint cycles, 378
virtual topology, 171, 259, 263, 296, 323, 376
reconfiguration, 308
virtual topology connection matrix, 272
virtual topology design problem, 259
virtual tree, 173

W-partition, 408
waveband, 369, 435
waveguide grating router (WGR), 51, 63
wavelength, 5
wavelength add/drop multiplexer (WADM), 9, 392
wavelength amplifier (WAMP), 392
wavelength color, 272
wavelength conversion, 69–79, 343
Best-Fit (BF) Strategy, 359
conversion gain, 352
converter bank, 345
cross-gain modulation (XGM), 75
cross-phase modulation (XPM), 75
degrees of sparseness, 356
limited range conversion, 357, 361
sparse nodal conversion, 356, 358
sparse switch-output conversion, 356, 361
difference frequency generation (DFG), 75

First-Fit (FF) algorithm, 358
four-wave mixing (FWM), 74
limited-range conversion, 355
multifiber networks, 354
network control, 348
dynamic routing, 348
static routing, 349
network design, 347
network management, 349
NSFNET performance, 358
optoelectronic, 73
performance benefits, 350
ring performance, 357
sparse conversion, 354
switch designs, 345
technologies, 73
transparency, 344
using coherent effects, 74
using cross modulation, 75
using semiconductr laser, 76
wavelength converter, 8, 72, 89, 291
characteristics, 72
wavelength crossconnect, 431
wavelength distributed data interface (WDDI), 397
wavelength interchanging crossconnect (WIXC), 13, 78, 393
wavelength reuse, 11, 13
wavelength router, 62, 393
nonreconfigurable, 62
reconfigurable, 66
routing matrix, 63
wavelength selective crossconnect (WSXC), 13, 66, 393
wavelength terminal multiplexer (WTM), 392
wavelength utilization, 313
wavelength-continuity constraint, 8, 69, 300, 324, 342
wavelength-convertible network, 341
wavelength-convertible switch, 13, 78
dedicated, 78
share-per-link, 78
share-per-node, 78
share-with-local, 79
wavelength-division multiplexing (WDM), 4, 5, 259, 463
wavelength-routed network, 8, 15, 23, 69, 291, 341, 381
wavelength-routing switch (WRS), 13, 66, 260, 341, 431
WDDI
delay analysis, 416
delay-based algorithm, 415
flow-based algorithms, 407
MIN-CROSS, 406
MIN-DELAY, 407
MIN-DIFF, 407, 412
stability, 416
throughput analysis, 416
WDM connection management, 381
connection setup, 384
connection switch table (CST), 383
connection takedown and update, 384
CST update protocol, 383
fault recovery, 384
link failure, 385
topology update, 383
wavelength failure, 384
WDM crossconnect, 341
WDM economics, 8, 17
WDM network design, 79
white cycle, 432, 443, 445, 446, 455

wide-sense NBN, 374
world wide web (WWW), 4

xDM vs. xDMA, 5

zero-dispersion wavelength, 500